# Global Lorentzian Geometry

## MONOGRAPHS AND TEXTBOOKS IN PURE AND APPLIED MATHEMATICS

1. *K. Yano,* Integral Formulas in Riemannian Geometry (1970) *(out of print)*
2. *S. Kobayashi,* Hyperbolic Manifolds and Holomorphic Mappings (1970) *(out of print)*
3. *V. S. Vladimirov,* Equations of Mathematical Physics (A. Jeffrey, editor; A. Littlewood, translator) (1970)
4. *B. N. Pshenichnyi,* Necessary Conditions for an Extremum (L. Neustadt, translation editor; K. Makowski, translator) (1971)
5. *L. Narici, E. Beckenstein, and G. Bachman,* Functional Analysis and Valuation Theory (1971)
6. *D. S. Passman,* Infinite Group Rings (1971)
7. *L. Dornhoff,* Group Representation Theory (in two parts). Part A: Ordinary Representation Theory. Part B: Modular Representation Theory (1971, 1972)
8. *W. Boothby and G. L. Weiss (eds.),* Symmetric Spaces: Short Courses Presented at Washington University (1972)
9. *Y. Matsushima,* Differentiable Manifolds (E. T. Kobayashi, translator) (1972)
10. *L. E. Ward, Jr.,* Topology: An Outline for a First Course (1972) *(out of print)*
11. *A. Babakhanian,* Cohomological Methods in Group Theory (1972)
12. *R. Gilmer,* Multiplicative Ideal Theory (1972)
13. *J. Yeh,* Stochastic Processes and the Wiener Integral (1973) *(out of print)*
14. *J. Barros-Neto,* Introduction to the Theory of Distributions (1973) *(out of print)*
15. *R. Larsen,* Functional Analysis: An Introduction (1973)
16. *K. Yano and S. Ishihara,* Tangent and Cotangent Bundles: Differential Geometry (1973)
17. *C. Procesi,* Rings with Polynomial Identities (1973)
18. *R. Hermann,* Geometry, Physics, and Systems (1973)
19. *N. R. Wallach,* Harmonic Analysis on Homogeneous Spaces (1973)
20. *J. Dieudonné,* Introduction to the Theory of Formal Groups (1973)
21. *I. Vaisman,* Cohomology and Differential Forms (1973)
22. *B.-Y. Chen,* Geometry of Submanifolds (1973)
23. *M. Marcus,* Finite Dimensional Multilinear Algebra (in two parts) (1973, 1975)
24. *R. Larsen,* Banach Algebras: An Introduction (1973)
25. *R. O. Kujala and A. L. Vitter (eds),* Value Distribution Theory: Part A; Part B. Deficit and Bezout Estimates by Wilhelm Stoll (1973)
26. *K. B. Stolarsky,* Algebraic Numbers and Diophantine Approximation (1974)
27. *A. R. Magid,* The Separable Galois Theory of Commutative Rings (1974)
28. *B. R. McDonald,* Finite Rings with Identity (1974)
29. *J. Satake,* Linear Algebra (S. Koh, T. Akiba, and S. Ihara, translators) (1975)

30. *J. S. Golan,* Localization of Noncommutative Rings (1975)
31. *G. Klambauer,* Mathematical Analysis (1975)
32. *M. K. Agoston,* ALgebraic Topology: A First Course (1976)
33. *K. R. Goodearl,* Ring Theory: Nonsingular Rings and Modules (1976)
34. *L. E. Mansfield,* Linear Algebra with Geometric Applications: Selected Topics (1976)
35. *N. J. Pullman,* Matrix Theory and Its Applications (1976)
36. *B. R. McDonald,* Geometric Algebra Over Local Rings (1976)
37. *C. W. Groetsch,* Generalized Inverses of Linear Operators: Representation and Approximation (1977)
38. *J. E. Kuczkowski and J. L. Gersting,* Abstract Algebra: A First Look (1977)
39. *C. O. Christenson and W. L. Voxman,* Aspects of Topology (1977)
40. *M. Nagata,* Field Theory (1977)
41. *R. L. Long,* Algebraic Number Theory (1977)
42. *W. F. Pfeffer,* Integrals and Measures (1977)
43. *R. L. Wheeden and A. Zygmund,* Measure and Integral: An Introduction to Real Analysis (1977)
44. *J. H. Curtiss,* Introduction to Functions of a Complex Variable (1978)
45. *K. Hrbacek and T. Jech,* Introduction to Set Theory (1978)
46. *W. S. Massey,* Homology and Cohomology Theory (1978)
47. *M. Marcus,* Introduction to Modern Algebra (1978)
48. *E. C. Young,* Vector and Tensor Analysis (1978)
49. *S. B. Nadler, Jr.,* Hyperspaces of Sets (1978)
50. *S. K. Sehgal,* Topics in Group Rings (1978)
51. *A. C. M. van Rooij,* Non-Archimedean Functional Analysis (1978)
52. *L. Corwin and R. Szczarba,* Calculus in Vector Spaces (1979)
53. *C. Sadosky,* Interpolation of Operators and Singular Integrals: An Introduction to Harmonic Analysis (1979)
54. *J. Cronin,* Differential Equations: Introduction and Quantitative Theory (1980)
55. *C. W. Groetsch,* Elements of Applicable Functional Analysis (1980)
56. *I. Vaisman,* Foundations of Three-Dimensional Euclidean Geometry (1980)
57. *H. I. Freedman,* Deterministic Mathematical Models in Population Ecology (1980)
58. *S. B. Chae,* Lebesgue Integration (1980)
59. *C. S. Rees, S. M. Shah, and Č. V. Stanojević,* Theory and Applications of Fourier Analysis (1981)
60. *L. Nachbin,* Introduction to Functional Analysis: Banach Spaces and Differential Calculus (R. M. Aron, translator) (1981)
61. *G. Orzech and M. Orzech,* Plane Algebraic Curves: An Introduction Via Valuations (1981)
62. *R. Johnsonbaugh and W. E. Pfaffenberger,* Foundations of Mathematical Analysis (1981)

63. *W. L. Voxman and R. H. Goetschel,* Advanced Calculus: An Introduction to Modern Analysis (1981)
64. *L. J. Corwin and R. H. Szczarba,* Multivariable Calculus (1981)
65. *V. I. Istrăţescu,* Introduction to Linear Operator Theory (1981)
66. *R. D. Järvinen,* Finite and Infinite Dimensional Linear Spaces: A Comparative Study in Algebraic and Analytic Settings (1981)
67. *J. K. Beem and P. E. Ehrlich,* Global Lorentzian Geometry (1981)

*Other Volumes in Preparation*

# Global Lorentzian Geometry

JOHN K. BEEM

PAUL E. EHRLICH

Department of Mathematics
University of Missouri
Columbia, Missouri

MARCEL DEKKER, INC. New York and Basel

Library of Congress Cataloging in Publication Data

Beem, John K., [date]
Global Lorentzian geometry.

(Pure and applied mathematics; 67)
Bibliography: p.
Indexes.
1. Geometry, Differential. 2. General relativity (Physics) I. Ehrlich, Paul E., [date]. II. Title. III. Title: Lorentzian geometry. IV. Series: Pure and applied mathematics (M. Dekker); 67.
QA649.B42 516.3'6 81-3272
ISBN 0-8247-1369-9 AACR2



MARCEL DEKKER, INC.
270 Madison Avenue, New York, New York 10016

Current printing (last digit):
10 9 8 7 6 5 4 3 2 1

PRINTED IN THE UNITED STATES OF AMERICA

# PREFACE

This book is about Lorentzian geometry, the mathematical theory used in general relativity, treated from the viewpoint of global differential geometry. Our goal is to help bridge the gap between modern differential geometry and the mathematical physics of general relativity by giving an invariant treatment of global Lorentzian geometry. The growing importance in physics of this approach is clearly illustrated by the recent Hawking-Penrose singularity theorems described in the text of Hawking and Ellis (1973).

The Lorentzian distance function is used as a unifying concept in our book. Furthermore, we frequently compare and contrast the results and techniques of Lorentzian geometry to those of Riemannian geometry to alert the reader to the basic differences between these two geometries.

This book has been written especially for the mathematician who has a basic acquaintance with Riemannian geometry and wishes to learn Lorentzian geometry. Accordingly, this book is written using the notation and methods of modern differential geometry. For readers less familiar with this notation, we have included Appendix A which gives the local coordinate representations for the symbols used.

The basic prerequisites for this book are a working knowledge of general topology and differential geometry. Thus this book should be accessible to advanced graduate students in either mathematics or mathematical physics.

In writing this monograph, both authors profited greatly from the opportunity to lecture on part of this material during the spring semester, 1978, at the University of Missouri-Columbia. The second author also gave a series of lectures on this material in Ernst Ruh's seminar in differential geometry at Bonn University during the summer semester, 1978, and would like to thank Professor Ruh for giving him the opportunity to speak on this material. We would like to thank C. Ahlbrandt, D. Carlson, and M. Jacobs for several helpful conversations on Section 2.4 and the calculus of variations. We would like to thank M. Engman, S. Harris, K. Nomizu, T. Powell, D. Retzloff, and H. Wu for helpful comments on our preliminary version of this monograph. We also thank S. Harris for contributing Appendix D to this monograph and J.-H. Eschenburg for calling our attention to the Diplomarbeit of Bölts (1977). To anyone who has read either of the excellent books of Gromoll, Klingenberg, and Meyer (1975) on Riemannian manifolds or of Hawking and Ellis (1973) on general relativity, our debt to these authors in writing this work will be obvious. It is also a pleasure for both authors to thank the Research Council of the University of Missouri-Columbia and for the second author to thank the Sonderforschungsbereich Theoretische Mathematik 40 of the Mathematics Department, Bonn University, and to acknowledge an NSF Grant MCS77-18723(02) held at the Institute for Advanced Study, Princeton, New Jersey, for partial financial support while we were working on this monograph. Finally it is a pleasure to thank Diane Coffman, DeAnna Williamson, and Debra Retzloff for the patient and cheerful typing of the manuscript.

John K. Beem<br>
Paul E. Ehrlich

CONTENTS

Chapter 1

# INTRODUCTION: RIEMANNIAN THEMES IN LORENTZIAN GEOMETRY

Recent progress on causality theory, singularity theory, and black holes in general relativity described in the influential text of Hawking and Ellis (1973) has resulted in a resurgence of interest in global Lorentzian geometry. Indeed, a better understanding of global Lorentzian geometry was required for the development of singularity theory. For example, it was necessary to know that causally related points in globally hyperbolic subsets of space-times could be joined by a nonspacelike geodesic segment maximizing the Lorentzian arc length among all nonspacelike curves joining the two given points. In addition, much work done in the 1970s on foliating asymptotically flat Lorentzian manifolds by families of maximal hypersurfaces has been motivated by general relativity [cf. Choquet-Bruhat, Fisher, and Marsden (1979) for a partial list of references].

All of these results naturally suggest that a systematic study of global Lorentzian geometry should be made. The development of "modern" global Riemannian geometry as described in any of the standard texts [cf. Bishop and Crittenden (1964), Gromoll, Klingenberg, and Meyer (1975), Helgason (1962), Hicks (1965)] supports the idea that a comprehensive treatment of global Lorentzian geometry should be grounded in three fundamental topics: geodesic and metric completeness, the Lorentzian distance function, and a Morse index theory valid for nonspacelike geodesic segments in an arbitrary Lorentzian manifold.

Geodesic completeness, or more accurately, geodesic incompleteness, has played a crucial role in the development of singularity theory in general relativity and has been thoroughly explored within this framework. However the Lorentzian distance function has not been as well investigated, although it has been of some use in the study of singularities [cf. Hawking (1967), Hawking and Ellis (1973), Tipler (1977a), Beem and Ehrlich (1979a)]. Some of the properties of the Lorentzian distance function needed in general relativity are briefly described in Hawking and Ellis (1973, pp. 215-217). Further results relating Lorentzian distance to causality and the global behavior of nonspacelike geodesics have been given in Beem and Ehrlich (1979b).

Uhlenbeck (1975), Everson and Talbot (1976), and Woodhouse (1976) have studied Morse index theory for globally hyperbolic space-times and we have sketched [cf. Beem and Ehrlich (1979 c,d)] a Morse index theory for nonspacelike geodesics in arbitrary space-times. But no complete treatment of this theory for arbitrary space-times has been published previously.

It is the purpose of this monograph to first review known results on geodesic and metric completeness. Then we give a detailed treatment of the Lorentzian distance function and of the Morse index theory for nonspacelike geodesics in arbitrary space-times. Finally we show how these concepts may be applied to global Lorentzian geometry and singularity theory in general relativity.

The Lorentzian distance function has many similarities with the Riemannian distance function but also many differences. Since the Lorentzian distance function is not so well known, we now review the main properties of the Riemannian distance function, then compare and contrast the corresponding results for the Lorentzian distance function.

For the rest of this portion of the introduction, we will let $(N,g_0)$ denote a Riemannian manifold and $(M,g)$ denote a Lorentzian manifold, respectively.

Thus N is a smooth paracompact manifold equipped with a positive definite inner product $g_0|_p : T_pN \times T_pN \longrightarrow \mathbb{R}$ on each tangent space $T_pN$. In addition, if X and Y are arbitrary smooth vector fields on N, the function $N \longrightarrow \mathbb{R}$ given by $p \longrightarrow g_0(X(p),Y(p))$ is required to be a smooth function. The Riemannian structure $g_0 : TN \times TN \longrightarrow \mathbb{R}$ then defines the Riemannian distance function

$$d_0 : N \times N \longrightarrow [0,\infty)$$

as follows. Let $\Omega_{p,q}$ denote the set of piecewise smooth curves in N from p to q. Given $c \in \Omega_{p,q}$, $c : [0,1] \longrightarrow N$, there is a finite partition $0 = t_1 < t_2 < \cdots < t_k = 1$ such that $c \mid [t_i,t_{i+1}]$ is smooth for each i. The Riemannian arc length of c with respect to $g_0$ is defined as

$$L_0(c) = \sum_{i=1}^{k-1} \int_{t_i}^{t_{i+1}} \sqrt{g_0(c'(t),c'(t))}\, dt$$

The Riemannian distance $d_0(p,q)$ between p and q is then defined to be

$$d_0(p,q) = \inf\{L_0(c) : c \in \Omega_{p,q}\} \geq 0$$

For any Riemannian metric $g_0$ for N, the function $d_0 : N \times N \longrightarrow [0,\infty)$ has the following properties:

(1) $d_0(p,q) = d_0(q,p)$ for all $p,q \in N$.

(2) $d_0(p,q) \leq d_0(p,r) + d_0(r,q)$ for all $p,q,r \in N$.

(3) $d_0(p,q) = 0$ iff $p = q$.

More surprisingly,

(4) $d_0 : N \times N \longrightarrow [0,\infty)$ is continuous and the family of metric balls

$$B(p,\varepsilon) = \{q \in N : d(p,q) < \varepsilon\}$$

for all $p \in N$ and $\varepsilon > 0$ forms a basis for the given manifold topology.

Thus the metric topology and the given manifold topology coincide. Furthermore, by a result of Whitehead (1932), given any $p \in N$, there exists an $R > 0$ such that for any $\varepsilon$ with $0 < \varepsilon < R$, the metric ball $B(p,\varepsilon)$ is geodesically convex. Thus for any $\varepsilon$ with $0 < \varepsilon < R$, the set $B(p,\varepsilon)$ is diffeomorphic to the n disk, $n = \dim(N)$, and the set $\{q \in N : d(p,q) = \varepsilon\}$ is diffeomorphic to $S^{n-1}$.

Removing the origin from $\mathbb{R}^2$ equipped with the usual Euclidean metric and setting $p = (-1,0)$, $q = (1,0)$, one calculates that $d_0(p,q) = 2$, but finds no curve $c \in \Omega_{p,q}$ with $L_0(c) = d_0(p,q)$ and also no smooth geodesic from p to q.

Thus the following questions arise naturally. Given a manifold N, find conditions on a Riemannian metric $g_0$ for N such that

(i) All geodesics in N may be extended to be defined on all of $\mathbb{R}$.

(ii) The pair $(N,d_0)$ is a complete metric space in the sense that all Cauchy sequences converge.

(iii) Given any two points $p,q \in N$, there is a smooth geodesic segment $c \in \Omega_{p,q}$ with $L_0(c) = d_0(p,q)$.

A distance realizing geodesic segment as in (iii) is called a *minimal* geodesic segment. The word *minimal* is used here since the definition of Riemannian distance implies that $L_0(\gamma) \geqslant d_0(p,q)$ for all $\gamma \in \Omega_{p,q}$. More generally, one may define an arbitrary piecewise smooth curve $\gamma \in \Omega_{p,q}$ to be *minimal* if $L_0(\gamma) = d_0(p,q)$. Using the variation theory of the arc length functional, it may be shown that if $\gamma \in \Omega_{p,q}$ is minimal, then $\gamma$ may be reparameterized to a smooth geodesic segment.

The question of finding criteria on $g_0$ such that (i), (ii), or (iii) hold was resolved by H. Hopf and W. Rinow in their famous paper (1931). In modern terminology the Hopf-Rinow theorem asserts the following:

HOPF-RINOW THEOREM For any Riemannian manifold $(N,g_0)$ the following are equivalent:

(a) Metric completeness: $(N,d_0)$ is a complete metric space.

(b) Geodesic completeness: For any $v \in TN$, the geodesic $c(t)$ in N with $c'(0) = v$ is defined for all positive and negative real numbers $t \in \mathbb{R}$.

(c) For some $p \in N$, the exponential map $\exp_p$ is defined on the entire tangent space $T_pN$ to N at p.

(d) Finite compactness: Every subset K of N that is $d_0$ bounded (i.e., $\sup\{d_0(p,q) : p,q \in K\} < \infty$) has compact closure.

Furthermore, if any of (a) through (d) holds, then

(e) Given any $p,q \in N$, there exists a smooth geodesic segment c from p to q with $L_0(c) = d_0(p,q)$.

A Riemannian manifold $(N,g_0)$ is said to be *complete* provided any one (and hence all) of conditions (a) through (d) is satisfied. It should be stressed that the Hopf-Rinow theorem guarantees the equivalence of metric and geodesic completeness and also that *all* Riemannian metrics for a compact smooth manifold are complete. Unfortunately, *none* of these statements are valid for arbitrary Lorentzian manifolds.

A remaining question for noncompact but paracompact manifolds is the existence of complete Riemannian metrics. This was settled by Nomizu and Ozeki's (1961) proof that given any Riemannian metric $g_0$ for N, there is a *complete* Riemannian metric for N globally conformal to $g_0$. Since any paracompact, connected smooth manifold N admits a Riemannian metric by a partition of unity argument, N also admits a complete Riemannian metric.

We now turn our attention to the Lorentzian manifold (M,g). A Lorentzian metric g for the smooth paracompact manifold M is the assignment of a nondegenerate bilinear form $g|_p : T_pM \times T_pM \longrightarrow \mathbb{R}$ with diagonal form (-, +, ..., +) to each tangent space. It is well known that if M is compact and $\chi(M) \neq 0$, then M admits *no* Lorentzian metrics. On the other hand, any noncompact manifold admits a Lorentzian metric. Geroch (1968a) and Marante (1972) have also shown that a smooth Hausdorff manifold which admits a Lorentzian metric is paracompact.

Nonzero tangent vectors are classified as *timelike*, *spacelike*, *nonspacelike*, or *null*, respectively, according to whether $g(v,v) < 0$, resp., $> 0$, $\leq 0$, $= 0$. [Some authors use the convention (+, -, ..., -) for the Lorentzian metric and hence all of the inequality signs in the above definition are reversed for them.] A Lorentzian manifold

(M,g) is said to be *time oriented* if M admits a continuous, nowhere vanishing timelike vector field X. This vector field is used to separate the nonspacelike vectors at each point into two classes, called *future directed* and *past directed*. A *space-time* is then a Lorentzian manifold (M,g) together with a choice of time orientation. We will usually work with space-times below.

In order to define the Lorentzian distance function and discuss its properties, we need to introduce some concepts from elementary causality theory. It is standard to write $p \ll q$ if there is a future-directed piecewise smooth timelike curve in M from p to q, and $p \leqslant q$ if $p = q$ or if there is a future directed piecewise smooth nonspacelike curve in M from p to q. The *chronological past* and *future* of p are then given respectively by $I^-(p) = \{q \in M : q \ll p\}$ and $I^+(p) = \{q \in M : p \ll q\}$. The *causal past* and *future* of p are defined as $J^-(p) = \{q \in M : q \leqslant p\}$ and $J^+(p) = \{q \in M : p \leqslant q\}$. The sets $I^-(p)$ and $I^+(p)$ are always open in any space-time, but the sets $J^-(p)$ and $J^+(p)$ are neither open nor closed in general (cf. Figure 1.1).

The *causal structure* of the space-time (M,g) may be defined as the collection of past and future sets at all points of M together with their properties. It may be shown that two strongly causal Lorentzian metrics $g_1$ and $g_2$ for M determine the same past and future sets at all points iff the two metrics are globally conformal [i.e., $g_1 = \Omega g_2$ for some smooth function $\Omega : M \longrightarrow (0,\infty)$]. Letting C(M,g) denote the set of Lorentzian metrics globally conformal to g, it follows that properties suitably defined using the past and future sets hold simultaneously either for all metrics in C(M,g) or for no metrics in C(M,g). Thus all of the basic properties of elementary causality theory depend only on the conformal class C(M,g) and not on the choice of Lorentzian metric representing C(M,g).

Perhaps the two most elementary properties to require of the conformal structure C(M,g) are either that (M,g) be chronological or that (M,g) be causal. A space-time (M,g) is said to be

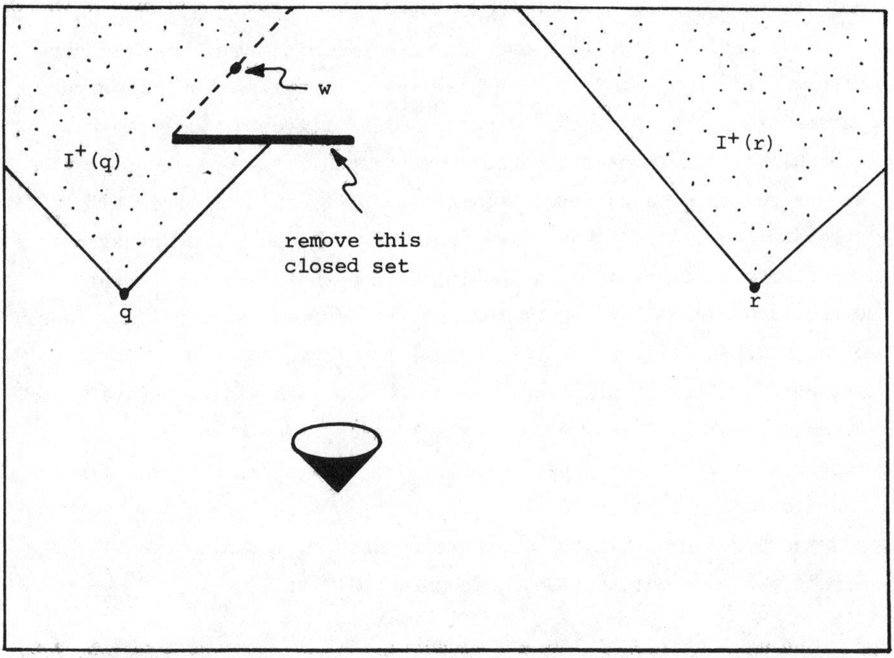


*Figure 1.1* The chronological (resp., causal) future of a point consists of all points which can be reached from that point by future-directed timelike (resp., nonspacelike) curves. In this example the causal future $J^+(r)$ of r is the closure of the chronological future $I^+(r)$ of r. On the other hand, the set $J^+(q)$ is not the closure of $I^+(q)$. In particular, w is in the closure of $I^+(q)$ but is not in $J^+(q)$.

*chronological* if $p \notin I^+(p)$ for all $p \in M$. This means that (M,g) contains no closed timelike curves. The space-time (M,g) is said to be *causal* if there is no pair of distinct points $p,q \in M$ with $p \leqslant q \leqslant p$. This is equivalent to requiring that (M,g) contain no closed nonspacelike curves.

Already at this stage, a basic difference emerges between Lorentzian and Riemannian geometry. On physical grounds, the space-times of general relativity are usually assumed to be chronological. But it is easy to show that if M is compact, (M,g) contains a closed timelike curve. Thus the space-times usually considered in general relativity are assumed to be noncompact.

In general relativity each point of a Lorentzian manifold corresponds to an event. Thus the existence of a closed timelike curve raises the possibility that a person might traverse some path and meet himself at an earlier age. More generally, closed nonspacelike curves generate paradoxes involving causality and are thus said to "violate causality." Even if a space-time has no closed nonspacelike curves, it may contain a point p such that there are future-directed nonspacelike curves leaving arbitrarily small neighborhoods of p and then returning. This behavior is said to be a violation of strong causality at p. Space-times with no such violation are *strongly causal*. The strongly causal space-times form an important subclass of the causal space-times. For this class of space-times the Alexandrov topology with basis $\{I^+(p) \cap I^-(q) : p,q \in M\}$ for M and the given manifold topology are related as follows [cf. Kronheimer and Penrose (1967), Penrose (1972)].

THEOREM The following are equivalent:

(a) (M,g) is strongly causal.

(b) The Alexandrov topology induced on M agrees with the given manifold topology.

(c) The Alexandrov topology is Hausdorff.

We are ready at last to define the *Lorentzian distance function* $d = d(g) : M \times M \longrightarrow [0,\infty]$ of an arbitrary space-time. If $c : [0,1] \longrightarrow M$ is a piecewise smooth nonspacelike curve differentiable except at $0 = t_1 < t_2 < \cdots < t_k = 1$, the length $L(c) = L_g(c)$ of c is given by the formula

$$L(c) = \sum_{i=1}^{k-1} \int_{t_i}^{t_{i+1}} \sqrt{-g(c'(t),c'(t))}\, dt$$

If $p \ll q$, there are timelike curves from p to q (very close to piecewise null curves) of arbitrarily small length. Hence the infimum of Lorentzian arc lengths of all piecewise smooth curves

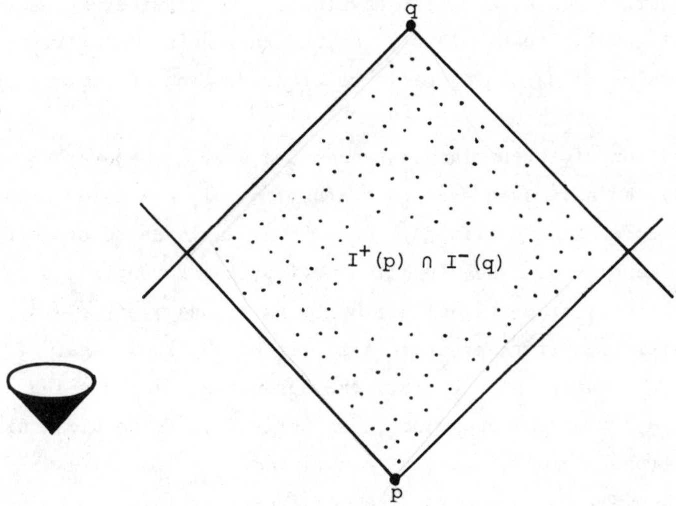


*Figure 1.2* Sets of the form $I^+(p) \cap I^-(q)$ with $p,q \in M$ arbitrary form a basis of the Alexandrov topology. This topology is always at least as coarse as the original topology on M. The Alexandrov topology agrees with the original topology iff (M,g) is strongly causal.

joining any two chronologically related points $p \ll q$ is zero. On the other hand, if $p \ll q$ and p, q lie in a geodesically convex neighborhood U, the future-directed timelike geodesic segment in U from p to q has the largest Lorentzian arc length among all nonspacelike curves in U from p to q. Thus the following definition for d(p,q) is natural: Fixing $p \in M$, set $d(p,q) = 0$ if $q \notin J^+(p)$, and otherwise calculate d(p,q) for $q \in J^+(p)$ as the supremum of Lorentzian arc length of all future-directed nonspacelike curves from p to q. Thus if $q \in J^+(p)$ and $\gamma$ is any future-directed nonspacelike curve from p to q, we have $L(\gamma) \leq d(p,q)$. Hence, unlike the Riemannian distance function, the Lorentzian distance function is not a priori finite valued. Indeed, a so-called totally vicious space-time may be characterized in terms of its Lorentzian distance function by the property that $d(p,q) = \infty$ for all $p,q \in M$. Also, if (M,g) is nonchronological and $p \in I^+(p)$, it follows that $d(p,p) = \infty$.

The Reissner-Nordström space-times, physically important examples of exact solutions to the Einstein equations in general relativity, also contain pairs of chronologically related, distinct points $p \ll q$ with $d(p,q) = \infty$.

By definition of Lorentzian distance, $d(p,q) = 0$ whenever $q \in M - J^+(p)$. We have even seen that $d(p,p) = \infty$ is possible. Thus for arbitrary Lorentzian manifolds, there is no analogue to property (3) of the Riemannian distance function. Also, the Lorentzian distance function tends from its definition to be a nonsymmetric distance. In particular, for any space-time, it may be shown that if $0 < d(p,q) < \infty$, then $d(q,p) = 0$. But the Lorentzian distance function does possess a useful analogue to property (2) of the Riemannian distance function. Namely, $d(p,q) \geqslant d(p,r) + d(r,q)$ for all $p,q,r \in M$ with $p \leqslant r \leqslant q$. The reversal of inequality sign is to be expected since nonspacelike geodesics in a Lorentzian manifold locally maximize rather than minimize arc length.

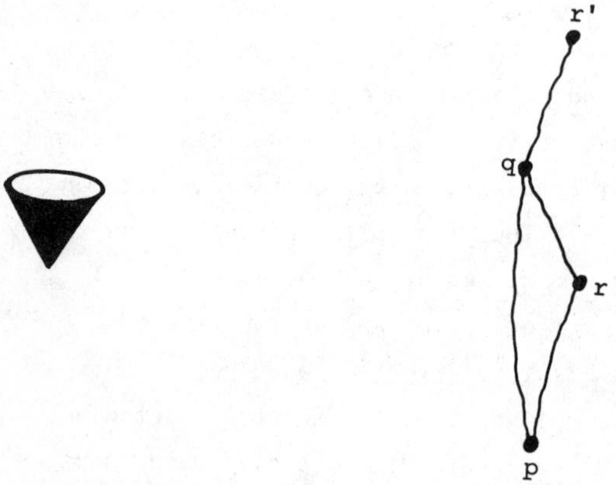


*Figure 1.3* If r is in the causal future of p and q is in the causal future of r, then the distance function satisfies the reverse triangle inequality $d(p,q) \geqslant d(p,r) + d(r,q)$. The reverse triangle inequality will not be valid in general for some point r' which is not causally between p and q.

Since $d(p,q) > 0$ iff $q \in I^+(p)$ and $d(q,p) > 0$ iff $q \in I^-(p)$, the distance function determines the chronology of $(M,g)$. On the other hand, conformally changing the metric changes distance but not the chronology, so that the chronology does not determine the distance function. Clearly, the distance function does not determine the sets $J^+(p)$ or $J^-(p)$ since $d(p,q) = 0$ not only for $q \in J^+(p) - I^+(p)$, but also for $q \in M - J^+(p)$.

Property (4) of the Riemannian distance function is the continuity of this function for all Riemannian metrics. For space-times, on the other hand, the Lorentzian distance function may fail to be upper semicontinuous. Indeed, the continuity of $d : M \times M \longrightarrow [0,\infty]$ has the following consequence for the causal structure of $(M,g)$ (cf. Theorem 3.24). If $(M,g)$ is a distinguishing space-time and d is continuous, then $(M,g)$ is causally continuous (cf. Chapter 2 for definitions of these concepts). Hence, it is necessary to accept the lack of continuity and lack of finiteness of the Lorentzian distance function for arbitrary space-times. The Lorentzian distance function is, at least, lower semicontinuous where it is finite. This may be combined with the upper semicontinuity in the $C^0$ topology of the Lorentzian arc length functional for strongly causal space-times to construct distance realizing nonspacelike geodesics in certain classes of space-times (cf. Sections 7.1 and 7.2).

With these remarks in mind, it is natural to ask if some class of space-times for which the Lorentzian distance function is finite valued and/or continuous may be found. The globally hyperbolic space-times turn out to be such a class. Here a space-time $(M,g)$ is said to be *globally hyperbolic* if it is strongly causal and satisfies the condition that $J^+(p) \cap J^-(q)$ is compact for all $p,q \in M$. It has been most useful in proving singularity theorems in general relativity to know that if $(M,g)$ is globally hyperbolic, its Lorentzian distance function is finite valued and continuous. Oddly enough, the finiteness of the distance function, rather than its continuity, characterizes globally hyperbolic space-times in the following sense (cf. Theorem 3.30). We say that a space-time (M,g) satisfies the *finite distance condition* provided that $d(g)(p,q) < \infty$ for all $p,q \in M$. It may then be shown that the strongly causal Lorentzian manifold (M,g) is globally hyperbolic iff (M,g') satisfies the finite distance condition for all $g' \in C(M,g)$.

Motivated by the definition of minimal curve in Riemannian geometry, we make the following definition for space-times.

DEFINITION A future directed nonspacelike curve $\gamma$ from p to q is said to be *maximal* if $L(\gamma) = d(p,q)$.

It may be shown (cf. Theorem 3.13), just as for minimal curves in Riemannian spaces, that if $\gamma$ is a maximal curve from p to q, then $\gamma$ may be reparameterized to a nonspacelike geodesic segment. This result may be used to construct geodesics in strongly causal space-times as limit curves of appropriate sequences of "almost maximal" nonspacelike curves (cf. Sections 7.1, 7.2).

In view of (e) of the Hopf-Rinow theorem for Riemannian manifolds, it is reasonable to look for a class of space-times satisfying the property that if $p \leqslant q$, there is a maximal geodesic segment from p to q. Using the compactness of $J^+(p) \cap J^-(q)$, one can show that globally hyperbolic space-times always contain maximal geodesics joining any two causally related points.

We are finally led to consider what can be said about Lorentzian analogues to the rest of the Hopf-Rinow theorem. Here, however, every conceivable thing goes wrong. Thus much of the difficulty in Lorentzian geometry from the viewpoint of global Riemannian geometry or its richness from the viewpoint of singularity theory in general relativity stems from the lack of a sufficiently strong analogue to the Hopf-Rinow theorem.

We now give a basic definition which corresponds to (b) of the Hopf-Rinow theorem.

DEFINITION A space-time (M,g) is said to be *timelike* (resp. *null*, *spacelike*, *nonspacelike*) *complete* if all timelike (resp. null, spacelike, nonspacelike) geodesics may be defined for all values $-\infty < t < \infty$ of an affine parameter t.

A space-time which is nonspacelike incomplete thus has a timelike or null geodesic which cannot be defined for all values of an affine parameter. Such space-times are said to be *singular* in the theory of general relativity.

It is first important to note that global hyperbolicity does *not* imply any of these forms of geodesic completeness. This may be seen by fixing points p and q in Minkowski space with $p \ll q$ and equipping $M = I^+(p) \cap I^-(q)$ with the Lorentzian metric it inherits as an open subset of Minkowski space. This space-time M is globally hyperbolic. Since geodesics in M are just the restriction of geodesics in Minkowski space to M, it follows that *every* geodesic in M is incomplete!

It was once hoped that timelike completeness might imply null completeness, etc. However, a series of examples have been given by Kundt, Geroch, and Beem of globally hyperbolic space-times for which timelike geodesic completeness, null geodesic completeness, and spacelike geodesic completeness are all logically inequivalent. Thus, there are globally hyperbolic space-times that are spacelike and timelike complete but null incomplete!

Metric completeness and geodesic completeness [(a) iff (b) of the Hopf-Rinow theorem] are unrelated for arbitrary Lorentzian manifolds. There are also Lorentzian metrics which are timelike geodesically complete, but also contain points p, q with $p \ll q$ such that *no* timelike geodesic from p to q exists (cf. Figure 5.1)

On the brighter side, there is some relationship between (a) and (d) of the Hopf-Rinow theorem for globally hyperbolic space-times. Since $d(p,q) = 0$ if $q \notin J^+(p)$, convergence of arbitrary sequences in (M,g) with respect to the Lorentzian distance function does not make sense. But timelike Cauchy completeness may be defined (cf. Section 5.3). It can be shown for globally hyperbolic space-times that timelike Cauchy completeness and a type of finite compactness are equivalent.

In addition, some results analogous to the Nomizu-Ozeki theorem mentioned above for Riemannian metrics have been obtained. For instance, given any strongly causal space-time (M,g), there is a conformal factor $\Omega : M \longrightarrow (0,\infty)$ such that the space-time $(M,\Omega g)$ is timelike and null geodesically complete (cf. Theorem 5.5). It is still unknown, however, whether this result can be strengthened to include spacelike geodesic completeness as well.

It should now be clear that while there are similarities between the Lorentzian and the Riemannian distance function, especially for globally hyperbolic space-times, there are also striking differences. In spite of these differences, the Lorentzian distance function has many uses similar to those of the Riemannian distance function.

In Chapter 7 the Lorentzian distance function is used in constructing maximal nonspacelike geodesics. These maximal geodesics play a key role in the proof of singularity theorems (cf. Chapter 11). In Chapter 8 the Lorentzian distance function is used to define and study the Lorentzian cut locus.

In Chapter 9 a Morse index theory is developed for both timelike and null geodesics. A number of global results for Lorentzian manifolds are obtained in Chapter 10 using the index theory and the Lorentzian distance function.

# Chapter 2

# LORENTZIAN MANIFOLDS AND CAUSALITY

Sections 2.1 and 2.2 give a brief review of elementary causality theory basic to this monograph as well as to general relativity. Then Section 2.3 describes an important relationship between the limit curve topology and the $C^0$ topology for sequences of nonspacelike curves in strongly causal space-times. Namely, if $\gamma : [a,b] \longrightarrow M$ is a future-directed, nonspacelike limit curve of a sequence $\{\gamma_n\}$ of future-directed nonspacelike curves, then a subsequence converges to $\gamma$ in the $C^0$ topology. This result is useful for constructing maximal geodesics in strongly causal space-times using the Lorentzian distance function (cf. Chapter 7 and Chapter 11, Section 4).

In Section 2.4 we study the causal structure of two-dimensional Lorentzian manifolds. In particular, we show that if $(M,g)$ is a space-time homeomorphic to $\mathbb{R}^2$, then $(M,g)$ is stably causal.

Section 2.5 gives a brief discussion of the theory of Lorentzian submanifolds and the second fundamental form needed for our discussion of singularity theory in Chapter 11.

An important splitting theorem of Geroch (1970) guarantees that a globally hyperbolic space-time may be written as a topological (although not necessarily *metric*) product $\mathbb{R} \times S$ where $S$ is a Cauchy hypersurface. This result suggests that product space-times of the form $(\mathbb{R} \times M, -dt^2 \oplus g)$ with $(M,g)$ a Riemannian manifold should be studied. While this class of space-times includes Minkowski space and the Einstein static universe, it fails to include the physically important exterior Schwarzschild and Robertson-Walker solutions to Einstein's equations.

In Section 2.6 we study a more general class of product space-times, the so-called *warped products*, which are space-times $M_1 \times_f M_2$ with metrics of the form $g_1 \oplus fg_2$. This class of metrics, studied for Riemannian manifolds by Bishop and O'Neill (1969), and for pseudo-Riemannian manifolds by O'Neill (1981), includes products, the exterior Schwarzchild space-times, and Robertson-Walker space-times. The following result, which may be regarded as a "metric converse" to Geroch's splitting theorem, is typical of the results of this section. Let $(\mathbb{R} \times M, -dt^2 \oplus g)$ be a Lorentzian product manifold with (M,g) an arbitrary Riemannian manifold. Then the following are equivalent:

(a) (M,g) is a complete Riemannian manifold.

(b) $(\mathbb{R} \times M, -dt^2 \oplus g)$ is globally hyperbolic.

(c) $(\mathbb{R} \times M, -dt^2 \oplus g)$ is geodesically complete.

## 2.1 LORENTZIAN MANIFOLDS AND CONVEX NORMAL NEIGHBORHOODS

Let M be a smooth connected paracompact Hausdorff manifold and let $\pi : TM \longrightarrow M$ denote the tangent bundle of M. A *Lorentzian metric* g for M is a smooth symmetric tensor field of type (0,2) on M such that for each $p \in M$, the tensor $g|_p : T_pM \times T_pM \longrightarrow \mathbb{R}$ is a nondegenerate inner product of signature (-, +, ..., +). In other words, a matrix representation of g at p will have one negative eigenvalue and all other eigenvalues will be positive.

A *Lorentzian manifold* (M,g) is a manifold M together with a Lorentzian metric g for M. All noncompact manifolds admit Lorentzian metrics. However, a compact manifold admits a Lorentzian metric iff its Euler characteristic vanishes [cf. Steenrod (1951, p. 207)]. The space of all Lorentzian metrics for M will be denoted by *Lor(M)*.

A nonzero vector $v \in TM$ is said to be *timelike* (resp., *nonspacelike, null, spacelike*) if $g(v,v) < 0$ (resp., $\leq 0$, $= 0$, $> 0$). A continuous vector field X on M is said to be *timelike* if $g(X(p),X(p)) < 0$ for all points $p \in M$. In general, a Lorentzian manifold does not necessarily have globally defined timelike vector fields. If $(M,g)$ does admit a timelike vector field $X : M \longrightarrow TM$, then $(M,g)$ is said to be *time oriented by* X. The timelike vector field X divides all nonspacelike tangent vectors into two separate classes, called future and past directed. Namely, a nonspacelike tangent vector $v \in T_pM$ is said to be *future* (resp., *past*) *directed* if $g(X(p),v) < 0$ [resp., $g(X(p),v) > 0$]. A Lorentzian manifold $(M,g)$ is said to be *time orientable* if $(M,g)$ admits a time orientation by some timelike vector field X. In this case, $(M,g)$ admits two distinct time orientations defined by X and $-X$, respectively.

Traditionally, a time-oriented Lorentzian manifold is called a *space-time*. More precisely,

DEFINITION 2.1 A *space-time* $(M,g)$ is a connected $C^\infty$ Hausdorff manifold of dimension $\geqslant 2$ which has a countable basis, a Lorentzian metric g of signature $(-, +, \ldots, +)$, and a time orientation.

We now show how to construct a time-oriented two sheeted Lorentzian covering manifold $\pi : (\tilde{M},\tilde{g}) \longrightarrow (M,g)$ for any Lorentzian manifold $(M,g)$ which is not time orientable. To this end, first let $(M,g)$ be an arbitrary Lorentzian manifold. Fix a base point $p_0 \in M$. Give a time orientation to $T_{p_0}M$ by choosing a timelike tangent vector $v_0 \in T_{p_0}M$ and defining a nonspacelike $w \in T_{p_0}M$ to be future (resp., past) directed if $g(v_0,w) < 0$ [resp., $g(v_0,w) > 0$]. Now let q be any point of M. Piecewise smooth curves $\gamma : [0,1] \longrightarrow M$ with $\gamma(0) = p_0$, $\gamma(1) = q$, may be divided into two equivalence classes as follows. Given $\gamma_1,\gamma_2 : [0,1] \longrightarrow M$ with $\gamma_1(0) = \gamma_2(0) = p_0$ and $\gamma_1(1) = \gamma_2(1) = q$, let $V_1$ (resp., $V_2$) be the unique parallel field (cf. Appendix A) along $\gamma_1$(resp., $\gamma_2$) with $V_1(0) = V_2(0) = v_0$. We say that $\gamma_1$ and $\gamma_2$ are equivalent if $g(V_1(1),V_2(1)) < 0$. If $\gamma_1$ and $\gamma_2$ are homotopic curves from $p_0$ to q, then $\gamma_1$ and $\gamma_2$ are equivalent. But equivalent curves are not necessarily homotopic. Given $\gamma : [0,1] \longrightarrow M$ with $\gamma(0) = p_0$, let $[\gamma]$ denote the equivalence class of

$\gamma$. Let $\tilde{M}$ consist of all such equivalence classes of piecewise smooth curves $\gamma : [0,1] \longrightarrow M$ with $\gamma(0) = p_0$. Define $\pi : \tilde{M} \longrightarrow M$ by $\pi([\gamma]) = \gamma(1)$. If $(M,g)$ is time orientable, then $\tilde{M} = M$. Otherwise, $\pi : \tilde{M} \longrightarrow M$ is a two-sheeted covering [cf. Markus (1955, p. 412)].

Suppose now that the Lorentzian manifold $(M,g)$ is *not* time orientable. It is standard from covering space theory to give the set $\tilde{M}$ a topology and differentiable structure such that $\pi : \tilde{M} \longrightarrow M$ is a two-sheeted covering manifold. Define a Lorentzian metric $\tilde{g}$ for M by $\tilde{g} = \pi^* g$, i.e., $\tilde{g}(v,w) = g(\pi_* v, \pi_* w)$. Then the map $\pi : (\tilde{M},\tilde{g}) \longrightarrow (M,g)$ is a local isometry.

In order to show that $(\tilde{M},\tilde{g})$ is time orientable, it is useful to establish a preliminary lemma. Fix a basepoint $\tilde{p}_0 \in \pi^{-1}(p_0)$ for $\tilde{M}$. Let $\tilde{v}_0 \in T_{\tilde{p}_0}\tilde{M}$ be the unique timelike tangent vector in $T_{\tilde{p}_0}\tilde{M}$ with $\pi_* \tilde{v}_0 = v_0$.

LEMMA 2.2 Let $\tilde{q} \in \tilde{M}$ be arbitrary and let $\tilde{\gamma}_1, \tilde{\gamma}_2 : [0,1] \longrightarrow \tilde{M}$ be two piecewise smooth curves with $\tilde{\gamma}_1(0) = \tilde{\gamma}_2(0) = p_0$ and $\tilde{\gamma}_1(1) = \tilde{\gamma}_2(1) = \tilde{q}$. If $\tilde{V}_1$, $\tilde{V}_2$ are the parallel vector fields along $\tilde{\gamma}_1$ and $\tilde{\gamma}_2$, respectively, with $\tilde{V}_1(0) = \tilde{V}_2(0) = \tilde{v}_0$, then $\tilde{g}(\tilde{V}_1(1),\tilde{V}_2(1)) < 0$.

*Proof.* Let $\gamma_1 = \pi \circ \tilde{\gamma}_1$ and $\gamma_2 = \pi \circ \tilde{\gamma}_2$. Since $\pi : (\tilde{M},\tilde{g}) \longrightarrow (M,g)$ is a local isometry, the vector fields $V_1 = \pi_*(\tilde{V}_1)$ and $V_2 = \pi_*(\tilde{V}_2)$ are parallel fields along $\gamma_1$ and $\gamma_2$, respectively, with $V_1(0) = V_2(0) = \pi_* \tilde{v}_0 = v_0$. Also, $g(V_1(1),V_2(1)) = g(\pi_* \tilde{V}_1(1), \pi_* \tilde{V}_2(1)) = \tilde{g}(\tilde{V}_1(1),\tilde{V}_2(1))$.

Now suppose that $\tilde{g}(\tilde{V}_1(1),\tilde{V}_2(1)) < 0$ is false. Since $\tilde{V}_1(1)$ and $\tilde{V}_2(1)$ are timelike tangent vectors, it follows that $\tilde{g}(\tilde{V}_1(1),\tilde{V}_2(1)) > 0$. Thus $g(V_1(1),V_2(1)) > 0$ at $q = \pi(\tilde{q})$. By definition of the equivalence relation on piecewise smooth curves from $p_0$ to q, we have $[\gamma_1] \neq [\gamma_2]$. From the construction of M, we know that $\tilde{\gamma}_1(1) = [\gamma_1]$ and $\tilde{\gamma}_2(1) = [\gamma_2]$. Thus $\tilde{\gamma}_1(1) \neq \tilde{\gamma}_2(1)$, in contradiction. □

Now we are ready to show

THEOREM 2.3 Suppose that $(M,g)$ is not time orientable. Then the two-sheeted Lorentzian covering manifold $(\tilde{M},\tilde{g})$ of $(M,g)$ constructed above is time orientable and hence is a space-time.

*Proof*. Let $\tilde{p}_0$ and $\tilde{v}_0$ be as above. Given any $\tilde{q} \in \tilde{M}$, let $\sigma : [0,1] \longrightarrow M$ be a smooth curve with $\sigma(0) = \tilde{p}_0$, $\sigma(1) = \tilde{q}$. Let $\tilde{V}$ be the unique parallel vector field along $\sigma$ with $\tilde{V}(0) = \tilde{v}_0$. Set $F^+(\tilde{q}) = \{\text{timelike } w \in T_{\tilde{q}}\tilde{M} : \tilde{g}(\tilde{V}(1),w) < 0\}$. By Lemma 2.2, the definition of $F^+(\tilde{q})$ is independent of the choice of $\sigma$. Hence $\tilde{q} \longrightarrow F^+(\tilde{q})$ consistently assigns a future cone to each tangent space $T_{\tilde{q}}\tilde{M}$, $\tilde{q} \in \tilde{M}$.

Now let $h$ be an auxiliary positive definite Riemannian metric for $\tilde{M}$. We may define a continuous nowhere zero timelike vector field $X$ on $\tilde{M}$ by choosing $X(\tilde{q})$ to be the vector in $F^+(\tilde{q})$ which is the unique $h$ unit vector in $F^+(\tilde{q})$ having a negative eigenvalue of $\tilde{g}$ with respect to $h$. That is, we may find a continuous function $\lambda : \tilde{M} \longrightarrow (-\infty,0)$ and a continuous timelike vector field $X$ on $\tilde{M}$ satisfying $X(\tilde{q}) \in F^+(\tilde{q})$, $h(X(\tilde{q}),X(\tilde{q})) = 1$ and $\tilde{g}(X(\tilde{q}),v) = \lambda(\tilde{q})h(X(\tilde{q}),v)$ for all $v \in T_{\tilde{q}}\tilde{M}$ and $\tilde{q} \in M$. □

Implicit in the proof of Theorem 2.3 is an alternate definition for a Lorentzian manifold $(M,g)$ to be time orientable. Namely, $(M,g)$ is time orientable if, fixing any base point $p_0 \in M$ and timelike tangent vector $v_0 \in T_{p_0}M$, the following condition is satisfied for all $q \in M$. Let $\gamma_1,\gamma_2 : [0,1] \longrightarrow M$ be any two smooth curves from $p$ to $q$. Hence a consistent choice of future timelike vectors $V_i(0) = v_0$ for $i = 1, 2$, then $g(V_1(1),V_2(1)) < 0$. This condition means that parallel translation of the future cone determined by $v_0$ at $p_0$ to any other point $q$ of $M$ is independent of the choice of path from $p$ to $q$. Hence a consistent choice of future timelike vectors for each tangent space may be made by parallel translation from $p_0$.

A Lorentzian manifold $(M,g)$ has a uniquely defined torsion free affine connection $\nabla$ which is compatible with the metric $g$. This means that

$$\nabla_X Y - \nabla_Y X = [X,Y]$$

and

$$X(g(Y,Z)) = g(\nabla_X Y,Z) + g(Y,\nabla_X Z)$$

for all smooth vector fields X, Y, Z on M. This connection $\nabla$ may be defined for Lorentzian manifolds in the same way that the Levi-Civita connection is defined for Riemannian manifolds. A discussion of the connection $\nabla$, *curvature tensor* R, *Ricci curvature* Ric and *scalar curvature* $\tau$ of g is given in Appendix A. Local coordinate representations are also included in this appendix.

A smooth curve in (M,g) is said to be *timelike* (resp., *nonspacelike*, *null*, *spacelike*) if its tangent vector is always timelike (resp., nonspacelike, null, spacelike). As in the Riemannian case, a *geodesic* $c : (a,b) \longrightarrow M$ is a smooth curve whose tangent vector moves by parallel displacement, i.e., $\nabla_{c'} c'(t) = 0$ for all $t \in (a,b)$. The tangent vector field $c'(t)$ of a geodesic c satisfies $g(c'(t),c'(t)) = \text{constant}$ for all $t \in (a,b)$ since

$$\frac{d}{dt}(g(c'(t),c'(t))) = 2g(\nabla_{c'} c'(t),c'(t)) = 0.$$

Consequently, a geodesic which is timelike, (resp. null, spacelike) for some value of its parameter is timelike (resp. null, spacelike) for all values of its parameter.

An *affine parameter* t for a geodesic curve c is any parameter for c such that $\nabla_{c'} c'(t) = 0$ for all values of the parameter. For timelike and spacelike geodesics, an affine parameter corresponds to a constant speed parameterization of the geodesic. For null geodesics, affine parameterizations are the closest analogue one may obtain to arc length parameterizations of nonnull geodesics.

The *exponential map* $\exp_p : T_pM \longrightarrow M$ is defined for Lorentzian manifolds just as for Riemannian manifolds. Given $v \in T_pM$, let $c_v(t)$ denote the unique geodesic in M with $c_v(0) = p$ and $c_v'(0) = v$. Then the exponential $\exp_p(v)$ of v is given by $\exp_p v = c_v(1)$ provided $c_v(1)$ is defined.

Let $v_1,\ldots,v_n$ be any basis for $T_pM$. For sufficiently small $(x_1,x_2,\ldots,x_n) \in \mathbb{R}^n$, the map

$$x_1v_1 + x_2v_2 + \cdots + x_nv_n \longrightarrow \exp_p(x_1v_1 + x_2v_2 + \cdots + x_nv_n)$$

is a diffeomorphism of a neighborhood of the origin of $T_pM$ onto a neighborhood U(p) of p in M. Thus assigning coordinates $(x_1,x_2,\ldots,x_n)$ to the point $\exp_p(x_1v_1 + \cdots + x_nv_n)$ in U(p) defines a coordinate chart for M, called *normal coordinates based at* p *for* U(p). The set U(p) is said to be a (simple) *convex neighborhood* of p if any two points in U(p) can be joined by a unique geodesic segment of (M,g) lying entirely in U(p). Whitehead (1932) has shown that any pseudo-Riemannian (hence Lorentzian) manifold has convex neighborhoods about each point, [cf. Hicks (1965, pp. 133-136)]. In fact, it may even be assumed that for each $q \in U(p)$, there are normal coordinates based at q containing U(p). We call such a neighborhood U(p) a *convex normal neighborhood* [cf. Hawking and Ellis (1973, p. 34)].

The next proposition is essential to the study of the local behavior of causality. A proof is given in Hawking and Ellis (1973, pp. 103-105).

PROPOSITION 2.4 Let U be a convex normal neighborhood of q. Then the points of U which can be reached by timelike (resp., nonspacelike) curves contained in U are those of the form $\exp_q(v)$, $v \in T_qM$, such that $g(v,v) < 0$ [resp., $g(v,v) \leq 0$].

## 2.2 CAUSALITY THEORY OF SPACE-TIMES

In a space-time (M,g) a (nowhere vanishing) nonspacelike vector field along a curve cannot continuously change from being future directed to being past directed. It follows that a smooth timelike, null, or nonspacelike curve in (M,g) is either always future directed or else always past directed.

We will use the standard notations $p \ll q$ if there is a smooth future-directed timelike curve from p to q, and $p \leqslant q$ if either $p = q$ or there is a smooth future-directed nonspacelike curve from p to q.

A *continuous* curve $\gamma : (a,b) \longrightarrow M$ is said to be a future-directed nonspacelike curve if for each $t_0 \in (a,b)$ there is an $\varepsilon > 0$ and a convex normal neighborhood $U(\gamma(t_0))$ of $\gamma(t_0)$ with $\gamma(t_0 - \varepsilon, t_0 + \varepsilon) \subset U(\gamma(t_0))$ such that given any $t_1$, $t_2$ with $t_0 - \varepsilon < t_1 < t_2 < t_0 + \varepsilon$, there is a smooth future-directed nonspacelike curve in $(U(\gamma(t_0)), g|U(\gamma(t_0)))$ from $\gamma(t_1)$ to $\gamma(t_2)$. It is necessary to use the convex normal neighborhood $U(\gamma(t_0))$ in this definition for the following reason. There exist space-times for which $p \ll q$ for all $(p,q) \in M \times M$. But, in these space-times, any continuous curve $\gamma$ satisfies both $\gamma(t_1) \ll \gamma(t_2)$ and $\gamma(t_2) \ll \gamma(t_1)$ for all $t_1$, $t_2$ in the domain of $\gamma$.

The *chronological future* $I^+(p)$ of p is the set $I^+(p) = \{q \in M : p \ll q\}$ and the *chronological past* is $I^-(p) = \{q \in M : q \ll p\}$. The *causal future* of p is $J^+(p) = \{q \in M : p \leqslant q\}$ and the *causal past* is $J^-(p) = \{q \in M : q \leqslant p\}$.

The relations $\ll$ and $\leqslant$ are clearly transitive. Moreover,

$$p \ll q \quad \text{and} \quad q \leqslant r \quad \text{implies} \quad p \ll r$$

and

$$p \leqslant q \quad \text{and} \quad q \ll r \quad \text{implies} \quad p \ll r$$

[cf. Penrose (1972, p. 14)]. If there is a future-directed timelike curve from p to q, there is a neighborhood U of q such that any point of U can be reached by a future-directed timelike curve. Consequently, it follows that

LEMMA 2.5 If p is any point of the space-time $(M,g)$, then $I^+(p)$ and $I^-(p)$ are open sets of M.

An example has been given in Chapter 1, Figure 1.1, to show that the sets $J^+(p)$, $J^-(p)$ are neither open nor closed in general.

It may happen that $p \in I^+(p)$. If so, there is a closed timelike curve through p and the space-time is said to have a causality violation. For example, on the cylinder $M = S^1 \times \mathbb{R}$ with the Lorentzian metric $ds^2 = -d\theta^2 + dt^2$, the circles t = constant are closed timelike curves. In this space-time, $I^+(p) = M$ for all $p \in M$. A number of causality conditions have been defined in general relativity in recent years because of the problems associated with examples of causality violations.

Space-times which do *not* contain any closed timelike curves [i.e., $p \notin I^+(p)$ for all $p \in M$] are said to be *chronological*. A space-time with no closed nonspacelike curves is said to be *causal*. Equivalently, a causal space-time contains no pair of distinct points p and q with $p \leqslant q \leqslant p$. The cylinder $M = S^1 \times \mathbb{R}$ with the Lorentzian metric $ds^2 = d\theta\, dt$ is an example of a chronological space-time that *fails* to be causal. The only closed nonspacelike curves in this example are the circles t = constant, which with proper parameterization, are null geodesics.

The chronological condition is the weakest causality condition which will be introduced. The next proposition guarantees that no compact space-time is either causal or chronological.

PROPOSITION 2.6 Any compact space-time (M,g) contains a closed timelike curve and thus fails to be chronological.

*Proof*. Since sets of the form $I^+(p)$ are open, it may be seen that $\{I^+(p) : p \in M\}$ forms an open cover of M. By compactness, we may extract a finite subcover $\{I^+(p_1), I^+(p_2), \ldots, I^+(p_k)\}$. Now $p_1 \in I^+(p_{i(1)})$ for some i(1) with $1 \leqslant i(1) \leqslant k$. Similarly, $p_{i(1)} \in I^+(p_{i(2)})$ for some index i(2) with $1 \leqslant i(2) \leqslant k$. Continuing inductively, we obtain an infinite sequence $\cdots p_{i(3)} \ll p_{i(2)} \ll p_{i(1)} \ll p_1$. Since k is finite, there are only a finite number of distinct $p_{i(j)}$'s. Thus there are repetitions on the list, and from the transitivity of $\ll$, it follows that $p_{i(n)} \in I^+(p_{i(n)})$ for some index $p_{i(n)}$. Thus (M,g) contains a closed timelike curve through $p_{i(n)}$. □

Recently, Tipler (1979) has proved that certain classes of compact space-times contain closed timelike *geodesics*, not just closed timelike curves. Since the proof uses the Lorentzian distance function as a tool, discussion of Tipler's result is postponed until Section 3.1, Theorem 3.15.

A space-time is said to be *distinguishing* if for all points $p,q \in M$, either $I^+(p) = I^+(q)$ or $I^-(p) = I^-(q)$ implies $p = q$. In a distinguishing space-time, distinct points have distinct chronological futures and chronological pasts. Thus points are distinguished both by their chronological futures and pasts.

A distinguishing space-time is said to be *causally continuous* if the set-valued functions $I^+$ and $I^-$ are outer continuous. Since $I^+$ and $I^-$ are always inner continuous [cf. Hawking and Sachs (1974, p. 291)], the causally continuous space-times are those distinguishing space-times for which both the chronological future and past of a point vary continuously with the point. Here $I^+$ is said to be *inner continuous* at $p \in M$ if for each compact set $K \subset I^+(p)$ there exists a neighborhood $U(p)$ of p such that $K \subset I^+(q)$ for each $q \in U(p)$. The set-valued function $I^+$ is *outer continuous at* $p \in M$ if for each compact set $K \subset M - \overline{I^+(p)}$ there exists some neighborhood $U(p)$ of p such that $K \subset M - \overline{I^+(q)}$ for each $q \in U(p)$. Inner and outer continuity of $I^-$ may be defined dually. An example of a space-time for which $I^-$ fails to be outer continuous is given in Figure 2.1. The concept of causal continuity was introduced by Hawking and Sachs (1974). For these space-times the causal structure may be extended to the causal boundary [cf. Budic and Sachs (1974)]. Furthermore, a metrizable topology may be defined on the causal completion of a causally continuous space-time [cf. Beem (1977)].

An open set U in a space-time is said to be *causally convex* if no nonspacelike curve intersects U in a disconnected set. Given $p \in M$, the space-time $(M,g)$ is said to be *strongly causal at* p if p has arbitrarily small causally convex neighborhoods. Thus p has arbitrarily small neighborhoods such that no nonspacelike curve

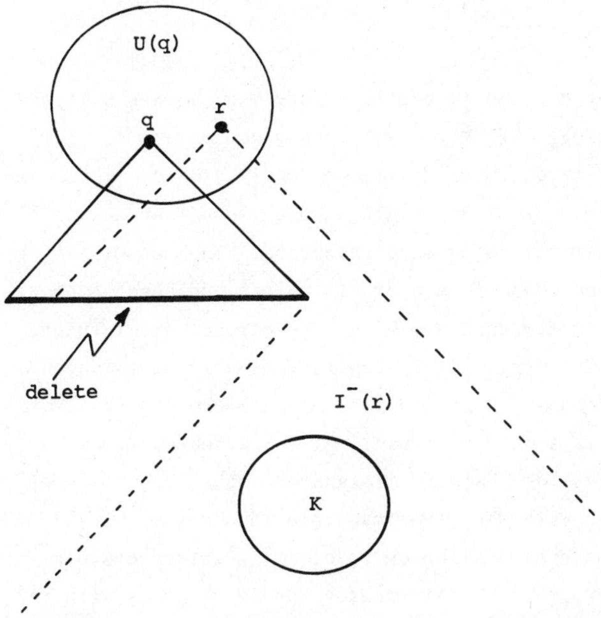


*Figure 2.1* A space-time which is *not* causally continuous is shown. The map $p \longrightarrow I^-(p)$ fails to be outer continuous at the point q. The compact set K is contained in $M - \overline{I^-(q)}$, yet each neighborhood U(q) of q contains some point r such that K is not contained in $M - \overline{I^-(r)}$

that leaves one of these neighborhoods ever returns. The space-time (M,g) is *strongly causal* if it is strongly causal at every point. It may be shown that the set of points of an arbitrary space-time (M,g) at which (M,g) is strongly causal is an open subset of M [cf. Penrose (1972, p. 30)]. It is not hard to show that strongly causal space-times are distinguishing.

Strongly causal space-times may be characterized in terms of the Alexandrov topology for M. The *Alexandrov topology* on an arbitrary space-time (M,g) is the topology given M by taking as a basis all sets of the form $I^+(p) \cap I^-(q)$, $p,q \in M$ (cf. Figure 1.2). The given manifold topology on M is always at least as fine as the Alexandrov topology since $I^+(p) \cap I^-(q)$ is an open set in the given topology by Lemma 2.5. The following relationship has been obtained between the given manifold topology and the Alexandrov topology [cf. Penrose (1972, p. 34)].

PROPOSITION 2.7 The Alexandrov topology for (M,g) agrees with the given manifold topology iff (M,g) is strongly causal.

*Proof.* Assume first that (M,g) is strongly causal. Thus each $p \in M$ has some convex normal neighborhood U(p) such that no nonspacelike curve intersects U(p) more than once. The set U(p) is a convex normal neighborhood of each of its points and hence Proposition 2.4 implies that for each $q \in U(p)$, the chronological future (resp., past) of q in $(U(p),g|U(p))$ consists of all points which can be reached by geodesic segments in U of the form $\exp_q(tv)$ for $0 \leq t \leq 1$, where v is a future- (resp., past-) directed timelike vector at q. This yields that the Alexandrov topology on $(U(p),g|U(p))$ agrees with the given manifold topology on U(p). Using the fact that no nonspacelike curve of (M,g) intersects U(p) more than once, it follows that the Alexandrov topology agrees with the given manifold topology.

Now assume that strong causality fails to hold at $p \in M$. Then there is a convex normal neighborhood V(p) of p such that if W(p) is any neighborhood of p with $W(p) \subset V(p)$, a nonspacelike curve starts in W(p), leaves V(p), and returns to W(p). It follows that all neighborhoods of p in the Alexandrov topology contain points outside of V(p). Thus the Alexandrov topology differs from the given manifold topology. □

In order to study causality breakdowns and geodesic incompleteness in general relativity, it is helpful to formulate the concept of *inextendibility* for nonspacelike curves. This may be done as follows. Let $\gamma : [a,b] \longrightarrow M$ be a curve in M. The point $p \in M$ is said to be the *endpoint* of $\gamma$ corresponding to t = b if

$$\lim_{t \to b^-} \gamma(t) = p$$

If $\gamma : [a,b] \longrightarrow M$ is a future- (resp., past-) directed nonspacelike curve with endpoint p corresponding to t = b, the point p is called a *future* (resp., *past*) *endpoint* of $\gamma$. A nonspacelike curve is said to be *future* (resp., *past*) *inextendible* if it has no future (resp., past) endpoint.

CONVENTION 2.8 A nonspacelike curve $\gamma : (a,b) \longrightarrow M$ is said to be *inextendible* if it is both future and past inextendible.

Causal space-times exist that contain inextendible nonspacelike curves having compact closure. An example given by Carter is shown in Figure 2.2 [cf. Hawking and Ellis (1973, p. 195)]. An inextendible nonspacelike curve which has compact closure and hence is contained in a compact set is said to be *imprisoned*. Thus Carter's example shows that imprisonment can occur in causal space-times.

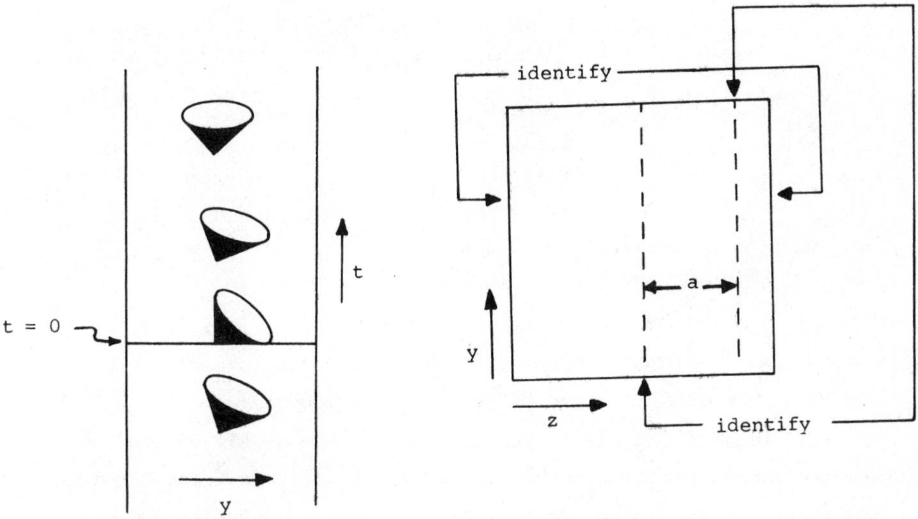


*Figure 2.2* A causal space-time (M,g) is shown which has imprisoned nonspacelike curves that are inextendible. Let a be an irrational number and let $M = \mathbb{R} \times S^1 \times S^1 = \{(t,y,z) \in \mathbb{R}^3 : (t,y,z) \sim (t, y, z + 1) \text{ and } (t,y,z) \sim (t, y + 1, z + a)\}$. The Lorentzian metric is given by $ds^2 = (\cosh t - 1)^2(dy^2 - dt^2) - dt\, dy + dz^2$.

Let $\gamma : [a,b) \longrightarrow M$ be a future-directed nonspacelike curve. Then $\gamma$ is said to be *future imprisoned* in the compact set K if there is some $t_0 < b$ such that $\gamma(t) \in K$ for all $t_0 < t < b$. The curve $\gamma$ is said to be *partially future imprisoned* in the compact set K if there exists an infinite sequence $t_n \uparrow b$ with $\gamma(t_n) \in K$ for each n.

If (M,g) is strongly causal and K is a compact subset of M, then K may be covered with a finite number of convex normal neighborhoods $\{U_i\}$ such that no nonspacelike curve which leaves some $U_i$ ever returns to that $U_i$. From this, it follows that

PROPOSITION 2.9 If (M,g) is strongly causal, then no inextendible nonspacelike curve can be partially future (or past) imprisoned in any compact set.

We now discuss another important class of space-times in general relativity, stably causal space-times. For this purpose as well as for later use, it is helpful to define the fine $C^r$ topologies on Lor(M).

Recall that Lor(M) denotes the space of all Lorentzian metrics on M. The *fine $C^r$ topologies* on Lor(M) may be defined by using a fixed countable covering $\mathcal{B} = \{B_i\}$ of M by coordinate neighborhoods with the property that each compact subset of M intersects only finitely many of the $B_i$'s. Such a coordinate cover is said to be locally finite. Let $\delta : M \longrightarrow (0,\infty)$ be a continuous function. Then $g_1, g_2 \in \text{Lor}(M)$ are said to be $\delta : M \longrightarrow (0,\infty)$ close in the $C^r$ topology, written $|g_1 - g_2|_r < \delta$, if for each $p \in M$ all of the corresponding coefficients and derivatives up to order r of the two metric tensors $g_1$ and $g_2$ are $\delta(p)$ close at p when calculated in the fixed coordinates of all $B_i \in \mathcal{B}$ which contain p. The sets $\{g_1 \in \text{Lor}(M) : |g_1 - g_2|_r < \delta\}$ with $g_2 \in \text{Lor}(M)$ arbitrary and $\delta : M \longrightarrow (0,\infty)$ an arbitrary continuous function form a basis for the fine $C^r$ topology on Lor(M). This topology may be shown to be independent of the choice of coordinate cover $\mathcal{B}$.

The $C^r$ topologies for $r = 0, 1, 2$ may be given the following interpretations.

REMARK 2.10 (a) Two Lorentzian metrics for M which are close in the fine $C^0$ topology have light cones which are close.
(b) Two Lorentzian metrics for M which are close in the fine $C^1$ topology have geodesic systems which are close (cf. Section 6.2).
(c) Two Lorentzian metrics for M which are close in the fine $C^2$ topology have curvature tensors which are close.

A space-time $(M,g)$ is said to be *stably causal* if there is a fine $C^0$ neighborhood $U(g)$ of g in Lor(M) such that each $g_1 \in U(g)$ is causal. Thus a stably causal space-time remains causal under small $C^0$ perturbations.

Stably causal space-times may be characterized in terms of a partial ordering $<$ for Lor(M) defined using light cones to compare Lorentzian metrics. Explicitly, if A is a subset of M, one defines $g_1 \leqslant_A g_2$ if for each $p \in A$ and $v \in T_pM$ with $v \neq 0$, $g_1(v,v) \leqslant 0$ implies $g_2(v,v) \leqslant 0$. One also defines $g_1 <_A g_2$ if for each $p \in A$ and $v \in T_pM$ with $v \neq 0$, $g_1(v,v) \leqslant 0$ implies $g_2(v,v) < 0$. We will write $g_1 < g_2$ (resp., $g_1 \leqslant g_2$) for $g_1 <_M g_2$ (resp., $g_1 \leqslant_M g_2$). Thus $g_1 < g_2$ means that at every point of M the light cone of $g_1$ is smaller than the light cone of $g_2$. It may be shown that $(M,g)$ is stably causal iff there exists some causal $g_1 \in \text{Lor}(M)$ with $g < g_1$.

A $C^0$ function $f : M \longrightarrow \mathbb{R}$ is a *global time function* if f is strictly increasing along each future-directed nonspacelike curve. A space-time $(M,g)$ admits a global time function iff it is stably causal [cf. Hawking (1968), Seifert (1977)]. However, there is generally no natural choice of a time function for a stably causal space-time.

Let $f : M \longrightarrow \mathbb{R}$ be a smooth function such that the gradient $\nabla f$ is always timelike. If $\gamma : (a,b) \longrightarrow M$ is a future-directed nonspacelike curve with nonvanishing tangent vector $\gamma'(t)$, then

$g(\nabla f(\gamma(t)),\gamma'(t)) = \gamma'(t)(f)$ must either be always positive or always negative. Thus f must be either strictly increasing or strictly decreasing along $\gamma$. It follows that f must be strictly increasing or decreasing along all future-directed nonspacelike curves. Hence f or -f is a smooth global time function for M. Furthermore, $\nabla f$ must be orthogonal to each of the level surfaces $f^{-1}(c) = \{p \in M : f(p) = c\}$, $c \in \mathbb{R}$, of f. These level surfaces are hypersurfaces orthogonal to a timelike vector field and are *spacelike*, i.e., g restricted to each of these hypersurfaces is a positive definite metric. Since the gradient of f is nonvanishing and df is an exact 1-form, it follows that M is foliated by the level surfaces $\{f^{-1}(c) : c \in \mathbb{R}\}$. Each nonspacelike curve $\gamma$ of M can intersect a given level surface at most once since f must be strictly increasing or decreasing along $\gamma$.

One of the most important causality conditions which we will discuss in this section is global hyperbolicity. Globally hyperbolic space-times have the important property that any pair of causally related points may be joined by a nonspacelike geodesic segment of maximal length.

DEFINITION 2.11 A strongly causal space-time (M,g) is said to be *globally hyperbolic* if for each pair of points $p,q \in M$, the set $J^+(p) \cap J^-(q)$ is compact.

A distinguishing space-time (M,g) is *causally simple* if $J^+(p)$ and $J^-(p)$ are closed subsets of M for all $p \in M$. It then follows that

PROPOSITION 2.12 A globally hyperbolic space-time is causally simple.

*Proof.* Suppose $q \in \overline{J^+(p)} - J^+(p)$ for some $p \in M$. Choose any $r \in I^+(q)$. Since $I^-(r)$ is open and $q \in \overline{J^+(p)}$, it may be seen that $r \in I^+(p)$ by taking a subsequence $\{q_n\} \subset J^+(p)$ with $q_n \longrightarrow q$ and using the fact that $p \leq q_n$, $q_n \ll r$ implies $p \ll r$. Consequently, $q \in \overline{J^+(p) \cap J^-(r)} - J^+(p) \cap J^-(r)$. But this is impossible since $J^+(p) \cap J^-(r)$ is compact, hence closed. $\square$

Globally hyperbolic space-times may be characterized using Cauchy surfaces. A *Cauchy surface* S is a subset of M which every inextendible nonspacelike curve intersects exactly once. It may be shown that a space-time is globally hyperbolic iff it admits a Cauchy surface [cf. Hawking and Ellis (1973, pp. 211-212)]. Furthermore, Geroch (1970) has established the following important structure theorem for globally hyperbolic space-times [cf. Sachs and Wu (1977b, p. 1155)].

THEOREM 2.13 If (M,g) is a globally hyperbolic space-time of dimension n, then M is homeomorphic to $\mathbb{R} \times S$ where S is an (n - 1)-dimensional topological submanifold of M, and for each t, $\{t\} \times S$ is a Cauchy surface.

The proof of this theorem uses a function $f : M \longrightarrow \mathbb{R}$ given by $f(p) = m(J^+(p))/m(J^-(p))$ where m is a measure on M with $m(M) = 1$. The level sets of f may be seen to be Cauchy surfaces as desired, but f is not necessarily smooth.

A time function $f : M \longrightarrow \mathbb{R}$ will be said to be a *Cauchy time function* if each level set $f^{-1}(c)$, $c \in \mathbb{R}$, is a Cauchy surface for M. In studying globally hyperbolic space-times, it is helpful to use Cauchy time functions rather than arbitrary time functions.

In a complete Riemannian manifold, any two points may be joined by a geodesic of minimal length. Avez (1963) and Seifert (1967) have obtained a Lorentzian analogue of this result for globally hyperbolic space-times.

THEOREM 2.14 Let (M,g) be globally hyperbolic and $p \leqslant q$. Then there is a nonspacelike geodesic from p to q whose length is greater than or equal to that of any other future-directed nonspacelike curve from p to q.

It should be emphasized that the geodesic in Theorem 2.14 is not necessarily unique. This result will also be discussed from the viewpoint of the Lorentzian distance function in Section 3.2.

An interesting list of 71 assertions on causality to be proved or disproved (together with answers) has been given in Geroch and

Horowitz (1979, pp. 289-293). We now give a diagram indicating the relations between the causality conditions discussed above [cf. Hawking and Sachs (1974, p. 295), Carter (1971b)].

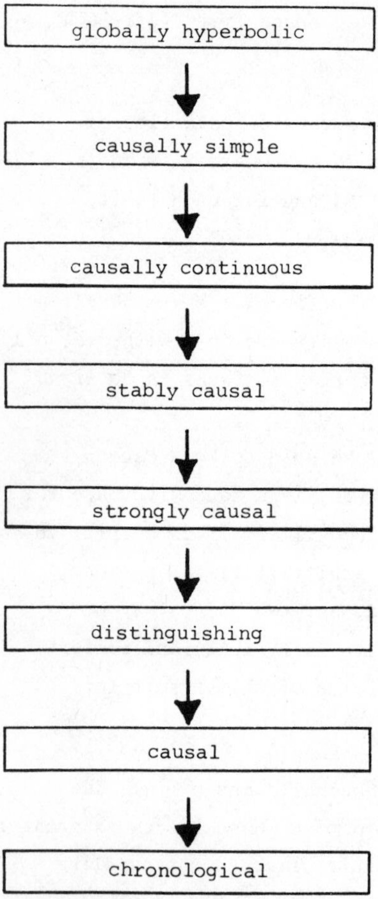


*Figure 2.3* This diagram illustrates the strengths of the causality conditions used in this book. Global hyperbolicity is the most restrictive causality assumption that we will use.

## 2.3 LIMIT CURVES AND THE $C^0$ TOPOLOGY ON CURVES

Two different forms of convergence for a sequence of nonspacelike curves $\{\gamma_n\}$ have been useful in Lorentzian geometry and general relativity [cf. Penrose (1972), Hawking and Ellis (1973)]. The first type of convergence uses the concept of a limit curve of a sequence of curves, while the second type uses the $C^0$ topology on curves. For arbitrary space-times, neither of these types of convergence is stronger than the other. However, we will show that for strongly causal space-times, these two forms of convergence are closely related. This relationship will be useful in constructing maximal geodesics in strongly causal space-times (cf. Sections 7.1, 7.2).

DEFINITION 2.15 A curve $\gamma$ is a *limit curve of the sequence* $\{\gamma_n\}$ if there is a subsequence $\{\gamma_m\}$ such that for all p in the image of $\gamma$, each neighborhood of p intersects all but a finite number of curves of the subsequence $\{\gamma_m\}$. The subsequence $\{\gamma_m\}$ is said to *distinguish* the limit curve $\gamma$.

In general, a sequence of curves $\{\gamma_n\}$ may have no limit curves or may have many limit curves. This is true even if the curves $\{\gamma_n\}$ are nonspacelike. Furthermore, even in causal space-times, a limit curve of a sequence of nonspacelike limit curves is not necessarily nonspacelike. For example, the curve $\gamma(u) = (0,0,u)$ in Carter's example (cf. Figure 2.2) is not nonspacelike although it is a limit curve of any sequence $\{\gamma_n\}$ of inextendible null geodesics contained in the set $t = 0$.

On the other hand, for strongly causal space-times we have

LEMMA 2.16 Let (M,g) be a strongly causal space-time. If $\gamma$ is a limit curve of the sequence $\{\gamma_n\}$ of nonspacelike curves, then $\gamma$ is nonspacelike.

*Proof.* Cover $\gamma$ by a locally finite collection $\{U_k\}$ of convex normal neighborhoods such that for each k, no nonspacelike curve which leaves $U_k$ ever returns. Since the causal relation $\leqslant$ is transitive, it suffices to show that $\gamma \cap U_k$ is nonspacelike for each k.

Let $\{\gamma_m\}$ be a subsequence of $\{\gamma_n\}$ which distinguishes $\gamma$. Given any pair of points $p,q \in \gamma \cap U_k$, we may find sequences $\{p_m\}$ and $\{q_m\}$ with $p_m, q_m \in \gamma_m$ for each m and $p_m \longrightarrow p$, $q_m \longrightarrow q$. The points $p_m$ and $q_m$ are causally related in $U_k$ for all sufficiently large m by construction of $U_k$ and the assumption that each $\gamma_m$ is nonspacelike. Taking limits, it follows that p and q are causally related in $U_k$. Since this holds for each pair $p,q \in \gamma \cap U_k$, it follows that the curve $\gamma \cap U_k$ is nonspacelike. □

The concept of a limit curve is closely related to the Hausdorff closed limit. Let $\{A_n\}$ be an arbitrary sequence of subsets (not necessarily curves) of M. The *Hausdorff upper and lower limits* of $\{A_n\}$ are defined, respectively, by [cf. Busemann (1955, p. 10)]

$$\lim\sup \{A_n\} = \{p \in M : \text{each neighborhood of } p \text{ intersects infinitely many of the sets } A_n\}$$

and

$$\lim\inf \{A_n\} = \{p \in M : \text{each neighborhood of } p \text{ intersects all but a finite number of the sets } A_n\}$$

The Hausdorff upper and lower limits always exist, although they may be empty, and are also always closed subsets of M. Clearly, $\lim\inf \{A_n\} \subset \lim\sup \{A_n\}$. If these two limits are equal, then the *Hausdorff closed limit* of $\{A_n\}$, denoted by $\lim \{A_n\}$, is defined to be $\lim \{A_n\} = \lim\inf \{A_n\} = \lim\sup \{A_n\}$.

A limit curve $\gamma$ of the sequence of curves $\{\gamma_n\}$ is contained in the Hausdorff upper limit $\lim\sup \{\gamma_n\}$. Further, a curve $\gamma$ is a limit curve of the sequence $\{\gamma_n\}$ iff $\gamma \subset \lim\inf \{\gamma_m\}$ for some subsequence $\{\gamma_m\}$ of $\{\gamma_n\}$.

We now turn to the proof (Proposition 2.18) of the existence of nonspacelike limit curves for sequences $\{\gamma_n\}$ of nonspacelike curves having points of accumulation. This result, an essential tool of causality theory in general relativity, is a consequence of Arzela's theorem (Theorem 2.17) which may be invoked since nonspacelike curves satisfy a local Lipschitz condition.

Let U be a convex normal neighborhood of (M,g) with compact closure $\overline{U}$ contained in a chart (V,x) having local coordinates $x = (x_1,\ldots,x_n)$ such that $f = x_1 : U \longrightarrow \mathbb{R}$ has a timelike gradient $\nabla f$ on U. Then f is a time function on U and whenever c is in the image of f, the level set $f^{-1}(c)$ is a spacelike hypersurface in U. For sufficiently small U there is some constant $K_0 > 0$ such that $g < g_0$ on U where $g_0$ is the flat metric on U given by

$$g_0 = -K_0\, dx_1^2 + \sum_{j=2}^{n} dx_j^2$$

Furthermore, each nonspacelike curve $\gamma$ in U joining $p,q \in U$ with $f(p) < f(q)$ can be reparameterized such that $\gamma$ is given in local coordinates by $\gamma(t) = (t,x_2(t),\ldots,x_n(t))$ for all t with $f(p) \leqslant t \leqslant f(q)$. Since $\gamma$ is nonspacelike for $g_0$ as well as g, $\gamma$ satisfies a Lipschitz condition of the form

$$\|\gamma(t_1) - \gamma(t_2)\|_2 \leqslant K_1|t_1 - t_2| \tag{2.1}$$

where $K_1 = (K_0 + 1)^{1/2}$. Here for $p,q \in U$, we use the given local coordinates to define

$$\|p - q\|_2 = \sqrt{\sum_{i=1}^{n} [x_i(p) - x_i(q)]^2}$$

and the constant $K_1$ depends on g, U and the choice of local coordinate chart (V,x). This Lipschitz condition implies that $\gamma$ is differentiable almost everywhere and that $|x_i'| \leqslant K_1$ along $\gamma$ for all $i = 1, 2, \ldots, n$.

Now let the space-time $(M,g)$ be given an auxiliary complete Riemannian metric $h$ with distance function $d_0$. By the Hopf-Rinow theorem, the closed balls $\{q \in M : d_0(p,q) \leq r\}$ are compact for all fixed $p \in M$ and $0 \leq r < \infty$. If the nonspacelike curve $\gamma(t)$ in $U$ is parameterized as $\gamma(t) = (t, x_2(t), \ldots, x_n(t))$ as above, then the length $L_0(\gamma|[t_1,t_2])$ of $\gamma$ from $t_1$ to $t_2$ with respect to $h$ is given by

$$L_0(\gamma|[t_1,t_2]) = \int_{t_1}^{t_2} \sqrt{\sum_{i,j} h_{ij} x_i' x_j'}\, dt$$

where $h_{ij}$ are the components of $h$ with respect to the local coordinates $x_1, \ldots, x_n$. Since $|x_i'| \leq K_1$, the length $L_0(\gamma|[t_1,t_2])$ satisfies

$$L_0(\gamma|[t_1,t_2]) \leq nH^{1/2}K_1|t_1 - t_2| \tag{2.2}$$

where $H$ is the supremum of $|h_{ij}|$ on the compact set $\overline{U}$ for $1 \leq i,j \leq n$. Thus any nonspacelike curve from the level set $f^{-1}(t_1)$ to the level set $f^{-1}(t_2)$ which lies in $U$ has length bounded by $nH^{1/2}K_1|t_1 - t_2|$. Furthermore, covering $(M,g)$ by a locally finite cover of sets with the properties of $U$ and $(V,x)$ above, it follows that any nonspacelike curve of $(M,g)$ defined on a compact interval of $\mathbb{R}$ must have finite length with respect to $h$. Thus every nonspacelike curve of $(M,g)$ may be given a parameterization which is an arc length parameterization with respect to $h$. Also, an inextendible curve $\gamma$ which has an arc length parameterization with respect to $h$ must be defined on all of $\mathbb{R}$ because $d_0$ is complete (cf. Lemma 2.52).

We now state a version of Arzela's theorem which may be established using standard techniques [cf. Munkres (1975, Section 7.5)].

THEOREM 2.17 Let $X$ be a locally compact Hausdorff space with a countable basis and let $(M,h)$ be a complete Riemannian manifold with distance function $d_0$. Assume that the sequence $\{f_n\}$ of functions $f_n : X \longrightarrow M$ is equicontinuous and that for each $x_0 \in X$ the set $\cup \{f_n(x_0)\}$ is bounded with respect to $d_0$. Then there exists a continuous function $f : X \longrightarrow M$ and a subsequence of $\{f_n\}$ which converges to f uniformly on each compact subset of X.

Using Arzela's theorem, we may now obtain the next proposition given in Hawking and Ellis (1973, p. 185), which guarantees the existence of limit curves for a sequence $\{\gamma_n\}$ of nonspacelike curves having points of accumulation.

PROPOSITION 2.18 Let $\{\gamma_n\}$ be a sequence of (future) inextendible nonspacelike curves in (M,g). If p is an accumulation point of the sequence $\{\gamma_n\}$, then there is a nonspacelike limit curve $\gamma$ of the sequence $\{\gamma_n\}$ such that $p \in \gamma$ and $\gamma$ is (future) inextendible.

*Proof.* We will give the proof only for inextendible curves since the proof for future-inextendible curves is similar.

Let h be an auxiliary complete Riemannian metric for M with distance function $d_0$ as above and give each $\gamma_n$ an arc length parameterization with respect to h. Then the domain of each $\gamma_n$ is $\mathbb{R}$ as each curve is assumed to be inextendible. Shifting parameterizations if necessary, we may then choose a subsequence $\{\gamma_m\}$ of $\{\gamma_n\}$ such that $\gamma_m(0) \longrightarrow p$ as $m \longrightarrow \infty$ since p is an accumulation point of the sequence $\{\gamma_n\}$. Using the fact that each $\gamma_m$ has an arc length parameterization with respect to h, we obtain

$$d_0(\gamma_m(t_1),\gamma_m(t_2)) \leqslant |t_1 - t_2| \tag{2.3}$$

for each m and $t_1,t_2 \in \mathbb{R}$. Thus the curves $\{\gamma_m\}$ form an equicontinuous family. Furthermore, since $\gamma_m(0) \longrightarrow p$ there exists an N such that $d_0(\gamma_m(0),p) < 1$ whenever $m \geqslant N$. This implies that for each fixed $t_0 \in \mathbb{R}$ the curve $\gamma_m \mid [-t_0,t_0]$ of the subsequence lies in the compact set $\{q \in M : d_0(p,q) \leqslant t_0 + 1\}$ whenever $m \geqslant N$. Hence the family $\{\gamma_m\}$ satisfies the hypotheses of Theorem 2.17 and we thus obtain a (continuous) curve $\gamma : \mathbb{R} \longrightarrow M$ and a subsequence $\{\gamma_k\}$ of the subsequence $\{\gamma_m\}$ such that $\{\gamma_k\}$ converges to $\gamma$ uniformly on each compact subset of $\mathbb{R}$. Clearly, $\gamma_k(0) \longrightarrow p = \gamma(0)$. The convergence of $\{\gamma_k\}$ to $\gamma$ also yields the inequality

$d_0(\gamma(t_1),\gamma(t_2)) \leqslant |t_1 - t_2|$ for all $t_1,t_2 \in \mathbb{R}$. It remains to show that $\gamma$ is nonspacelike and inextendible.

To show that $\gamma$ is nonspacelike, fix $t_1 \in \mathbb{R}$ and let U be a convex normal neighborhood of (M,g) containing $\gamma(t_1)$. Choose $\delta > 0$ such that the set $\{q \in M : d_0(\gamma(t_1),q) < \delta\}$ is contained in U. If $t_1 < t_2 < t_1 + \delta$, then (2.3) and the uniform convergence on compact subsets yields that for all large k the set $\gamma_k[t_1,t_2]$ lies in U. Using $\gamma_k(t_1) \longrightarrow \gamma(t_1)$, $\gamma_k(t_2) \longrightarrow \gamma(t_2)$, $\gamma_k(t_1) \leqslant_U \gamma_k(t_2)$ for all large k, and the fact that U is a convex normal neighborhood, we obtain that $\gamma(t_1) \leqslant_U \gamma(t_2)$. Thus $\gamma \mid [t_1,t_2]$ is a future-directed nonspacelike curve in U, [cf. Hawking and Ellis (1973, Lemma 4.5.1)]. It follows that $\gamma$ is a future-directed nonspacelike curve in (M,g).

It remains to show that $\gamma$ is inextendible. We will give the proof only of the future inextendibility since the past inextendibility may be proven similarly. To this end, assume that $\gamma$ is not future inextendible. Then $\gamma(t) \longrightarrow q_0 \in M$ as $t \longrightarrow \infty$. Let U' be a convex normal neighborhood of $q_0$ such that $\overline{U'}$ is a compact set contained in a chart (V,x) of M with local coordinates $(x_1,\ldots,x_n)$ such that $f = x_1 : U' \longrightarrow \mathbb{R}$ is a time function for U'. An inequality of the form of (2.2) shows that if $\gamma \mid [t_1,\infty) \subset U'$, then no nonspacelike curve in U' from the level set $f^{-1}(f(\gamma(t_1)))$ to the level set $f^{-1}(f(q_0))$ can have arc length with respect to h greater than some number $\delta' > 0$. On the other hand, for sufficiently large k we must have $\gamma_k[t_1 + 1, t_1 + \delta' + 2] \subset f^{-1}([f(\gamma(t_1)),f(q_0)])$. Since the length $L_0(\gamma_k[t_1 + 1, t_1 + \delta' + 2]) = \delta' + 1$ for all k, this yields a contradiction. □

Even if the inextendible nonspacelike curves of the sequence $\{\gamma_n\}$ are parameterized by arc length with respect to a complete Riemannian metric h, the limit curve $\gamma$ obtained in the proof of Proposition 2.18 need not be parameterized by arc length. This is a consequence of the fact that the Riemannian length functional, while lower semicontinuous, is not upper semicontinuous in the topology of uniform convergence on compact subsets. Even though the curve $\gamma$ constructed in the proof of Proposition 2.18 need not be parameterized by arc length, the curve $\gamma$ will still be defined on all of $\mathbb{R}$ provided each $\gamma_n$ is inextendible. Furthermore, if $(M,g)$ is strongly causal, the Hopf-Rinow theorem and Proposition 2.9 imply that $d_0(\gamma(0),\gamma(t)) \longrightarrow \infty$ as $|t| \longrightarrow \infty$. Here $d_0$ denotes the complete Riemannian distance function induced on M by h, as above.

In the globally hyperbolic case, a stronger version of Proposition 2.18 may be obtained.

COROLLARY 2.19 Let $(M,g)$ be globally hyperbolic. Suppose that $\{p_n\}$ and $\{q_n\}$ are sequences in M converging to $p,q \in M$, respectively, with $p \leqslant q$, $p \neq q$, and $p_n \leqslant q_n$ for each n. Let $\gamma_n$ be a future-directed nonspacelike curve from $p_n$ to $q_n$ for each n. Then there exists a future-directed nonspacelike limit curve $\gamma$ of the sequence $\{\gamma_n\}$ which joins p to q.

*Proof*. Let H be an auxiliary complete Riemannian metric on M with length functional $L_0$. Choose a finite cover of the compact set $J^+(p) \cap J^-(q)$ by convex normal neighborhoods $U_1, \ldots, U_k$, each of which has compact closure and such that no nonspacelike curve which leaves any $U_i$ ever returns to that $U_i$. As in the proof of Proposition 2.18, there exists a number $N_i$ for each i such that each nonspacelike curve $\gamma : [a,b] \longrightarrow U_i$ has length less than $N_i$ with respect to h [cf. equation (2.2)]. Thus if $U = U_1 \cup \cdots \cup U_k$ and $N = N_1 + \cdots + N_k$, every nonspacelike curve $\gamma : [a,b] \longrightarrow U$ must satisfy $L_0(\gamma) \leqslant N$.

Extend each given nonspacelike curve $\gamma_n$ to a future inextendible nonspacelike curve, also denoted by $\gamma_n$. We may assume that each $\gamma_n : [0,\infty) \longrightarrow M$ has been parameterized by arc length with respect to h [cf. equation (2.2)]. Thus if $U = U_1 \cup \cdots \cup U_k$ and Proposition 2.18, there exists a future-inextendible nonspacelike limit curve $\gamma : [0,\infty) \longrightarrow M$ with $\gamma(0) = p$ and a subsequence $\{\gamma_m\}$ of $\{\gamma_n\}$ such that $\gamma_m \longrightarrow \gamma$ uniformly on compact subsets of $[0,\infty)$. Using $\gamma_m(t_m) = q_m$ for $0 < t_m \leqslant N$ and $q_m \longrightarrow q$, we conclude that $\gamma$ passes through q for some parameter value $\tau$ which satisfies

$0 < \tau \leqslant N$. It follows that $\gamma \mid [0,\tau]$ is a nonspacelike limit curve of $\{\gamma_m | [0,t_m]\}$ which joins p to q. □

We now consider convergence in the $C^0$ topology [cf. Penrose (1972, p. 49)].

DEFINITION 2.20 Let $\gamma$ and all curves of the sequence $\{\gamma_n\}$ be defined on the closed interval [a,b]. The sequence $\{\gamma_n\}$ is said to *converge to* $\gamma$ *in the* $C^0$ *topology on curves* if $\gamma_n(a) \longrightarrow \gamma(a)$, $\gamma_n(b) \longrightarrow \gamma(b)$, and given any open set V containing $\gamma$, there is an integer N such that $\gamma_n \subset V$ for all $n \geqslant N$.

Any space-time contains a sequence $\{\gamma_n\}$ that has a limit curve $\gamma$, yet $\{\gamma_n\}$ does not converge to $\gamma$ in the $C^0$ topology. For let $\alpha,\beta : [0,1] \longrightarrow M$ be any two future-directed timelike curves with $\alpha([0,1]) \cap \beta([0,1]) = \emptyset$. Set

$$\gamma_n = \begin{cases} \alpha & \text{if } n = 2m \\ \beta & \text{if } n = 2m - 1 \end{cases}$$

Then $\{\gamma_n\}$ does not converge to either $\alpha$ or $\beta$ in the $C^0$ topology. However, the subsequences $\{\gamma_{2n}\}$ (resp., $\{\gamma_{2n-1}\}$) of $\{\gamma_n\}$ converge to $\alpha$ (resp., $\beta$) in the $C^0$ topology. A space-time which is not strongly causal may also contain a sequence $\{\gamma_n\}$ of nonspacelike curves which has a nonspacelike limit curve $\gamma$, yet no subsequence $\{\gamma_m\}$ of $\{\gamma_n\}$ converges to $\gamma$ in the $C^0$ topology on curves. This is illustrated in Figure 2.4 [cf. Hawking and Ellis (1973, p. 193) for a discussion of the causal properties of this example].

Conversely, a sequence of nonspacelike curves $\{\gamma_n\}$ may converge in the $C^0$ topology to some nonspacelike curve $\gamma$, but fail to have $\gamma$ as a limit curve. This may be seen on the cylinder $M = S^1 \times \mathbb{R}$ with the Lorentzian metric $ds^2 = d\theta\, dt$. Let $\gamma_n$ be the segment on the generator $\theta = 0$ given by $\gamma_n(t) = (0,t)$ for $0 \leqslant t \leqslant 1$ for all n. If $\gamma$ is the piecewise smooth nonspacelike curve obtained by going around the circle on the null geodesic

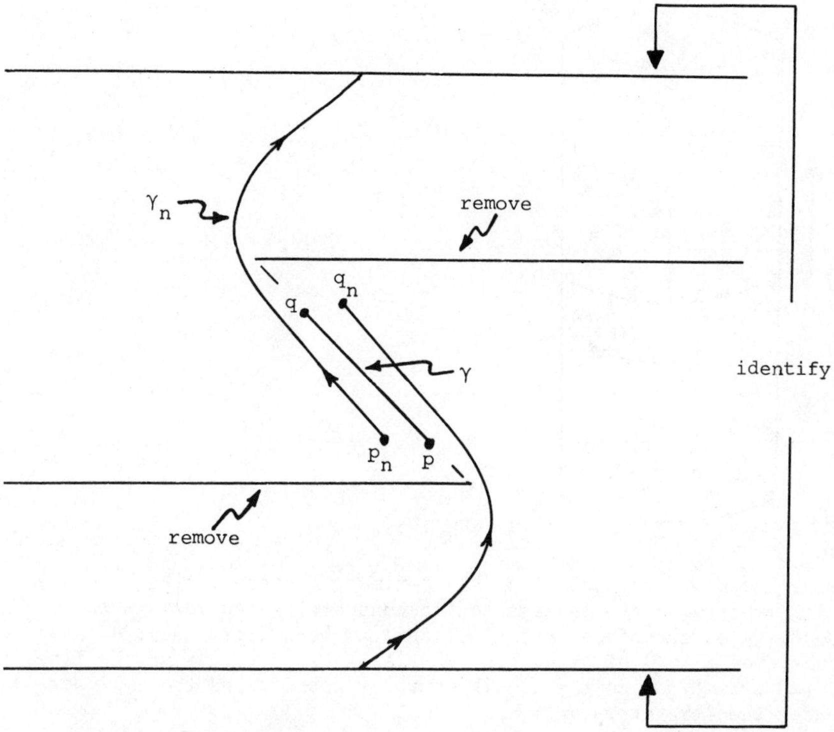


*Figure 2.4* A causal space-time (M,g) in which a sequence $\{\gamma_n\}$ of nonspacelike curves may have a limit curve $\gamma$, yet fail to have a subsequence which converges to $\gamma$ in the $C^0$ topology on curves, may be formed from a subset of Minkowski space as shown.

t = 0, then up the generator $\theta = 0$ from t = 0 to t = 1, then $\{\gamma_n\}$ converges to $\gamma$ in the $C^0$ topology, but $\gamma$ is not a limit curve of $\{\gamma_n\}$ (cf. Figure 2.5).

In strongly causal space-times, however, for nonspacelike sequences of curves, these two types of convergence are almost equivalent [cf. Beem and Ehrlich (1979a, p. 164)]. More precisely,

PROPOSITION 2.21 Let (M,g) be a strongly causal space-time. Suppose that $\{\gamma_n\}$ is a sequence of nonspacelike curves defined on [a,b] such that $\gamma_n(a) \longrightarrow p$ and $\gamma_n(b) \longrightarrow q$. A nonspacelike curve $\gamma : [a,b] \longrightarrow M$ with $\gamma(a) = p$ and $\gamma(b) = q$ is a limit curve of

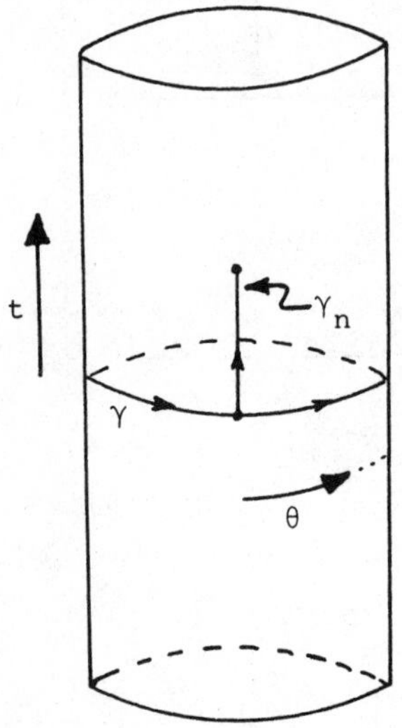


*Figure 2.5* In chronological space-times, a nonspacelike sequence of curves $\{\gamma_n\}$ may converge to the nonspacelike curve $\gamma$ in the $C^0$ topology on curves and yet $\gamma$ may fail to be a limit curve of any subsequence $\{\gamma_m\}$ of $\{\gamma_n\}$. The curves $\gamma_n$ are segments on the line $\theta = 0$ from $t = 0$ to $t = 1$. The curve $\gamma$ goes around the cylinder once, then traverses $\gamma_n$.

$\{\gamma_n\}$ iff there is a subsequence $\{\gamma_m\}$ of $\{\gamma_n\}$ which converges to $\gamma$ in the $C^0$ topology on curves.

*Proof.* ($\Rightarrow$) We may assume without loss of generality that $\gamma$ and $\{\gamma_n\}$ are all future directed curves. Let V be any open set with $\gamma \subset V$. Cover the compact image of $\gamma$ with convex normal neighborhoods $W_1, \ldots, W_k$ such that each $W_i \subset V$ and no nonspacelike curve which leaves $W_i$ ever returns to $W_i$. There exists a subdivision $a = t_0 < t_1 < \cdots < t_j = b$ of $[a,b]$ such that for all $0 \leq i \leq j - 1$, each pair $\gamma(t_i)$, $\gamma(t_{i+1})$ lies in some $W_h$. Here $h = h(i)$ and $1 \leq h(i) \leq k$ for all i. Let $\{\gamma_m\}$ be a subsequence that distinguishes $\gamma$ as a limit curve. For each m, let $p(0,m) = \gamma_m(a)$ and $p(j,m) = \gamma_m(b)$. Furthermore, for each fixed i with $0 < i < j$, choose $p(i,m) \in \gamma_m$ such that $\{p(i,m)\}$ converges to $\gamma(t_i)$. Since $\gamma(t_{i+1})$ lies in the causal future of $\gamma(t_i)$ and M is strongly causal, the point $p(i + 1, m)$ lies in the causal future of $p(i,m)$ for all m larger than some $N_1$. Also, there is some $N_2$ such that $p(i,m)$ and $p(i + 1, m)$ lie in $W_{h(i)}$ for all $0 \leqslant i \leqslant j - 1$ and $m \geqslant N_2$. Let $N = \max\{N_1, N_2\}$. The portion of $\gamma_m$ joining $p(i,m)$ to $p(i + 1, m)$ must lie entirely in $W_{h(i)}$ for $m \geqslant N$ because no nonspacelike curve can leave $W_h$ and return. It follows that $\gamma_m \subset W_1 \cup \cdots \cup W_k \subset V$ for all $m \geqslant N$ as required.

($\Leftarrow$) Let $\{\gamma_m\}$ be a subsequence converging to $\gamma$ in the $C^0$ topology on curves. Set $A = \{t_0 \in [a,b]$ : each point of $\gamma \mid [a,t_0]$ is a limit point of the given subsequence$\}$. We wish to show that $A = [a,b]$. Clearly, $\gamma_m(a) \longrightarrow \gamma(a)$ implies $a \in A$. If $\tau = \sup\{t_0 : t_0 \in A\}$, then for each $a \leqslant t < \tau$ the point $\gamma(t)$ is a limit point of the subsequence $\{\gamma_m\}$. To show $\tau \in A$ we assume $\tau > a$ and let $\{t_k\}$ be a sequence with $t_k \longrightarrow \tau^-$. Each neighborhood $U(\gamma(\tau))$ of $\gamma(\tau)$ is also a neighborhood of $\gamma(t_k)$ for sufficiently large k and hence must intersect all but a finite number of curves of the subsequence $\{\gamma_m\}$. Thus $\gamma(\tau)$ is a limit point of $\{\gamma_m\}$ and A must be a closed subinterval of $[a,b]$. Assume that $\tau < b$. Using the strong causality of $(M,g)$ we may find a convex normal neighborhood V of $\gamma(\tau)$ such that no nonspacelike curve of $(M,g)$ which leaves V ever returns. Letting V be sufficiently small we may assume that $(V,g|V)$ is globally hyperbolic and that $f : V \longrightarrow \mathbb{R}$ is a Cauchy time function for $(V,g|V)$ with $f(V) = \mathbb{R}$ and $f(\gamma(\tau)) = 0$. We may also assume $\gamma(b) \notin V$. Then each inextendible nonspacelike curve of $(M,g)$ which has a nonempty intersection with V must intersect each Cauchy surface $f^{-1}(s)$ exactly once. Fix s with $0 < s < \infty$ and define $x(s) = \gamma \cap f^{-1}(s)$. This intersection exists because $\gamma(\tau) \in V$ and $\gamma(b) \notin V$. Since $\gamma(\tau)$ and $\gamma(b)$ are limit points of $\{\gamma_m\}$, the curves $\gamma_m$ must have a nonempty intersection with $f^{-1}(s)$ for all sufficiently large m. Set $x_m(s) = \gamma_m \cap f^{-1}(s)$ for all such m. In order to verify that $x_m(s) \longrightarrow x(s)$, we observe that for each neighborhood W of $\gamma$ the points $x_m(s)$ must lie in $W \cap f^{-1}(s)$ for all large m (cf. Figure 2.6). This shows that each $x(s)$ is a limit point of the subsequence $\{\gamma_m\}$. Consequently, the set A contains numbers greater than $\tau$, in contradiction to the definition of $\tau$. We conclude $A = [a,b]$, which shows $\gamma$ is a limit curve of the subsequence $\{\gamma_m\}$. □

Let $\gamma$ be a nonspacelike curve in a strongly causal space-time $(M,g)$. Choose a compact subset $K$ of $M$ such that $\gamma \subset \text{Int}(K)$. Let the nonspacelike curves which are contained in $K$ be given the $C^0$ topology. It is known [cf. Penrose (1972, p. 54)] that the

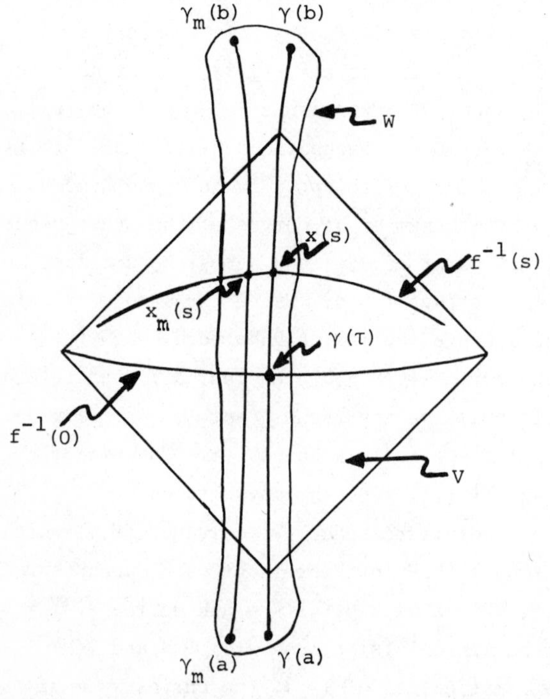


*Figure 2.6* In the proof of Proposition 2.21 the globally hyperbolic neighborhood $V$ of $\gamma(\tau)$ has a Cauchy time function $f : V \longrightarrow \mathbb{R}$ with $f(\gamma(\tau)) = 0$. For all large m the curves $\gamma_m$ must intersect the Cauchy surface $f^{-1}(s)$ at a single point $x_m(s)$. If W is any neighborhood of $\gamma$, then $x_m(s) \in W \cap f^{-1}(s)$ for all large m. We may choose W such that $W \cap f^{-1}(s)$ is as small a neighborhood of $x(s) = \gamma \cap f^{-1}(s)$ in $f^{-1}(s)$ as we wish. Thus $x_m(s) \longrightarrow x(s)$ and $x(s)$ must be a limit curve of the subsequence $\{\gamma_m\}$.

Lorentzian arc length functional $L(\gamma)$ [cf. Eq. (3.1)] is upper semicontinuous with respect to the $C^0$ topology on curves [cf. Busemann (1967, p. 10)]. This is the analogue of the well-known result that the Riemannian arc length functional is lower semicontinuous.

REMARK 2.22 Let $(M,g)$ be strongly causal and let $\gamma$ be a given nonspacelike curve in $(M,g)$. If the sequence $\{\gamma_n\}$ of nonspacelike curves converges to $\gamma$ in the $C^0$ topology on curves, then

$$L(\gamma) \geqslant \limsup L(\gamma_n)$$

## 2.4 TWO-DIMENSIONAL SPACE-TIMES

In this section, we consider the topological and causal structures of two-dimensional Lorentzian manifolds. Using the pair of null vector fields generated by the tangent vectors to the two null geodesics passing through each point of M, we show that the universal covering manifold of any two-dimensional Lorentzian manifold is homeomorphic to $\mathbb{R}^2$. We then show that any two-dimensional Lorentzian manifold homeomorphic to $\mathbb{R}^2$ is stably causal. In particular, every simply connected, two-dimensional Lorentzian manifold is causal. Thus *no* Lorentzian metric for $\mathbb{R}^2$ has any closed nonspacelike curves. It should also be noted that two- (but not higher) dimensional Lorentzian manifolds have the property that $(M,-g)$ is also Lorentzian. This is sometimes useful in obtaining results about all geodesics in $(M,g)$ from results valid in higher dimensions just for nonspacelike geodesics.

Let $(M,g)$ be an arbitrary two-dimensional Lorentzian manifold and fix a point $p \in M$. Choose a convex normal neighborhood $U(p)$ based at p and consider the following method of assigning local coordinates to points in $U(p)$ sufficiently close to p. Let the two null geodesics $\gamma_1$ and $\gamma_2$ through p be given parameterizations $\gamma_1 : (-\varepsilon_1,\varepsilon_1) \longrightarrow U(p)$ and $\gamma_2 : (-\varepsilon_2,\varepsilon_2) \longrightarrow U(p)$ with

$\gamma_1(0) = \gamma_2(0) = p$. For each point $q \in U(p)$ sufficiently close to $p$, the two null geodesics through $q$ will intersect $\gamma_1$ and $\gamma_2$ in $U(p)$ at unique points $\gamma_1(t_0)$ and $\gamma_2(s_0)$ repsectively. Assign coordinates $(t_0,s_0)$ to $q$. In these coordinates, the null geodesics near $p$ are contained in sets of the form $t = t_0$ or $s = s_0$. We have established

LEMMA 2.23 Let $(M,g)$ be a two-dimensional Lorentzian manifold. Then each $p \in M$ has local coordinates $x = (x_1,x_2)$ with $x(p) = 0$ such that each null geodesic in this neighborhood is contained in a set of the form $x_1$ = constant or $x_2$ = constant.

Suppose $X$ is a future directed timelike vector field on $M$. Then at each $p \in M$, there are two uniquely defined future-directed null vectors $n_1,n_2 \in T_pM$ such that $X(p) = n_1 + n_2$. Clearly, a sufficiently small neighborhood $U(p)$ of $p$ may be found such that $n_1$ and $n_2$ may be extended to continuous null vector fields $X_1$, $X_2$ defined on $U(p)$ with $X(q) = X_1(q) + X_2(q)$ for all $q \in U(p)$. If $M$ is simply connected, we now show $X_1$ and $X_2$ can be extended to all of $M$.

PROPOSITION 2.24 Let $(M,g)$ be a simply connected two-dimensional Lorentzian manifold. Then two smooth nonvanishing null vector fields $X_1$ and $X_2$ may be defined on $M$ such that $X_1$ and $X_2$ are linearly independent at each point of $M$.

*Proof*. Since $M$ is simply connected, $(M,g)$ is time orientable. Thus we may choose a smooth future-directed timelike vector field $X$ on $M$.

Fix a base point $p_0 \in M$ and let $X(p_0) = n_1 + n_2$ as above. Given any other point $q \in M$, let $\gamma : [0,1] \longrightarrow M$ be a curve from $p_0$ to $q$. There is exactly one way to define continuous null vector fields $X_1$ and $X_2$ along $\gamma$ such that $X_1(0) = n_1$, $X_2(0) = n_2$, and $X(\gamma(t)) = X_1(t) + X_2(t)$ for all $t \in [0,1]$. If $\eta : [0,1] \longrightarrow M$ is any other curve from $p_0$ to $q$, then $\gamma$ and $\eta$ are homotopic since $M$ is assumed to be simply connected. Hence if $Y_1$ and $Y_2$ were null vector fields along $\eta$ with $Y_1(0) = n_1$ and $Y_2(0) = n_2$, we would have

$Y_1(1) = X_1(1)$ and $Y_2(1) = X_2(1)$ by standard homotopy arguments. Thus this construction produces a pair of continuous vector fields $X_1$ and $X_2$ on M which are linearly independent at each point. □

COROLLARY 2.25 Let (M,g) be any two-dimensional Lorentzian manifold. Then the universal Lorentzian covering manifold $(\tilde{M},\tilde{g})$ of (M,g) is homeomorphic to $\mathbb{R}^2$.

*Proof.* Since $\tilde{M}$ is simply connected and two dimensional, $\tilde{M}$ is homeomorphic to $\mathbb{R}^2$ or $S^2$. But since the Euler characteristic of $S^2$ is nonzero, $S^2$ does not admit any nowhere zero continuous vector fields. □

Recall that an *integral curve* for a smooth vector field X on M is a smooth curve $\gamma$ such that $\gamma'(t) = X(\gamma(t))$ for all t in the domain of $\gamma$ [cf. Kobayashi and Nomizu (1963, p. 12)]. The following result is well known [cf. Hartman (1964, p. 156)].

PROPOSITION 2.26 Let X be a smooth nonvanishing vector field on $\mathbb{R}^2$ and let $\gamma : (a,b) \longrightarrow \mathbb{R}^2$ be a maximal integral curve of X. Then $\gamma(t)$ does not remain in any compact subset of $\mathbb{R}^2$ as $t \longrightarrow a^+$ (or $t \longrightarrow b^-$).

Now assume that (M,g) is a Lorentzian manifold homeomorphic to $\mathbb{R}^2$. Let $X_1$, $X_2$ be the null vector fields on M given by Proposition 2.24. Clearly, each null geodesic of (M,g) may be reparameterized to an integral curve of $X_1$ or $X_2$. Equivalently, the integral curves of $X_1$ and $X_2$ are said to be *null pregeodesics*. Suppose $\gamma : (a,b) \longrightarrow M$ is an inextendible null geodesic which may be reparameterized to an integral curve of $X_1$. If $\gamma(t_1) = \gamma(t_2)$ for some $t_1 \neq t_2$, since both $\gamma'(t_1)$ and $\gamma'(t_2)$ are scalar multiples of $X_1(\gamma(t_1))$, it follows from geodesic uniqueness that $\gamma$ is a smooth closed geodesic. However, this is impossible by Proposition 2.26. Thus Proposition 2.26 has the following corollary.

COROLLARY 2.27 If (M,g) is a Lorentzian manifold homeomorphic to $\mathbb{R}^2$, then (M,g) contains no closed null geodesics. Moreover, every inextendible null geodesic $\gamma : (a,b) \longrightarrow M$ is injective and hence contains no loops.

A family F of inextendible null geodesics is said to cover a manifold M *simply* if each point $p \in M$ lies on exactly one null geodesic of F. Suppose (M,g) is a Lorentzian manifold homeomorphic to $\mathbb{R}^2$. Then the integral curves of the null vector field $X_1$ (resp., $X_2$) given in Proposition 2.24 may be reparameterized to define a family $F_1$ (resp., $F_2$) of geodesics on M. Each family $F_i$ covers M since $X_i(p) \neq 0$ for i = 1, 2 and all $p \in M$. Furthermore, since exactly one integral curve of $X_i$ passes through any $p \in M$, each family $F_i$ covers M simply. Consequently Proposition 2.24 implies [cf. Beem and Woo (1969, p. 51)]

PROPOSITION 2.28 Let (M,g) be a Lorentzian manifold homeomorphic to $\mathbb{R}^2$. Then the inextendible null geodesics of (M,g) may be partitioned into two families $F_1$ and $F_2$ such that each of these families covers M simply.

Let $\gamma : (a,b) \longrightarrow M$ be an inextendible timelike curve and let $c : (\alpha,\beta) \longrightarrow M$ be an inextendible null geodesic. Obviously, in arbitrary two-dimensional Lorentzian manifolds, $\gamma$ and c may intersect more than once. However, if M is homeomorphic to $\mathbb{R}^2$, $\gamma$ and c intersect in at most one point [cf. Beem and Woo (1969, p. 52), Smith (1960b)].

PROPOSITION 2.29 Let (M,g) be a Lorentzian manifold homeomorphic to $\mathbb{R}^2$. Then each timelike curve intersects a given null geodesic at most once.

*Proof.* Let $c_0$ be an inextendible future-directed null geodesic in M, which we may assume belongs to the family $F_1$ defined by the null vector field $X_1$ as above. Suppose that $\sigma$ is a future-directed timelike curve in M which intersects $c_0$ twice (possibly at the same point). We may then find $a,b \in \mathbb{R}$ with $a < b$ such that $\sigma(a)$, $\sigma(b)$ lie on $c_0$ and $\sigma(t) \notin c_0$ for $a < t < b$. Since $\sigma$ is timelike, $\sigma$ is locally 1-1. Hence if $\sigma \mid [a,b]$ is not 1-1, $\sigma$ contains at worst closed timelike loops. Using one of these loops, it is possible to find $\alpha,\beta \in \mathbb{R}$ with $a < \alpha < \beta < b$ and a second null geodesic $c_1 \in F_1$ such that $\sigma \mid [\alpha,\beta]$ is 1-1, $\sigma(\alpha)$ and $\sigma(\beta)$ lie on $c_1$, and $\sigma(t) \notin c_1$ for $\alpha < t < \beta$. We will show below that if $\gamma : [a,b] \longrightarrow M$ is an injective future-directed timelike curve, there is no $c \in F_1$ such that $\gamma(a)$ and $\gamma(b)$ lie on $c$ but $\gamma(t) \notin c$ for $a < t < b$. If the original timelike curve $\sigma \mid [a,b]$ is injective, this argument applied to $\sigma \mid [a,b]$ yields the desired contradiction. If $\sigma \mid [a,b]$ is not injective but intersects $c_0$ at $\sigma(a)$ and $\sigma(b)$, then this argument applied to $c_1$ and $\sigma \mid [\alpha,\beta]$ yields the desired contradiction.

Thus the theorem will be established if we show that it is impossible to find an injective future-directed timelike curve $\gamma : [a,b] \longrightarrow M$ with $\gamma(a)$ and $\gamma(b)$ on $c_0$ and $\gamma(t) \notin c_0$ for $a < t < b$. Traversing $\gamma$ from $\gamma(a)$ to $\gamma(b)$ and then the portion of $c_0$ from $\gamma(b)$ to $\gamma(a)$ yields a closed Jordan curve which encloses a set $W$ with $\overline{W}$ compact (cf. Figure 2.7). Let $U$ be a convex normal neighborhood based on $\gamma(a)$. Choose $t_1$ with $a < t_1 < b$ and $\gamma(t_1) \in U$. Let $c_1$ be the null geodesic in $F_1$ passing through $\gamma(t_1)$. Since $c_1$ may be reparameterized to be an integral curve of $X_1$ and $c_1$ enters $\overline{W}$ at $\gamma(t_1)$, it follows by Proposition 2.26 that $c_1$ leaves $W$ at some point $\gamma(t_1')$ with $t_1' > t_1$. As $c_0$ intersects $\gamma$ at $\gamma(b)$, we must have $t_1' < b$. In particular, $[t_1,t_1'] \subset (a,b)$. Hence we have found a closed interval $[t_1,t_1'] \subset (a,b)$ such that $\gamma([t_1,t_1']) \subset \overline{W}$ and $\gamma$ intersects the null geodesic $c_1 \in F_1$ at $\gamma(t_1)$ and $\gamma(t_1')$ (cf. Figure 2.7).

We may now form a second closed Jordan curve by traversing $\gamma$ from $t_1$ to $t_1'$ followed by the portion of $c_1$ from $\gamma(t_1')$ to $\gamma(t_1)$. Repeating the argument of the preceding paragraph, we obtain a closed interval $[t_2,t_2'] \subset (t_1,t_1')$ such that the timelike curve $\gamma \mid [t_2,t_2']$ intersects a null geodesic $c_2$ in the family $F_1$ at $\gamma(t_2)$ and $\gamma(t_2')$, and such that $\gamma(t_2)$ is contained in a convex normal neighborhood of $\gamma(t_1)$. Inductively, we can construct a nested sequence of intervals $[t_{k+1},t'_{k+1}] \subset (t_k,t'_k)$ such that $\gamma(t_{k+1})$ lies in a convex normal neighborhood of $\gamma(t_k)$ and $\gamma \mid [t_{k+1},t'_{k+1}]$ intersects a null geodesic $c_{k+1} \in F_1$ at $\gamma(t_{k+1})$ and $\gamma(t'_{k+1})$. Moreover, the intervals $[t_k,t'_k]$ may be chosen such that $\cap_{k=1}^{\infty} [t_k,t'_k] = \{t_0\}$ for some $t_0 \in (a,b)$. We thus have constructed two sequences $t_k \uparrow t_0$ and $t'_k \downarrow t_0$ such that the timelike curve $\gamma$ intersects a null geodesic in $F_1$ at both $\gamma(t_k)$ and $\gamma(t'_k)$ for each $k \geqslant 1$. But this is impossible by Proposition 2.4. Hence the geodesic $\gamma$ : $[a,b] \longrightarrow M$ intersects $c_0$ at most once. □

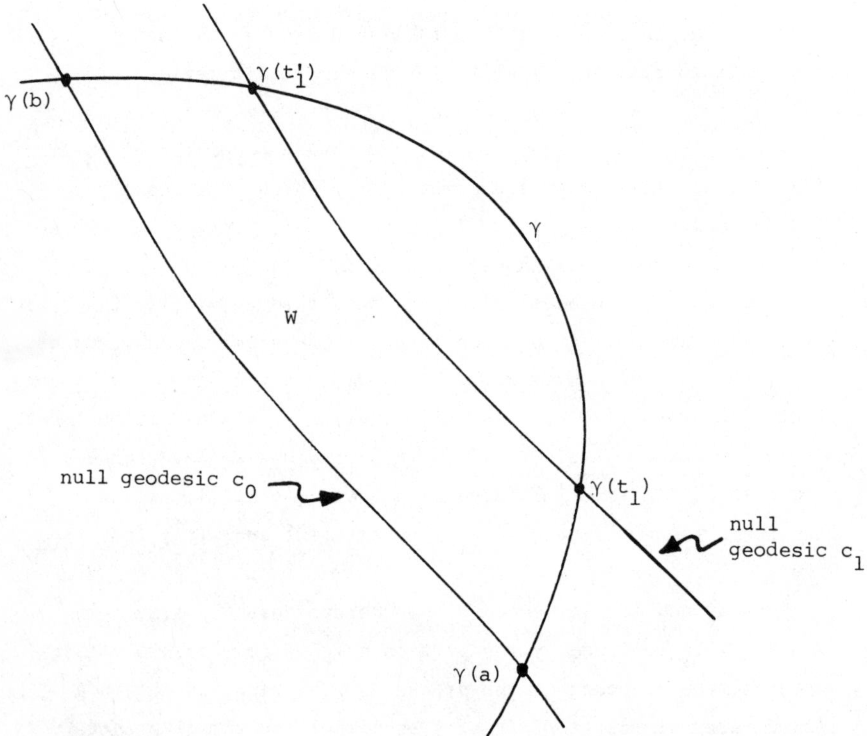


*Figure 2.7* In a Lorentzian manifold homeomorphic to $\mathbb{R}^2$, the timelike curve $\gamma$ is assumed to cross the null geodesic $c_0$ at $\gamma(a)$ and $\gamma(b)$. The null geodesic $c_1$ enters W at $\gamma(t_1)$ and first leaves at $\gamma(t'_1)$, $t'_1 > t_1$.

THEOREM 2.30 Let (M,g) be a Lorentzian manifold homeomorphic to $\mathbb{R}^2$. Then (M,g) is stably causal.

*Proof.* Recall that $g \in \mathrm{Lor}(M)$ is stably causal if there is a fine $C^0$ neighborhood U of g in Lor(M) such that all metrics in U are causal. Since strongly causal implies causal, it will thus follow that all metrics in Lor(M) are stably causal if all metrics in Lor(M) are strongly causal. Thus to prove the theorem, it is enough to show that if g is any Lorentzian metric for M, then (M,g) is strongly causal.

Thus suppose that g is a Lorentzian metric for M such that (M,g) is not strongly causal. Then there is some $p \in M$ such that strong causality fails at p. Let (U,x) be a chart about p guaranteed by Lemma 2.23 such that the null geodesics in U lie on the sets $x_1$ = constant and $x_2$ = constant. Since strong causality fails at p, there are arbitrarily small neighborhoods V of p with $V \subset U$ and timelike curves which begin at p, leave V, and then return to V. By Proposition 2.29, there are no closed timelike curves through p. Thus if $\gamma$ is a future directed timelike curve with $\gamma(0) = p$ which leaves V and then returns, we have $\gamma(t) \neq p$ for all $t > 0$. Since the null geodesics in U through p are given by $x_1 = 0$ and by $x_2 = 0$ in the local coordinates $x = (x_1, x_2)$ for U, it follows that $\gamma$ may be deformed to intersect one of the null geodesics through p on returning to V (cf. Figure 2.8). Hence $\gamma$ intersects a null geodesic in $F_1$ or $F_2$ twice, contradicting Proposition 2.29. □

COROLLARY 2.31 No Lorentzian metric for $\mathbb{R}^2$ contains any closed nonspacelike curves.

A different proof that any simply connected Lorentzian 2-manifold is strongly causal may be found in O'Neill [1981]. For $n \geqslant 3$, Lorentzian metrics which are not chronological and hence not strongly causal may be constructed on $\mathbb{R}^n$.

Every two-dimensional Lorentzian manifold (M,g) has an associated Lorentzian manifold (M,-g). The timelike curves of (M,-g) are the spacelike curves of (M,g) and vice versa. Using $M = \mathbb{R}^2$ and applying Corollary 2.31 to (M,-g) we obtain

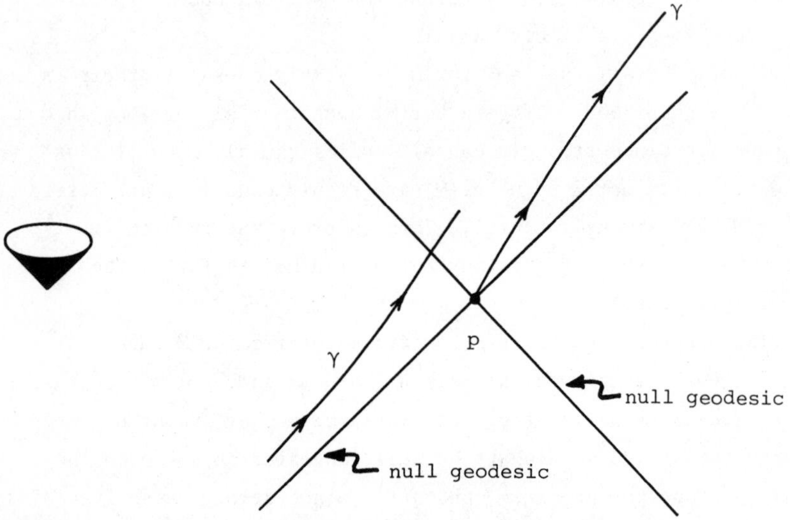


*Figure 2.8* (M,g) is a two-dimensional space-time such that strong causality fails at p. There is a future-directed timelike curve γ which starts at p, later returns close to p, and crosses one of the null geodesics through p.

COROLLARY 2.32 No Lorentzian metric for $\mathbb{R}^2$ contains any closed spacelike curves.

If (M,g) is two dimensional and both (M,g) and (M,-g) are stably causal, then using techniques given in Beem (1976a) one may show there is some smooth conformal factor $\Omega : M \longrightarrow (0,\infty)$ such that $(M,\Omega g)$ is geodesically complete. This yields

COROLLARY 2.33 Let (M,g) be a Lorentzian manifold homeomorphic to $\mathbb{R}^2$. Then there is a smooth conformal factor $\Omega : M \longrightarrow (0,\infty)$ such that $(M,\Omega g)$ is geodesically complete.

There are examples of two-dimensional space-times such that no global conformal change makes them nonspacelike geodesically complete (cf. Section 5.2). Thus Corollary 2.33 cannot be extended to all two-dimensional space-times by covering space arguments.

## 2.5 THE SECOND FUNDAMENTAL FORM

Let N be a smooth submanifold of the Lorentzian manifold (M,g). If $i : N \longrightarrow M$ denotes the inclusion map, by identifying $i_{*p}(T_pN)$ with $T_pN$, we may regard $T_pN$ as being a subspace of $T_pM$. Let $g_0 = i^* g$ denote the pullback of the Lorentzian metric g for M to a symmetric tensor field on N. Under the identification of $T_pN$ and $i_{*p}(T_pN)$, we may also identify $g_0$ at p and $g \mid T_pN \times T_pN$ for all $p \in N$. This identification will be used throughout this section.

DEFINITION 2.34 The submanifold N of (M,g) is said to be *nondegenerate* if for each $p \in N$ and nonzero $v \in T_pN$, there exists some $w \in T_pN$ with $g(v,w) \neq 0$. If in addition, $g \mid T_pN \times T_pN$ is positive definite for each $p \in N$, then N is said to be a *spacelike submanifold*. If $g \mid T_pN \times T_pN$ is a Lorentzian metric for each $p \in N$, then N is said to be a *timelike submanifold*.

For the rest of this section, we will suppose that N is a nondegenerate submanifold. Thus for each $p \in N$, there is a well-defined subspace $T_p^{\perp}N$ of $T_pM$ given by

$$T_p^{\perp}N = \{v \in T_pM : g(v,w) = 0 \text{ for all } w \in T_pN\}$$

which has the property that $T_p^{\perp}N \cap T_pN = \{0\}$. Consequently, there is a well-defined orthogonal projection map $P : T_pM \longrightarrow T_pN$. The connection $\nabla$ on (M,g) may be projected to a connection $\nabla^0$ on N by defining $\nabla^0_X Y = P(\nabla_X Y)$ for vector fields X, Y tangent to N. It is easily verified that $\nabla^0$ is the unique torsion free connection on $(N,g_0)$ satisfying

$$X(g_0(Y,Z)) = g_0(\nabla^0_X Y, Z) + g_0(Y, \nabla^0_X Z)$$

for all vector fields X, Y, Z on N. The second fundamental form, which measures the difference between $\nabla$ and $\nabla^0$, may be defined just as for Riemannian submanifolds [cf. Hermann (1968, p. 319), Bölts (1977, p. 25, pp. 51-52)].

DEFINITION 2.35 Let N be a nondegenerate submanifold of (M,g). Given $n \in T_p^\perp N$, define the *second fundamental form* $S_n : T_pN \times T_pN \longrightarrow \mathbb{R}$ *in the direction* n as follows. Given $x,y \in T_pN$, extend to local vector fields X, Y tangent to N and put

$$S_n(x,y) = g(\nabla_X Y\big|_p, n) = g(\nabla_X Y\big|_p - \nabla_X^0 Y\big|_p, n)$$

Define the *second fundamental form* $S : T_p^\perp N \times T_pN \times T_pN \longrightarrow \mathbb{R}$ by $S(n,x,y) = S_n(x,y)$. Given $n \in T_p^\perp N$, the *second fundamental form operator* $L_n : T_pN \longrightarrow T_pN$ is defined by $g(L_n(x),y) = S_n(x,y)$ for all $x,y \in T_pN$.

It may be checked that this definition of $S_n(x,y)$ is independent of the choice of extensions X, Y for $x,y \in T_pN$ and also that $S_n : T_pN \times T_pN \longrightarrow \mathbb{R}$ is a symmetric bilinear map. Furthermore, $S : T_p^\perp N \times T_pN \times T_pN \longrightarrow \mathbb{R}$ is trilinear for each $p \in N$.

LEMMA 2.36 Let N be a nondegenerate submanifold of (M,g). The second fundamental form S = 0 on N iff $\nabla_X Y = \nabla_X^0 Y$ for all vector fields X, Y tangent to N.

*Proof*. Obviously, Definition 2.35 implies that if $\nabla_X Y = \nabla_X^0 Y$ for all vector fields tangent to N, then S = 0.

Now suppose S = 0. Let $p \in N$ be arbitrary. We then have $g(\nabla_X Y\big|_p - \nabla_X^0 Y\big|_p, n) = 0$ for all $n \in T_p^\perp N$ and vector fields X, Y tangent to N. Since $g \mid T_pN \times T_pN$ is nondegenerate, $g \mid T_p^\perp N \times T_p^\perp N$ is also nondegenerate. Thus $\nabla_X Y\big|_p$ and $\nabla_X^0 Y\big|_p$ have the same projection onto $T_p^\perp N$. Since $T_pM = T_pN \oplus T_p^\perp N$, we have $\nabla_X Y\big|_p = \nabla_X^0 Y\big|_p$ as required. □

The second fundamental form may be used to characterize totally geodesic nondegenerate submanifolds of (M,g). A submanifold N of (M,g) is said to be *geodesic* at $p \in N$ if each geodesic γ of (M,g) with γ(0) = p and $\gamma'(0) \in T_pN$ is contained in N in some neighborhood of p. The submanifold N is said to be *totally geodesic* if it is geodesic at each of its points. The following proposition is the

Lorentzian analogue of a well-known Riemannian result [cf. Hermann (1968, p. 338), Cheeger and Ebin (1975, p. 23)].

PROPOSITION 2.37 Let N be a nondegenerate submanifold of (M,g). Then N is totally geodesic iff the second fundamental form S = 0 on N.

*Proof.* Given that S = 0 on N, Lemma 2.36 implies that $\nabla_X Y = \overset{0}{\nabla}_X Y$ for all vector fields X, Y tangent to N. Let $c : (-\varepsilon,\varepsilon) \longrightarrow M$ be a geodesic in (M,g) with $c'(0) = v \in T_pN$ for some $p \in N$. Also let $\gamma : (-\delta,\delta) \longrightarrow N$ be the geodesic in $(N,g_0)$ with $\gamma'(0) = v$. Since $\nabla_{\gamma'}\gamma' = \overset{0}{\nabla}_{\gamma'}\gamma' = 0$, the curve $\gamma$ is also a geodesic in (M,g). Set $\eta = \min(\varepsilon,\delta)$. From the uniqueness of the geodesic in (M,g) with the given initial direction v, we have $c(t) = \gamma(t)$ for all $t \in (-\eta,\eta)$. Hence $\gamma \mid (-\eta,\eta) \subset N$ as required.

Conversely suppose N is totally geodesic in (M,g). Let $p \in N$ be arbitrary. Given $n \in T_p^{\perp}N$ and $x \in T_pN$, let $c : J \longrightarrow N$ be the geodesic (in both M and N) with $c'(0) = x$. Extend $c'(t)$ to a vector field X tangent to N near p. We then have $S(n,x,x) = g(\nabla_X X|_p, n) = g(\nabla_{c'}c'(0),n) = g(0,n) = 0$. By polarization, it follows that $S(n,x,y) = 0$ for all $x,y \in T_pN$. Hence S = 0 on N. □

As will be seen in Chapter 11, the second fundamental form plays an important role in singularity theory in general relativity.

## 2.6 WARPED PRODUCTS

If (M,g) and (H,h) are two Riemannian manifolds, there is a natural product metric $g_0$ defined on the product manifold $M \times H$ such that $(M \times H, g_0)$ is again a Riemannian manifold. Bishop and O'Neill (1969) studied a larger class of Riemannian manifolds including products which they called *warped products*. If (M,g) and (H,h) are two Riemannian manifolds and $f : M \longrightarrow (0,\infty)$ is any smooth function, the product manifold $M \times H$ equipped with the metric $g \oplus fh$ is said to be a *warped product* and $f : M \longrightarrow (0,\infty)$ is called the *warping function*. Following Bishop and O'Neill, we will denote the Riemannian manifold $(M \times H,\ g \oplus fh)$ by $M \times_f H$. Bishop and O'Neill (1969, p. 23) showed that $M \times_f H$ is a complete Riemannian manifold iff both $(M,g)$ and $(H,h)$ are complete Riemannian manifolds. Utilizing this result, they were able to construct a wide variety of complete Riemannian manifolds of everywhere negative sectional curvature using warped products.

In this section, we will use warped product metrics to construct Lorentzian manifolds and will then study the causal structure and completeness properties of this class of Lorentzian manifolds. The theory for Lorentzian manifolds differs from the Riemannian theory somewhat, since the product of two Lorentzian manifolds $(M,g)$ and $(N,h)$ has signature $(-, -, +, \ldots, +)$ and hence is not Lorentzian. Nevertheless, warped product Lorentzian metrics may be constructed from products of Lorentzian and Riemannian manifolds. In particular, this product construction may be used to produce examples of bi-invariant Lorentzian metrics for Lie groups (cf. Section 4.5). A treatment of warped products of pseudo-Riemannian (not necessarily Lorentzian) manifolds including a calculation of their Riemannian and Ricci curvature tensors is given in O'Neill (1981).

Throughout this section, we will let $\pi : M \times H \longrightarrow M$ and $\eta : M \times H \longrightarrow H$ denote the projection maps given by $\pi(m,h) = m$ and $\eta(m,h) = h$ for $(m,h) \in M \times H$ respectively

DEFINITION 2.38 Let $(M,g)$ be an n-dimensional manifold $(n \geqslant 1)$ with a signature of $(-, +, \ldots, +)$ and let $(H,h)$ be a Riemannian manifold. Let $f : M \longrightarrow (0,\infty)$ be a smooth function. The *Lorentzian warped product* $M \times_f H$ is the manifold $\overline{M} = M \times H$ equipped with the Lorentzian metric $\overline{g}$ defined for $v,w \in T_{\overline{p}}\overline{M}$ by

$$\overline{g}(v,w) = g(\pi_* v, \pi_* w) + f(\pi(\overline{p}))h(\eta_* v, \eta_* w)$$

DEFINITION 2.39 A warped product $M \times_f H$ with $f = 1$ is said to be a *Lorentzian product* and will be denoted by $M \times H$.

REMARK 2.40 One may also obtain Lorentzian manifolds by considering warped products $H \times_f M$, where $(H,h)$ is a Riemannian manifold, $(M,g)$ is a Lorentzian manifold, and $f : H \longrightarrow (0,\infty)$ is a smooth function. The universal covering manifold of anti de Sitter space (cf. Section 4.3) is an example of a space-time important in general relativity which may be written as a warped product of the form $H \times_f M$ with H Riemannian and M Lorentzian but *not* as a warped product of the form $M \times_f H$ of Definition 2.38. We will only treat warped products of the form $M \times_f H$ in this book.

We begin our study of the causal properties of warped products with the following lemma.

LEMMA 2.41 The warped product $M \times_f H$ of $(M,g)$ and $(H,h)$ may be time oriented iff either $(M,g)$ is time oriented (if $\dim M \geq 2$) or $(M,g)$ is a one-dimensional manifold with a negative definite metric.

*Proof.* Suppose that $M \times_f H$ is time orientable. If $\dim M = 1$, then $(M,g)$ has a negative definite metric by Definition 2.38. Now consider the case $\dim M \geq 2$. Since $M \times_f H$ is time orientable, there exists a continuous timelike vector field X for $M \times_f H$. Since $f > 0$ and h is positive definite, we then have $g(\pi_* X, \pi_* X) \leq \bar{g}(X,X) < 0$. Thus the vector field $\pi_* X$ provides a time orientation for $(M,g)$.

Conversely, suppose first that $\dim M \geq 2$ and $(M,g)$ is time oriented by the timelike vector field V. Then V may be lifted to a timelike vector field $\bar{V}$ on $M \times H$ which satisfies $\pi_* \bar{V} = V$ and $\eta_* \bar{V} = 0$. Explicitly, fixing $\bar{p} = (m,b) \in M \times H$, there is a natural isomorphism

$$T_{\bar{p}}(M \times_f H) = T_{\bar{p}}(M \times H) \cong T_m M \times T_b H$$

Thus we may define $\bar{V}$ at $\bar{p}$ by setting $\bar{V}(\bar{p}) = (V(m), 0_b)$ using this isomorphism to identify $T_{\bar{p}}(M \times H)$ and $T_m M \times T_b H$. It is immediate from Definition 2.38 that $\bar{g}(\bar{V},\bar{V}) = g(V,V) < 0$. Hence $\bar{V}$ time orients $M \times_f H$ as required.

Now consider the case dim M = 1. It is then known that M is diffeomorphic to $S^1$ or $\mathbb{R}$. In either case, let T be a smooth vector field on M with g(T,T) = -1. Defining $\overline{T}(\overline{p}) = (T(\pi(\overline{p})), 0_{\eta(\overline{p})})$ as above, we have $\eta_* \overline{T} = 0$ so that $\overline{T}$ time orients $\overline{M}$. Note also in the case that $M = S^1$, the integral curves of $\overline{T}$ in $\overline{M}$ are closed timelike curves. Thus $\overline{M}$ is not chronological. □

LEMMA 2.42 Let (H,h) be an arbitrary Riemannian manifold and let M = (a,b) with $-\infty \leqslant a < b \leqslant +\infty$ given the negative definite metric $-dt^2$. For any smooth function f : M ⟶ (0,∞), the warped product $(M \times_f H, \overline{g})$ is stably causal.

*Proof.* The projection map $\pi : M \times H \longrightarrow M \subset \mathbb{R}$ serves as a time function. □

From the hierarchy of causality conditions given in Figure 2.3, we then obtain

COROLLARY 2.43 Let (H,h) be an arbitrary Riemannian manifold and let M = (a,b) with $-\infty \leqslant a < b \leqslant +\infty$ given the negative definite metric $-dt^2$. For any smooth function f : M ⟶ (0,∞), the warped product $(M \times_f H, \overline{g})$ is chronological, causal, distinguishing, and strongly causal.

In the proof of Lemma 2.41 above, we have seen that if $M = S^1$, then $(S^1 \times_f H, \overline{g})$ fails to be chronological, and hence fails to be causal, distinguishing, or strongly causal.

We now list some elementary properties of warped products that follow directly from Definition 2.38. A *homothetic map* F : $(M_1, g_1) \longrightarrow (M_2, g_2)$ is a diffeomorphism such that $F^*(g_2) = cg_1$ for some constant c. We remark that some authors only require homothetic maps to be smooth and not necessarily one-to-one.

REMARK 2.44 Let $M \times_f H$ be a Lorentzian warped product.

(a) For each b ∈ H, the restriction $\pi \mid \eta^{-1}(b) : \eta^{-1}(b) \longrightarrow M$ is an isometry of $\eta^{-1}(b)$ onto M.

(b) For each $m \in M$, the restriction $\eta \mid \pi^{-1}(m) : \pi^{-1}(m) \longrightarrow H$ is a homothetic map of $\pi^{-1}(m)$ with homothetic factor $1/f(m)$.

(c) If $v \in T(M \times H)$, then $g(\pi_* v, \pi_* v) \leqslant \bar{g}(v,v)$. Thus $\pi_* : T_p(M \times H) \longrightarrow T_{\pi(p)}M$ maps nonspacelike vectors to nonspacelike vectors and $\pi$ maps nonspacelike curves of $M \times_f H$ to nonspacelike curves of $M$.

(d) The map $\pi$ is length nondecreasing on nonspacelike curves since $|g(\pi_* v, \pi_* v)| \geqslant |\bar{g}(v,v)|$ if $v \in T(M \times H)$ is nonspacelike (cf. Section 3.1, formula (3.1) for the definition of Lorentzian arc length).

(e) For each $(m,b) \in M \times H$, the submanifolds $\pi^{-1}(m)$ and $\eta^{-1}(b)$ of $M \times_f H$ are nondegenerate in the sense of Definition 2.34.

(f) If $\phi : H \longrightarrow H$ is an isometry, then the map $\Phi = 1 \times \phi :$ $M \times_f H \longrightarrow M \times_f H$ given by $\Phi(m,b) = (m,\phi(b))$ is an isometry of $M \times_f H$.

(g) If $\psi : M \longrightarrow M$ is an isometry of $M$ such that $f \circ \psi = f$, then the map $\Psi = \psi \times 1 : M \times_f H \longrightarrow M \times_f H$ given by $\Psi(m,b) = (\psi(m),b)$ is an isometry of $M \times_f H$. Thus if $X$ is a Killing vector field on $M$ (i.e., $L_X g = 0$) with $X(f) = 0$, then the natural lift $\bar{X}$ of $X$ to $M \times_f H$ given by $\bar{X}(p) = (X(\pi(p)), 0_{\eta(p)})$ is a Killing vector field on $M \times_f H$.

LEMMA 2.45 Let $M \times_f H$ be a Lorentzian warped product. Then for each $b \in H$, the leaf $\eta^{-1}(b)$ is totally geodesic.

*Proof*. Since the map $\pi : M \times_f H \longrightarrow M$ is length nondecreasing on nonspacelike curves and since nonspacelike geodesics are locally length maximizing, it follows that any nonspacelike geodesic of $\eta^{-1}(b)$ (in the metric induced by the inclusion $\eta^{-1}(b) \subset M \times_f H$) is a geodesic in the ambiant manifold $M \times_f H$. Thus the second fundamental form vanishes on all nonspacelike vectors in $T(\eta^{-1}(b))$. Since any tangent vector in $T(\eta^{-1}(b))$ may be written as a linear combination of nonspacelike vectors in $T(\eta^{-1}(b))$, it follows that the second fundamental form vanishes identically. Hence, $\eta^{-1}(b)$ is totally geodesic by Proposition 2.37. □

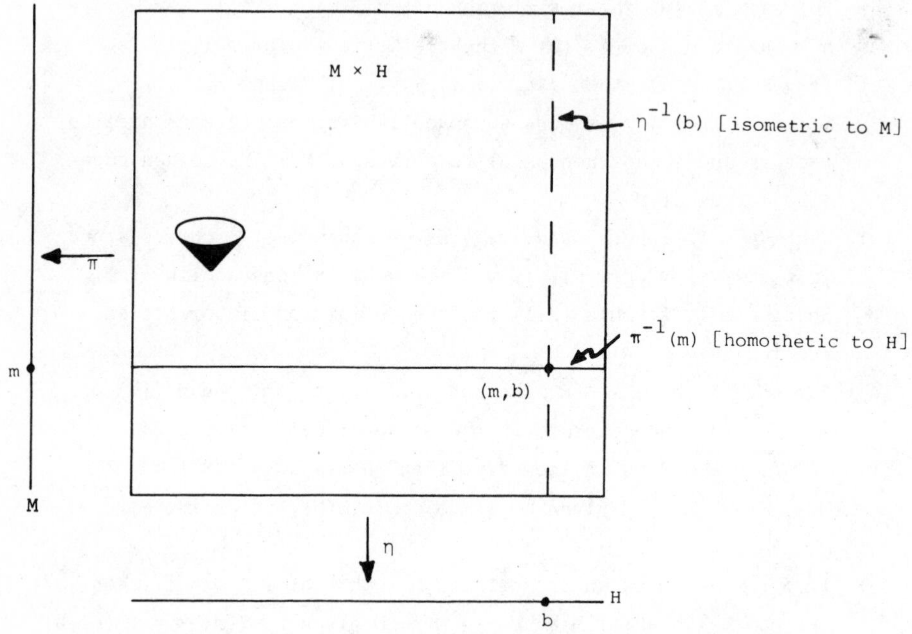


*Figure 2.9* Let (m,b) be a point of the warped product $M \times_f H$. Then the projection map $\pi$ restricted to $\eta^{-1}(b)$ is an isometry onto M and the projection map $\eta$ restricted to $\pi^{-1}(m)$ is a homothetic map onto H.

In view of Corollary 2.43, we may now restrict our attention to studying the basic causal properties of time-oriented Lorentzian warped products $(M \times_f H, \bar{g})$ with $\dim M \geq 2$.

LEMMA 2.46 Let $p = (p_1, p_2)$ and $q = (q_1, q_2)$ be two points in $M \times_f H$ with $p \ll q$ (resp., $p \leq q$) in $(M \times_f H, \bar{g})$. Then $p_1 \ll q_1$ (resp., $p_1 \leq q_1$) in $(M,g)$.

*Proof.* If $\gamma$ is a future-directed timelike (resp., nonspacelike) curve in $M \times_f H$ from p to q, then $\pi \circ \gamma$ is a future-directed timelike (resp., nonspacelike) curve in M from $p_1$ to $q_1$. □

While $\pi : M \times_f H \longrightarrow M$ takes nonspacelike curves to nonspacelike curves, $\pi$ does *not* preserve null curves. Indeed, it follows from Definition 2.38 that if $\gamma$ is any smooth null curve with $\eta_*\dot{\gamma}(t) \neq 0$ for all t, then $g(\pi_*\dot{\gamma}(t), \pi_*\dot{\gamma}(t)) < 0$ for all t.

For points p, q in the same leaf $\eta^{-1}(b)$ of $M \times_f H$, Lemma 2.46 may be strengthened as follows.

LEMMA 2.47 If $p = (p_1, b)$ and $q = (q_1, b)$ are points in the same leaf $\eta^{-1}(b)$ of $M \times_f H$, then $p \ll q$ (resp., $p \leq q$) in $(M \times_f H, \bar{g})$ iff $p_1 \ll q_1$ (resp., $p_1 \leq q_1$) in $(M,g)$.

*Proof.* By Lemma 2.46, it only remains to show that if $p_1 \ll q_1$ (resp., $p_1 \leq q_1$) in $(M,g)$, then $p \ll q$ (resp., $p \leq q$) in $(M \times_f H, \bar{g})$. But if $\gamma_1 : [0,1] \longrightarrow M$ is a future-directed timelike (resp., nonspacelike) curve in M from $p_1$ to $q_1$, then $\gamma(t) = (\gamma_1(t), b)$, $0 \leq t \leq 1$, is a future-directed timelike (resp., nonspacelike) curve in $M \times_f H$ from p to q. □

Lemma 2.47 implies that each leaf $\eta^{-1}(b)$, $b \in H$, has the same chronology and causality as $(M,g)$. In particular, Lemmas 2.46 and 2.47 imply that $(M \times_f H, \bar{g})$ has a closed timelike (resp., nonspacelike) curve iff $(M,g)$ has a closed timelike (resp., nonspacelike curve). Hence

PROPOSITION 2.48 Let $(M,g)$ be a space-time and let $(H,h)$ be a Riemannian manifold. Then the Lorentzian warped product $(M \times_f H, \bar{g})$ is chronological (resp., causal) iff $(M,g)$ is chronological (resp., causal).

A similar result holds for strong causality.

PROPOSITION 2.49 Let $(M,g)$ be a space-time and let $(H,h)$ be a Riemannian manifold. Then the Lorentzian warped product $(M \times_f H, \bar{g})$ is strongly causal iff $(M,g)$ is strongly causal.

*Proof.* We first show that if $(M,g)$ is not strongly causal at $p_1$, then $(M \times_f H, \bar{g})$ is not strongly causal at $p = (p_1, b)$ for any $b \in H$. Since $(M,g)$ is not strongly causal at $p_1$, there is an open neighborhood $U_1$ of $p_1$ in M and a sequence $\{\gamma_k : [0,1] \longrightarrow M\}$ of future-directed nonspacelike curves with $\gamma_k(0) \to p_1$, $\gamma_k(1) \to p_1$ as $k \to \infty$, but $\gamma_k(1/2) \notin U_1$ for all k. Define $\sigma_k : [0,1] \to M \times H$ by $\sigma_k(t) = (\gamma_k(t), b)$. Let $V_1$ be any open neighborhood of b in H and set $U = U_1 \times V_1$ in $M \times H$. Then U is an open neighborhood of $p = (p_1, b)$ in $M \times_f H$ and $\{\sigma_k\}$ is a sequence of nonspacelike future-directed curves in $M \times_f H$ with $\sigma_k(0) \to p$, $\sigma_k(1) \to p$ as $k \to \infty$, but $\sigma_k(1/2) \notin U$ for all k. Thus $(M \times_f H, \overline{g})$ is not strongly causal at p.

Conversely, suppose that strong causality fails at $p = (p_1, q_1)$ of $(M \times_f H, \overline{g})$. Let $(x_1, \ldots, x_i)$ be local coordinates on M near $p_1$ such that g has the form diag$\{-1, +1, \ldots, +1\}$ at $p_1$ and let $(x_{i+1}, \ldots, x_n)$ be local coordinates on H near $q_1$ such that fh has the form diag$\{+1, \ldots, +1\}$ at $q_1$. Then $(x_1, \ldots, x_i, x_{i+1}, \ldots, x_n)$ are local coordinates for $M \times_f H$ near p. Furthermore, $F_1 = x_1$ and $F_2 = x_1 \circ \pi$ are (locally defined) time functions for M near $p_1$ and for $M \times_f H$ near p, respectively. The failure of strong causality at p implies the existence of a sequence $\gamma_k : [0,1] \to M \times_f H$ of future-directed nonspacelike curves with $\gamma_k(0) \to p$ and $\gamma_k(1) \to p$ as $k \to \infty$, but $F_2(\gamma_k(1/2)) \geq \varepsilon > 0$ for all k and some parameterization of $\gamma_k$. Choose a neighborhood W of $p_1$ in M such that W is covered by the local coordinates $(x_1, \ldots, x_i)$ above and such that $\sup\{F_1(r) : r \in W\} \leq \varepsilon/2$. The curves $\pi \circ \gamma_k$ are then future-directed nonspacelike curves in M with $\pi \circ \gamma_k(0) \to p_1$, $\pi \circ \gamma_k(1) \to p_1$ and $\pi \circ \gamma_k(1/2) \notin W$. Hence W and $\{\pi \circ \gamma_k\}$ show that strong causality fails at $p_1$ in (M,g) as required. □

In Proposition 2.51, we prove the equivalence of stable causality for $(M \times_f H, \overline{g})$ and (M,g) for $\dim M \geq 2$. From this proposition and the last two propositions it follows that the basic causal properties of $(M \times_f H, \overline{g})$ are determined by those of (M,g).

Before giving Proposition 2.51 we state the following

REMARK 2.50 If $g < g_1$ on M, then there is a smooth conformal factor $\Omega : M \to (0,\infty)$ such that $\Omega g_1(v,v) < g(v,v)$ for all nontrivial vectors which are nonspacelike with respect to g.

PROPOSITION 2.51 Let $(M,g)$ be a space-time and $(H,h)$ a Riemannian manifold. Then the Lorentzian warped product $(M \times_f H, \bar{g})$ is stably causal iff $(M,g)$ is stably causal.

*Proof.* In this proof we will use the identification $T_p(M \times H) \cong T_{p_1}M \times T_bH$ for all $p = (p_1,b) \in M \times H$.

Assuming that $(M \times_f H, \bar{g})$ is stably causal, there exists $\bar{g}_1 \in \mathrm{Lor}(M \times H)$ such that $\bar{g} < \bar{g}_1$ and $\bar{g}_1$ is causal. If b is a fixed point of H, then we may assume without loss of generality that $\bar{g}_1 \mid \eta^{-1}(b)$ is nondegenerate since $\bar{g} \mid \eta^{-1}(b)$ is nondegenerate. Setting $\tilde{g}_1 = \bar{g}_1 \mid \eta^{-1}(b)$ and using $\pi \mid \eta^{-1}(b)$ to identify $\eta^{-1}(b)$ with M, we obtain a metric $g_1 \in \mathrm{Lor}(M)$ such that $\pi \mid \eta^{-1}(b)$ is an isometry of $(\eta^{-1}(b),\tilde{g}_1)$ onto $(M,g_1)$. Notice that since $(M \times H, \bar{g}_1)$ is causal, the space-time $(\eta^{-1}(b),\tilde{g}_1)$ is causal and hence also $(M,g_1)$. To show $g < g_1$ on M, we choose a nonzero vector $v_1 \in T_{p_1}M$ such that $g(v_1,v_1) \leq 0$. If $0_b$ denotes the zero vector in $T_bH$, then $g(v,v) = g(v_1,v_1) \leq 0$, where $v = (v_1,0_b) \in T_{p_1}M \times T_bH$. Since $\bar{g} < \bar{g}_1$, we obtain $\bar{g}_1(v,v) = g_1(v_1,v_1) < 0$. Hence $g < g_1$ and $(M,g)$ is stably causal.

Conversely, we now assume that $(M,g)$ is stably causal. Let $g_1 \in \mathrm{Lor}(M)$ be a causal metric with $g < g_1$. By Remark 2.50 we may also assume that $g_1(v_1,v_1) < g(v_1,v_1)$ for all vectors $v_1 \neq 0$ which are nonspacelike with respect to g. Since Proposition 2.48 implies that $\bar{g}_1 = g_1 \oplus fh$ is a causal metric on $M \times H$, it suffices to show that $\bar{g} < \bar{g}_1$. To this end, let $v = (v_1,v_2)$ be a nontrivial vector of $T_p(M \times H)$ which is nonspacelike with respect to $\bar{g}$. Then since $\bar{g}(v,v) = g(v_1,v_1) + f(\pi(v))h(v_2,v_2) \leq 0$ and $f(\pi(v))h(v_2,v_2) > 0$ with $v_2 \neq 0$, the nontriviality of v implies that $v_1 \neq 0$ and $g(v_1,v_1) \leq 0$. Thus $\bar{g}_1(v,v) = g_1(v_1,v_1) + f(\pi(v))h(v_2,v_2) < g(v_1,v_1) + f(\pi(v))h(v_2,v_2) \leq 0$ which shows that $\bar{g} < \bar{g}_1$ and establishes the proposition. □

Geroch's splitting theorem (cf. Theorem 2.13) guarantees that any globally hyperbolic space-time may be written as a topological product $\mathbb{R} \times S$ where S is a Cauchy hypersurface. Geroch's result suggests investigating conditions on $(M,g)$ and $(H,h)$ which imply that the warped product $(M \times_f H, \overline{g})$ is globally hyperbolic. These conditions are given for $\dim M = 1$ and $\dim M \geqslant 2$ in Theorem 2.53 and 2.55, respectively. In order to prove these results, it is first necessary to show that a curve in a complete Riemannian manifold which is inextendible in one direction must have infinite length.

LEMMA 2.52 Let $(H,h)$ be a complete Riemannian manifold. If $\gamma : [0,1) \longrightarrow H$ is a curve of finite length in $(H,h)$, then there exists $p \in H$ such that $\gamma(t) \longrightarrow p$ as $t \longrightarrow 1^-$.

*Proof.* Let $d_0$ denote the Riemannian distance function induced on $H$ by the Riemannian metric $h$. Let $L = L_0(\gamma)$ be the Riemannian arc length of $\gamma$ and set $K = \{q \in H : d_0(\gamma(0), q) \leqslant L\}$. The Hopf-Rinow theorem [cf. Hicks (1965, pp. 163-164)] implies that $K$ is compact. Fix a sequence $\{t_n\}$ in $[0,1)$ with $t_n \longrightarrow 1$. Since $d(\gamma(0),\gamma(t)) \leqslant L(\gamma|[0,t]) \leqslant L$ for $t \in [0,1)$, we have $\gamma[0,1) \subset K$. Thus by the compactness of $K$, the sequence $\{\gamma(t_n)\}$ has a limit point $p \in K$. If $\lim_{t\to 1^-} \gamma(t) \neq p$, there would then exist an $\varepsilon > 0$ such that $\gamma$ leaves the ball $\{m \in M : d(p,m) \leqslant \varepsilon\}$ infinitely often. But this would imply that $\gamma$ has infinite length, in contradiction. □

The next theorem may be obtained from Corollary 2.43 and Lemma 2.52. The proof which is similar to that of Theorem 2.55 will be omitted.

THEOREM 2.53 Let $(H,h)$ be a Riemannian manifold and $M = (a,b)$ with $-\infty \leqslant a < b \leqslant +\infty$ given the negative definite metric $-dt^2$. Then the Lorentzian warped product $(M \times_f H, \overline{g})$ is globally hyperbolic iff $(H,h)$ is complete.

Theorem 2.53 may be regarded as a "metric converse" to Geroch's splitting theorem. If $f = 1$ is assumed, so that the warped product $(M \times_f H, \overline{g})$ is simply a metric product $(M \times H, g \oplus h)$, Theorem 2.53 may be strengthened to include geodesic completeness (cf. Definition 5.2 for the definition of geodesic completeness).

THEOREM 2.54 Suppose that $(H,h)$ is a Riemannian manifold and that $\mathbb{R} \times H$ is given the product Lorentzian metric $-dt^2 \oplus h$. Then the following are equivalent:

(a) $(H,h)$ is geodesically complete.

(b) $(\mathbb{R} \times H, -dt^2 \oplus h)$ is geodesically complete.

(c) $(\mathbb{R} \times H, -dt^2 \oplus h)$ is globally hyperbolic.

*Proof.* We know that (a) iff (c) from Theorem 2.53. Thus it remains to show (a) iff (b). But this is a consequence of the fact that all geodesics of $\mathbb{R} \times H$ are either (up to parameterization) of the form $(\lambda t, c(t))$, $(\lambda_0, c(t))$, or $(\lambda t, h_0)$ where $\lambda, \lambda_0 \in \mathbb{R}$ are constants, $h_0 \in H$, and $c : J \longrightarrow H$ is a unit speed geodesic in H. □

Suppose that a space-time $(M,g)$ of dimension $\geqslant 3$ has everywhere nonnegative nonspacelike Ricci curvatures and satisfies the "generic condition" that all inextendible nonspacelike geodesics contain a point with nonzero curvatures at that point (cf. Definition 11.7 for a precise formulation of the generic condition). Then if $(M,g)$ has a compact Cauchy surface, the space-time $(M,g)$ is geodesically incomplete. Thus Theorem 2.53 may *not* be strengthened for arbitrary warped products to include geodesic completeness. The "big bang" Robertson-Walker cosmological models (cf. Section 4.4) are examples of globally hyperbolic warped products which are not geodesically complete.

On the other hand, let $(\mathbb{R} \times H, -dt^2 \oplus h)$ be a product space-time of the form considered in Theorem 2.54. Fix any $b_0 \in H$. Then $\gamma(t) = (t, b_0)$ is a timelike geodesic with $R(\gamma'(t), v) = 0$ for all $v \in T_{\gamma(t)}(\mathbb{R} \times H)$ for each $t \in \mathbb{R}$. Thus $(\mathbb{R} \times H, -dt^2 \oplus h)$ *fails* to satisfy the generic condition.

If $\dim M = 1$ and M is homeomorphic to $\mathbb{R}$, we have just given necessary and sufficient conditions for the warped product $M \times_f H$ to be globally hyperbolic. If $M = S^1$, we remarked above that $(M \times_f H, \bar{g})$ is nonchronological no matter which Riemannian metric h is chosen for H. Thus no warped product space-time $(S^1 \times_f H, \bar{g})$ is globally hyperbolic.

We now consider the case $\dim M \geqslant 2$.

THEOREM 2.55 Let $(M,g)$ be a space-time and let $(H,h)$ be a Riemannian manifold. Then the Lorentzian warped product $(M \times_f H, \bar{g})$ is globally hyperbolic iff both of the following conditions are satisfied:

(1) $(M,g)$ is globally hyperbolic.

(2) $(H,h)$ is a complete Riemannian manifold.

*Proof.* ($\Rightarrow$) Suppose first that $(M \times_f H, \bar{g})$ is globally hyperbolic. Fixing $b \in H$, we may identify $(M,g)$ with the closed submanifold $\eta^{-1}(b) = M \times \{b\}$ since the projection map $\pi : \eta^{-1}(b) \longrightarrow M$ is an isometry. Lemma 2.47 implies that under this identification, the set $J^+(p_1) \cap J^-(q_1)$ in $M$ corresponds to $\eta^{-1}(b) \cap J^+((p_1,b)) \cap J^-((q_1,b))$ in $M \times_f H$ for any $p_1,q_1 \in M$. Since $\eta^{-1}(b)$ is closed and $(M \times_f H, \bar{g})$ is globally hyperbolic, $\eta^{-1}(b) \cap J^+((p_1,b)) \cap J^-((q_1,b))$ is compact in $M \times_f H$. Hence $J^+(p_1) \cap J^-(q_1)$ is compact in $M$. Because $(M \times_f H, \bar{g})$ is globally hyperbolic, it is also strongly causal. Thus $(M,g)$ is strongly causal by Proposition 2.49. Hence $(M,g)$ is globally hyperbolic as required.

Now we show that $(M \times_f H, \bar{g})$ globally hyperbolic implies that $(H,h)$ is a complete Riemannian manifold. We will suppose that $(H,h)$ is incomplete and derive a contradiction to the global hyperbolicity of $(M \times_f H, \bar{g})$. For this purpose, fix any pair of points $p_1,q_1 \in M$ with $p_1 \ll q_1$ and let $\gamma_1 : [0,L] \longrightarrow M$ be a unit speed future-directed timelike curve in $M$ from $p_1$ to $q_1$. Set $\alpha = \sup\{f(\gamma_1(t)) : t \in [0,L]\}$ where $f : M \longrightarrow (0,\infty)$ is the given warping function. Since $\gamma_1([0,L])$ is a compact subset of $M$, we have $0 < \alpha < \infty$.

Assuming that $(H,h)$ is not complete, by the Hopf-Rinow theorem there exists a geodesic $c : [0,\beta) \longrightarrow H$ with $h(c'(t),c'(t)) = 1/\alpha$ which is not extendible to $t = \beta < \infty$. By changing $c(0)$ and reparameterizing $c$ if necessary, we may suppose that $0 < \beta < L/2$. Define a future-directed nonspacelike curve $\bar{\gamma} : [0,\beta) \longrightarrow M \times H$ and a past-directed nonspacelike curve $\tilde{\gamma} : [0,\beta) \longrightarrow M \times H$ by $\bar{\gamma}(t) = (\gamma_1(t),c(t))$ and $\tilde{\gamma}(t) = (\gamma_1(L - t), c(t))$, respectively. For each $t$

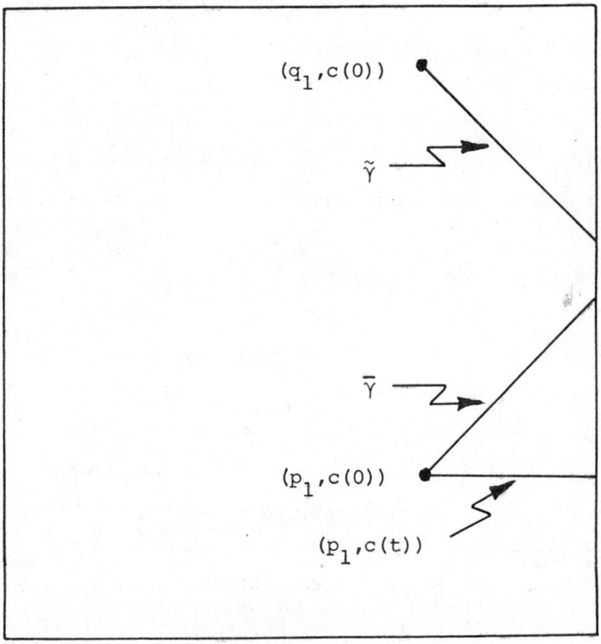


*Figure 2.10* In the proof of Theorem 2.55, the curve $c : [0,\beta) \longrightarrow H$ is a geodesic which is not extendible to $t = \beta < \infty$. The curves $\bar{\gamma}(t) = (\gamma_1(t), c(t))$ and $\tilde{\gamma}(t) = (\gamma_1(L - t), c(t))$ are inextendible nonspacelike curves in $(M \times_f H, \bar{g})$ and hence do not have compact closure.

with $0 \leqslant t < \beta$, we have $\gamma_1(t) \ll \gamma_1(L - t)$ in $(M,g)$ since $t < L - t$. Hence by Lemma 2.47, we have $(\gamma_1(t),c(t)) \ll (\gamma_1(L - t), c(t))$ in $M \times_f H$. Thus $(p_1,c(1)) \leqslant \bar{\gamma}(t) \ll \tilde{\gamma}(t) \leqslant (q_1,c(0))$ for all $0 \leqslant t < \beta$ (cf. Figure 2.10). It follows that $\bar{\gamma}([0,\beta))$ is contained in $J^+((p_1,c(0))) \cap J^-((q_1,c(0)))$. Since $c = \eta \circ \bar{\gamma}$ does not have compact closure in H, the curve $\bar{\gamma} : [0,\beta) \longrightarrow M \times_f H$ does not have compact closure in $J^+((p_1,c(0))) \cap J^-((q_1,c(0)))$. But since $(M \times_f H, \bar{g})$ is globally hyperbolic, the set $J^+((p_1,c(0))) \cap J^-((q_1,c(0)))$ is compact, in contradiction.

($\Leftarrow$) Suppose now that $(M,g)$ is globally hyperbolic. Assuming that $(M \times_f H, \bar{g})$ is not globally hyperbolic, we must show that $(H,h)$ is not complete. Since $(M,g)$ is strongly causal, the warped product

$(M \times_f H, \bar{g})$ is also strongly causal by Proposition 2.49. Hence since $(M \times_f H, \bar{g})$ is not globally hyperbolic, there exist distinct points $(p_1,b_1)$ and $(p_2,b_2)$ in $M \times_f H$ such that $J^+((p_1,b_1)) \cap J^-((p_2,b_2))$ is noncompact. There is then a future-directed nonspacelike curve $\gamma : [0,1) \longrightarrow J^+((p_1,b_1)) \cap J^-((p_2,b_2))$ which is future inextendible in $(M \times_f H, \bar{g})$. Let $\gamma(t) = (u_1(t),u_2(t))$, where $u_1 : [0,1) \longrightarrow M$ and $u_2 : [0,1) \longrightarrow H$. Then $u_1 : [0,1) \longrightarrow M$ is a future-directed nonspacelike curve contained in $J^+(p_1) \cap J^-(p_2)$. Since $(M,g)$ is globally hyperbolic, $J^+(p_1) \cap J^-(p_2)$ is compact. Hence if we set $\alpha_0 = \inf\{f(m) : m \in J^+(p_1) \cap J^-(p_2)\}$, then $\alpha_0 > 0$. Also since $(M,g)$ is strongly causal, no future-directed future-inextendible nonspacelike curve may be future imprisoned in the compact set $J^+(p_1) \cap J^-(p_2)$ (cf. Proposition 2.9). Hence there exists a point $r \in J^+(p_1) \cap J^-(p_2)$ with $\lim_{t \to 1^-} u_1(t) = r$. We may then extend $u_1$ to a continuous curve $u_1 : [0,1] \longrightarrow M$ by setting $u_1(1) = r$. Since the curve $\gamma = (u_1,u_2)$ was inextendible to $t = 1$, it follows that $u_2(t)$ cannot converge to any point of $H$ as $t \longrightarrow 1^-$. By Lemma 2.52, either $(H,h)$ is incomplete or $u_2$ has infinite length. As $u_1 : [0,1] \longrightarrow M$ is a nonspacelike curve defined on a compact interval, $u_1$ has finite length in $(M,g)$. Since $f(u_1(t)) \geq \alpha_0 > 0$ for all $t \in [0,1]$ and

$$\bar{g}(\gamma'(t),\gamma'(t)) = g(u_1'(t),u_1'(t)) + f(u_1(t))h(u_2'(t),u_2'(t)) \leq 0$$

it follows that $u_2$ has finite length in $(H,h)$. Thus $(H,h)$ is incomplete as required. □

Cauchy surfaces may be constructed for globally hyperbolic Lorentzian warped products as follows.

THEOREM 2.56 Let $(H,h)$ be a complete Riemannian manifold. Let $(M \times_f H, \bar{g})$ be the Lorentzian warped product of $(M,g)$ and $(H,h)$.

(1) If $M = (a,b)$ with $-\infty \leq a < b \leq +\infty$ is given the metric $-dt^2$, then $\{p_1\} \times H$ is a Cauchy surface of $(M \times_f H, \bar{g})$ for each $p_1 \in M$.

(2) If $(M,g)$ is globally hyperbolic with Cauchy surface $S_1$, then $S_1 \times H$ is a Cauchy surface of $(M \times_f H, \bar{g})$.

*Proof.* Since the proofs of (1) and (2) are similar, we shall only give the proof of (2). In this case, $S_1 \times H$ is an achronal subset of $(M \times_f H, \bar{g})$. To show $S_1 \times H$ is a Cauchy surface, we must show that every inextendible nonspacelike curve in $M \times_f H$ meets $S_1 \times H$. Now given $(p_1,p_2) \in M \times H - S_1 \times H$, either every future-directed, future-inextendible nonspacelike curve in $(M,g)$ beginning at $p_1$ meets $S_1$, or every past-directed, past-inextendible nonspacelike curve starting at $p_1$ meets $S_1$. Since the two cases are similar, we will suppose the former holds and then show that every future-directed, future-inextendible nonspacelike curve $\gamma : [0,1) \longrightarrow M \times_f H$ with $\gamma(0) = (p_1,p_2)$ meets $S_1 \times H$.

Thus suppose that $\gamma : [0,1) \longrightarrow M \times_f H$ is a future-directed, future-inextendible nonspacelike curve with $\gamma(0) = (p_1,p_2)$ which does not meet $S_1 \times H$. Decompose $\gamma(t) = (u_1(t),u_2(t))$ with $u_1 : [0,1) \longrightarrow M$, $u_2 : [0,1) \longrightarrow H$. Since $S_1$ is a Cauchy surface for $(M,g)$ and $(M,g)$ is globally hyperbolic, the set $J^+(p_1) \cap J^-(S_1)$ is compact [cf. Beem and Ehrlich (1979a, p. 163)]. As in the proof of Theorem 2.55, the strong causality of $(M,g)$ implies that there exists a point $r \in J^+(p_1) \cap J^-(S_1)$ with $\lim_{t\to 1^-} u_1(t) = r$. Since $J^+(p_1) \cap J^-(S_1)$ is compact, the warping function $f : M \longrightarrow (0,\infty)$ achieves a minimum $\alpha_0 > 0$ on $J^+(p_1) \cap J^-(S_1)$. As in the proof of Theorem 2.55, this then implies that $u_2 : [0,1) \longrightarrow H$ has finite length. Since $(H,h)$ is complete, by Lemma 2.52 there exists a point $b \in H$ with $\lim_{t\to 1^-} u_2(t) = b$. Setting $\gamma(1) = (r,b)$, we have then extended $\gamma$ to a nonspacelike future-directed curve $\gamma : [0,1] \longrightarrow M \times_f H$, contradicting the inextendibility of $\gamma$. Hence $\gamma$ must meet $S_1 \times H$ as required. □

We now consider the nonspacelike geodesic completeness of Lorentzian warped products $\bar{M} = (a,b) \times_f H$, $\bar{g} = -dt^2 \oplus fh$. Here a space-time is said to be *null* (resp., *timelike*) *geodesically incomplete* if some future-directed null (resp., timelike) geodesic cannot be extended to be defined for arbitrary negative and positive values of an affine parameter, cf. Definitions 5.2 and 5.3. Since we are using the metric $-dt^2$ on $(a,b)$, the curve $c(t) = (t,y_0)$ with $y_0 \in H$ fixed is a unit speed timelike geodesic in $(\overline{M},\overline{g})$ no matter which warping function is chosen. Consequently, if $a > -\infty$ or $b < +\infty$, then $(\overline{M},\overline{g})$ is timelike geodesically incomplete for all possible warping functions f. Moreover, if a and b are both finite and if $\gamma$ is any timelike geodesic in $\overline{M} = (a,b) \times_f H$, then $L(\gamma) \leq b - a < \infty$. Thus if a and b are finite, all timelike geodesics are past and future incomplete. Nonetheless, if the warping function f is chosen suitably, $(\overline{M},\overline{g})$ may be null geodesically complete even if a and b are both finite. This will be clear from the proof of Theorem 2.57 below.

If $\overline{M} = \mathbb{R} \times_f H$ with $\overline{g} = -dt^2 \oplus fh$, then any timelike geodesic of the form $c(t) = (t,y_0)$ is past and future timelike complete. On the other hand, warped product space-times $\overline{M} = \mathbb{R} \times_f H$ may be constructed for which all nonspacelike geodesics except for those of the form $t \longrightarrow (t,y_0)$ are future incomplete. One such example may be given as follows. Busemann and Beem (1966) studied the space-time $\overline{M} = \{(x,y) \in \mathbb{R}^2 : y > 0\}$ with the Lorentzian metric $ds^2 = y^{-2}(dx^2 - dy^2)$. Busemann and Beem (1966, p. 245) noted that all timelike geodesics except for those of the form $t \longrightarrow (t,y_0)$ are future incomplete. Setting $t = \ln y$, this space-time is transformed into the Lorentzian warped product $\mathbb{R} \times_f \mathbb{R}$ with $\overline{g} = -dt^2 \oplus f\,dt^2$, where $f(t) = e^{-2t}$. Since the map $F : (\overline{M},ds^2) \longrightarrow (\mathbb{R} \times_f \mathbb{R}, \overline{g})$ given by $F(x,y) = (x, \ln y)$ is a global isometry, all timelike geodesics of $(\mathbb{R} \times_f \mathbb{R}, \overline{g})$ except for those of the form $t \longrightarrow (t,y_0)$ are future incomplete. It will also follow from Theorem 2.57 below that all null geodesics are future incomplete. Similarly, if $(\mathbb{R}^n,h)$ denotes $\mathbb{R}^n$ with the usual Euclidean metric $h = dx_1^2 + \cdots + dx_n^2$, then $(\mathbb{R} \times_f \mathbb{R}^n, \overline{g})$ with $\overline{g} = -dt^2 \oplus fh$ and $f(t) = e^{-2t}$ is a space-time with all nonspacelike geodesics future incomplete except for those of the form $t \longrightarrow (t,y_0)$.

In order to study geodesic completeness, it is necessary to determine the Levi-Civita connection for a Lorentzian warped product metric. For this purpose, we will consider the general warped product $(M \times_f H, g \oplus fh)$ where $f : M \longrightarrow (0,\infty)$, $(H,h)$ is Riemannian and $(M,g)$ is equipped with a metric of signature $(-, +, \ldots, +)$. Let $\nabla^1$ denote the Levi-Civita connection for $(M,g)$ and $\nabla^2$ denote the Levi-Civita connection for $(H,h)$. Given vector fields $X_1$, $Y_1$ on $M$ and $X_2$, $Y_2$ on $H$ we may lift them to $M \times H$ and obtain the vector fields $X = (X_1,0) + (0,X_2) = (X_1,X_2)$ and $Y = (Y_1,0) + (0,Y_2) = (Y_1,Y_2)$ on $M \times H$. Recall that the connection $\overline{\nabla}$ for $(M \times_f H, g \oplus fh)$ is related to the metric $\overline{g} = g \oplus fh$ by the formula

$$2\overline{g}(\overline{\nabla}_X Y,Z) = X\overline{g}(Y,Z) + Y\overline{g}(X,Z) - Z\overline{g}(X,Y) + \overline{g}([X,Y],Z)$$
$$- \overline{g}([X,Z],Y) - \overline{g}([Y,Z],X)$$

[cf. Cheeger and Ebin (1975, p. 2)]. Using this formula and setting $\phi = \ln f$, we obtain the following formula for $\overline{\nabla}$ for X and Y as above:

$$\overline{\nabla}_X Y = \nabla^1_{X_1} Y_1 + \nabla^2_{X_2} Y_2 + \tfrac{1}{2}[X_1(\phi)Y_2 + Y_1(\phi)X_2 - \overline{g}(X_2,Y_2)\text{grad } \phi] \quad (2.4)$$

Here grad $\phi$ denotes the gradient of the function $\phi$ on $(M,g)$ and we are identifying the vector $\nabla^1_{X_1} Y_1|_p \in T_pM$ with the vector $(\nabla^1_{X_1} Y_1|_p, 0_q) \in T_{(p,q)}(M \times H)$, etc.

We are now ready to obtain the following criterion for null geodesic incompleteness of Lorentzian warped products $\overline{M} = (a,b) \times_f H$ [cf. Beem, Ehrlich, and Powell (1980)]. Throughout the rest of this section, let $\omega_0$ denote an interior point of $(a,b)$.

THEOREM 2.57 Let $\overline{M} = (a,b) \times_f H$ be a Lorentzian warped product with Lorentzian metric $\overline{g} = -dt^2 \oplus fh$ where $-\infty \leqslant a < b \leqslant +\infty$, $(H,h)$ is an arbitrary Riemannian manifold, and $f : (a,b) \longrightarrow (0,\infty)$. Set $S(t) = \sqrt{f(t)}$. Then if $\lim_{t\to a^+} \int_t^{\omega_0} S(s)\, ds$ [resp., $\lim_{t\to b^-} \int_{\omega_0}^t S(s)\, ds$] is finite, every future-directed null geodesic in $(\overline{M},\overline{g})$ is past (resp., future) incomplete.

*Proof*. Let $\gamma_0$ be an arbitrary future-directed null geodesic in $(\overline{M},\overline{g})$. We may reparameterize $\gamma_0$ to be of the form $\gamma(t) = (t,c(t))$, where $\gamma$ is a smooth null pregeodesic. Accordingly, there exists a smooth function $g(t)$ such that

$$\overline{\nabla}_{\gamma'}\gamma'\Big|_t = g(t)\gamma'(t) = g(t)\frac{\partial}{\partial t}\Big|_t + g(t)c'(t)$$

[cf. Hawking and Ellis (1973, p. 33)]. On the other hand, since $\gamma'(t) = \partial/\partial t|_t + c'(t)$ and $\overline{g}(\gamma',\gamma') = -1 + \overline{g}(c',c') = 0$, we obtain using formula (2.4) that

$$\overline{\nabla}_{\gamma'}\gamma'\Big|_t = \nabla^1_{\partial/\partial t}\frac{\partial}{\partial t}\Big|_t + \nabla^2_{c'}c'\Big|_t + \frac{\partial}{\partial t}(\phi)c'(t) - \frac{1}{2}\overline{g}(c'(t),c'(t))\operatorname{grad}\phi$$

$$= \nabla^2_{c'}c'\Big|_t + \frac{f'(t)}{f(t)}c'(t) + \frac{f'(t)}{2f(t)}\frac{\partial}{\partial t}\Big|_t$$

Equating terms with a $\partial/\partial t$ component, we obtain the formula

$$g(t) = \frac{f'(t)}{2f(t)} \tag{2.5}$$

Thus $\overline{\nabla}_{\gamma'}\gamma'\Big|_t = (1/2)[\ln f(t)]'\gamma'(t) = [\ln S(t)]'\gamma'(t)$. If we define $p : (a,b) \longrightarrow \mathbb{R}$ by

$$p(t) = \int_{\omega_0}^{t} S(s)\, ds$$

then $p'(t) = S(t) > 0$ so that $p^{-1}$ exists. Moreover, from the classical theory of projective transformations, we know that $\gamma_1(t) = \gamma \circ p^{-1}(t) = (p^{-1}(t), c \circ p^{-1}(t))$ is a null geodesic [cf. Spivak (1970, pp. 6-35 ff.)]. Let

$$A = \lim_{t\to a^+} p(t) \qquad B = \lim_{t\to b^-} p(t)$$

Since $p$ is monotone increasing, we have that $p : (a,b) \longrightarrow (A,B)$ is a bijection. Hence $p^{-1} : (A,B) \longrightarrow (a,b)$ and thus $\gamma_1 = \gamma \circ p^{-1} : (A,B) \longrightarrow \overline{M}$. Therefore if $A$ is finite, $\gamma_1$ is past incomplete and if $B$ is finite, $\gamma_1$ is future incomplete as required. □

It is immediate from Theorem 2.57 that if a and b are finite and the warping function $f : (a,b) \longrightarrow (0,\infty)$ is bounded, then $(\overline{M},\overline{g})$ is past and future null geodesically incomplete. Thus assuming that a and b are finite, 1-parameter families $(\overline{M},\overline{g}(s)) = (\overline{M}, -dt^2 \oplus f(s)h)$ of past and future null geodesically incomplete space-times may easily be constructed. Choosing the 1-parameter family of functions $f(s) : (a,b) \longrightarrow (0,\infty)$ suitably, the curve $s \longrightarrow g(s) = -dt^2 \oplus f(s)h$ of metrics will *not* be a continuous curve in $\mathrm{Lor}(\overline{M})$ in the fine $C^r$ topologies. Thus the space-times $(\overline{M},\overline{g}(0))$ and $(\overline{M},\overline{g}(s))$ may be far apart in $\mathrm{Lor}(\overline{M})$ for $s \neq 0$.

Notice that if the Riemannian manifold $(H,h)$ is geodesically incomplete, then $\overline{M} = (a,b) \times_f H$ may be null geodesically incomplete even if both integrals in Theorem 2.57 diverge. On the other hand, if the completeness of $(H,h)$ is assumed, the following necessary and sufficient condition for the null geodesic incompleteness of $\overline{M} = (a,b) \times_f H$ may be obtained from the proof of Theorem 2.57.

REMARK 2.58 Let $\overline{M} = (a,b) \times_f H$ be a Lorentzian warped product with Lorentzian metric $\overline{g} = -dt^2 \oplus fh$, where $(H,h)$ is a complete Riemannian manifold and $-\infty \leq a < b \leq +\infty$. Let $S(t) = \sqrt{f(t)}$ as above. Then $(\overline{M},\overline{g})$ is past (resp., future) null geodesically incomplete iff

$$\lim_{t\to a^+} \int_t^{\omega_0} S(s)\,ds \text{ is finite (resp., } \lim_{t\to b^-} \int_{\omega_0}^{t} S(s)\,ds \text{ is finite).}$$

In singularity theory in general relativity, conditions on the curvature tensor of $(\overline{M},\overline{g})$ which will be discussed in Section 11.2, called the *generic condition* and the *strong energy condition*, are considered. These two conditions guarantee that if a nonspacelike geodesic $\gamma$ may be extended to be defined for all positive and negative values of an affine parameter and $\dim \overline{M} \geq 3$, then $\gamma$ contains a pair of conjugate points. Hence these curvature conditions may be combined with geometric or physical assumptions such as $(\overline{M},\overline{g})$ is causally disconnected or $(\overline{M},\overline{g})$ contains a closed trapped set to show that $(\overline{M},\overline{g})$ is nonspacelike geodesically incomplete (cf. Section 11.4).

Since $(\overline{M},\overline{g})$ satisfies the generic condition and strong energy condition if all nonspacelike Ricci curvatures are positive, it is thus of interest to consider conditions on the warping function f of a Lorentzian warped product which guarantee that $(\overline{M},\overline{g})$ has everywhere positive nonspacelike Ricci curvatures. The assumption $\dim \overline{M} \geqslant 3$ made in singularity theory is necessary for null conjugate points to exist since no null geodesic in any two-dimensional Lorentzian manifold contains a pair of conjugate points.

We now give the formula for the curvature tensor R and Ricci curvature tensor Ric for the warped product space-time $(M \times_f H, \overline{g})$ where $\overline{g} = g \oplus fh$. As above, let $\nabla^1$ (resp., $\nabla^2$) denote the covariant derivative of $(M,g)$, [resp., $(H,h)$]. Also let $\phi = \ln f$ and recall that grad $\phi$ denotes the gradient of $\phi$ on $(M,g)$. As before, we will decompose tangent vectors $x \in T_{\overline{p}}(M \times H)$ as $x = (x_1,x_2)$. Let $R^1$ (resp., $R^2$) denote the curvature tensor of $(M,g)$, [resp., $(H,h)$]. Given tangent vectors $x_1,y_1 \in T_pM$, define the Hessian tensors $H_\phi$ and $h_\phi$ by

$$H_\phi(x_1) = \nabla^1_{x_1} \operatorname{grad} \phi \tag{2.6}$$

and

$$h_\phi(x_1,y_1) = g(\nabla^1_{x_1} \operatorname{grad} \phi,\ y_1) \tag{2.7}$$

respectively. We will also write $\|\operatorname{grad} \phi\|^2 = g(\operatorname{grad} \phi, \operatorname{grad} \phi)$. Using the sign convention

$$R(X,Y)Z = \overline{\nabla}_X\overline{\nabla}_Y Z - \overline{\nabla}_Y\overline{\nabla}_X Z - \overline{\nabla}_{[X,Y]}Z$$

for the curvature tensor and substituting from formula (2.4), one obtains the formula

$$R(x,y)z = R^1(x_1,y_1)z_1 + R^2(x_2,y_2)z_2 + \tfrac{1}{2}[h_\phi(x_1,z_1)y_2 - h_\phi(y_1,z_1)x_2 + \overline{g}(x_2,y_2)H_\phi(y_1) - \overline{g}(y_2,z_2)H_\phi(x_1)]$$

$$+ \frac{1}{4}\{[x_1(\phi)z_1(\phi) + \bar{g}(x_2,z_2)\|\text{grad } \phi\|^2(p)]y_2$$

$$- [y_1(\phi)z_1(\phi) + \bar{g}(y_2,z_2)\|\text{grad } \phi\|^2(p)]x_2$$

$$+ [y_1(\phi)\bar{g}(x_2,z_2) - x_1(\phi)\bar{g}(y_2,z_2)]\text{grad } \phi(p)\} \qquad (2.8)$$

where $x,y,z \in T_{(p,q)}(M \times H)$.

Suppose now that dim M = m and dim H = n. To calculate the Ricci curvature at $\bar{p} = (p,q) \in M \times H$ let $\{e_1,\ldots,e_m\}$ be a basis for $T_pM$ with $g(e_1,e_1) = -1$, $g(e_j,e_j) = 1$ for $2 \leqslant j \leqslant m$ and $g(e_i,e_j) = 0$ if $i \neq j$. Also let $\{e_{m+1},\ldots,e_{n+m}\}$ be a $\bar{g}$-orthonormal basis for $T_qH$. Then for any $x,y \in T_{\bar{p}}(M \times H)$, we have

$$\text{Ric}(x,y) = -\bar{g}(R(e_1,x)y,e_1) + \sum_{j=2}^{n+m} \bar{g}(R(e_j,x)y,e_j)$$

The d'Alembertian $\Box\phi$ of $\phi$ may also be calculated as

$$\Box\phi(p) = -h_\phi(e_1,e_1) + \sum_{j=2}^{m} h_\phi(e_j,e_j)$$

Using (2.8), it then follows that

$$\text{Ric}(x,y) = \text{Ric}^1(x_1,y_1) + \text{Ric}^2(x_2,y_2)$$

$$- \bar{g}(x_2,y_2)\left[\frac{1}{2}\Box\phi(p) + \frac{\dim H}{4}\|\text{grad } \phi(p)\|^2\right]$$

$$- \frac{\dim H}{2} h_\phi(x_1,y_1) - \frac{\dim H}{4} x_1(\phi)y_1(\phi) \qquad (2.9)$$

where $x = (x_1,y_1)$, $y = (y_1,y_2) \in T_{(p,q)}(M \times H)$ and $\text{Ric}^1$ and $\text{Ric}^2$ denote the Ricci curvature tensors of (M,g) and (H,h), respectively.

We now specialize to the case $\bar{M} = (a,b) \times_f H$ with warped product metric $\bar{g} = -dt^2 \oplus fh$. In this case, $\Box\phi(t) = -\phi''(t)$ and $\|\text{grad } \phi(t)\|^2 = -[\phi'(t)]^2$. Thus we obtain from (2.9) for $\bar{v} = (0,v) \in T_{(t,q)}(\mathbb{R} \times H)$ that

$$\text{Ric}(\bar{v},\bar{v}) = \text{Ric}^2(v,v) + \bar{g}(v,v)\left\{\frac{1}{2}\phi''(t) + \frac{\dim H}{4}[\phi'(t))^2]\right\} \qquad (2.10)$$

If $x = \partial/\partial t|_t + v \in T_{(t,q)}(\mathbb{R} \times H)$ with $v \in T_q H$, we obtain

$$\mathrm{Ric}(x,x) = \mathrm{Ric}^2(v,v) + \bar{g}(v,v)\left\{\frac{1}{2}\phi''(t) + \frac{\dim H}{4}[\phi'(t)]^2\right\}$$
$$+ \left\{-\frac{\dim H}{2}\phi''(t) - \frac{\dim H}{4}[\phi'(t)]^2\right\} \tag{2.11}$$

Both bracketed terms in formulas (2.10) and (2.11) will be positive provided that

$$-[\phi'(t)]^2 \dim H < 2\phi''(t) < -[\phi'(t)]^2 \tag{2.12}$$

for all $t \in (a,b)$. Thus if $\mathrm{Ric}^2(v,v) \geqslant 0$ for all $v \in TH$ and condition (2.12) holds, the space-time $(\bar{M},\bar{g})$ will have everywhere positive Ricci curvatures. A globally hyperbolic family of such space-times is provided by warped products $\bar{M} = (0,\infty) \times_f H$, where $(H,h)$ is a complete Riemannian manifold of nonnegative Ricci curvature and $\bar{g} = -dt^2 \oplus fh$ with $f(t) = t^r$ for a fixed constant $r \in \mathbb{R}$ with $2/\dim H < r < 2$. If $(H,h)$ is taken to be $\mathbb{R}^3$ with the usual Euclidean metric and $r = 4/3$, we recover the Einstein-de Sitter universe of cosmology theory [cf. Hawking and Ellis (1973, p. 138), Sachs and Wu (1977a, Proposition 6.2.7 ff.)].

We may also obtain the following condition on $\phi = \ln f$ for positive nonspacelike Ricci curvature if the Ricci tensor of $(H,h)$ is bounded from below.

PROPOSITION 2.59 Let $\bar{M} = (a,b) \times_f H$ with $n = \dim H \geqslant 2$ and $\bar{g} = -dt^2 \oplus fh$, $\phi = \ln f$. Suppose that $\mathrm{Ric}^2(v,v) \geqslant \lambda h(v,v)$ for some constant $\lambda \in \mathbb{R}$ for all $v \in TH$. Then if

$$2\phi''(t) < \min\{-(\phi'(t))^2, 4(n-1)^{-1}\lambda e^{-\phi(t)}\} \tag{2.13}$$

for all $t \in (a,b)$, the Lorentzian warped product $(\bar{M},\bar{g})$ has everywhere positive nonspacelike Ricci curvature.

*Proof.* It suffices to show that $\mathrm{Ric}(x,x) > 0$ for all nonspacelike tangent vectors $x$ of the form $x = \partial/\partial t + v \in T(M \times H)$,

$v \in TH$. Since $\bar{g}(x,x) \leqslant 0$ and $\bar{g}(\partial/\partial t, \partial/\partial t) = -1$, we have $\beta = \bar{g}(v,v) \leqslant 1$. Hence $0 \leqslant \beta \leqslant 1$. Then $h(v,v) = \beta e^{-\phi}$ and we obtain from (2.11) that

$$\operatorname{Ric}(x,x) \geqslant \beta e^{-\phi}\lambda + \left(\frac{\beta}{2} - \frac{n}{2}\right)\phi'' + \frac{n}{4}(\beta - 1)(\phi')^2 \tag{2.14}$$

Thus $\operatorname{Ric}(x,x) > 0$ provided $\phi'' < G(\beta)$ for all $\beta \in [0,1]$, where

$$G(\beta) = \frac{4\beta e^{-\phi}\lambda - n(1 - \beta)(\phi')^2}{2(n - \beta)}$$

Calculating $G''(\beta)$, one finds that $G'(\beta)$ does not change sign in $[0,1]$. Thus $G(\beta)$ obtains its minimum on $[0,1]$ for $\beta = 0$ or $\beta = 1$. Hence $\operatorname{Ric}(x,x) > 0$ provided that $\phi'' < \min\{G(0),G(\beta)\}$ which yields inequality (2.13). □

We now consider the scalar curvature of warped product manifolds of the form $\bar{M} = \mathbb{R} \times_f H$, $\bar{g} = dt^2 \oplus fh$. We will let $n = \dim H$ below. Given $(t,p) \in \bar{M}$, choose $e_j \in T_pH$ for $1 \leqslant j \leqslant n$ such that if $\bar{e}_j = (0,e_j) \in T_{(t,p)}\bar{M}$, then $\{\overline{\partial/\partial t} = (\partial/\partial t, 0_p), \bar{e}_1,\ldots,\bar{e}_n\}$ forms a $\bar{g}$-orthonormal basis for $T_{(t,p)}\bar{M}$. Hence $\{\sqrt{f(t)}\, e_1,\ldots,\sqrt{f(t)}\, e_n\}$ forms an h-orthonormal basis for $T_pH$. Thus if $\tau : \bar{M} \longrightarrow \mathbb{R}$ and $\tau_H : H \longrightarrow \mathbb{R}$ denote the scalar curvature functions of $(\bar{M},\bar{g})$ and $(H,h)$, respectively, we have

$$\tau(t,p) = -\operatorname{Ric}\left(\overline{\frac{\partial}{\partial t}}, \overline{\frac{\partial}{\partial t}}\right) + \sum_{j=1}^{n} \operatorname{Ric}(\bar{e}_j,\bar{e}_j)$$

and

$$\tau_H(p) = f(t) \sum_{j=1}^{n} \operatorname{Ric}^2(e_j,e_j)$$

Now formulas (2.10) and (2.11) above simplify to

$$\operatorname{Ric}\left(\overline{\frac{\partial}{\partial t}}, \overline{\frac{\partial}{\partial t}}\right) = -\frac{n}{2}\phi''(t) - \frac{n}{4}[\phi'(t)]^2 \tag{2.15}$$

and

$$\mathrm{Ric}(\overline{e_j},\overline{e_j}) = \mathrm{Ric}^2(e_j,e_j) + \frac{1}{2}\phi''(t) + \frac{n}{4}[\phi'(t)]^2 \tag{2.16}$$

for $1 \leqslant j \leqslant n$. Consequently, we obtain the formula

$$\tau(t,p) = \frac{1}{f(t)}\,\tau_H(p) + n\phi''(t) + \frac{1}{4}(n^2 + n)[\phi'(t)]^2$$

Recalling that $\phi(t) = \ln f(t)$, this may be rewritten as

$$\tau(t,p) = \frac{1}{f(t)}\,\tau_H(p) + n\,\frac{f''(t)}{f(t)} + \frac{1}{4}(n^2 - 3n)\left[\frac{f'(t)}{f(t)}\right]^2 \tag{2.17}$$

where dim $H = n$ as above. In particular, in the case that $n = 3$ as in general relativity, we obtain the simpler formula

$$\tau(t,p) = \frac{1}{f(t)}\,\tau_H(p) + 3\,\frac{f''(t)}{f(t)} \tag{2.18}$$

EXAMPLE 2.60 With the formulas of this section in hand, we are now ready to give an example of a 1-parameter family $\overline{g}_\lambda$ of nonisometric Einstein metrics for $\mathbb{R}^{n+1}$ such that for $\lambda = 0$, $(\mathbb{R}^{n+1}, g_0)$ is Minkowski space-time of dimension $n + 1$. Let $(\mathbb{R}^n, h)$ be Euclidean n-space with the usual Euclidean metric $h = dx_1{}^2 + dx_2{}^2 + \cdots + dx_n{}^2$ and putting $M_\lambda = \mathbb{R}^{n+1} = \mathbb{R} \times_f \mathbb{R}^n$ with the Lorentzian metric $\overline{g}_\lambda = -dt^2 \oplus e^{\lambda t}h$, i.e., $f(t) = e^{\lambda t}$. By Theorem 2.57, for all $\lambda > 0$, the space-time $(\mathbb{R}^{n+1}, \overline{g}_\lambda)$ is future null geodesically complete but past null geodesically incomplete and for all $\lambda < 0$, the space-time $(\mathbb{R}^{n+1}, \overline{g}_\lambda)$ is past null geodesically complete but future null geodesically incomplete. Using formulas (2.15), (2.16), and (2.17), we obtain

$$\mathrm{Ric}(\overline{g}_\lambda) = \frac{n\lambda^2}{4}\,\overline{g}_\lambda \tag{2.19}$$

and

$$\tau_{\overline{g}_\lambda} = \frac{1}{4}(n^2 + n)\lambda^2 \tag{2.20}$$

Thus if $\lambda \neq 0$, $(\overline{M}_\lambda, \overline{g}_\lambda)$ is an Einstein space-time with constant positive scalar curvature.

EXAMPLE 2.61 Let $\overline{M}_\lambda = (0,\infty) \times_f \mathbb{R}^3$, where $\overline{g}_\lambda = -dt^2 \oplus fh$ with $f(t) = \lambda t$, $\lambda > 0$, and h is the usual Euclidean metric on $\mathbb{R}^3$. It is then immediate from formula (2.18) that $\tau(g_\lambda) \equiv 0$ for all $\lambda > 0$. On the other hand, since $\phi(t) = \ln(\lambda t)$, it may be checked using formulas (2.15) and (2.16) that $(\overline{M}_\lambda, \overline{g}_\lambda)$ is neither Ricci flat nor Einstein for any $\lambda > 0$. Also we have for any $\lambda > 0$ that

$$\mathrm{Ric}\left(\frac{\partial}{\partial t}, \frac{\partial}{\partial t}\right) = \frac{3}{4}\, t^{-2}$$

for all $t > 0$. Hence the space-times $(\overline{M}_\lambda, \overline{g}_\lambda)$ are "inextendible across" $\{0\} \times \mathbb{R}^3$ (cf. Section 5.5). Also, $(\overline{M}_\lambda, \overline{g}_\lambda)$ is future null geodesically complete by Theorem 2.57.

Chapter 3

# LORENTZIAN DISTANCE

With the basic properties of Riemannian metrics (cf. Chapter 1) in mind, it is the aim of this chapter to study the corresponding properties of Lorentzian distance and to show how the Lorentzian distance is related to the causal structure of the given space-time. We also show that Lorentzian distance preserving maps of a strongly causal space-time onto itself are diffeomorphisms which preserve the metric tensor.

While there are many similarities between the Riemannian and Lorentzian distance functions, many basic differences will also be apparent from this chapter. Nonetheless, a duality between "minimal" for Riemannian manifolds and "maximal" for Lorentzian manifolds will be noticed in this and subsequent chapters.

## 3.1 BASIC CONCEPTS AND DEFINITIONS

Let (M,g) be a Lorentzian manifold of dimension $\geqslant 2$. Given $p,q \in M$, with $p \leqslant q$, let $\Omega_{p,q}$ denote the path space of all future directed nonspacelike curves $\gamma : [0,1] \longrightarrow M$ with $\gamma(0) = p$ and $\gamma(1) = q$. The Lorentzian arc length functional $L = L_g : \Omega_{p,q} \longrightarrow \mathbb{R}$ is then defined as follows [cf. Hawking and Ellis (1973, p. 105)]. Given a piecewise smooth curve $\gamma \in \Omega_{p,q}$, choose a partition $0 = t_0 < t_1 < t_2 < \cdots < t_{n-1} < t_n = 1$ such that $\gamma \mid (t_i, t_{i+1})$ is smooth for each $i = 0, 1, 2, \ldots, n - 1$. Then define

$$L(\gamma) = L_g(\gamma) = \sum_{i=0}^{n-1} \int_{t=t_i}^{t_{i+1}} \sqrt{-g(\gamma'(t),\gamma'(t))}\, dt \tag{3.1}$$

It may be checked as in elementary differential geometry [cf. O'Neill (1966, pp. 51-52)] that this definition of Lorentzian arc length is independent of the parameterization of $\gamma$. Since an arbitrary nonspacelike curve satisfies a local Lipschitz condition, it is differentiable almost everywhere. Hence the Lorentzian arc length $L(\gamma)$ of $\gamma$ may still be defined using (3.1). Alternate but equivalent definitions of $L(\gamma)$ for arbitrary nonnull nonspacelike curves may be given by approximating $\gamma$ by $C^1$ timelike curves [cf. Hawking and Ellis (1973, p. 214)] or by approximating $\gamma$ by sequences of broken nonspacelike geodesics [cf. Penrose (1972, p. 53)]. The Lorentzian arc length of an arbitrary null curve may be set equal to zero.

Now fix $p,q \in M$ with $p \ll q$. If $\gamma$ is any timelike curve from p to q, then $L(\gamma) > 0$. On the other hand, $\gamma$ may be approximated by a sequence $\{\gamma_n\}$ of piecewise smooth "almost null" curves $\gamma_n : [0,1] \longrightarrow M$ with $\gamma_n(0) = p$, $\gamma_n(1) = q$ such that $\gamma_n \longrightarrow \gamma$ in the $C^0$ topology, but $L(\gamma_n) \longrightarrow 0$, (cf. Figure 3.1). This construction shows, moreover, that given any $p,q \in M$ with $p \ll q$, there are curves $\gamma \in \Omega_{p,q}$ with arbitrarily small Lorentzian arc length. Hence the infimum of Lorentzian arc lengths of all piecewise smooth curves joining any two chronologically related points $p \ll q$ is always zero. On the other hand, if p and q lie in a geodesically convex neighborhood U, the future-directed timelike geodesic segment in U from p to q has the largest Lorentzian arc length among all nonspacelike curves in U from p to q.

Thus it is natural to make the following definition of the Lorentzian distance function $d = d(g) : M \times M \longrightarrow \mathbb{R} \cup \{\infty\}$ of $(M,g)$.

DEFINITION 3.1 Given $p \in M$, if $q \notin J^+(p)$, set $d(p,q) = 0$. If $q \in J^+(p)$, set $d(p,q) = \sup \{L_g(\gamma) : \gamma \in \Omega_{p,q}\}$.

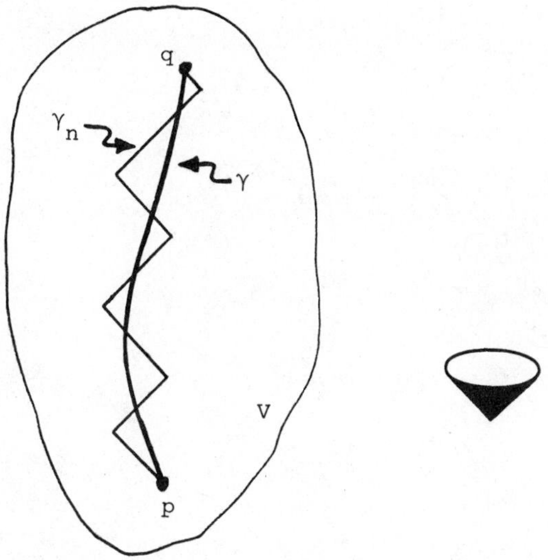


*Figure 3.1* The timelike curve $\gamma$ from p to q is approximated by a sequence of curves $\gamma_n$ with $\gamma_n \longrightarrow \gamma$ in the $C^0$ topology, but $L(\gamma_n) \longrightarrow 0$.

From the definition, it is immediate that

$$d(p,q) > 0 \Leftrightarrow q \in I^+(p) \tag{3.2}$$

Thus the Lorentzian distance function determines the chronological past and future of any point. However, the Lorentzian distance function in general fails to determine the causal past and future sets of p since $d(p,q) = 0$ does *not* imply $q \in J^+(p) - I^+(p)$. But at least if $q \in J^+(p) - I^+(p)$, then $d(p,q) = 0$.

We emphasize that the Lorentzian distance $d(p,q)$ need *not* be finite. One way that $d(p,q) = \infty$ may occur is that timelike curves from p to q may attain arbitrarily large arc lengths by approaching certain boundary points of the space-time. In Figure 3.2, two points with $d(p,q) = \infty$ are shown in a Reissner-Nordström space-time with $e^2 = m^2$ [cf. Hawking and Ellis (1973, p. 160)].

A second way Lorentzian distance may become infinite is through causality violations. Recall that a space-time is said to be totally vicious if $I^+(p) \cap I^-(p) = M$ for all $p \in M$.

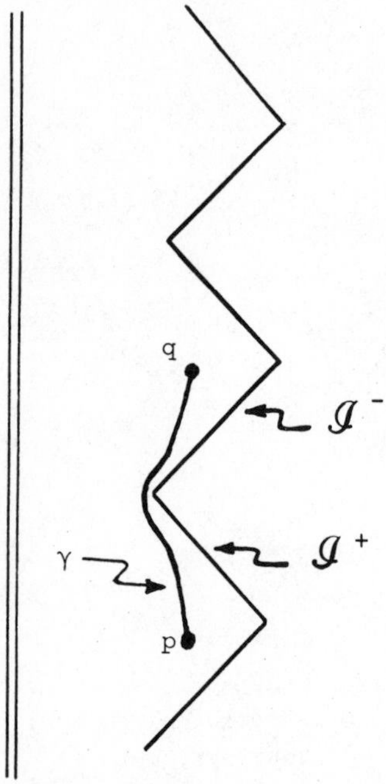


*Figure 3.2* A Reissner-Nordström space-time with $e^2 = m^2$ is shown. By taking timelike curves $\gamma$ from p to q close to $\mathscr{I}^+$ and $\mathscr{I}^-$, we can make $L(\gamma)$ arbitrarily large. Thus $d(p,q) = \infty$.

LEMMA 3.2 Let $(M,g)$ be an arbitrary space-time.

(a) If $p \in I^+(p)$, then $d(p,p) = \infty$. Thus for each $p \in M$, either $d(p,p) = 0$ or $d(p,p) = \infty$.

(b) $(M,g)$ is totally vicious iff $d(p,q) = \infty$ for all $p,q \in M$.

*Proof.* (a) Suppose $p \in I^+(p)$. Then we may find a closed timelike curve $\gamma : [0,1] \longrightarrow M$ with $\gamma(0) = \gamma(1) = p$. Since $\gamma$ is timelike, $L(\gamma) > 0$. If $\sigma_n \in \Omega_{p,p}$ is the timelike curve obtained by traversing $\gamma$ n times, then $L(\sigma_n) = nL(\gamma) \longrightarrow \infty$ as $n \longrightarrow \infty$. Thus $d(p,p) = \infty$.

(b) Suppose $(M,g)$ is totally vicious. Fix $p,q \in M$ and let $n > 0$ be any positive integer. Since $p \in I^+(p)$, we may find $\gamma_1 \in \Omega_{p,p}$ with $L(\gamma_1) \geq n$ by part (a). Since $q \in I^+(p)$, there is a timelike curve $\gamma_2$ from p to q. Then $\gamma = \gamma_1 * \gamma_2 \in \Omega_{p,q}$ is a timelike curve with length $L(\gamma) = L(\gamma_1) + L(\gamma_2) > n$. Hence $d(p,q) = \infty$.

Conversely, suppose $d(p,q) = \infty$ for all $p,q \in M$. Fixing $r \in M$, we have $d(r,p) > 0$ and $d(p,r) > 0$ for all $p \in M$. Thus by (3.2), it follows that $I^+(r) \cap I^-(r) = M$. □

By Definition 3.1, if $I^+(p) \neq M$, there are points $q \in M$ with $d(p,q) = 0$ but $p \neq q$. Hence unlike the Riemannian distance function, the Lorentzian distance function usually fails to be nondegenerate. Indeed, we have seen that $d(p,p) > 0$ is possible. But if $(M,g)$ is chronological, $d(p,p) = 0$ for all $p \in M$. Also, the Lorentzian distance function tends to be nonsymmetric. More precisely, it may be shown for arbitrary space-times that

REMARK 3.3 If $p \neq q$ and $d(p,q)$ and $d(q,p)$ are both finite, then either $d(p,q) = 0$ or $d(q,p) = 0$. Equivalently, if $d(p,q) > 0$ and $d(q,p) > 0$, then $d(p,q) = d(q,p) = \infty$.

*Proof.* If $d(p,q) > 0$ and $d(q,p) > 0$, we may find future-directed timelike curves $\gamma_1$ from p to q and $\gamma_2$ from q to p, respectively. Let $\gamma_n = \gamma_1 * (\gamma_2 * \gamma_1)^n \in \Omega_{p,q}$. As $n \longrightarrow \infty$, $L(\gamma_n) \longrightarrow \infty$, whence $d(p,q) = \infty$. Similarly, $d(q,p) = \infty$. □

A further consequence of Definition 3.1 is that if $\gamma : [0,\infty) \longrightarrow (M,g)$ is any future-directed, future-complete timelike geodesic in an arbitrary space-time $(M,g)$, then $\lim_{t\to\infty} d(\gamma(0),\gamma(t)) \geq \lim_{t\to\infty} L(\gamma|[0,t]) = \infty$. Complete Riemannian manifolds $(N,g_0)$, on the other hand, may contain (nonclosed) geodesics $\sigma : [0,\infty) \longrightarrow (N,g_0)$ for which $\sup\{d_0(\sigma(0),\sigma(t)) : t \geq 0\}$ is finite. Further assumptions are needed for Riemannian manifolds to guarantee that $\lim_{t\to\infty} d(\sigma(0),\sigma(t)) = \infty$ for all geodesics $\sigma : [0,\infty) \longrightarrow (N,g_0)$ [cf. Cheeger and Ebin (1975, pp. 53 and 151)].

While the Lorentzian distance function fails to be symmetric and nondegenerate, at least a reverse triangle inequality holds (cf. Figure 1.3). Explicitly,

$$\text{If } p \leq r \leq q, \text{ then } d(p,q) \geq d(p,r) + d(r,q) \tag{3.3}$$

We now discuss some properties of the Lorentzian distance that make it a useful tool in general relativity and Lorentzian geometry. First, the Lorentzian distance function is lower semicontinuous where it is finite [cf. Hawking and Ellis (1973, p. 215)].

LEMMA 3.4 If $d(p,q) < \infty$ and $p_n \longrightarrow p$, $q_n \longrightarrow q$, then $d(p,q) \leqslant \liminf d(p_n,q_n)$. Also, if $d(p,q) = \infty$ and $p_n \longrightarrow p$, $q_n \longrightarrow q$, then $\lim_{n\to\infty} d(p_n,q_n) = \infty$.

*Proof.* First consider the case $d(p,q) < \infty$. If $d(p,q) = 0$, there is nothing to prove. If $d(p,q) > 0$, then $q \in I^+(p)$ and the lower semicontinuity follows from the following fact. Given any $\varepsilon > 0$, a timelike curve $\gamma$ of length $d(p,q) - \varepsilon/2$ from p to q and sufficiently small neighborhoods $U_1$ of p and $U_2$ of q may be found such that $\gamma$ may be deformed to give a timelike curve of length $\geqslant d(p,q) - \varepsilon$ from any point r of $U_1$ to any point s of $U_2$.

Suppose now that $d(p,q) = \infty$, but $\liminf d(p_n,q_n) = R < \infty$. Since $d(p,q) = \infty$ there exists a timelike curve $\gamma$ from p to q of length $L(\gamma) > R + 2$. This implies that there exist neighborhoods $U_1$ and $U_2$ of p and q, respectively, such that $\gamma$ can be deformed to give a timelike curve of length $\geqslant R + 1$ from any point r of $U_1$ to any point s of $U_2$. This contradicts $\liminf d(p_n,q_n) = R$. □

In general, the Lorentzian distance function fails to be upper semicontinuous. We give an example of a space-time (M,g) containing an infinite sequence $p_n \longrightarrow p$ and $q \in I^+(p)$ such that $d(p_n,q) = \infty$ for all large n, but $d(p,q) < \infty$ (Figure 3.3).

For globally hyperbolic space-times, on the other hand, the Lorentzian distance function is finite and continuous just like the Riemannian distance function.

LEMMA 3.5 If (M,g) is a globally hyperbolic space-time, then the Lorentzian distance function d is finite and continuous on $M \times M$.

*Proof.* To prove the finiteness of d, cover the compact set $J^+(p) \cap J^-(q)$ with a finite number of convex normal neighborhoods $B_1, B_2, \ldots, B_m$ such that no nonspacelike curve which leaves any $B_i$

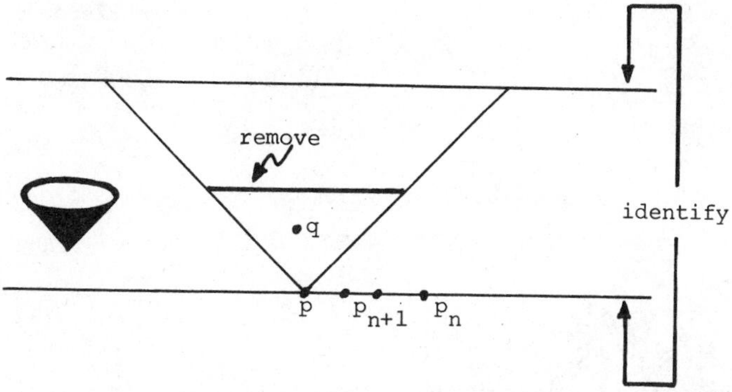


*Figure 3.3* Let M be $\{(x,y) \in \mathbb{R}^2 : 0 \leq y \leq 2\} - \{(x,1) : -1 \leq x \leq 1\}$ with the identification $(x,0) \sim (x,2)$ and the flat Lorentzian metric $ds^2 = dx^2 - dy^2$. Let $p = (0,0)$, $q = (0,1/2)$ and $p_n \longrightarrow p$ as shown. Then $p_n \in I^+(p_n)$ and hence $d(p_n,p_n) = \infty$ for all n. For large n we have $q \in I^+(p_n)$ and thus $d(p_n,q) = \infty$. On the other hand, $d(p,q) = 1/2$ which yields $d(p,q) < \liminf d(p_n,q)$. This space-time is not causal. However, the distance function may also fail to be upper semicontinuous in causal space-times (cf. Figure 3.6).

ever returns and such that every nonspacelike curve in each $B_i$ has length at most 1. Since any nonspacelike curve $\gamma$ from p to q can enter each $B_i$ no more than once, $L(\gamma) \leq m$. Hence $d(p,q) \leq m$.

If d failed to be upper semicontinuous at $(p,q) \in M \times M$, we could find a $\delta > 0$ and sequences $\{p_n\}$ and $\{q_n\}$ converging to p and q, respectively, such that $d(p_n,q_n) \geq d(p,q) + 2\delta$ for all n. By definition of $d(p_n,q_n)$, we may then find a future-directed nonspacelike curve $\gamma_n$ from $p_n$ to $q_n$ with $L(\gamma_n) \geq d(p,q) + \delta$ for each n. By Corollary 2.19, the sequence $\{\gamma_n\}$ has a nonspacelike limit curve $\gamma$ from p to q. By Proposition 2.21, a subsequence $\{\gamma_m\}$ of $\{\gamma_n\}$ converges to $\gamma$ in the $C^0$ topology. Hence $L(\gamma) \geq d(p,q) + \delta$ by Remark 2.22. But this contradicts the definition of Lorentzian distance. Thus d is upper semicontinuous at (p,q). □

We now define the following condition [cf. Beem and Ehrlich (1977, Condition 4)].

DEFINITION 3.6 The space-time $(M,g)$ satisfies the *finite distance condition* if $d(g)(p,q) < \infty$ for all $p,q \in M$.

Lemma 3.5 then has the following

COROLLARY 3.7 If $(M,g)$ is globally hyperbolic, then $(M,g)$ satisfies the finite distance condition and $d(g) : M \times M \longrightarrow \mathbb{R}$ is continuous.

If $(M,g)$ is globally hyperbolic, all metrics in the conformal class $C(M,g)$ are globally hyperbolic. Hence all metrics in $C(M,g)$ satisfy the finite distance condition. We will examine the converse of this statement in Section 3.3, Theorem 3.30.

Since the given topology of a smooth manifold coincides with the metric topology induced by any Riemannian metric, it is natural to consider the sets $\{m \in I^+(p) : d(p,m) < \varepsilon\}$ for a Lorentzian manifold. However, as Minkowski space shows, these sets do not form a basis for the given manifold topology (cf. Figure 3.4). Indeed, this same example shows that no matter how small $\varepsilon > 0$ is chosen, the sets $\{m \in J^+(p) : d(p,m) \leqslant \varepsilon\}$ may fail to be compact and fail to be geodesically convex as well as failing to be diffeomorphic to the closed n disk.

The sphere of radius $\varepsilon$ for the point $p \in M$ is given by $K(p,\varepsilon) = \{q \in M : d(p,q) = \varepsilon\}$. This set need not be compact. However, the reverse triangle inequality and (3.2) imply that $K(p,\varepsilon)$ is achronal for all finite $\varepsilon > 0$ and all $p \in M$.

In arbitrary space-times the future (resp., past) inner ball $B^+(p,\varepsilon) = \{q \in I^+(p) : d(p,q) < \varepsilon\}$ (resp., $B^-(p,\varepsilon) = \{q \in I^-(p) : d(q,p) < \varepsilon\}$) need not be open. On the other hand, when the distance function $d : M \times M \longrightarrow \mathbb{R} \cup \{\infty\}$ is continuous, these inner balls must be open. In Section 3.3 we will show that for distinguishing space-times with continuous distance functions, the past and future inner balls form a subbasis for the manifold topology.

A different subbasis for the topology of any strongly causal space-time $(M,g)$ with a possibly discontinuous distance function

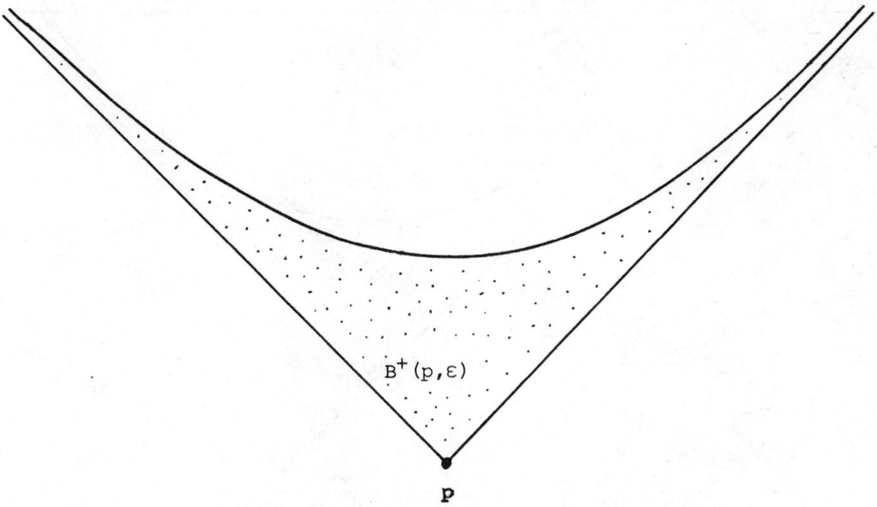


*Figure 3.4* The set $B^+(p,\varepsilon) = \{q \in I^+(p) : d(p,q) < \varepsilon\}$ in Minkowski space-time does not have compact closure, is not geodesically convex and does not contain p. Furthermore, sets of the form $B^+(p,\varepsilon)$ do not form a basis for the manifold topology.

But in general, if (M,g) is a distinguishing space-time with a continuous Lorentzian distance function, then a subbasis for the manifold topology is given by sets of the form $B^+(p,\varepsilon)$ and $B^-(p,\varepsilon)$ (cf. Proposition 3.31). Hence these sets do form a *sub*basis for the given topology of Minkowski space-time.

$d = d(g) : M \times M \longrightarrow \mathbb{R} \cup \{\infty\}$ may be obtained by using the outer balls $O^+(p,\varepsilon)$ and $O^-(p,\varepsilon)$ rather than the inner balls.

DEFINITION 3.8 The *outer ball* $O^+(p,\varepsilon)$ [resp., $O^-(p,\varepsilon)$] of $I^+(p)$ [resp., $I^-(p)$] is given by

$$O^+(p,\varepsilon) = \{q \in M : d(p,q) > \varepsilon\}$$

and

$$O^-(p,\varepsilon) = \{q \in M : d(q,p) > \varepsilon\}$$

respectively (cf. Figure 3.5).

Since the Lorentzian distance function is lower semicontinuous where it is finite, the outer balls $O^+(p,\varepsilon)$ and $O^-(p,\varepsilon)$ are open in

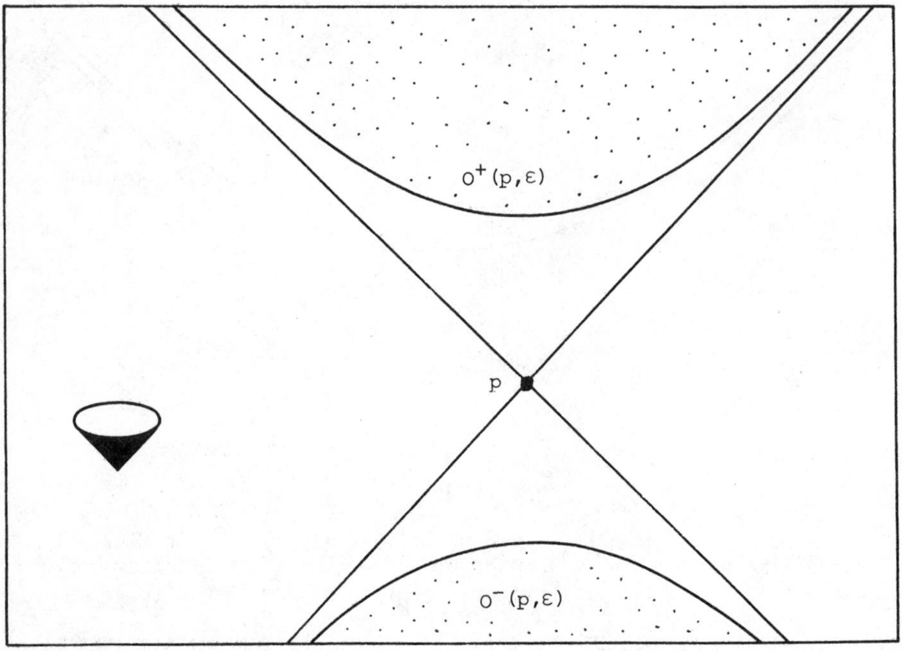


*Figure 3.5* The outer balls $O^+(p,\varepsilon) = \{q \in M : d(p,q) > \varepsilon\}$ and $O^-(p,\varepsilon) = \{q \in M : d(q,p) > \varepsilon\}$ are open in arbitrary space-times. Furthermore, $O^+(p,\varepsilon)$ and $O^-(p,\varepsilon)$ are always subsets of $I^+(p)$ and $I^-(p)$, respectively. If (M,g) is strongly causal, the outer balls $O^+(p,\varepsilon)$ and $O^-(p,\varepsilon)$ with $p \in M$ and $\varepsilon > 0$ arbitrary form a subbasis for the manifold topology.

arbitrary space-times. The reverse triangle inequality implies that these sets also have the property that if $m,n \in O^+(p,\varepsilon)$, [resp., $m,n \in O^-(p,\varepsilon)$] and $m \leq n$, then any future-directed nonspacelike curve from m to n lies in $O^+(p,\varepsilon)$, [resp., $O^-(p,\varepsilon)$]. Moreover,

THEOREM 3.9 Let (M,g) be strongly causal. Then the collection $\{O^+(p,\varepsilon_1) \cap O^-(q,\varepsilon_2) : p,q \in M, \varepsilon_1,\varepsilon_2 > 0\}$ forms a basis for the given manifold topology.

*Proof*. Let $m \in M$ be given and let U be any open neighborhood containing m. We may find a local causality neighborhood $U_1$ with $m \in U_1 \subset U$, i.e., no nonspacelike curve which leaves $U_1$ ever returns. Choose $p_1,p_2 \in U_1$ with $p_1 \ll m \ll p_2$ such that $I^+(p_1) \cap I^-(p_2) \subset U_1$.

By the chronology assumptions on $p_1$ and $p_2$, we have $d(p_1,m) > 0$ and $d(m,p_2) > 0$. Choose constants $\varepsilon_1$, $\varepsilon_2$ with $0 < \varepsilon_1 < d(p_1,m)$ and $0 < \varepsilon_2 < d(m,p_2)$. Then $m \in O^+(p_1,\varepsilon_1) \cap O^-(p_2,\varepsilon_2)$. Since $O^+(p_1,\varepsilon_1) \subset I^+(p_1)$ and $O^-(p_2,\varepsilon_2) \subset I^-(p_2)$, we also have $O^+(p_1,\varepsilon_1) \cap O^-(p_2,\varepsilon_2) \subset I^+(p_1) \cap I^-(p_2) \subset U_1 \subset U$ as required. □

For complete Riemannian manifolds, any two points may be joined by a minimal (distance-realizing) geodesic segment. We now examine the dual of this property for space-times.

In Hawking and Ellis (1973, p. 110), a timelike geodesic $\gamma$ from p to q is said to be maximal if the index form of $\gamma$ is negative semidefinite. This definition implies that if the geodesic $\gamma$ is *not* maximal, there exist variations of $\gamma$ which yield curves from p to q "close" to $\gamma$ having longer Lorentzian arc length than $\gamma$. If $\gamma$ is maximal in this sense, however, no small variation of $\gamma$ keeping p and q fixed will produce timelike curves $\sigma$ from p to q with $L(\sigma) > L(\gamma)$. Nonetheless, there may still exist a timelike geodesic $\sigma_1$ in M from p to q ("far" from $\gamma$) with $d(p,q) = L(\sigma_1) > L(\gamma)$. Thus maximality as defined by Hawking and Ellis does not imply "maximality in the large." To study "maximality in the large," we adopt in analogy to the concept of minimality in Riemannian geometry, a definition of maximality valid for all *curves* in the path space $\Omega_{p,q}$ [cf. Beem and Ehrlich (1977, Definition 1)]. The motivation for our definition is Theorem 3.13 below [cf. Beem and Ehrlich (1979a, p. 166)] and its applications, particularly, the construction of geodesics as limit curves of sequences of "almost maximal" curves in Chapter 7 and the definition of the Lorentzian cut locus in Chapter 8.

DEFINITION 3.10 Let $p,q \in M$ with $p \leqslant q$, $p \neq q$. The curve $\gamma \in \Omega_{p,q}$ is said to be *maximal* if $L(\gamma) = d(p,q)$.

An immediate consequence of the reverse triangle inequality (3.3) is

REMARK 3.11 If $\gamma : [0,1] \longrightarrow M$ in $\Omega_{p,q}$ is maximal, then for all s, t with $0 \leq s < t \leq 1$, we have $d(\gamma(s),\gamma(t)) = L(\gamma|[s,t])$.

The following result, stated somewhat differently in Penrose (1972, Prop. 7.2), is the analogue of the principle in Riemannian geometry that "locally" geodesics minimize arc length [cf. Bishop and Crittenden (1964, p. 149, Theorem 2)].

PROPOSITION 3.12 Let U be a convex normal neighborhood centered at $p \in M$. For $q \in J^+(p)$, let $\overline{pq}$ denote the unique nonspacelike geodesic $c : [0,1] \longrightarrow U$ in U with $c(0) = p$, $c(1) = q$. If $\gamma$ is any future-directed nonspacelike curve in U from p to q with $L(\gamma) = d(p,q)$, then $\gamma$ coincides with $\overline{pq}$ up to parameterization.

*Proof.* If $q \in I^+(p)$ and $d(p,q) > 0$, Penrose (1972, p. 53) shows using a synchronous coordinate system that if $\gamma$ is any causal trip in U from p to q other than $\overline{pq}$, then $L(\gamma) < L(\overline{pq}) = d(p,q)$. This may be obtained equivalently using the Gauss lemma (cf. Corollary 9.19 of Section 9.1). Hence the result is established if $d(p,q) > 0$.

Suppose now that $d(p,q) = 0$ and let $\gamma$ be any nonspacelike curve in U from p to q. Then $L(\gamma) \leqq d(p,q) = 0$. Thus $\gamma : [0,1] \longrightarrow M$ is a null curve. Suppose that $\gamma(t) \notin \text{Int}(\overline{pq})$. Let $\gamma_1$ be the unique null geodesic in U from p to $\gamma(t)$ and let $\gamma_2$ be the unique null geodesic in U from $\gamma(t)$ to q. By Proposition 2.19 of Penrose (1972, p. 15), $\gamma_1 * \gamma_2$ is either a smooth null geodesic or $p \ll q$. Since $d(p,q) = 0$, $p \ll q$ is impossible. Hence $\gamma_1 * \gamma_2$ is a smooth null geodesic which by convexity of U must coincide with $\overline{pq}$ up to parameterization. □

Proposition 3.12 has the following important consequence.

THEOREM 3.13 If $\gamma \in \Omega_{p,q}$ satisfies $L(\gamma) = d(p,q)$, then $\gamma$ may be reparameterized to be a smooth geodesic.

*Proof.* Fix any point $\gamma(t)$ on $\gamma$. We may find $\delta > 0$ such that a convex neighborhood centered at $\gamma(t + \delta)$ contains $\gamma([t - \delta, t + \delta])$. By Remark 3.11 the curve $\gamma \mid [t - \delta, t + \delta]$ is maximal. Thus

Proposition 3.12 implies that $\gamma \mid [t - \delta, t + \delta]$ may be reparameterized to be a smooth geodesic. As t was arbitrary, the theorem now follows. □

As an illustration of the use of Definition 3.10 and Theorem 3.13, we give a simple proof of a basic result in elementary causality theory [cf. Penrose (1972, Proposition 2.20)] that is usually obtained by different methods.

COROLLARY 3.14 If $p \leqslant q$, but $p \ll q$ is false, there is a null geodesic from p to q.

*Proof.* The causality assumptions on p and q imply that $d(p,q) = 0$. Now let $\gamma$ be a future-directed nonspacelike curve from p to q. By definition of Lorentzian distance, $d(p,q) \geqslant L(\gamma) \geqslant 0$. Thus $L(\gamma) = d(p,q) = 0$ and $\gamma$ is maximal. By Theorem 3.13, $\gamma$ may be reparameterized to a smooth geodesic $c : [0,1] \longrightarrow M$ from p to q. Since $L(c) \leqslant d(p,q) = 0$, the geodesic c must be a null geodesic. □

As a second application of the elementary properties of the distance function, we give a proof of the existence of a smooth closed timelike geodesic on any compact space-time having a regular cover with a compact Cauchy surface. Using infinite dimensional Morse theory, it may be shown [cf. Klingenberg (1978)] that any compact Riemannian manifold admits at least one smooth closed geodesic. However, the method of proof relies crucially on the positive definiteness of the metric and thus is not applicable to Lorentzian manifolds. Nonetheless, one may obtain the following theorem of Tipler by direct methods [cf. Tipler (1979) for a stronger result].

THEOREM 3.15 Let (M,g) be a compact space-time with a regular covering space which is globally hyperbolic and has a compact Cauchy surface. Then (M,g) contains a closed timelike geodesic.

*Proof.* Since M is compact, there exists a closed future-directed timelike curve $\gamma : [0,1] \longrightarrow M$. Set $p = \gamma(0) = \gamma(1)$. Let $\pi : \tilde{M} \longrightarrow M$ denote the given covering manifold and let

$\tilde{\gamma} : [0,1] \longrightarrow \tilde{M}$ be a lift of $\gamma$, i.e., $\pi \circ \tilde{\gamma}(t) = \gamma(t)$ for all $t \in [0,1]$. Then $\tilde{\gamma}$ is a future-directed timelike curve in $\tilde{M}$. Put $p_1 = \tilde{\gamma}(0)$, $p_2 = \tilde{\gamma}(1)$. Then the global hyperbolicity of M implies $p_1$ and $p_2$ are distinct points which cannot lie on any common Cauchy surface. Since $\pi : \tilde{M} \longrightarrow M$ is regular, there must be a deck transformation $\psi : \tilde{M} \longrightarrow \tilde{M}$ taking $p_1$ to $p_2$ [cf. Wolf (1974, pp. 35-38, p. 60)]. Choose a compact Cauchy surface $S_1$ of $\tilde{M}$ containing $p_1$ and define $S_2 = \psi(S_1)$. Since $(\tilde{M},g)$ is globally hyperbolic, the distance function $d = d(\tilde{g}) : \tilde{M} \times \tilde{M} \longrightarrow \mathbb{R} \cup \{\infty\}$ is finite valued and continuous. Thus we may define a continuous function $f : S_1 \longrightarrow \mathbb{R}$ by $f(s) = d(s,\psi(s))$. Since $f(p_1) > 0$, we have $A = \sup\{d(s,\psi(s)) : s \in S_1\} > 0$. Moreover, since $S_1$ is compact, $A < \infty$ and there exists an $r_1 \in S_1$ with $d(r_1,\psi(r_1)) = A$. Let $\tilde{c} : [0,1] \longrightarrow M$ be a timelike geodesic segment with $\tilde{c}(0) = r_1$, $\tilde{c}(1) = \psi(r_1)$, and $L(c) = d(r_1,\psi(r_1)) = A$. This geodesic exists since (M,g) is globally hyperbolic. Because $\tilde{g} = \pi^* g$, it follows that $c = \pi \circ \tilde{c} : [0,1] \longrightarrow M$ is a timelike geodesic. Since $\tilde{c}(0) = r_1$ and $\tilde{c}(1) = \psi(r_1)$, we also have $c(0) = c(1)$. If c were not smooth at $c(0)$, we could deform c to a timelike curve $\sigma : [0,1] \longrightarrow M$ with $L_g(\sigma) > L(c)$, $\sigma(0) = \sigma(1) \in \pi(S_1)$, which lifts to a curve $\tilde{\sigma} : [0,1] \longrightarrow \tilde{M}$ with $\tilde{\sigma}(0) \in S_1$, $\tilde{\sigma}(1) = \psi(\tilde{\sigma}(0)) \in S_2$. But then $L_{\tilde{g}}(\tilde{\sigma}) = L_g(\sigma) > L_g(c) = L_{\tilde{g}}(\tilde{c}) = A$, in contradiction.

## 3.2 DISTANCE-PRESERVING AND HOMOTHETIC MAPS

Myers and Steenrod (1939) and Palais (1957) have shown that if f is a distance-preserving map of a Riemannian manifold $(N_1,g_1)$ onto a Riemannian manifold $(N_2,g_2)$, then f is a diffeomorphism which preserves the metric tensors, i.e., $f^* g_2 = g_1$. In particular, every distance-preserving map of $(N_1,g_1)$ onto itself is a smooth isometry. In this section, we give similar results for Lorentzian manifolds following Beem (1978a).

Recall that a diffeomorphism $f : (M_1,g_1) \longrightarrow (M_2,g_2)$ of the Lorentzian manifold $(M_1,g_1)$ onto the Lorentzian manifold $(M_2,g_2)$ is said to be *homothetic* if there exists a constant $c > 0$ such that $g_2(f_*v, f_*w) = cg_1(v,w)$ for all $v,w \in T_pM_1$ and all $p \in M_1$. In particular, if $c = 1$, then f is a (smooth) isometry. The group of homothetic transformations is important in general relativity since it has been shown to be the group of transformations which for a large class of space-times preserves the causal structure [cf. Zeeman (1964), (1967), Göbel (1976)].

We will let $d_1$ denote the Lorentzian distance function of $(M_1, g_1)$ and $d_2$ denote the Lorentzian distance function of $(M_2, g_2)$ below. The distance analogue of a smooth homothetic map is defined as follows.

DEFINITION 3.16 A map $f : (M_1, g_1) \longrightarrow (M_2, g_2)$ is said to be *distance homothetic* if there exists a constant $c > 0$ such that $d_2(f(p), f(q)) = cd_1(p,q)$ for all $p,q \in M$. If $c = 1$, then f is said to be *distance preserving*.

It is important to note that for arbitrary Lorentzian manifolds, distance preserving does *not* imply continuity. For if $(M,g)$ is a totally vicious space-time, we have seen that $d(p,q) = \infty$ for all $p,q \in M$ [cf. Lemma 3.2(b)]. Hence any set theoretic bijection $f : M \longrightarrow M$ is distance preserving, but need not be continuous.

THEOREM 3.17 Let $(M_1, g_1)$ be a strongly causal space-time and let $(M_2, g_2)$ be an arbitrary space-time. If $f : (M_1, g_1) \longrightarrow (M_2, g_2)$ is a distance homothetic map (*not* assumed to be continuous) of $M_1$ onto $M_2$, then f is a smooth homothetic map. That is, f is a diffeomorphism and there exists a constant $c > 0$ such that $f^*g_2 = cg_1$. In particular, every map of a strongly causal space-time $(M,g)$ onto itself which preserves Lorentzian distance is an isometry.

COROLLARY 3.18 If $(M,g)$ is a strongly causal space-time, the space of distance homothetic maps of $(M,g)$ equipped with the compact-open topology is a Lie group.

*Proof of Corollary 3.18* Since $(M,g)$ is strongly causal, this group coincides by Theorem 3.17 with the space of smooth, homothetic maps of M onto itself which preserve the time orientation. But this second group is a Lie group. □

The proof of Theorem 3.17 will be broken up into a series of lemmas.

LEMMA 3.19 Let $(M_1,g_1)$ and $(M_2,g_2)$ be space-times and let $f : (M_1,g_1) \longrightarrow (M_2,g_2)$ be an onto (but not necessarily continuous) map. If f is distance homothetic, then

(a) $p \ll q$ iff $f(p) \ll f(q)$.

(b) $f(I^+(p) \cap I^-(q)) = I^+(f(p)) \cap I^-(f(q))$.

*Proof*. First (a) holds since $d_2(f(p),f(q)) = cd_1(p,q)$ and $p \ll q$ [resp., $f(p) \ll f(q)$] iff $d_1(p,q) > 0$ [resp., $d_2(f(p),f(q)) > 0$]. Since (a) implies $p \ll r \ll q$ iff $f(p) \ll f(r) \ll f(q)$, statement (b) follows. □

The importance of (b) stems from the fact that if (M,g) is strongly causal, the sets $\{I^+(p) \cap I^-(q) : p,q \in M\}$ form a basis for the topology of M. Recall that a map $f : M_1 \longrightarrow M_2$ is said to be *open* if f maps each open set in $M_1$ to an open set in $M_2$.

LEMMA 3.20 Let $(M_1,g_1)$ be strongly causal and let $(M_2,g_2)$ be an arbitrary space-time. If f is a distance homothetic (not necessarily continuous) map of $(M_1,g_1)$ onto $(M_2,g_2)$, then f is open and one-to-one.

*Proof*. The openness of f is immediate from part (b) of Lemma 3.19. It remains to show that f is one-to-one. Assume there are distinct points p and q of $M_1$ with $f(p) = f(q)$. Let U(p) be an open neighborhood of p with $q \notin U(p)$ and such that no nonspacelike curve intersects U(p) more than once. Choose $r_1,r_2 \in U(p)$ with $r_1 \ll p \ll r_2$. Clearly, $q \notin I^+(r_1) \cap I^-(r_2)$. It follows from Lemma 3.19 that $f(r_1) \ll f(p) = f(q) \ll f(r_2)$ implies $r_1 \ll q \ll r_2$. This yields $q \in I^+(r_1) \cap I^-(r_2)$ which is a contradiction. □

Applying Lemma 3.20 to f and $f^{-1}$, we obtain

PROPOSITION 3.21 Let $(M_1,g_1)$ be strongly causal and let $(M_2,g_2)$ be an arbitrary space-time. Let f be a not necessarily continuous map of $M_1$ onto $M_2$. If f is distance homothetic, then f is a homeomorphism and $(M_2,g_2)$ is strongly causal.

*Proof.* The relation $f^{-1}$ is a function since f is one-to-one by Lemma 3.20. Furthermore, $f^{-1}$ is continuous since Lemma 3.20 shows f is an open map.

In order to complete the proof it is sufficient to show $M_2$ is strongly causal since Lemma 3.20 will then imply $f^{-1}$ is an open map whence f is continuous. Given $p' \in M_2$, let $p = f^{-1}(p')$. If $r' \ll p' \ll q'$, then Lemma 3.19 applied to the distance homothetic map $f^{-1}$ yields $f^{-1}(r') \ll p \ll f^{-1}(q')$. Let $U'(p')$ be an open neighborhood of p'. Choose $V'(p') \subset U'(p')$ with the closure of $V'(p')$ a compact set contained in an open convex normal neighborhood $W'(p')$ of p'. We may assume that $(W'(p'), g_2|W'(p'))$ is globally hyperbolic. Let $\{r_n'\}$ and $\{q_n'\}$ be sequences in $V'(p')$ such that $r_n' \to p'$, $q_n' \to p'$, and $r_n' \ll p' \ll q_n'$ for all n. Assume the strong causality of $M_2$ fails at p'. This means that for each n, the set $I^+(r_n') \cap I^-(q_n')$ cannot be contained in the convex normal neighborhood $W'(p')$ because otherwise, the sets $I^+(r_n') \cap I^-(q_n')$ would give arbitrarily small neighborhoods of p' which each nonspacelike curve intersects at most once. Choose a sequence of points $\{z_n'\}$ contained in the boundary of $V'(p')$ with $z_n' \in I^+(r_n') \cap I^-(q_n')$ for each n. The sequence $\{z_n'\}$ has an accumulation point z because the closure of $V'(p')$ is compact. Furthermore, $f^{-1}(z_n') \in f^{-1}(I^+(r_n') \cap I^-(q_n')) = I^+(f^{-1}(r_n')) \cap I^-(f^{-1}(q_n'))$. The continuity of $f^{-1}$ implies that $f^{-1}(r_n') \to p$ and $f^{-1}(q_n') \to p$. The strong causality of $M_1$ yields that the sets $I^+(f^{-1}(r_n')) \cap I^-(f^{-1}(q_n'))$ are approaching the point p. Thus, $f^{-1}(z_n') \to p$ which means $f^{-1}(z) = p = f^{-1}(p')$. This contradicts the one-to-one property of $f^{-1}$. Consequently, $M_2$ must be strongly causal and the proposition is established. □

Consider the strongly causal space-time M. Given $p \in M$ let U(p) be a convex normal neighborhood of p. The set U(p) may be chosen so small that whenever $q,r \in U(p)$ with $q \leqslant r$, the distance $d(q,r)$ is the length of the unique geodesic segment $\alpha(q,r)$ from q to r which lies in U(p). Furthermore, U(p) may be chosen such that if $q,z,r \in U(p)$ with $q \ll z \ll r$, then the reverse triangle inequality $d(q,r) \geqslant d(q,z) + d(z,r)$ is valid with strict equality iff z is on the geodesic segment from q to r in U(p). Thus, timelike geodesics in a strongly causal space-time are characterized by the space-time distance function, and it follows that distance homothetic maps take timelike geodesics to timelike geodesics.

LEMMA 3.22 If f is a distance homothetic map defined on a strongly causal space-time, then f maps null geodesics to null geodesics.

*Proof.* Let U(p) be a convex normal neighborhood of p as in the above paragraph, chosen sufficiently small such that f(U(p)) lies in a convex normal neighborhood of f(p). Let $\alpha(q,r)$ be a null geodesic in U(p). Choose $q_n \longrightarrow q$ and $r_n \longrightarrow r$ with $q_n \ll r_n$ for all n. Proposition 3.21 then implies that $f(q_n) \longrightarrow f(q)$ and $f(r_n) \longrightarrow f(r)$. The map f takes the timelike geodesic $\alpha(q_n,r_n)$ with endpoints $q_n$ and $r_n$ to the timelike geodesic $\alpha(f(q_n),f(r_n))$. Since the geodesics $\alpha(q_n,r_n)$ converge to $\alpha(q,r)$ and the geodesics $\alpha(f(q_n),f(r_n))$ converge to $\alpha(f(q),f(r))$, it follows that f maps $\alpha(q,r)$ to $\alpha(f(q),f(r))$. □

*Proof of Theorem 3.17* The fact that f is a diffeomorphism follows from a result proved by Hawking, King, and McCarthy (1976) which states that a homeomorphism which maps null geodesics to null geodesics must be a diffeomorphism. Since $M_1$ and $M_2$ are strongly causal, for each $p \in M_1$ there exists a convex normal neighborhood $U_1(p)$ such that for $q \in U_1(p)$ with $p \ll q$ the lengths of the timelike geodesics $\alpha(p,q)$ joining p to q and $\alpha(f(p),f(q))$ joining f(p) to f(q) are given by $d_1(p,q)$ and $d_2(f(p),f(q))$, respectively. Using $d_2(f(p),f(q)) = cd_1(p,q)$, it follows that f maps $g_1$ onto the tensor $c^{-2}g_2$. □

It is well known that if a complete Riemannian manifold is not locally flat, then it admits no homothetic maps that are not isometries [cf. Kobayashi and Nomizu (1963, p. 242, Lemma 2)]. An essential step in the proof consists of showing for arbitrary complete Riemannian manifolds that any homothetic map which is not an isometry has a unique fixed point. This may be done by using the triangle inequality for the Riemannian distance function and the metric completeness of any geodesically complete Riemannian manifold.

In view of Theorem 3.17 above, it is then of interest to consider the analogous question of the existence of nonisometric homothetic maps of a Lorentzian manifold [cf. Beem (1978b)]. We will use the standard terminology of *proper* homothetic map for a homothetic map which is not an isometry below.

We first note that $\mathbb{R}^2$ with the Lorentzian metric $ds^2 = dx\, dy$ provides an example of a globally hyperbolic, geodesically complete space-time that admits a fixed-point free, proper homothetic map. For fixing any $\beta \neq 0$ and choosing any $c > 0$, the map $f(x,y) = (x + \beta, cy)$ is a fixed-point free homothetic map with homothetic constant c. Thus the existence of a fixed point for a proper homothetic map must be assumed for geodesically complete Lorentzian manifolds unlike the Riemannian case.

Now suppose f is a proper homothetic map of a space-time (M,g) such that $f(p) = p$ for some $p \in M$. Then $f_{*p} : T_pM \longrightarrow T_pM$ has at least one nonspacelike eigenvector [cf. Beem (1978, p. 319, Lemma 3)]. This eigenvector may be null, however. For example, composing the Lorentzian "boost" isometry

$$F(x,y) = (x \cosh t + y \sinh t,\ x \sinh t + y \cosh t)$$

with $t > 0$ fixed of $(\mathbb{R}^2,\ ds^2 = dx^2 - dy^2)$ and a dilation $T(x,y) = (cx, cy)$, $c > 0$, $c \neq 1$, yields a proper homothetic map f of Minkowski space-time fixing the origin such that $f_{*(0,0)}$ has null vectors for eigenvectors.

But if $f_{*_p} : T_pM \longrightarrow T_pM$ is a proper homothetic map which has a timelike eigenvector with eigenvalue $\lambda < 1$, it may be shown that $(M,g)$ is Minkowski space-time [cf. Beem (1978b, p. 319, Proposition 4)]. Also if f is a homothetic map with a fixed point p such that all eigenvalues of $f_{*_p}$ are real and all have absolute value less than 1, then $(M,g)$ is Minkowski space-time [cf. Beem (1978b, p. 316, Theorem 1)].

We now give an example of a nonflat space-time admitting a global homothetic flow. Let $M = \mathbb{R}^3$ with the metric $g = ds^2 = e^{xz}\, dx\, dy + dz^2$. Thus if

$$v = a\frac{\partial}{\partial x} + b\frac{\partial}{\partial y} + c\frac{\partial}{\partial z} \qquad w = \bar{a}\frac{\partial}{\partial x} + \bar{b}\frac{\partial}{\partial y} + \bar{c}\frac{\partial}{\partial z}$$

are tangent vectors at $(x,y,z)$, we have

$$g(v,w) = e^{xz}\frac{\bar{a}b + a\bar{b}}{2} + c\bar{c}$$

It may then be checked that while $(M,g)$ is not flat, the map $\phi_t : (\mathbb{R}^3, ds^2) \longrightarrow (\mathbb{R}^3, ds^2)$ given by

$$\phi_t(x,y,z) = (e^t x,\ e^{-3t} y,\ e^{-t} z)$$

is a proper homothety with $g(\phi_{t_*} v, \phi_{t_*} w) = e^{-2t} g(v,w)$ for each fixed nonzero t.

We now show, however, that this space-time is null geodesically incomplete. Let $X = \partial/\partial x$, $Y = \partial/\partial y$, and $Z = \partial/\partial z$. Then all inner products vanish except for $g(X,Y) = e^{xz}/2$, $g(Z,Z) = 1$, and $[X,Y] = [X,Z] = 0$. Hence using the formula

$$\begin{aligned} 2g(\nabla_U V, W) &= Ug(V,W) + Vg(U,W) - Wg(U,V) \\ &\quad + g([U,V],W) - g([U,W],V) - g([V,W],U) \end{aligned}$$

we obtain the following formulas for the Levi-Civita connection of $(\mathbb{R}^3, ds^2)$:

$$\nabla_X X = zX \qquad \nabla_Y Y = \nabla_Z Z = 0$$

$$\nabla_X Y = \frac{-x}{4} e^{xz} Z \qquad \nabla_X Z = \frac{x}{2} X \qquad \nabla_Y Z = \frac{x}{2} Y$$

Thus the only nonzero Christoffel symbols are $\Gamma^1_{11} = z$, $\Gamma^3_{12} = \Gamma^3_{21} = -x/4\ e^{xz}$, $\Gamma^1_{13} = \Gamma^1_{31} = \Gamma^2_{23} = \Gamma^2_{32} = x/2$. Hence if $\gamma(t) = (x(t), y(t), z(t))$ is a geodesic, the usual system of second order differential equations

$$\frac{d^2 x_k}{dt^2} + \sum_{i,j} \Gamma^k_{ij} \circ \gamma(t) \frac{dx_i}{dt} \frac{dx_j}{dt} = 0$$

for $\gamma$ reduces to the following system:

$$x'' + z(x')^2 + xx'z' = 0$$
$$y'' + xy'z' = 0$$
$$z'' - \frac{e^{xz}}{2} xx'y' = 0$$

It may finally be checked that $\gamma : (-1,\infty) \longrightarrow (\mathbb{R}^3, ds^2)$ given by $\gamma(t) = (\ln(1 + t), 0, 1)$ satisfies this system of differential equations and hence is the unique null geodesic in $(\mathbb{R}^3, ds^2)$ with $\gamma'(0) = \partial/\partial x|_{(0,0,1)}$. Thus this space-time is null geodesically incomplete.

## 3.3 THE LORENTZIAN DISTANCE FUNCTION AND CAUSALITY

In this section we study the relationship between the continuity and finiteness of the Lorentzian distance function $d = d(g) : M \times M \longrightarrow \mathbb{R} \cup \{\infty\}$ for $(M,g)$ and the causal structure of $(M,g)$. The most elementary properties, extending Lemma 3.2 above, are summarized in the following lemma. Recall that Lor(M) denotes the space of all Lorentzian metrics for M. The $C^0$ topology on Lor(M) was defined in Section 2.2.

LEMMA 3.23

(a) $d(p,q) > 0$ iff $q \in I^+(p)$.

(b) The space-time $(M,g)$ is totally vicious iff $d(p,q) = \infty$ for all $p,q \in M$.

(c) The space-time $(M,g)$ is chronological iff $d$ is identically zero on the diagonal $\Delta(M) = \{(p,p) : p \in M\}$ of $M \times M$.

(d) The space-time $(M,g)$ is future (resp., past) distinguishing iff for each pair of distinct $p,q \in M$, there is some $x \in M$ such that exactly one of $d(p,x)$ and $d(q,x)$ (resp., $d(x,p)$ and $d(x,q)$) is zero.

(e) The space-time $(M,g)$ is stably causal iff there exists a neighborhood $U$ of $g$ in the fine $C^0$ topology on Lor(M) such that $d(g')(p,p) = 0$ for all $g' \in U$ and $p \in M$.

*Proof*. Similar to Lemma 3.2 and Remark 3.3. □

Recall that the Lorentzian distance function in general fails to be upper semicontinuous. Thus the continuity of $d(g)$ should have implications for the causal structure of $(M,g)$. An example is the following result first stated in Beem and Ehrlich (1977, p. 1130). Here $d$ is regarded as being continuous at $(p,q) \in M \times M$ with $d(p,q) = \infty$ because $d(p_n,q_n) \to \infty$ for all sequences $p_n \to p$ and $q_n \to q$, cf. Lemma 3.4.

THEOREM 3.24 Let $(M,g)$ be a distinguishing space-time. If $d = d(g) : M \times M \to \mathbb{R} \cup \{\infty\}$ is continuous, then $(M,g)$ is causally continuous.

*Proof*. We need only show that $I^+$ and $I^-$ are outer continuous. Assume $I^+$ is not outer continuous. There is then some compact set $K \subset M - \overline{I^+(p)}$ and some sequence $p_n \to p$ such that $K \cap \overline{I^+(p_n)} \neq \phi$ for all $n$. Let $q_n \in K \cap \overline{I^+(p_n)}$ and let $\{q_m\}$ be a subsequence of $\{q_n\}$ such that $q_m$ converges to some point $q$ of the compact set $K$. Then $q_m \to q$ and $q_m \in \overline{I^+(p_m)}$ imply there must be a sequence $\{q_m'\}$ converging to $q$ such that $q_m' \in I^+(p_m)$ for each $m$. Since $M - \overline{I^+(p)}$ is an open neighborhood of $q$, there is some $r \in M - \overline{I^+(p)}$

with $q \ll r$. For sufficiently large m we then have $q'_m \ll r$ and hence $p_m \ll q'_m \ll r$. Thus $d(p_m,r) \geq d(p_m,q'_m) + d(q'_m,r)$. Using the lower semicontinuity of distance and the causality relation $q \ll r$, we obtain $0 < d(q,r) \leq \liminf d(q'_m,r)$. Consequently, $d(p_m,r) \geq d(q,r)/2 > 0$ for all sufficiently large m. However, since $r \notin I^+(p)$, we have $d(p,r) = 0$, and hence $d(p,r) \neq \lim d(p_m,r)$. Thus if d is continuous, then $I^+$ is outer continuous. A similar argument shows that $I^-$ is outer continuous. Thus continuity of d implies that (M,g) is causally continuous. □

On the other hand, causal continuity does not imply continuity of the Lorentzian distance function. Let (M,g) denote Minkowski space-time with a single point removed. The space-time $(M,\Omega g)$ will be causally continuous for any smooth conformal factor $\Omega : M \longrightarrow (0,\infty)$. However, $\Omega$ may be chosen such that $d = d(\Omega g)$ is *not* continuous (cf. Figure 3.6).

We now turn to a characterization of strongly causal space-times in terms of the Lorentzian distance function. The definition of convex normal neighborhood was given in Section 2.1. Given any space-time (M,g):

DEFINITION 3.25 A *local distance function* (D,U) *on* (M,g) is a convex normal neighborhood U together with the distance function $D : U \times U \longrightarrow \mathbb{R}$ induced on U by the space-time $(U,g|U)$.

More explicitly, if $p,q \in U$, then $D(p,q) = 0$ if there is no future-directed timelike geodesic segment in U from p to q. Otherwise, D(p,q) is the Lorentzian arc length of the unique future-directed timelike geodesic segment in U from p to q.

We will let $I^+(p,U)$ (resp., $J^+(p,U)$) denote the chronological (resp., causal) future of p with respect to the space-time $(U,g|U)$.

LEMMA 3.26 Let (M,g) be a space-time and let U be a convex normal neighborhood of (M,g). Assume that $D : U \times U \longrightarrow \mathbb{R}$ is the distance function for $(U,g|U)$. Then D is a continuous function on $U \times U$ and D is differentiable on $U^+ = \{(p,q) \in U \times U : q \in I^+(p,U)\}$.

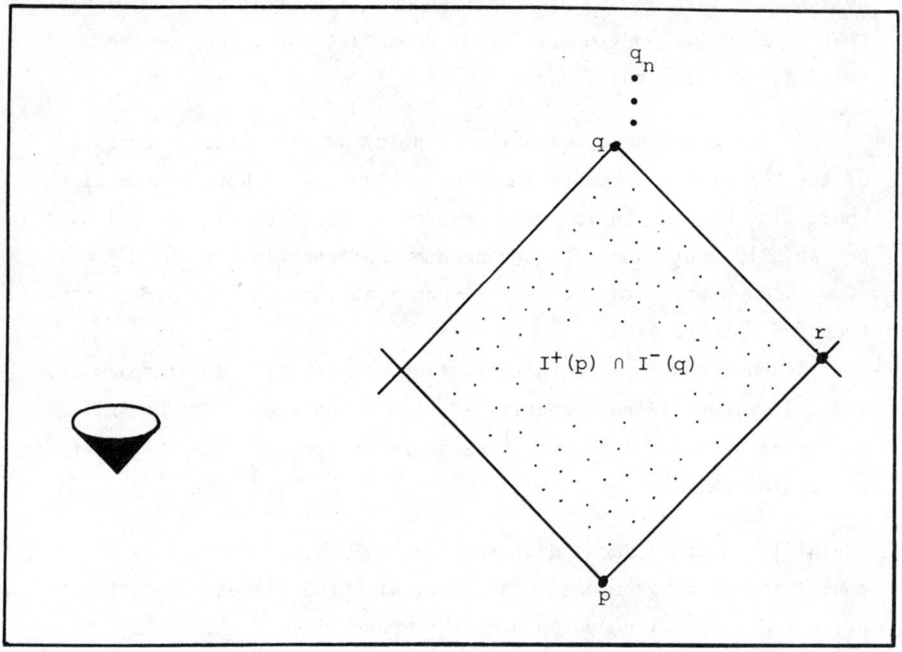


*Figure 3.6* Let (M,g) denote Minkowski space-time with the point r deleted. Choose $p,q \in M$ such that in Minkowski space-time the point r is on the boundary of $I^+(p) \cap I^-(q)$ as shown. Let $\{q_n\}$ be a sequence of points approaching q with $q \ll q_n$ for each n. There is a smooth conformal factor $\Omega : M \longrightarrow (0,\infty)$ such that $\Omega \equiv 1$ on $I^+(p) \cap I^-(q)$ and yet $d(\Omega g)(p,q_n) \geq 2d(g)(p,q)$ for each n. The function $\Omega$ will be unbounded near the deleted point r. Since $d(g)(p,q) = d(\Omega g)(p,q) < \liminf d(\Omega g)(p,q_n)$, the causally continuous space-time $(M,\Omega g)$ has a Lorentzian distance function which is discontinuous at $(p,q) \in M \times M$.

*Proof.* Given $p,q \in U$ with $q \in J^+(p,U)$, let $c_{pq} : [0,1] \longrightarrow U$ denote the unique nonspacelike geodesic segment with $c_{pq}(0) = p$ and $c_{pq}(1) = q$. We then have $D(p,q) = [-g(c'_{pq}(0), c'_{pq}(0))]^{1/2}$ and $[D(p,q)]^2 = -g(c'_{pq}(0), c'_{pq}(0))$. From the differentiable dependence of geodesics on endpoints in convex neighborhoods, it is immediate that D is continuous on $U \times U$ and that D is differentiable on $U^+$. □

Minkowski space-time shows that D fails to be differentiable across the null cones and thus fails to be smooth on all of $U \times U$.

It is not hard to see that the local distance function $(D,U)$ uniquely determines the Lorentzian metric g on U. Consequently if $\{U_\alpha\}$ is a covering of M by convex normal neighborhoods with associated local distance functions $\{(D_\alpha, U_\alpha)\}$, then $\{(D_\alpha, U_\alpha)\}$ uniquely determines g on M.

We now characterize strongly causal space-times in terms of local distance functions [cf. Beem and Ehrlich (1979c, Theorem 3.4)].

THEOREM 3.27 A space-time $(M,g)$ is strongly causal iff each point $r \in M$ has a convex normal neighborhood U such that the local distance function $(D,U)$ agrees on $U \times U$ with the distance function $d = d(g) : M \times M \longrightarrow \mathbb{R} \cup \{\infty\}$.

*Proof.* If $(M,g)$ is strongly causal and $r \in M$, then there is some convex normal neighborhood U of r such that no nonspacelike curve which leaves U ever returns. The local distance function for U then agrees with $d = d(g) \mid (U \times U)$.

Conversely, assume that strong causality breaks down at some point $r \in M$. Let U be a convex normal neighborhood of r such that $D(p,q) = d(p,q)$ for all $p,q \in U$. There exists a neighborhood $W \subset U$ of r such that any future-directed nonspacelike curve $\gamma : (0,1] \longrightarrow U$ with $\gamma(1) \in W$ and $\gamma$ past inextendible in U contains some point not in $J^+(W,U)$. Since strong causality fails to hold at r, there is a future-directed timelike curve $\gamma_1 : [0,1] \longrightarrow M$ with $r' = \gamma_1(0) \in W$, $\gamma_1(1/2) \notin U$ and $\gamma_1(1) \in W$. By construction of W, there is some point $p \in \gamma_1 \cap U$ with $p \notin J^+(r',U)$. Hence $D(r',p) = 0$. However, $d(r',p) > 0$, since $d(r',p)$ is at least as large as the length of $\gamma_1$ from r' to

p. Thus $D(r',p) \neq d(r',p)$. Taking the contrapositive then establishes the theorem. □

COROLLARY 3.28 If $(M,g)$ is strongly causal, then $d$ is continuous on some neighborhood of $\Delta(M) = \{(p,p) : p \in M\}$ in $M \times M$. Also given any point $m \in M$, there exists a convex normal neighborhood $U$ of $m$ such that $d \mid (U \times U)$ is finite valued.

We now give a characterization of globally hyperbolic space-times among all strongly causal space-times using the Lorentzian distance function. For this purpose, it is first necessary to show that the usual definition of globally hyperbolic may be weakened. In the proof of Lemma 3.29, we will use cl to denote closure.

LEMMA 3.29 Let $(M,g)$ be a strongly causal space time. If $J^+(p) \cap J^-(q)$ has compact closure for all $p,q \in M$, then $(M,g)$ is globally hyperbolic.

*Proof.* It is only necessary to show $J^+(p) \cap J^-(q)$ is always closed. Assume $r \in \mathrm{cl}(J^+(p) \cap J^-(q)) - J^+(p) \cap J^-(q)$. Choose a sequence $\{r_n\}$ of points in $J^+(p) \cap J^-(q)$ with $r_n \longrightarrow r$. For each $n$ let $\gamma_n : [0,1) \longrightarrow M$ be a future-directed, future-inextendible non-spacelike curve with $p = \gamma_n(0)$ and $q, r_n \in \gamma_n$. By Proposition 2.18, there is some future-directed, future-inextendible nonspacelike limit curve $\gamma : [0,1) \longrightarrow M$ of the sequence $\{\gamma_n\}$. Furthermore, $p = \gamma(0)$. The limit curve $\gamma$ cannot be future imprisoned in any compact subset of $M$ because $(M,g)$ is strongly causal (cf. Proposition 2.9). Consequently, there is some point $x$ on $\gamma$ with $x \notin \mathrm{cl}(J^+(p) \cap J^-(q))$. The definition of limit curve yields a subsequence $\{\gamma_m\}$ of $\{\gamma_n\}$ and points $x_m \in \gamma_m$ with $x_m \longrightarrow x$. Since $x \notin \mathrm{cl}(J^+(p) \cap J^-(q))$, we have $x_m \notin J^+(p) \cap J^-(q)$ for all large $m$. Using $\gamma_m \subset J^+(p)$, it follows that $x_m \notin J^-(q)$ for large $m$. Hence $q$ lies between $p$ and $x_m$ on $\gamma_m$ for large $m$. Let $\gamma[p,x]$ (resp., $\gamma_m[p,x_m]$) denote the portion of $\gamma$ (resp., $\gamma_m$) from $p$ to $x$ (resp., $x_m$). By Proposition 2.21 we may assume, by taking a subsequence of $\{\gamma_m[p,x_m]\}$ if necessary, that $\{\gamma_m[p,x_m]\}$ converges to $\gamma[p,x]$ in the

$C^0$ topology on curves. Hence $q \in \gamma_m[p,x_m]$ for large m implies $q \in \gamma[p,x]$. Also $r_m \longrightarrow r$ and $r_m \leqslant q$ which yield $r \in \gamma[p,q]$. Thus $r \in J^+(p) \cap J^-(q)$, in contradiction. □

Recall from Definition 3.6 above that a space-time $(M,g)$ is said to satisfy the *finite distance condition* iff $d(g)(p,q) < \infty$ for all $p,q \in M$. This condition may be used to characterize globally hyperbolic space-times among strongly causal space-times [cf. Beem and Ehrlich (1979b, Theorem 3.5)].

THEOREM 3.30 The strongly causal space-time $(M,g)$ is globally hyperbolic iff $(M,g')$ satisfies the finite distance condition for all $g' \in C(M,g)$.

*Proof.* It has already been remarked that if $(M,g)$ is globally hyperbolic, then all metrics in $C(M,g)$ satisfy the finite distance condition, (cf. Corollary 3.7).

Conversely, assume that $(M,g)$ is not globally hyperbolic. Lemma 3.29 implies that there exist $p,q \in M$ such that $J^+(p) \cap J^-(q)$ does not have compact closure. Let h be an auxiliary geodesically complete positive definite metric on M and let $d_0 : M \times M \longrightarrow \mathbb{R}$ be the Riemannian distance function induced on M by h. The Hopf-Rinow theorem implies that all subsets of M which are bounded with respect to $d_0$ have compact closure. Thus $J^+(p) \cap J^-(q)$ is not bounded. Hence, for each n we may choose $p_n \in J^+(p) \cap J^-(q)$ such that $d_0(p,p_n) > n$. Choose $p'$ and $q'$ with $p' \ll p \ll q \ll q'$. We wish to show there exists a conformal factor $\Omega$ such that $d(\Omega g)(p',q') = \infty$. For each $n > 1$ choose $\gamma_n$ to be a future-directed timelike curve from $p'$ to $p_n$ such that $\gamma_n[1/2,3/4] \subset \{r \in M : n - 1 < d_0(p,r) < n\}$. For each $n > 1$ let $\Omega_n : M \longrightarrow \mathbb{R}$ be a smooth function such that $\Omega_n(x) = 1$ if $x \notin \{r : n - 1 < d_0(p,r) < n\}$ and such that the length of $\gamma_n[1/2,3/4]$ is greater than n for the metric $\Omega_n g$. Let $\Omega = \Pi\Omega_n$. This infinite product is well defined on M since for each $x \in M$ at most one of the factors $\Omega_n$ is not unity. Then $d(\Omega g)(p',p_n) > n$ for each $n > 1$. Hence $d(\Omega g)(p',q') = \infty$ as $d(\Omega g)(p',q') \geqslant d(\Omega g)(p',p_n) + d(\Omega g)(p_n,q')$ for each n. □

We now turn to the proof that for distinguishing space-times with continuous distance functions, the future and past inner balls form a subbasis for the given manifold topology. Recall that

$$B^+(p,\varepsilon) = \{q \in I^+(p) : d(p,q) < \varepsilon\} = \{q \in M : 0 < d(p,q) < \varepsilon\}$$

and

$$B^-(p,\varepsilon) = \{q \in I^-(p) : d(q,p) < \varepsilon\} = \{q \in M : 0 < d(q,p) < \varepsilon\}$$

Thus defining $f_i : M \longrightarrow \mathbb{R}$ for $i = 1, 2$ by $f_1(q) = d(p,q)$ and $f_2(q) = d(q,p)$, we have $B^+(p,\varepsilon) = f_1^{-1}(0,\varepsilon)$ and $B^-(p,\varepsilon) = f_2^{-1}(0,\varepsilon)$. Hence if $(M,g)$ has a continuous distance function, the inner balls $B^+(p,\varepsilon)$ and $B^-(p,\varepsilon)$ of $M$ are open in the manifold topology.

PROPOSITION 3.31 Let $(M,g)$ be a distinguishing space-time with a continuous distance function. Then the collection $\{B^+(p,\varepsilon_1) \cap B^-(q,\varepsilon_2) : p,q \in M,\ \varepsilon_1,\varepsilon_2 > 0\}$ forms a basis for the given manifold topology of $M$.

*Proof.* The above arguments show that sets of the form $B^+(p,\varepsilon_1) \cap B^-(q,\varepsilon_2)$ are open in the manifold topology. Thus given an arbitrary point $r \in M$ and an arbitrary open neighborhood $U(r)$ of $r$ in the manifold topology, it is sufficient to show that there exist $p,q \in M$ and $\varepsilon_1,\varepsilon_2 > 0$ with $r \in B^+(p,\varepsilon_1) \cap B^-(q,\varepsilon_2) \subset U(r)$.

Theorem 3.24 yields that $(M,g)$ is causally continuous and hence also strongly causal. Thus we may choose a convex normal neighborhood $V$ of $r$ with $V \subset U(r)$ such that no nonspacelike curve which leaves $V$ ever returns and such that $d : V \times V \longrightarrow \mathbb{R} \cup \{\infty\}$ is finite valued (cf. Corollary 3.28). Fix $p,q \in V$ with $p \ll r \ll q$. Then $r \in I^+(p) \cap I^-(q) \subset V$ since no nonspacelike curve from $p$ to $q$ can leave $V$ and return. Letting $\varepsilon_1 = d(p,r) + 1$ and $\varepsilon_2 = d(r,q) + 1$, we obtain

$$r \in B^+(p,\varepsilon_1) \cap B^-(q,\varepsilon_2) \subset I^+(p) \cap I^-(q) \subset V \subset U(r)$$

which establishes the proposition. □

We conclude this section with a characterization of totally geodesic timelike submanifolds in terms of the Lorentzian distance function. Analogous result holds for submanifolds of (not necessarily complete) Riemannian manifolds [cf. Gromoll, Klingenberg, and Meyer (1975, p. 159)].

Let $(M,g)$ be an arbitrary strongly causal space-time. Suppose that $i : N \longrightarrow M$ is a smooth submanifold and set $\overline{g} = i^*g$. Recall that $(N,\overline{g})$ is said to be a *timelike submanifold* of $(M,g)$ if $\overline{g}|_p : T_pN \times T_pN \longrightarrow \mathbb{R}$ is a Lorentzian metric for each $p \in N$. As usual, we will identify $N$ and $i(N)$. Let $\overline{L}$, $L$ and $\overline{d}$, $d$ denote the arc-length functionals and Lorentzian distance functions of $(N,\overline{g})$ and $(M,g)$, respectively. Then if $\gamma$ is a smooth curve in $(N,\overline{g})$, we have $\overline{L}(\gamma) = L(\gamma)$. Note also that if $q \in I^+(p,N)$, then $p \ll q$ in $(M,g)$ and if $q \in J^+(p,N)$, then $p \leqslant q$ in $(M,g)$. Thus it follows immediately from the definitions of $d$ and $\overline{d}$ that

$$\overline{d}(m,n) \leqslant d(m,n) \qquad \text{for all } m,n \in N \tag{3.4}$$

With this remark in hand, we are ready to prove the following result.

PROPOSITION 3.32 Let $(N,\overline{g})$ be a totally geodesic, timelike submanifold of the strongly causal space-time $(M,g)$. Then given any $p \in N$, there exists a neighborhood $V$ of $p$ in $N$ such that $d \mid (V \times V) = \overline{d} \mid (V \times V)$.

*Proof.* First let $W$ be a convex neighborhood of $p$ in $(M,g)$ such that every pair of points $m,n \in W$ are joined by a unique geodesic of $(M,g)$ lying in $W$ and if $m \leqslant n$, then this geodesic is maximal in $(M,g)$. We may then choose a smaller neighborhood $V_0$ of $p$ in $M$ with $V_0 \subset W$ such that if $V = V_0 \cap N$, then $V$ is contained in a convex normal neighborhood $U$ of $p$ in $N$ with $U$ contained in $W$.

Suppose first that $m,n \in V$ and $n \in J^+(m,N)$. Since $V \subset U$, there exists a nonspacelike geodesic $\gamma$ of $(N,\overline{g})$ in $U$ from $m$ to $n$. Also as $N$ is totally geodesic, $\gamma$ is a nonspacelike geodesic in $(M,g)$. Because $\gamma$ is contained in $U \subset W$, $\gamma$ is maximal in $(M,g)$. We thus have $\overline{d}(m,n) \geqslant \overline{L}(\gamma) = L(\gamma) = d(m,n)$. In view of (3.4), we obtain $\overline{d}(m,n) = d(m,n)$ as required.

It remains to consider the case that $m,n \in V$ and $n \notin J^+(m,N)$. Thus $\bar{d}(m,n) = 0$ by definition. Suppose that $d(m,n) > 0$. Then there exists a timelike geodesic $\gamma_1$ of $(M,g)$ in $W$ from $m$ to $n$. On the other hand, since $m,n \in U$, there exists a geodesic $\gamma_2$ of $(N,\bar{g})$ from $m$ to $n$ lying in $U$ which must be spacelike since $n \notin J^+(m,N)$. Since $(N,\bar{g})$ is totally geodesic, $\gamma_2$ is also a spacelike geodesic of $(M,g)$ from $m$ to $n$ which lies in $U \subset W$. Thus we have distinct geodesics $\gamma_1$ and $\gamma_2$ in $W$ from $m$ to $n$, in contradiction. Hence $d(m,n) = 0 = \bar{d}(m,n)$ as required. □

We now prove the converse of Proposition 3.32.

PROPOSITION 3.33 Let $(N,\bar{g})$ be a timelike submanifold of the strongly causal space-time $(M,g)$. Suppose that for all $p \in N$, there exists a neighborhood $V$ of $p$ in $N$ such that $d \mid (V \times V) = \bar{d} \mid (V \times V)$. Then $(N,\bar{g})$ is totally geodesic in $(M,g)$.

*Proof.* It suffices to fix any $p \in N$ and show that the second fundamental form $S_n$ vanishes at $p$ (cf. Definition 2.35). Since any tangent vector in $T_pN$ may be written as a sum of nonspacelike tangent vectors, it is enough to show that $S_n(v,w) = 0$ for all nonspacelike tangent vectors in $T_pN$. Also as $S_n(-v,w) = -S_n(v,w)$, it suffices to show that $S_n(v,w) = 0$ for all future-directed nonspacelike tangent vectors in $T_pN$.

Thus let $v \in T_pN$ be a future-directed nonspacelike tangent vector. Let $\gamma$ denote the unique geodesic in $(N,\bar{g})$ with $\gamma'(0) = v$. Also let $V$ be a neighborhood of $p$ in $N$ on which the distance functions $\bar{d}$ and $d$ coincide. Choose $t > 0$ such that if $m = \gamma(t)$, then $m \in V$ and $\bar{d}(p,m) = \bar{L}(\gamma|[0,t]) < \infty$. We then obtain

$$d(p,m) \geqslant L(\gamma|[0,t]) = \bar{L}(\gamma|[0,t]) = \bar{d}(p,m)$$

But since $m \in V$, we have $d(p,m) = \bar{d}(p,m)$ whence $L(\gamma|[0,t]) = d(p,m)$. Hence $\gamma \mid [0,t]$ is a geodesic in $(M,g)$ by Theorem 3.13. Thus we have shown that if $v \in T_pN$ is any future-directed tangent vector, the geodesic in $(M,g)$ with initial direction $v$ is also a geodesic in

$(N,\bar{g})$ near p. Therefore $S_n(v,v) = 0$ for all future-directed nonspacelike tangent vectors. Since the sum of two nonparallel future-directed nonspacelike tangent vectors is future timelike, it follows by polarization that $S_n(v,w) = 0$ for all future-directed nonspacelike tangent vectors $v,w \in T_pN$ as required. □

Combining Propositions 3.32 and 3.33 yields the following characterization of totally geodesic, timelike submanifolds of strongly causal space-times in terms of the Lorentzian distance function.

THEOREM 3.34 Let (M,g) be a strongly causal space-time of dimension $\geqslant 2$ and suppose that $(N,i^*g)$ is a smooth timelike submanifold of (M,g), i.e., $\bar{g} = i^*g$ is a Lorentzian metric for N. Then $(N,\bar{g})$ is totally geodesic iff given any $p \in N$, there exists a neighborhood V of p in N such that the Lorentzian distance functions $\bar{d}$ of $(N,\bar{g})$ and d of (M,g) agree on $V \times V$.

Chapter 4

# EXAMPLES OF SPACE-TIMES

In this chapter we present a variety of examples of space-times. Some of these space-times are important for physical as well as mathematical reasons. In particular, Minkowski space-time, Schwarzschild space-times, Kerr space-times, and Robertson-Walker space-times all have significant physical interpretations.

Minkowski space-time is simultaneously the geometry of special relativity and the geometry induced on each fixed tangent space of an arbitrary Lorentzian manifold. Thus Minkowskian geometry plays the same role for Lorentzian manifolds that Euclidean geometry plays for Riemannian manifolds. Minkowski space-time is sometimes called *flat space-time*. But more generally, any Lorentzian manifold on which the curvature tensor is identically zero is flat.

The Schwarzschild space-times represent the spherically symmetric, empty space-times outside nonrotating, spherically symmetric bodies. Since suns and planets are assumed to be slowly rotating and approximately spherically symmetric, the Schwarzschild space-times may be used to model the gravitational fields outside of these bodies. These space-times may also be used to model the gravitational fields outside of dead (i.e., nonrotating) black holes. The usual coordinates for the Schwarzschild solution outside a massive body are $(t,r,\theta,\phi)$, where t represents a kind of time and r represents a kind of radius [cf. Sachs and Wu (1977a, Chapter 7)]. This metric has a special radius $r = 2m$ associated with it. Points with $r = 2m$ correspond to the surface of a black hole. It was once thought that the metric was singular at $r = 2m$. But it is now known that the usual form of the Schwarzschild metric with $r > 2m$ may be analytically extended to points with $0 < r < 2m$. In fact, there is a maximal analytic extension of Schwarzschild space-time [cf. Kruskal (1960)] which contains an alternative universe lying on the "other side" of the black hole.

The gravitational fields outside of rotating black holes apparently correspond to the Kerr space-times [cf. Hawking and Ellis (1973, pp. 161, 331), Carter (1971b)]. These space-times represent stationary, axisymmetric metrics outside of rotating objects. The Kerr and Schwarzschild space-times are asymptotically flat and correspond to universes which are empty, apart from one massive body. Thus while these metrics may be reasonable models near a given single massive body, they cannot be used as large scale models for a universe with many massive bodies.

The usual "big bang" cosmological models are based on the Robertson-Walker space-times. These space-times are foliated by a special set of spacelike hypersurfaces such that each hypersurface corresponds to an instant of time. The isometry group $I(M)$ of a Robertson-Walker space-time $(M,g)$ acts transitively on these hypersurfaces of constant time. Thus Robertson-Walker universes are spatially homogeneous. Furthermore, they are spatially isotropic in the sense that for each $p \in M$, the subgroup of $I(M)$ fixing $p$ is transitive on the directions at $p$ which are tangential to the hypersurface of constant time through $p$. In our discussion of Robertson-Walker space-times, we will use Lorentzian warped products $M_0 \times_f H$ described in Section 2.6. The cosmological assumptions made about Robertson-Walker universes imply that $(H,h)$ is an isotropic Riemannian manifold. Hence the classification of two-point homogeneous Riemannian manifolds yields a classification of all Robertson-Walker space-times. We also show how the results of Section 2.6 may be specialized to construct Lie groups with bi-invariant globally hyperbolic Lorentzian metrics.

## 4.1 MINKOWSKI SPACE-TIME

Minkowski space-time is the manifold $M = \mathbb{R}^n$ together with the metric

$$ds^2 = -dx_1^2 + \sum_{i=2}^{n} dx_i^2$$

This space-time is time oriented by the vector field $\partial/\partial x_1$. It is also globally hyperbolic and hence satisfies all of the causality conditions discussed in Section 2.2.

The geodesics of Minkowski space-time are just the straight lines of the underlying Euclidean space $\mathbb{R}^n$. The affine parameterizations of these geodesics in Minkowski space are even proportional to the usual Euclidean arc length parameterizations in $\mathbb{R}^n$. The null geodesics through a given point p in Minkowski space form an elliptic cone with vertex p. The future-directed null geodesics starting at p thus form one nappe of the null cone of p. This nappe forms the boundary in $\mathbb{R}^n$ of an open convex set which is exactly the chronological future $I^+(p)$ of p. In Minkowski space, the causal future $J^+(p)$ of p is the closure of $I^+(p)$. The future horismos $E^+(p) = J^+(p) - I^+(p)$ is the nappe of the null cone of p corresponding to the future (cf. Figure 4.1).

Minkowski space-time is a Lorentzian product (i.e., a warped product in the sense of Definition 2.38 with f = 1). If $\mathbb{R}$ is given the negative definite metric $-dt^2$ and $\mathbb{R}^{n-1}$ is given the usual Euclidean metric $g_0$, then $(\mathbb{R}^n = \mathbb{R} \times \mathbb{R}^{n-1}, -dt^2 \oplus g_0)$ is the n-dimensional Minkowski space-time.

Consider two points $p = (p_1,\ldots,p_n)$ and $q = (q_1,\ldots,q_n)$ in Minkowski space-time. The chronological relation $p \ll q$ holds whenever $p_1 < q_1$ and $(p_1 - q_1)^2 > (p_2 - q_2)^2 + \cdots + (p_n - q_n)^2$ in $\mathbb{R}$. If $p \ll q$, then the distance from p to q is given by

$$d(p,q) = \left[(p_1 - q_1)^2 - \sum_{i=2}^{n} (p_i - q_i)^2\right]^{1/2}$$

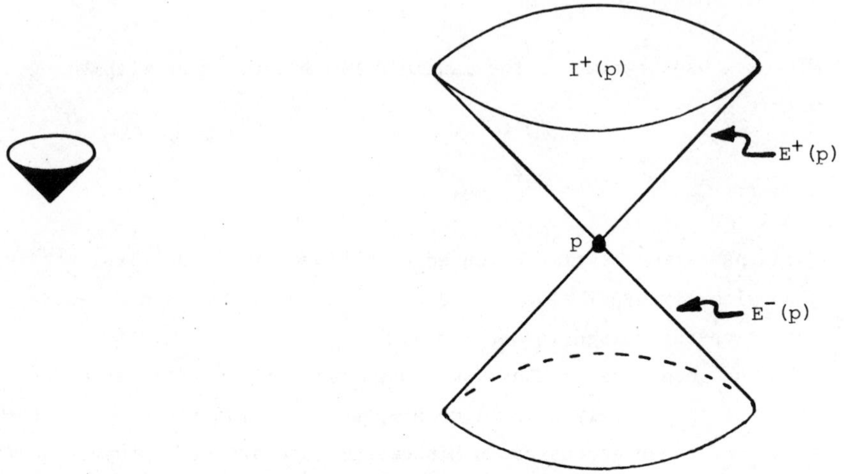


*Figure 4.1* Let (M,g) be Minkowski space-time. The null cone at p has a future nappe and past nappe. The future (resp., past) nappe is also the horismos $E^+(p)$ [resp., $E^-(p)$] of p. The chronological future $I^+(p)$ is an open convex set bounded by $E^+(p)$. In more general space-times, $I^+(p)$ may fail to be convex but is always open.

The "unit sphere" in Minkowski space-time centered at p is then $K(p,1) = \{q \in M : d(p,q) = 1\}$. But this set is actually one sheet of a hyperboloid of two sheets (cf. Figure 4.2).

If we remove a point from Minkowski space-time, then it is no longer causally simple and hence no longer globally hyperbolic (cf. Figure 4.3).

It is possible to conformally map all of Minkowski space-time onto a small open set about the origin. This is illustrated in Figure 4.4 [cf. Penrose (1968, p. 178), Hawking and Ellis (1973, p. 123)].

Minkowski space-time and many other important space-times may be represented by *Penrose diagrams*. We now indicate the conventions used in Penrose diagrams.

A Penrose diagram is a two-dimensional representation of a spherically symmetric space-time. The radial null geodesics are represented by null geodesics at ±45°. Dotted lines represent the

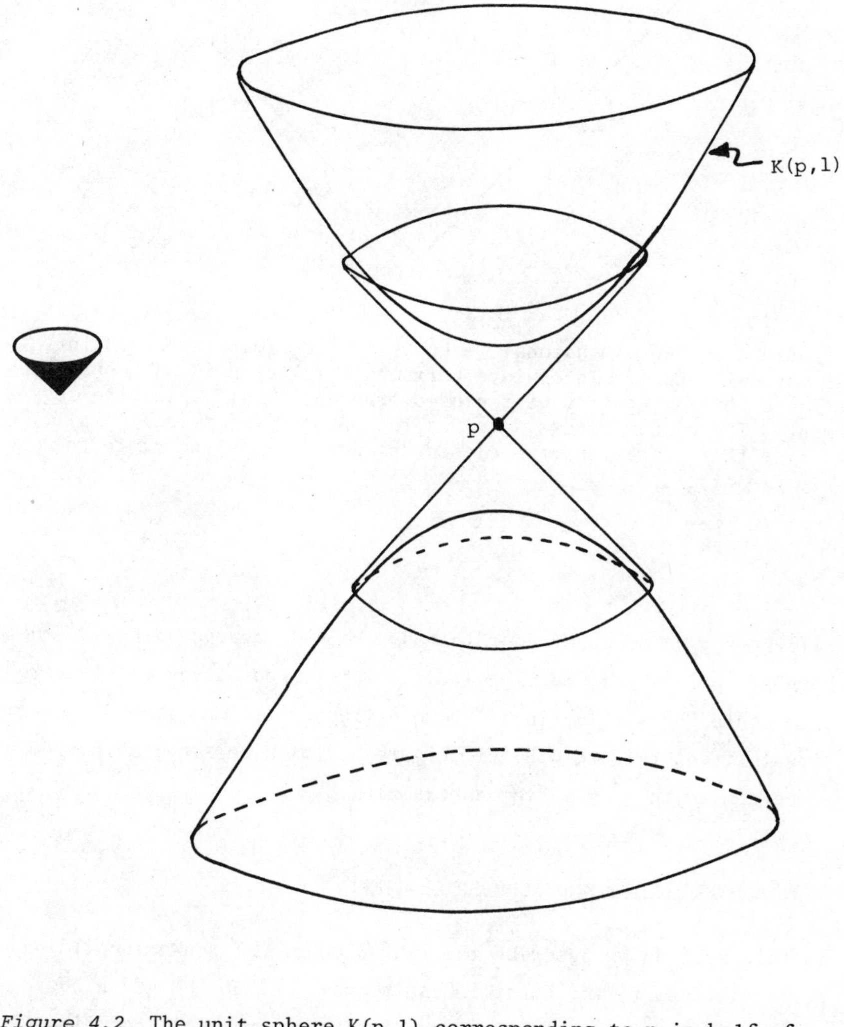


*Figure 4.2* The unit sphere K(p,1) corresponding to p is half of a hyperboloid of two sheets. It is *not* compact and p does *not* lie in the convex open set bounded by K(p,1).

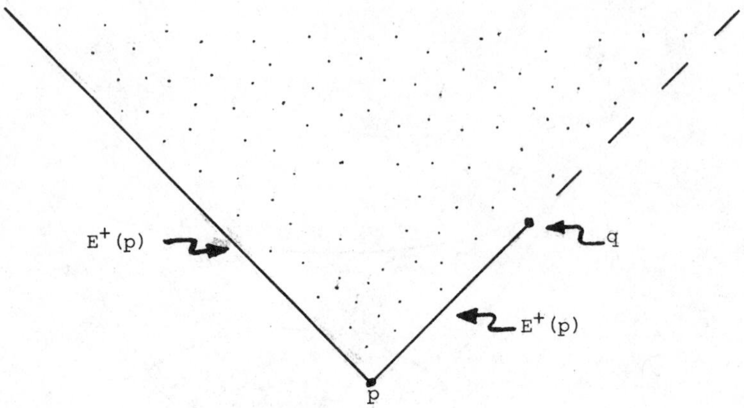


*Figure 4.3* Two-dimensional Minkowski space-time with one point q removed is shown. The future horismos $E^+(p)$ of p is an L-shaped figure consisting of a half-closed line and a half-open line segment. The causal future $J^+(p)$ is the union of $I^+(p)$ and $E^+(p)$. Notice that $J^+(p)$ is *not* a closed set nor is $J^+(p)$ equal to the closure of $I^+(p)$.

origin (r = 0) of polar coordinates. Points corresponding to smooth boundary points (cf. Section 11.5) which are not singularities are represented by single lines. Double lines represent irremovable singularities (Figure 4.5; cf. Figure 3.2 of a Reissner-Nördström space-time with $e^2 = m^2$ for an example).

## 4.2 SCHWARZSCHILD AND KERR SPACE-TIMES

In this section we describe the four-dimensional Schwarzschild and Kerr solutions to the Einstein equations. Let $\mathbb{R}^4$ be given coordinates $(t,r,\theta,\phi)$, where $(r,\theta,\phi)$ are the usual spherical coordinates on $\mathbb{R}^3$. Given a positive constant m, the exterior Schwarzschild space-time is defined on the subset $r > 2m$ of $\mathbb{R}^4$, a subset which is topologically $\mathbb{R}^2 \times S^2$. The Schwarzschild metric for the region $r > 2m$ is given in $(t,r,\theta,\phi)$ coordinates by the formula

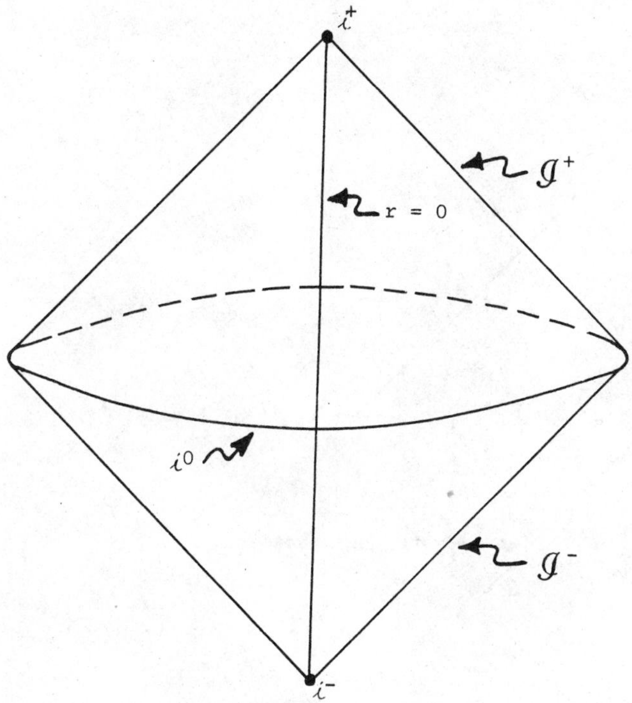


*Figure 4.4* Minkowski space-time is conformal to the open set enclosed by the two null cones indicated. The vertices $i^+$ and $i^-$ correspond to timelike infinity. All future-directed timelike geodesics go from $i^-$ to $i^+$. The sets $\mathscr{I}^+$ and $\mathscr{I}^-$ represent future and past null infinity. Topologically $\mathscr{I}^+$ and $\mathscr{I}^-$ are each $\mathbb{R} \times S^{n-2}$. The intersection of the two null cones is a set which is identified to a single point $i^0$. The point $i^0$ is called *spacelike infinity*.

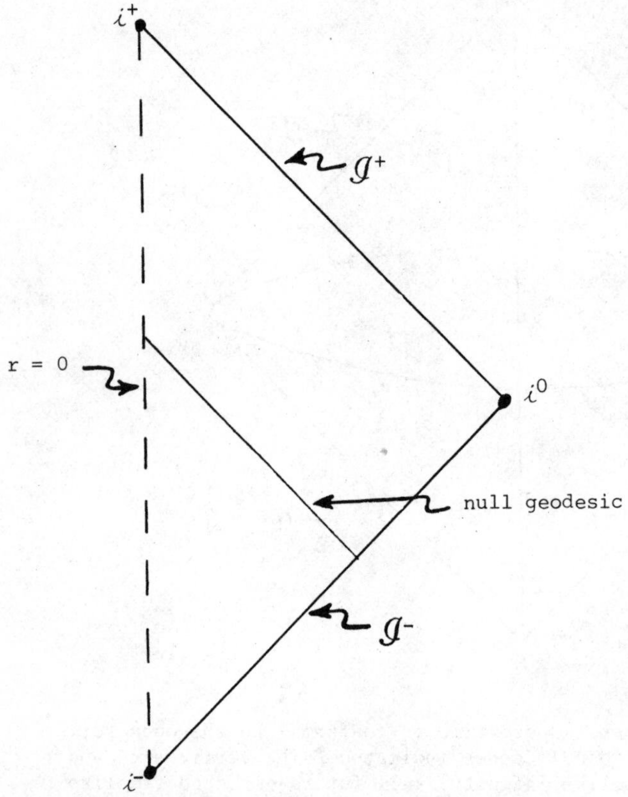


*Figure 4.5* The Penrose diagram for Minkowski space-time is shown.

$$ds^2 = -(1 - \frac{2m}{r})\, dt^2 + (1 - \frac{2m}{r})^{-1}\, dr^2 + r^2(d\theta^2 + \sin^2\theta\, d\phi^2)$$

Each element of the rotation group SO(3) for $\mathbb{R}^3$ induces a motion of the Schwarzschild solution. Namely, given $\psi \in SO(3)$, a motion $\bar{\psi}$ of Schwarzschild space-time may be defined by setting $\bar{\psi}(t,r,\theta,\phi) = (t,\psi(r,\theta,\phi))$. Thus at a fixed instant t in time, the exterior Schwarzschild space-time is spherically symmetric. The metric for this space-time is also invariant under the time translation $t \longrightarrow t + a$. The coordinate vector field $\partial/\partial t$ is a timelike Killing vector field which is a gradient and the metric is said to be static. This space-time is also Ricci flat (i.e., Ric = 0). Using the Einstein equations (cf. Appendix C), it follows that the energy momentum tensor for the exterior Schwarzschild space-time vanishes. Thus this space-time is empty.

The exterior Schwarzschild space-time may be regarded as a Lorentzian warped product (cf. Section 2.6). For let $M = \{(t,r) \in \mathbb{R}^2 : r > 2m\}$ be given the Lorentzian metric

$$g = -(1 - \frac{2m}{r})\, dt^2 + (1 - \frac{2m}{r})^{-1}\, dr^2$$

and let $H = S^2$ be given the usual Riemannian metric h of constant sectional curvature 1 induced by the inclusion $S^2 \longrightarrow \mathbb{R}^3$. Define $f : M \longrightarrow \mathbb{R}$ by $f(t,r) = r^2$. Then $(M \times_f H, \bar{g})$ is the exterior Schwarzschild space-time where $\bar{g} = g \oplus fh$.

Physically, the exterior Schwarzschild solution represents the gravitational field outside of a nonrotating spherically symmetric massive object. Comparison with the Newtonian theory [cf. Einstein (1916, p. 819), Pathria (1974, p. 217)] shows that m can be identified with the gravitational mass of the massive body. The solution is not valid in the interior of the body.

The above form of the exterior Schwarzschild metric appears to have a singularity at r = 2m. However, this is not a true singularity. The exterior Schwarzschild solution may be analytically continued across the surface r = 2m.

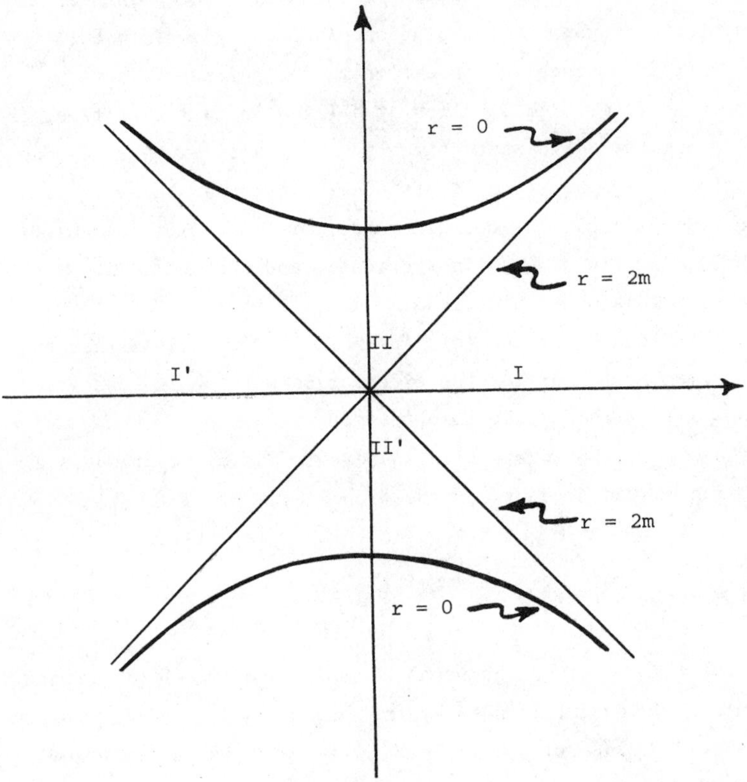


*Figure 4.6* The Kruskal diagram for the maximal analytic extension of the exterior Schwarzschild space-time is shown. The extended space-time is the connected nonconvex region I ∪ II ∪ I' ∪ II' bounded by the hyperbola corresponding to r = 0. The points of this hyperbola are the true singularities of this space-time. The lines at ±45° separate the space-time into four regions. Region I corresponds to the exterior Schwarzschild solution. Region II is the "interior" of a nonrotating black hole. Region I' is isometric to region I and corresponds to an alternative universe on the "other side" of the black hole. There is no nonspacelike curve from region I to region I'.

Kruskal (1960) investigated the maximal analytic extension of Schwarzschild space-time. Suppressing $\theta$ and $\phi$, the following two-dimensional representation of this maximal extension may be given (cf. Figure 4.6).

The gravitational field outside of a rotating black hole will not correspond to the Schwarzschild solution. The generally accepted solutions of the Einstein equations for rotating black holes are Kerr solutions. In Boyer and Lindquist coordinates $(t,r,\theta,\phi)$ the Kerr metrics are given by [cf. Hawking and Ellis (1973, p. 161)]

$$ds^2 = \rho^2\left(\frac{dr^2}{\Delta} + d\theta^2\right) + (r^2 + a^2)\sin^2\theta\, d\phi^2 - dt^2 + \frac{2mr}{\rho^2}(a \sin^2\theta\, d\phi - dt)^2$$

where $\rho^2 = r^2 + a^2\cos^2\theta$ and $\Delta = r^2 - 2mr + a^2$. The constant m represents the mass and the constant ma represents the angular momentum of the black hole [cf. Boyer and Price (1965), Boyer and Lindquist (1967)]. Tomimatsu and Sato (1973) have given a series of exact solutions which include the Kerr solutions as special cases.

## 4.3 SPACES OF CONSTANT CURVATURE

It is known that two Lorentzian manifolds of the same dimension which have constant sectional curvature k are locally isometric [cf. Wolf (1974, p. 69)]. Thus any Lorentzian manifold of constant sectional curvature zero is locally isometric to Minkowski space-time. In this section we will consider Lorentzian model spaces which have constant nonzero sectional curvature.

We first define $\mathbb{R}^n_s$ to be the standard pseudo-Euclidean space of signature $(-, \ldots, -, +, \ldots, +)$ where there are s negative eigenvalues and n - s positive eigenvalues. Hence the pseudo-Euclidean metric on $\mathbb{R}^n_s$ is given by

$$ds^2 = -\sum_{i=1}^{s} dx_i^2 + \sum_{i=s+1}^{n} dx_i^2$$

In particular, $\mathbb{R}_1^n$ is the n-dimensional Minkowski space-time. We also define for $r > 0$ [cf. Wolf (1974, Section 4.2)]

$$S_1^n = \{x \in \mathbb{R}_1^{n+1} : -x_1^2 + x_2^2 + \cdots + x_{n+1}^2 = r^2\}$$

and

$$H_1^n = \{x \in \mathbb{R}_2^{n+1} : -x_1^2 - x_2^2 + x_3^2 + \cdots + x_{n+1}^2 = -r^2\}$$

Topologically, $S_1^n$ is $\mathbb{R}^1 \times S^{n-1}$ and $H_1^n$ is $S^1 \times \mathbb{R}^{n-1}$ [cf. Wolf (1974, p. 68)]. The pseudo-Euclidean metric on $\mathbb{R}_1^{n+1}$ (resp., $\mathbb{R}_2^{n+1}$) induces a Lorentzian metric of constant sectional curvature $k = r^{-2}$ (resp., $k = -r^{-2}$) on $S_1^n$ (resp., $H_1^n$). The space-time $S_1^n$ is a Lorentzian analogue of the usual Riemannian spherical space of radius r and has positive curvature $r^{-2}$. The universal covering manifold $\tilde{H}_1^n$ of $H_1^n$ is topologically $\mathbb{R}^n$ and is thus a Lorentzian analogue of the usual Riemannian hyperbolic space of negative curvature $-r^{-2}$.

DEFINITION 4.1 Let $S_1^n$ and $H_1^n$ be defined as above. Then $S_1^n$ is called *de Sitter space-time* and the universal covering $\tilde{H}_1^n$ of $H_1^n$ is called (universal) *anti-de Sitter space-time*.

REMARK 4.2

(a) $S_1^n$ is simply connected for $n > 2$ and $\Pi_1(S_1^2) = \mathbb{Z}$.

(b) $S_1^n$ is globally hyperbolic and geodesically complete.

(c) $H_1^n$ is nonchronological since $\gamma(t) = (r \cos t, r \sin t, 0, \ldots, 0)$ is a closed timelike curve. Also $\tilde{H}_1^n$, while strongly causal, is not globally hyperbolic.

The de Sitter space-time represented in Figure 4.7 may be covered by global coordinates $(t,\chi,\theta,\phi)$ with $-\infty < t < \infty$, $0 \leqslant \chi \leqslant \pi$, $0 \leqslant \theta \leqslant \pi$, and $0 \leqslant \phi \leqslant 2\pi$. Here t is the coordinate on $\mathbb{R}$ and $(\chi,\theta,\phi)$ represent coordinates on $S^3$ [cf. Hawking and Ellis (1973, pp. 125, 136)]. In these coordinates, the metric for de Sitter space-time of constant positive sectional curvature $1/r^2$ is given by

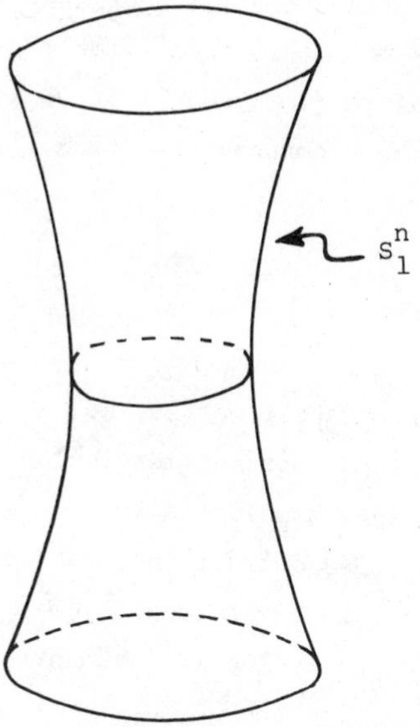


*Figure 4.7* The n-dimensional de Sitter space-time with positive constant sectional curvature $r^{-2}$ is the set $-x_1^2 + x_2^2 + \cdots + x_{n+1}^2 = r^2$ in Minkowski space-time $\mathbb{R}_1^{n+1}$. The geodesics of $S_1^n$ lie on the intersections of $S_1^n$ with the planes through the origin of $\mathbb{R}_1^{n+1}$.

$$ds^2 = -dt^2 + r^2 \cosh^2 t/r \, [d\chi^2 + \sin^2\chi(d\theta^2 + \sin^2\theta \, d\phi^2)]$$

This may be reinterpreted as a Lorentzian warped product metric (cf. Section 2.6) as follows. Let $f : \mathbb{R} \longrightarrow (0,\infty)$ be given by $f(t) = r^2 \cosh^2 t/r$ and let $S^3$ be given the usual complete Riemannian metric of constant sectional curvature 1. Then the de Sitter space-time described in local coordinates as above is the warped product $(\mathbb{R} \times S^3, -dt^2 \oplus fh)$.

Universal anti-de Sitter space-time of curvature $k = -1$ may be given coordinates $(t',r,\theta,\phi)$ for which the metric has the form

$$ds^2 = -\cosh^2(r)(dt')^2 + dr^2 + \sinh^2 r(d\theta^2 + \sin^2\theta \, d\phi^2)$$

[cf. Hawking and Ellis (1973, pp. 131, 136)]. Regarding $-(dt')^2$ as a negative definite metric on $\mathbb{R}$ and $dr^2 + \sinh^2 r(d\theta^2 + \sin^2\theta \, d\phi^2)$

as the complete Riemannian metric h of constant negative sectional curvature -1 on the hyperbolic 3-space $H = \mathbb{R}^3$, this space-time may be represented as a warped product of the form $(\mathbb{R} \times_f H, -f\, dt^2 \oplus h)$, where the warping function is defined on the Riemannian factor H (cf. Remark 2.40).

## 4.4 ROBERTSON-WALKER SPACE-TIMES

In this section we discuss Robertson-Walker space-times in the framework of Lorentzian warped products. These space-times include the Einstein static universe and the big bang cosmological models of general relativity. In order to give a precise definition of a Robertson-Walker space-time, it is necessary to first recall some concepts from the theory of two point homogeneous Riemannian manifolds and isotropic Riemannian manifolds.

Let (H,h) be a Riemannian manifold. Denote by I(H) the isometry group of (H,h) and by $d_0 : H \times H \longrightarrow \mathbb{R}$ the Riemannian distance function of (H,h).

DEFINITION 4.3 The Riemannian manifold (H,h) is said to be *homogeneous* if I(H) acts transitively on H, i.e., given any $p, q \in H$, there is an isometry $\phi \in I(H)$ with $\phi(p) = q$. Further, (H,h) is said to be *two-point homogeneous* if given any $p_1, q_1, p_2, q_2 \in H$ with $d_0(p_1, q_1) = d_0(p_2, q_2)$, there is an isometry $\phi \in I(H)$ with $\phi(p_1) = p_2$ and $\phi(q_1) = q_2$.

Since it is possible to choose $p_i = q_i$ for i = 1, 2, a two-point homogeneous Riemannian manifold is also homogeneous. Two-point homogeneous spaces were first studied by Busemann (1942) in the more general setting of locally compact metric spaces. Wang (1951, 1952) and Tits (1955) classified two-point homogeneous Riemannian manifolds.

Notice that in Definition 4.3, it is not required that (H,h) be a complete Riemannian manifold. Nonetheless, homogeneous

Riemannian manifolds have the important basic property of always being complete.

LEMMA 4.4 If $(H,h)$ is a homogeneous Riemannian manifold, then $(H,h)$ is complete.

*Proof.* By the Hopf-Rinow theorem, it suffices to show that $(H,h)$ is geodesically complete. Thus suppose that $c : [a,1) \longrightarrow H$ is a unit speed geodesic which is not extendible to $t = 1$. Choosing any $p \in H$, we may find a constant $\alpha > 0$ such that any unit speed geodesic starting at p has length $\geqslant \alpha$. Set $\delta = \min(\alpha/2, (1 - a)/2 > 0$. Since isometries preserve geodesics, it follows from the homogeneity of $(H,h)$ that any unit speed geodesic starting at $c(1 - \delta)$ may be extended to a geodesic of length $\geqslant 2\delta$. In particular, c may be extended to a geodesic $c : [a, 1 + \delta) \longrightarrow H$, in contradiction to the inextendibility of c to $t = 1$. □

REMARK 4.5 It is important to note that the conclusion of Lemma 4.4 is *false* in general for homogeneous Lorentzian manifolds [cf. Wolf (1974, p. 95), Marsden (1973)].

We now recall the concept of an isotropic Riemannian manifold. Given $p \in (H,h)$, the *isotropy group* $I_p(H)$ of $(H,h)$ at p is the closed subgroup $I_p(H) = \{\phi \in I(H) : \phi(p) = p\}$ of $I(H)$ consisting of all isometries of $(H,h)$ which fix p. Given any $\phi \in I_p(H)$, the differential $\phi_{*p}$ maps $T_pH$ onto $T_pH$ since $\phi(p) = p$. As $h(\phi_* v, \phi_* v) = h(v,v)$ for any $v \in T_pH$, the differential $\phi_{*p}$ also maps the unit sphere $S_pH = \{v \in T_pM : h(v,v) = 1\}$ in $T_pH$ onto itself.

DEFINITION 4.6 The Riemannian manifold $(H,h)$ is said to be *isotropic at* p if $I_p(H)$ acts transitively on the unit sphere $S_pH$ of $T_pH$, i.e., given any $v,w \in S_pH$, there is an isometry $\phi \in I_p(H)$ with $\phi_* v = w$. The Riemannian manifold $(H,h)$ is said to be *isotropic* if it is isotropic at every point.

We now show that the class of isotropic Riemannian manifolds coincides with the class of two point homogeneous Riemannian manifolds [cf. Wolf (1974, p. 289)].

PROPOSITION 4.7 A Riemannian manifold (H,h) is isotropic iff it is two-point homogeneous.

*Proof.* Recall that $d_0$ denotes the Riemannian distance function of (H,h). First suppose that (H,h) is isotropic. Then for each $p \in H$ and each inextendible geodesic $c : (a,b) \longrightarrow H$ with $c(0) = p$, there is an isometry $\phi \in I_p(H)$ with $\phi_* c'(0) = -c'(0)$. Hence by geodesic uniqueness, $\phi(c(t)) = c(-t)$ for all $t \in (a,b)$. This implies that the length of $c \mid (a,0]$ equals the length of $c \mid [0,b)$. Since p may be taken to be any point of the geodesic c, it follows that $a = -\infty$ and $b = +\infty$. Thus (H,h) is geodesically complete. Hence by the Hopf-Rinow theorem given any two points $p_1, p_2 \in H$, there is a geodesic segment $c_0$ of minimal length $d_0(p_1,p_2)$ from $p_1$ to $p_2$. Let p be the midpoint of $c_0$. As (H,h) is isotropic, there is an isometry $\phi \in I_p(H)$ which reverses $c_0$. It follows that $\phi(p_1) = p_2$. Hence (H,h) is homogeneous. It remains to show that if $p_1, q_1, p_2, q_2 \in H$ with $d_0(p_1,q_1) = d_0(p_2,q_2) > 0$ are given, we may find an isometry $\phi \in I(H)$ with $\phi(p_1) = p_2$ and $\phi(q_1) = q_2$. Choose minimal unit speed geodesics $c_1$ from $p_1$ to $q_1$ and $c_2$ from $p_2$ to $q_2$, respectively. Since (H,h) is homogeneous, we may first find an isometry $\psi \in I(H)$ with $\psi(p_1) = p_2$. Then as (H,h) is isotropic, we may find $\eta \in I_{p_2}(H)$ with $\eta_*((\psi \circ c_1)'(0)) = c_2'(0)$. It follows that $\phi = \eta \circ \psi$ is the required isometry.

Now suppose that (H,h) is two point homogeneous. Fix any $p \in M$ and let U be a convex normal neighborhood based at p. Choose $\alpha > 0$ such that $\exp_p(v) \in U$ for all $v \in T_pH$ with $h(v,v) \leq \alpha$. Now let $v,w \in T_pH$ be any pair of nonzero tangent vectors with $h(v,v) = h(w,w) < \alpha/2$. Set $q_1 = \exp_p v$ and $q_2 = \exp_p w$. Then $q_1, q_2 \in U$ and $d(p,q_1) = \sqrt{h(v,v)} = \sqrt{h(w,w)} = d(p,q_2)$. Since (H,h) is two point homogeneous, there is thus an isometry $\phi \in I(H)$ with $\phi(p) = p$ and $\phi(q_1) = q_2$. It follows that $\phi_* v = w$. The linearity of

$\eta_{*p} : T_pH \longrightarrow T_pH$ for any $\eta \in I_p(H)$ then implies that $I_p(H)$ acts transitively on $S_pH$. Thus $(H,h)$ is isotropic at p. As the same argument clearly holds for all $p \in H$, it follows that $(H,h)$ is isotropic as required. □

COROLLARY 4.8 Any isotropic Riemannian manifold is homogeneous and complete.

REMARK 4.9 (a) The two-point homogeneous Riemannian manifolds are well known [cf. Wolf (1974, pp. 290-296)]. In particular, the odd-dimensional two-point homogeneous (hence isotropic) Riemannian manifolds are just the odd-dimensional Euclidean, hyperbolic, spherical, and elliptic spaces [cf. Wang (1951, p. 473)].
(b) Astronomical observations indicate that the spatial universe is approximately spherically symmetric about the earth. This suggests that the spatial universe should be modeled as a three-dimensional isotropic Riemannian manifold. Hence the possibilities are limited to the Euclidean, hyperbolic, spherical, and elliptic spaces. On the other hand, if one only assumes local isotropy there are more possibilities [cf. Misner, Thorne, and Wheeler (1973, pp. 713-725)].
(c) Any three-dimensional isotropic Riemannian manifold $(H,h)$ has constant sectional curvature and also $\dim I(H) = 6$ [cf. Walker (1944)].

We are now ready to define Robertson-Walker space-times using Lorentzian warped products and isotropic Riemannian manifolds.

DEFINITION 4.10 A *Robertson-Walker* space-time $(M,g)$ is any Lorentzian manifold which can be written as a Lorentzian warped product $(M_0 \times_f H, g)$ with $M_0 = (a,b)$ for $-\infty \leqslant a < b \leqslant +\infty$ given the negative definite metric $-dt^2$, with $(H,h)$ an isotropic Riemannian manifold, and with warping function $f : M_0 \longrightarrow (0,\infty)$.

In the notation of Section 2.6, we thus have $g = -dt^2 \oplus fh$ and $M_0 \times_f H$ is also topologically the product $M_0 \times H$. Letting $d\sigma^2$

denote the Riemannian metric h for H and defining $S(t) = \sqrt{f(t)}$, the Lorentzian metric g for $M_0 \times_f H$ may be rewritten in the more familiar form

$$ds^2 = -dt^2 + S^2(t)\, d\sigma^2$$

The map $\pi : M_0 \times_f H \longrightarrow \mathbb{R}$ given by $\pi(t,x) = t$ is a smooth time function on $M_0 \times_f H$ so that the Lorentzian manifold $M_0 \times_f H$ of Definition 4.10 actually is a (stably causal) space-time. Also each level surface $\pi^{-1}(c)$ of $\pi : M_0 \times_f H \longrightarrow M_0 \subset \mathbb{R}$ is an isotropic Riemannian manifold which is homothetic to (H,h). Furthermore, the isometry group I(H) of (H,h) may be identified with a subgroup $\tilde{I}(H)$ of $I(M_0 \times_f H)$ as follows. Given $\phi \in I(H)$, define $\overline{\phi} \in \tilde{I}(H)$ by $\overline{\phi}(r,h) = (r,\phi(h))$ for all $(r,h) \in M_0 \times H$. With this definition, $\tilde{I}(H)$ restricted to the level surfaces $\pi^{-1}(c)$ of $\pi$ acts transitively on each level surface.

Since all isotropic Riemannian manifolds are complete, Theorem 2.53 implies that all Robertson-Walker space-times are globally hyperbolic. From Theorem 2.56 we also know that every level surface $\pi^{-1}(c) = \{c\} \times H$ is a Cauchy surface for $M_0 \times_f H$.

Next to Minkowski space $\mathbb{R}^n = \mathbb{R} \times \mathbb{R}^{n-1}$ itself, the Einstein static universe is the simplest example of a Robertson-Walker space-time.

EXAMPLE 4.11 (Einstein Static Universe) Let $M_0 = \mathbb{R}$ with the negative definite metric $-dt^2$ and let $H = S^{n-1}$ with the standard spherical Riemannian metric. If $f : \mathbb{R} \longrightarrow (0,\infty)$ is the trivial warping function f = 1, then the product Lorentzian manifold $M = M_0 \times H = M_0 \times_f H$ is the n-dimensional Einstein static universe. If n = 2, then M is the cylinder $\mathbb{R} \times S^1$ with flat metric $-dt^2 + d\theta^2$. If $n \geq 3$, then this metric for $M = \mathbb{R} \times S^{n-1}$ is not flat since $S^{n-1}$ has constant positive sectional curvature 1.

For the rest of this section, we restrict our attention to

four-dimensional Robertson-Walker space-times. By Remark 4.9, these are warped products $M_0 \times_f H$, where (H,h) is Euclidean, hyperbolic, spherical, or elliptic of dimension 3. In the first two cases, H is topologically $\mathbb{R}^3$. In the third case $H = S^3$ and in the last case, H is the real projective 3-space $\mathbb{R}P^3$. We thus have the following

COROLLARY 4.12 All four-dimensional Robertson-Walker space-times are topologically either $\mathbb{R}^4$, $\mathbb{R} \times S^3$, or $\mathbb{R} \times \mathbb{R}P^3$.

Also by Remark 4.9, the sectional curvature k of (H,h) is constant. If k is nonzero, the metric may be rescaled to be of the form $ds^2 = -dt^2 + S^2(t)\, d\sigma^2$ on M so that k is either identically +1 or -1. This is the form of the metric usually studied in general relativity.

In physics, cosmological models are constructed from four-dimensional Robertson-Walker space-times assumed to be filled with a perfect fluid. The Einstein equations (cf. Appendix C) are then used to find the form of the above warping function $S^2(t)$. Among the models this technique yields are the big bang cosmological models [cf. Hawking and Ellis (1973, pp. 134-138)]. These models depend on the energy density $\mu$ and pressure p of the perfect fluid as well as the value of the cosmological constant $\Lambda$ in the Einstein equations. In the big bang cosmological models, the inextendible nonspacelike geodesics are all past incomplete. The stability of this incompleteness under metric perturbations will be considered in Section 6.3. Astronomical observations of clusters of galaxies indicate that distant clusters of galaxies are receding from us. This expansion of the universe suggests the existence of a "big bang" in the past and also suggests that the universe is a warped product with a nontrivial warping function rather than simply a Lorentzian product. Observations of black body radiation support these ideas [cf. Hawking and Ellis (1973, Chapter 10)].

## 4.5 BI-INVARIANT LORENTZIAN METRICS ON LIE GROUPS

The purpose of this section is to show how Theorems 2.54 and 2.55 of Section 2.6 may be used to construct a large class of Lie groups admitting globally hyperbolic, bi-invariant Lorentzian metrics.

We first summarize some basic facts from the elementary theory of Lie groups. Details may be found in a lucid exposition by Milnor (1963, Part IV) or at a more advanced level in Helgason (1962, Chapter 2). A Lie group is a group G which is also an analytic manifold such that the mapping $(g,h) \longrightarrow gh^{-1}$ from $G \times G \longrightarrow G$ is analytic. This multiplication induces left and right translation maps $L_g$, $R_g$ for each $g \in G$ given respectively by $L_g(h) = gh$ and $R_g(h) = hg$. A Riemannian or Lorentzian metric $\langle\ ,\ \rangle$ for G is then said to be *left invariant* (resp., *right invariant*) if $\langle L_{g_*}v, L_{g_*}w\rangle = \langle v,w\rangle$ (resp., $\langle R_{g_*}v, R_{g_*}w\rangle = \langle v,w\rangle$) for all $g \in G$, $v,w \in TG$. A metric which is both left and right invariant is said to be *bi-invariant*. By an averaging procedure involving the Haar integral, any compact Lie group may be given a bi-invariant metric [cf. Milnor (1963, p. 112)]. In fact, the Haar integral may be used to produce a bi-invariant Riemannian metric for G from any left invariant Riemannian metric for G. Any Lie group may be equipped with a left invariant Riemannian (or Lorentzian) metric by starting with a positive definite inner product (resp., inner product of signature n - 2) $\langle\ ,\ \rangle|_e$ on the tangent space $T_eG$ to G to the identity element $e \in G$, then defining $\langle\ ,\ \rangle|_g : T_gG \times T_gG \longrightarrow \mathbb{R}$ by

$$\langle v,w\rangle|_g = \langle L_{g^{-1}_*}v, L_{g^{-1}_*}w\rangle|_e \tag{4.1}$$

Thus any compact Lie group is furnished with a large supply of bi-invariant Riemannian metrics.

On the other hand, while (4.1) equips any Lie group with left-invariant Lorentzian metrics, the standard Haar integral averaging procedure used for Riemannian metrics fails to preserve signature (-, +, ..., +), so it cannot be used to convert left-invariant Lorentzian metrics into bi-invariant Lorentzian metrics.

But we will see shortly that a large class of bi-invariant Lorentzian metrics may be constructed for noncompact Lie groups of the form $\mathbb{R} \times G$ where G is any Lie group admitting a bi-invariant Riemannian metric.

Before giving the construction, we need to discuss product Lie groups briefly. Let G and H be two Lie groups. The product manifold $G \times H$ is then turned into a Lie group by defining the multiplication by

$$(g_1,h_1) \times (g_2,h_2) = (g_1g_2,h_1h_2) \tag{4.2}$$

It is immediate from (4.2), that if $\sigma = (g,h) \in G \times H$, then the translation maps $L_\sigma, R_\sigma : G \times H \longrightarrow G \times H$ are given by $L_\sigma = (L_g, L_h)$, and $R_\sigma = (R_g, R_h)$, i.e., $L_\sigma(g_1,h_1) = (L_g g_1, L_h h_1)$, etc. Recall that $T_\sigma(G \times H) \cong T_g G \times T_h H$. It is straightforward to check that for any $\sigma \in G \times H$ and any tangent vector $\xi = (v,w) \in T_\sigma(G \times H) \cong T_g G \times T_h H$, that

$$L_{\sigma_*}\xi = (L_{g_*}v, L_{h_*}w) \tag{4.3}$$

and

$$R_{\sigma_*}\xi = (R_{g_*}v, R_{h_*}w) \tag{4.4}$$

Now if $\langle\ ,\ \rangle_1$ is a Lorentzian metric for G and $\langle\ ,\ \rangle_2$ is a Riemannian metric for H, the product metric $\langle\langle\ ,\ \rangle\rangle = \langle\ ,\ \rangle_1 \oplus \langle\ ,\ \rangle_2$ is a Lorentzian metric for $G \times H$. Explicitly, recalling Definition 2.38, we have for tangent vectors $\xi_1 = (v_1,w_1)$ and $\xi_2 = (v_2,w_2)$ in $T_\sigma(G \times H)$ the formula

$$\langle\langle \xi_1,\xi_2 \rangle\rangle = \langle v_1,v_2\rangle_1 + \langle w_1,w_2\rangle_2$$

It is then immediate from (4.3) and (4.4) that if $\langle\ ,\ \rangle_1$ is a bi-invariant Lorentzian metric for G and $\langle\ ,\ \rangle_2$ is a bi-invariant Riemannian metric for H, then $\langle\langle\ ,\ \rangle\rangle$ is a bi-invariant Lorentzian metric for $G \times H$. To summarize,

PROPOSITION 4.13 Let $(G, \langle\ ,\ \rangle_1)$ be a Lie group equipped with a bi-invariant Lorentzian metric and let $(H, \langle\ ,\ \rangle_2)$ be a Lie group equipped with a bi-invariant Riemannian metric. Then the product metric $\langle\langle\ ,\ \rangle\rangle = \langle\ ,\ \rangle_1 \oplus \langle\ ,\ \rangle_2$ is a bi-invariant Lorentzian metric for the product Lie group $G \times H$. Hence $(G \times H, \langle\langle\ ,\ \rangle\rangle)$ is a Lorentzian symmetric space and, in particular, is geodesically complete.

*Proof.* It is only necessary to prove the last statement which is a standard fact in Lie group theory. Recall that we must show that for each $\sigma \in G \times H$, there exists an isometry $I_\sigma : G \times H \longrightarrow G \times H$ which fixes $\sigma$ and reverses the geodesics through $\sigma$. That is, if $\gamma$ is a geodesic in $G \times H$ with $\gamma(0) = \sigma$, we must show that $I_\sigma(\gamma(t)) = \gamma(-t)$ for all t. This is equivalent to showing that $I_{\sigma_*} : T_\sigma(G \times H) \longrightarrow T_\sigma(G \times H)$ is the map $I_{\sigma_*}(\xi) = -\xi$ and also implies that $I_\sigma^2 = \text{Id}$.

We will follow the proof given in Milnor (1963, pp. 109, 112). First, if we denote the identity element of $G \times H$ by e, and define a map $I_e : G \times H \longrightarrow G \times H$ by $I_e(\sigma) = \sigma^{-1}$, then $I_{e_*} : T_e(G \times H) \longrightarrow T_e(G \times H)$ is given by $I_{e_*}(v) = -v$. Thus $I_{e_*} : T_e(G \times H) \longrightarrow T_e(G \times H)$ is an isometry of $T_e(G \times H)$. To see that $I_{e_*}$ is an isometry of any other tangent space $T_\sigma(G \times H) \longrightarrow T_{\sigma^{-1}}(G \times H)$ and hence that $I_e : G \times H \longrightarrow G \times H$ is an isometry, we simply note that

$$I_e = R_{\sigma^{-1}} I_e L_{\sigma^{-1}}$$

Since $\langle\langle\ ,\ \rangle\rangle$ is bi-invariant, all left and right translation maps are isometries. Then as

$$I_{e_*}\Big|_\sigma = R_{\sigma^{-1}*}\Big|_e I_{e_*}\Big|_e L_{\sigma^{-1}*}\Big|_\sigma$$

and $I_{e_*}\big|_e$ is an isometry of $T_e(G \times H)$, it follows that $I_{e_*} : T_\sigma(G \times H) \longrightarrow T_{\sigma^{-1}}(G \times H)$ is also an isometry. The map $I_e : G \times H \longrightarrow G \times H$ is thus the required geodesic symmetry at e.

We define the geodesic symmetry $I_\sigma$ for any $\sigma \in G$ by setting $I = R_\sigma I_e R_{\sigma^{-1}}$. Since $R_\sigma$, $R_{\sigma^{-1}}$ are isometries by the bi-invariance of << , >> and we have just shown that $I_e$ is an isometry, it follows that $I_\sigma : G \times H \longrightarrow G \times H$ is an isometry and obviously $I_\sigma(\sigma) = \sigma$ since $I_\sigma(h) = \sigma h^{-1}\sigma$. Finally, for any $\xi \in T_\sigma(G \times H)$, we have

$$\begin{aligned} I_{\sigma_*}\xi &= R_{\sigma_*}(I_{e_*}(R_{\sigma^{-1}_*}\xi)) \\ &= R_{\sigma_*}(-R_{\sigma^{-1}_*}\xi) \qquad \text{since } R_{\sigma^{-1}_*}\xi \in T_e(G \times H) \\ &= -R_{\sigma_*}R_{\sigma^{-1}_*}\xi = -(R_\sigma R_{\sigma^{-1}})_*\xi = -\xi \end{aligned}$$

Thus $I_\sigma$ reverses geodesics at $\sigma$ as required. We have therefore shown that $G \times H$ is a symmetric space.

It may be shown that any symmetric space is geodesically complete as follows. Let $\gamma$ be a geodesic in M and set $p = \gamma(0)$. Supposing that $q = \gamma(A)$ is defined, one may derive the formula [cf. Milnor (1963, p. 109)]

$$I_q I_p(\gamma(t)) = \gamma(t + 2A)$$

provided that $\gamma(t)$ and $\gamma(t + 2A)$ are defined.

Thus if $\gamma$ is defined originally on an interval $\gamma : [0,\lambda] \longrightarrow G \times H$, $\gamma$ may be extended to a geodesic $\tilde{\gamma} : [0,2\lambda] \longrightarrow G \oplus H$ by choosing $q := \gamma(\lambda/2)$ and putting $\tilde{\gamma}(t) = I_q I_p(\gamma(t - \lambda))$ for $t \in [\lambda,2\lambda]$. It is then clear that $\gamma$ may be defined on $(-\infty,\infty)$. Thus $(G \times H, << , >>)$ is geodesically complete. □

Now Proposition 4.13 has the apparent defect that the existence of Lie groups $(G,< , >_1)$ equipped with bi-invariant Lorentzian metrics is assumed. It will now be shown how such Lie groups may be constructed by taking products of the form $(\mathbb{R} \times G, -dt^2 \oplus < , >)$ where $(G,< , >)$ is a Lie group equipped with a Riemannian bi-invariant metric.

The Lie group structure on $(\mathbb{R}, -dt^2)$ we will use is that induced by the usual addition of real numbers. Accordingly, we will write $(a,b) \longmapsto a + b$ for the Lie group "multiplication" despite our use of the product notation above for the group operation. Here

$\mathbb{R}$ is the analytic manifold determined by the chart $t : \mathbb{R} \to \mathbb{R}$, $t(r) = r$. Let $\partial/\partial t$ denote the corresponding coordinate vector field on $\mathbb{R}$. The left and right translation maps $L_a, R_a : \mathbb{R} \to \mathbb{R}$ are given by $L_a(r) = a + r$ and $R_a(r) = r + a$. It is easy to check that if $v = \lambda\, \partial/\partial t|_r \in T_r\mathbb{R}$, then $L_{a_*}v, R_{a_*}v \in T_{a+r}(\mathbb{R})$ are given by $L_{a_*}v = R_{a_*}v = \lambda\, \partial/\partial t|_{a+r}$. Hence $-dt^2(L_{a_*}v, L_{a_*}v) = -dt^2(R_{a_*}v, R_{a_*}v) = -\lambda^2 = -dt^2(v,v)$ so that $-dt^2$ is left and right invariant.

Let $(G, \langle\ ,\ \rangle)$ be a Lie group with a bi-invariant Riemannian metric. By the proof given in Proposition 4.13, $G$ is a complete symmetric space. Also using (4.3) and (4.4), it is easily seen that the metric $\langle\langle\ ,\ \rangle\rangle = -dt^2 \oplus \langle\ ,\ \rangle$ is a bi-invariant Lorentzian metric for $\mathbb{R} \times G$. [Here if $\xi_1 = (\lambda_1 \partial/\partial t|_r, v_1)$ and $\xi_2 = (\lambda_2 \partial/\partial t|_r, v_2)$ with $v_1, v_2 \in T_gG$, the inner product $\langle\langle \xi_1, \xi_2 \rangle\rangle = -\lambda_1\lambda_2 + g(v_1, v_2)$]. Since $(G, \langle\ ,\ \rangle)$ is a complete Riemannian manifold, the product $(\mathbb{R} \times G, \langle\langle\ ,\ \rangle\rangle)$ is globally hyperbolic by Theorem 2.54. We have obtained

THEOREM 4.14 Let $(\mathbb{R}, -dt^2)$ be given the usual additive group structure and let $(G, \langle\ ,\ \rangle)$ be any Lie group equipped with a bi-invariant Riemannian metric. Then the product metric $\langle\langle\ ,\ \rangle\rangle = -dt^2 \oplus \langle\ ,\ \rangle$ is a bi-invariant Lorentzian metric for the product Lie group $\mathbb{R} \times G$. Thus $(\mathbb{R} \times G, \langle\langle\ ,\ \rangle\rangle)$ is a geodesically complete, globally hyperbolic space-time.

Much research inspired by E. Cartan's work was done on Lie groups, homogeneous spaces and symmetric spaces equipped with indefinite metrics before modern causality theory had assumed such a prominent role in general relativity. Thus most of this work was carried out not for Lorentzian metrics in particular, but rather for general pseudo-Riemannian metrics of arbitrary signature. Rather than attempting to give an exhaustive list of references, we refer the reader to the bibliography in Wolf's (1974) text. Much of this research has been concerned with the problem of classifying all geodesically complete pseudo-Riemannian manifolds of constant curvature (the "space-form problem"). Two recent papers dealing with pseudo-Riemannian Lie theory have been written by Kulkarni (1978) and Nomizu (1979). Nomizu's paper deals specifically with Lorentzian metrics, considering the existence of constant curvature left-invariant Lorentzian metrics on a certain class of noncommutative Lie groups.

Chapter 5

# COMPLETENESS AND EXTENDIBILITY

We mentioned in Chapter 1 that the Hopf-Rinow theorem guarantees the equivalence of geodesic and metric completeness for arbitrary Riemannian manifolds. Further, either of these conditions implies the existence of minimal geodesics. That is, given any two points $p,q \in M$, there is a geodesic from p to q whose arc length realizes the metric distance from p to q. If M is compact, it also follows from the Hopf-Rinow theorem that all Riemannian metrics for M are complete. In the noncompact case, Nomizu and Ozeki (1961) established that every noncompact smooth manifold admits a complete Riemannian metric. Extending their proof, Morrow (1970) showed that the complete Riemannian metrics for M are dense in the compact-open topology in the space of all Riemannian metrics for M [cf. Fegan and Millman (1978)].

In the first three sections of this chapter, we compare and contrast these results with the theory of geodesic and metric completeness for arbitrary Lorentzian manifolds. In Section 5.1, a standard example is given to show that geodesic completeness does not imply the existence of maximal geodesic segments joining causally related points. Then we recall that the class of globally hyperbolic space-times possesses this useful property. In Section 5.2, we consider forms of completeness such as nonspacelike geodesic completeness, bounded acceleration (b.a.) completeness, and b-completeness that have been studied in singularity theory in general relativity [cf. Clarke and Schmidt (1977), Ellis and Schmidt

(1977)]. We also state a corollary to Theorem 8 of Beem (1976a, p. 184) establishing the existence of nonspacelike complete metrics for all distinguishing space-times. In Section 5.3 we discuss Lorentzian metric completeness and finite compactness.

In the last four sections of this chapter, we discuss extensions and local extensions of space-times. Since extendibility is related to geodesic completeness, extendibility plays an important role in singularity theory in general relativity [cf. Clarke (1973, 1975, 1976), Hawking and Ellis (1973), Ellis and Schmidt (1977)]. In particular, one usually wants to avoid investigating space-times which are proper subsets of larger space-times since such proper subsets are always geodesically incomplete.

A space-time (M',g') is said to be an *extension* of a given space-time (M,g) if (M,g) may be isometrically imbedded as a proper open subset of (M',g'). A space-time which has no extension is either said to be *inextendible* [cf. Hawking and Ellis (1973)] or *maximal* [cf. Sachs and Wu (1977b, p. 29)].

A local extension is an extension of a certain type of subset of a given space-time. In general, local inextendibility (i.e., the nonexistence of local extensions) implies global inextendibility. Since questions of extendibility naturally relate to the boundary of space-time, in Section 5.4 we briefly describe the Schmidt b-boundary and the Geroch-Kronheimer-Penrose causal boundary. In Section 5.5, two types of local extensions are defined and studied. If a Lorentzian manifold has no local extensions of either of these two types, it is shown to be inextendible. We also give a local extension of Minkowski space-time which shows that while b-completeness forces a space-time to be (globally) inextendible, b-completeness does *not* prevent a space-time from having local extensions.

In Section 5.6, local extensions are related to curvature singularities. For example, if (M,g) is an analytic space-time such that each timelike geodesic $\gamma : [0,a) \longrightarrow M$ which is inextendible to $t = a$ is either complete (in the sense that $a = \infty$) or else corresponds to a curvature singularity, then (M,g) has no analytic local b-boundary extensions.

## 5.1 EXISTENCE OF MAXIMAL GEODESIC SEGMENTS

The purpose of this section is twofold. First we recall that for arbitrary Lorentzian manifolds, geodesic completeness does *not* imply the existence of maximal geodesic segments joining causally related pairs of points. Second we discuss the important and useful fact that distance realizing geodesics do exist for the class of globally hyperbolic space-times.

The universal covering manifold $(M,g)$ of two-dimensional anti-de Sitter space provides an example that geodesic completeness does not imply that every $p,q \in M$ with $p \ll q$ may be joined by a timelike geodesic $\gamma$ with $L(\gamma) = d(p,q)$ (cf. Figure 5.1). Recall that

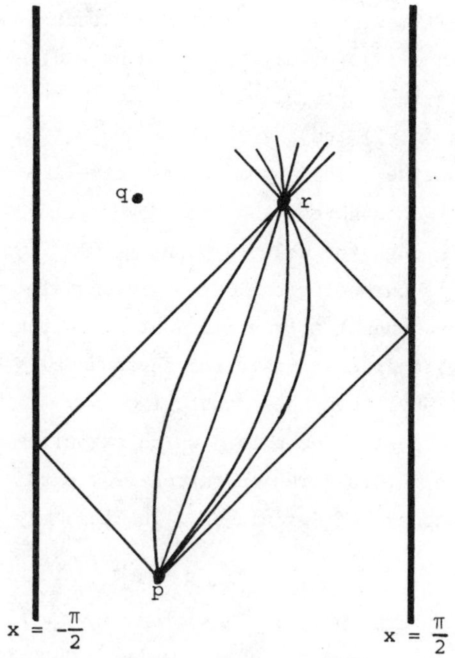


*Figure 5.1* The universal cover $M = \{(x,t) : -\pi/2 < x < \pi/2\}$ of two-dimensional anti-de Sitter space is shown. The metric is given by $ds^2 = \sec^2 x(-dt^2 + dx^2)$. The points p and q are chronologically related in M. Yet no maximal timelike geodesic in M joins p to q since all future-directed timelike geodesics emanating from p are focused at r.

if $\gamma$ is any future-directed timelike curve from p to q with $L(\gamma) = d(p,q)$, then $\gamma$ may be reparameterized to a timelike geodesic (cf. Theorem 3.13). Thus this same example shows that geodesically complete space-times exist which contain points $p \ll q$ such that $L(\gamma) < d(p,q)$ for all $\gamma \in \Omega_{p,q}$.

The space-time (M,g) may be represented by the strip $M = \{(x,t) \in \mathbb{R}^2 ; -\pi/2 < x < \pi/2\}$ in $\mathbb{R}^2$ with the Lorentzian metric $ds^2 = \sec^2 x\,(-dt^2 + dx^2)$, [cf. Penrose (1972, p. 7)]. The points p and q in Figure 5.1 satisfy $p \ll q$. Yet all future timelike geodesics emanating from p are focused again at the future timelike conjugate point r. Thus there is no timelike geodesic in M from p to q. Hence there is no maximal timelike geodesic or maximal timelike curve from p to q.

We now consider which space-times do have the property that every pair of points $p,q \in M$ with $q \in J^+(p)$ may be joined by a distance realizing geodesic. If $M = \mathbb{R}^2 - \{(0,0)\}$ with the Lorentzian metric $ds^2 = dx^2 - dy^2$, then $p = (0,-1)$ and $q = (0,1)$ are points in M with $d(p,q) = 2 > 0$ which cannot be joined by a maximal timelike geodesic. [The desired geodesic would have to be the curve $\gamma(t) = (0,t)$, $-1 \leqslant t \leqslant 1$ which passes through the deleted point (0,0).] On the other hand, this space-time is chronological, strongly causal, and stably causal. Thus it is reasonable to restrict our attention to the class of globally hyperbolic space-times. For these space-times, Avez (1963) and Seifert (1967) have shown that given any $p,q \in M$ with $p \leqslant q$, there is a geodesic from p to q which maximizes arc length among all nonspacelike future-directed curves from p to q (cf. Theorem 2.14). In the language of Definition 3.10, this may be stated as follows.

THEOREM 5.1 Let (M,g) be globally hyperbolic. Then given any $p,q \in M$ with $q \in J^+(p)$, there is a maximal geodesic segment $\gamma \in \Omega_{p,q}$, i.e., a future-directed nonspacelike geodesic $\gamma$ from p to q with $L(\gamma) = d(p,q)$.

We sketch Seifert's (1967, Theorem 1) proof of this result [cf. Penrose (1972, Chapter 6)]. Since (M,g) is globally hyperbolic, it may be shown that if $p \leqslant q$, the nonspacelike path space $\Omega_{p,q}$ is compact. On the other hand, since (M,g) is strongly causal, the arc-length functional $L : \Omega_{p,q} \longrightarrow \mathbb{R}$ is upper semicontinuous in the $C^0$ topology (cf. Section 2.3). Thus there exists a curve $\gamma_0 \in \Omega_{p,q}$ with $L(\gamma_0) = \sup\{L(\gamma) : \gamma \in \Omega_{p,q}\}$. It follows from the variational theory of arc length that if $\gamma_0$ is not a reparameterization of a smooth geodesic, a curve $\sigma \in \Omega_{p,q}$ with $L(\sigma) > L(\gamma)$ may be constructed, in contradiction. Alternately, if $L(\gamma_0) = \sup\{L(\gamma) : \gamma \in \Omega_{p,q}\}$, then $L(\gamma_0) = d(p,q)$ by definition of Lorentzian distance. Hence Theorem 3.13 implies that $\gamma_0$ is up to reparameterization a smooth geodesic.

In the case that $p \ll q$, the maximal curve $\gamma_0$ may also be constructed using the results of Section 2.3. Let $h : M \longrightarrow \mathbb{R}$ be a globally hyperbolic time function for (M,g). Choose $t_0$ with $h(p) < t_0 < h(q)$. Then $K = J^+(p) \cap J^-(q) \cap h^{-1}(t_0)$ is compact and any nonspacelike curve from p to q intersects K. By definition of Lorentzian distance, we may find a curve $\gamma_n \in \Omega_{p,q}$ with

$$d(p,q) \geqslant L(\gamma_n) \geqslant d(p,q) - \frac{1}{n}$$

for each positive integer n. Let $r_n \in \gamma_n \cap K$. Since K is compact, a subsequence $r_{n(j)}$ converges to $r \in K$. By Corollary 2.19, there is a nonspacelike limit curve $\gamma_0$ passing through r and joining p to q of the sequence $\{\gamma_{n(j)}\}$. Since (M,g) is strongly causal, a subsequence of $\{\gamma_{n(j)}\}$ converges to $\gamma_0$ in the $C^0$ topology by Proposition 2.21. Using Remark 2.22 and condition (1), we obtain $L(\gamma_0) \geqslant d(p,q)$. Hence by definition of distance, $L(\gamma_0) = d(p,q)$ and $\gamma_0$ may be reparameterized to a smooth geodesic by Theorem 3.13. If $p \leqslant q$ and $d(p,q) = 0$, we already know that there is a maximal null geodesic segment from p to q by Corollary 3.14.

In connection with Theorem 5.1, it should be noted that global hyperbolicity is not a necessary condition for the existence of maximal geodesic segments joining all pairs of causally related points. For let $M = \{(x,y) \in \mathbb{R}^2 : 0 < x < 10,\ 0 < y < 10\}$ be equipped with the Lorentzian metric it inherits as an open subset of Minkowski space. Since the geodesics in M are just Euclidean straight line segments, it is readily seen that maximal geodesics exist joining any pair of causally related points. On the other hand, if $p = (1,1)$ and $q = (1,9)$, then $J^+(p) \cap J^-(q)$ is noncompact. Hence this space-time, while strongly causal, fails to be globally hyperbolic.

## 5.2 GEODESIC COMPLETENESS

We showed in Theorem 3.9 that for strongly causal space-times, the Lorentzian distance function may be used to construct a subbasis for the given manifold topology. Nonetheless, the sets $\{q \in M : d(p,q) < R\}$ fail to form a basis for the given manifold topology. Thus geodesic completeness rather than metric completeness of space-times has usually been considered in general relativity.

Let (M,g) be an arbitrary Lorentzian manifold.

DEFINITION 5.2 A geodesic c in (M,g) with affine parameter t is said to be *complete* if the geodesic can be extended to be defined for $-\infty < t < \infty$. A past and future inextendible geodesic is said to be *incomplete* if it cannot be extended to arbitrarily large positive and negative values of an affine parameter. *Future* or *past incomplete* geodesics may be defined similarly.

An affine parameter for the curve c is a parameterization such that in this parameterization, c(t) satisfies the geodesic differential equation $\nabla_{c'} c'(t) = 0$ for all t [cf. Kobayashi and Nomizu (1963, p. 138)]. It is necessary to use the concept of an affine parameter since null geodesics, which have zero arc length, cannot be parameterized by arc length. If s and t are two affine parameters for c, it follows from the geodesic differential equations that there exist constants $a,b \in \mathbb{R}$ such that $s(t) = at + b$ for all t in the domain of c. Hence completeness or incompleteness as defined in Definition 5.2 is independent of the choice of affine parameter. In particular, if c is an inextendible timelike geodesic parameterized by arc length [i.e., $g(c'(t),c'(t)) = -1$ for all $t \in$ domain c], then c is incomplete if $L(c) < \infty$. Even if $L(c) = \infty$ it may happen that c is incomplete. This occurs, for example, when the domain of c is of the form $(a,\infty)$, where $a > -\infty$.

Certain exact solutions to the Einstein equations in general relativity, like the extended Schwarzschild solution, contain nonspacelike geodesics which become incomplete on running into black holes. Even though the existence of incomplete, inextendible nonspacelike geodesics does not force a space-time to contain a black hole, these examples suggest that nonspacelike geodesic incompleteness might be used as a first order test for "singular space-times" [cf. Hawking and Ellis (1973, Chapter 8), Clarke and Schmidt (1977), Ellis and Schmidt (1977)]. Thus it is standard to make the following definitions in general relativity. Recall that a geodesic is said to be inextendible if it is both past and future inextendible.

DEFINITION 5.3 A space-time (M,g) is said to be *timelike* (resp., *null, nonspacelike, spacelike*) *geodesically complete* if all timelike (resp., null, nonspacelike, spacelike) inextendible geodesics are complete. The space-time (M,g) is said to be *geodesically complete* if all inextendible geodesics are complete. Also (M,g) is said to be *timelike* (resp., *null, nonspacelike, spacelike*) *geodesically incomplete* if some timelike (resp., null, nonspacelike, spacelike) geodesic is incomplete. A nonspacelike incomplete space-time is said to be a geodesically *singular space-time*.

It was once hoped that timelike geodesic completeness might imply null geodesic completeness, etc. However Kundt (1963) gave an example of a space-time that is timelike and null geodesically complete, but not spacelike complete. Then Geroch (1968b, p. 531) gave an example of a space-time conformal to Minkowski 2-space and thus globally hyperbolic which is timelike incomplete, but null and spacelike complete. Also Geroch remarked that modifications of

Kundt's and his examples give space-times that are (i) incomplete in any two ways, but complete in the third way, (ii) spacelike incomplete but null and timelike complete, and (iii) timelike incomplete, but spacelike and null complete. Then Beem (1976c) gave an example of a globally hyperbolic space-time that is null incomplete, but spacelike and timelike complete. These results may be summarized as follows.

THEOREM 5.4 Timelike geodesic completeness, null geodesic completeness, and spacelike geodesic completeness are all logically inequivalent.

We now describe Geroch's example of a space-time which is null and spacelike complete but timelike incomplete to illustrate the constructions used in the proof of Theorem 5.4. Let $(\mathbb{R}^2, g_0)$ be Minkowski 2-space with global coordinates $(x,t)$ and the usual Lorentzian metric $g_0 = ds^2 = dx^2 - dt^2$. Conformally change the metric $g_0$ to a new metric $g = \phi g_0$ for $\mathbb{R}^2$, where $\phi : \mathbb{R}^2 \longrightarrow (0,\infty)$ is a smooth function with the following properties (cf. Figure 5.2):

(1) $\phi(x,t) = 1$ if $x \leqslant -1$ or $x \geqslant 1$.

(2) $\phi(x,t) = \phi(-x,t)$ for all $(x,t) \in \mathbb{R}^2$.

(3) On the t axis, $\phi(0,t)$ goes to zero like $t^{-4}$ as $t \longrightarrow \infty$.

Since $g$ is conformal to $g_0$, the space-time $(\mathbb{R}^2, g)$ is globally hyperbolic, and null geodesics still have as images straight lines making angles of 45° with the positive or negative x axis. By property (2), the reflection $F(x,t) = (-x,t)$ is an isometry of $(\mathbb{R}^2, g)$. Since the fixed point set of an isometry is totally geodesic, the t axis may be parameterized as a timelike geodesic. By condition (3), this geodesic is incomplete as $t \longrightarrow \infty$. Thus $(\mathbb{R}^2, g)$ is timelike incomplete. But every null or spacelike geodesic which enters the region $-1 \leqslant x \leqslant 1$ eventually leaves, then remains outside this region. Thus condition (1) implies that $(\mathbb{R}^2, g)$ is null and spacelike complete.

We now consider the converse problem of constructing geodesically complete Lorentzian metrics for paracompact smooth manifolds. In

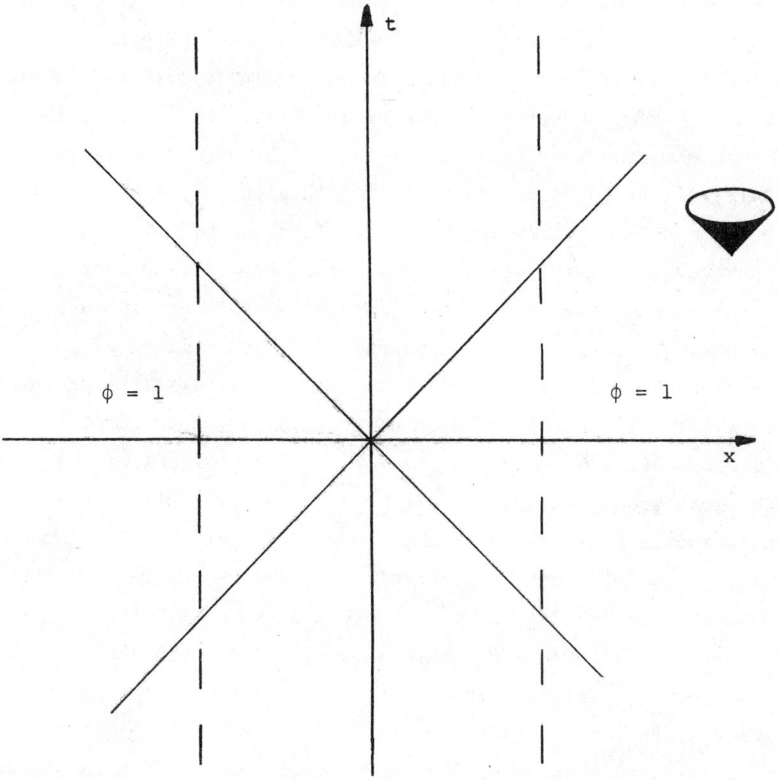


*Figure 5.2* Geroch's example of a space-time globally conformal to Minkowski 2-space, which is null and spacelike geodesically complete, but timelike geodesically incomplete, is shown. Here the positive t axis may be parameterized to be an incomplete timelike geodesic since $\phi(0,t) \longrightarrow 0$ like $t^{-4}$ as $t \longrightarrow \infty$.

order to preserve the causal structure of the given space-time, we restrict our attention to global conformal changes rather than arbitrary metric deformations.

For Riemannian metrics Nomizu and Ozeki (1961) showed that an arbitrary metric can be made complete by a global conformal change. On the other hand, space-times exist with the property that no global conformal factor will make these space-times nonspacelike geodesically complete. A two-dimensional example with this property has been given by Misner (1967). In this example there are inextendible null geodesics which are future incomplete and future trapped in a compact set [cf. Hawking and Ellis (1973, pp. 171-172)]. Any conformal change of this example will leave these null geodesics pointwise fixed and future incomplete. Thus one may not establish an analogue of the Nomizu and Ozeki result for arbitrary space-times.

However, the existence of nonspacelike complete Lorentzian metrics has been shown for space-times satisfying certain causality conditions. Seifert (1971, p. 258) has shown that if (M,g) is stably causal, then M is conformal to a space-time with all future-directed (or all past-directed) nonspacelike geodesics complete. Also Clarke (1971) has shown that a strongly causal space-time may be made null geodesically complete by a conformal factor. Beem (1976a) studied space-times with the property that for each compact subset K of M, no future-inextendible nonspacelike curve is future imprisoned in K. [Recall that the nonspacelike curve $\gamma$ is said to be future imprisoned in K if there exists $t_0 \in \mathbb{R}$ such that $\gamma(t) \in K$ for all $t \geqslant t_0$.] If (M,g) is a causal space-time satisfying this condition, then there exists a conformal factor $\Omega : M \longrightarrow (0,\infty)$ such that $(M,\Omega g)$ is null and timelike geodesically complete [Beem (1976a, p. 184, Theorem 8)]. This imprisonment condition is satisfied if (M,g) is stably causal, strongly causal, or distinguishing. Hence we may state the following result.

THEOREM 5.5 If (M,g) is distinguishing, strongly causal, stably causal, or globally hyperbolic, then there exists a smooth conformal factor $\Omega : M \longrightarrow (0,\infty)$ such that the space-time $(M,\Omega g)$ is timelike and null geodesically complete.

It is an open question as to whether Theorem 5.5 can be strengthened to include spacelike geodesic completeness as well (cf. Corollary 2.33 for space-times homeomorphic to $\mathbb{R}^2$).

Suppose that a space-time is defined to be nonsingular if it is geodesically complete. Then "no regions have been deleted from the space-time manifold" of a nonsingular space-time [Geroch (1968b, Property 1)]. But Geroch (1968b, Property 2) suggested a second condition that nonsingular space-times should satisfy: namely, "observers who follow 'reasonable' (in some sense) world lines should have an infinite total proper time." Here a "world line" is a timelike curve in (M,g). Then Geroch (1968b, pp. 534-540) constructed a geodesically complete space-time which contains a smooth timelike *curve* of bounded acceleration, but having finite length. Thus this example fails to satisfy Geroch's Property 2 even though all timelike *geodesics* have infinite length by the geodesic completeness.

Accordingly, in addition to geodesically incomplete space-times, further kinds of singular space-times have been studied in general relativity. In the rest of this section, we will discuss two of these additional types of completeness, b.a. completeness ("bounded acceleration completeness") and b-completeness ("bundle completeness").

The concept of b.a. completeness stems from the preceding example of Geroch. For the purpose of stating Definition 5.6, we recall that any $C^2$ timelike curve may be reparameterized to a $C^2$ timelike curve $\gamma : J \longrightarrow M$ with $g(\gamma'(t),\gamma'(t)) = -1$ for all $t \in J$.

DEFINITION 5.6 A $C^2$ timelike curve $\gamma : J \longrightarrow M$ with $g(\gamma'(t),\gamma'(t)) = -1$ for all $t \in J$ is said to have *bounded acceleration* if there exists a constant $B > 0$ such that $|g(\nabla_{\gamma'}\gamma'(t),\nabla_{\gamma'}\gamma'(t))| \leq B$ for all $t \in J$.

Here $\nabla$ is the unique torsion free connection for M defined by the metric g (cf. Section 2.1). In particular, if $\gamma$ is a geodesic, then $\gamma$ has zero hence bounded acceleration. The requirement that $\gamma$ be $C^2$ makes it possible to calculate $\nabla_{\gamma'}\gamma'$.

We may now define b.a. completeness.

DEFINITION 5.7 A space-time (M,g) is said to be *b.a. complete* if all future- (resp., past-) directed, future- (resp., past-) inextendible unit speed $C^2$ timelike curves with bounded acceleration have infinite length. If there exists a future- (or past-) directed, future- (or past-) inextendible unit speed $C^2$ timelike curve with bounded acceleration but finite length, then (M,g) is said to be *b.a. incomplete*.

Geroch's example (1968b, pp. 534-540) shows that geodesic completeness does not imply b.a. completeness. Further Beem (1976c, p. 509) has given an example to show that even for globally hyperbolic space-times, geodesic completeness does not imply b.a. completeness. Trivially, b.a. completeness implies timelike geodesic completeness. On the other hand, the example of Geroch given in Figure 5.2 may be modified by changing the sign of the metric tensor to show that b.a. completeness does not imply spacelike geodesic completeness.

A stronger form of completeness, *b-completeness*, does imply geodesic completeness and hence overcomes this last objection to b.a. completeness. B-completeness was first studied for Lorentzian manifolds by Schmidt (1971). Intuitively, b-completeness is defined as follows [cf. Hawking and Ellis (1973, p. 259)]. First the concept of an affine parameter is extended from geodesics to all $C^1$ curves. Then a space-time is said to be b-complete if every $C^1$ curve of finite length in such a parameter has an endpoint.

We now give a brief discussion of b-completeness. First it is necessary to discuss the concept of a *generalized affine parameter* for any $C^1$ curve $\gamma : J \longrightarrow M$. Recall that a smooth vector field V along $\gamma$ is a smooth map $V : J \longrightarrow TM$ such that $V(t) \in T_{\gamma(t)}M$ for all $t \in J$. Such a smooth vector field V along $\gamma$ is said to be a parallel field along $\gamma$ if V satisfies the differential equation $\nabla_{\gamma'}V(t) = 0$ for all $t \in J$, (cf. Appendix A).

A generalized affine parameter $\mu = \mu(\gamma, E_1, E_2, \ldots, E_n)$ may be constructed for $\gamma : J \longrightarrow M$ as follows. Choosing any $t_0 \in J$, let $\{e_1, \ldots, e_n\}$ be any basis for $T_{\gamma(t_0)}M$. Let $E_i$ be the unique parallel field along $\gamma$ with $E_i(t_0) = e_i$ for $1 \leq i \leq n$. Then $\{E_1(t), E_2(t), \ldots, E_n(t)\}$ forms a basis for $T_{\gamma(t)}M$ for each $t \in J$. We may thus write $\gamma'(t) = \Sigma_{i=1}^n V^i(t)E_i(t)$, with $V^i : J \longrightarrow \mathbb{R}$ for $1 \leq i \leq n$. Then the generalized affine parameter $\mu = \mu(\gamma, E_1, \ldots, E_n)$ is given by

$$\mu(t) = \int_{t_0}^{t} \sqrt{\sum_{i=1}^{n} [V^i(s)]^2}\, ds \qquad t \in J$$

The assumption that $\gamma$ is $C^1$ was necessary in order to obtain the vector fields $\{E_1, \ldots, E_n\}$ by parallel translation. It may be checked that $\gamma$ has finite arc length in the generalized affine parameter $\mu = \mu(\gamma, E_1, \ldots, E_n)$ iff $\gamma$ has finite arc length in any other generalized affine parameter $\mu = \mu(\gamma, \overline{E}_1, \ldots, \overline{E}_n)$ calculated from any other basis $\{\overline{E}_i\}_{i=1}^n$ for $TM|_\gamma$ obtained by parallel translation along $\gamma$ [cf. Hawking and Ellis (1973, p. 259)]. Hence the concept of finite arc length with respect to a generalized affine parameter is independent of the particular choice of generalized affine parameter. It thus makes sense to make the following definition.

DEFINITION 5.8 The space-time $(M,g)$ is said to be *b-complete* if every $C^1$ curve of finite arc length as measured by a generalized affine parameter has an endpoint in $M$.

Suppose $\gamma : J \longrightarrow M$ is any smooth geodesic. Taking $E_1(t) = \gamma'(t)$ in the above construction, for any choice of $E_2, \ldots, E_n$ we have $\mu(\gamma, E_1, E_2, \ldots, E_n)(t) = t$. Hence b-completeness implies geodesic completeness. It is also known that b-completeness implies b.a. completeness. Geroch's example (cf. Figure 5.2) with the sign of the metric tensor changed shows that there are globally hyperbolic space-times which are b.a. complete but not b-complete. Thus b.a. completeness does not imply b-completeness.

## 5.3 METRIC COMPLETENESS

The Hopf-Rinow theorem for Riemannian manifolds $(N,g_0)$ implies that the following are equivalent:

(1) N with the Riemannian distance function $d_0 : N \times N \longrightarrow [0,\infty)$ is a complete metric space, i.e., all Cauchy sequences converge.

(2) $(N,d_0)$ is finitely compact, i.e., all $d_0$-bounded sets have compact closure.

(3) $(N,g_0)$ is geodesically complete.

Here a set K in a Riemannian manifold $(N,g_0)$ is said to be *bounded* if $\sup\{d_0(p,q) : p,q \in K\} < \infty$. By the triangle inequality, this is equivalent to the condition that K be contained inside a closed metric ball of finite radius.

In Section 5.2, we considered the geodesic completeness of Lorentzian manifolds. In this section, we shall consider Lorentzian analogues of conditions (1) and (2) above. From the very definition of Lorentzian distance [i.e., $d(p,q) = 0$ if $q \notin J^+(p)$], it is clear that attention should be restricted to timelike Cauchy sequences.

Busemann (1967) studied general Hausdorff spaces having a partial ordering with properties similar to those of the chronological partial ordering $p \ll q$ of a space-time. Also Busemann supposed that these spaces, which he called *timelike spaces*, were equipped with a function which behaves just like the Lorentzian distance function of a chronological space-time restricted to the set $\{(p,q) \in M \times M : p \leqslant q\}$. For this class of nondifferentiable spaces, Busemann observed that the length of continuous curves could be defined and, moreover, the length functional is upper semicontinuous in a topology of uniform convergence [cf. Busemann (1967, p. 10)]. Busemann's aim in studying timelike spaces was to develop a geometric theory for indefinite metrics analogous to the theory of metric G-spaces [cf. Busemann (1955)]. In particular, Busemann studied finite compactness and metric completeness for timelike spaces in the spirit of (1) and (2) of the Hopf-Rinow theorem.

Beem (1976b) observed that Busemann's definitions of finite compactness and metric completeness for timelike G-spaces may be adapted to causal space-times. First, timelike Cauchy completeness may be defined for causal space-times as follows.

DEFINITION 5.9 The causal space-time $(M,g)$ is said to be *timelike Cauchy complete* if any sequence $\{x_n\}$ of points with $x_n \ll x_{n+m}$ for $n,m = 1, 2, 3, \ldots$ and $d(x_n, x_{n+m}) \leqslant B_n$ [or else $x_{n+m} \ll x_n$ and $d(x_{n+m}, x_n) \leqslant B_n$] for all $m \geqslant 0$, where $B_n \longrightarrow 0$ as $n \longrightarrow \infty$, is a convergent sequence.

For Riemannian manifolds, finite compactness may be defined by requiring that all closed metric balls be compact. On the other hand, we have noted above (cf. Figure 3.4) that the subsets $\{q \in J^+(p) : d(p,q) \leqslant \varepsilon\}$ of a space-time are generally noncompact. Thus the Riemannian definition must be modified. One possibility is the following [cf. Busemann (1967, p. 22)].

DEFINITION 5.10 The causal space-time $(M,g)$ is said to be *finitely compact* if for each fixed constant $B > 0$ and each sequence of points $\{x_n\}$ with either $p \ll q \leqslant x_n$ and $d(p,x_n) \leqslant B$ for all n, or $x_n \leqslant q \ll p$ and $d(x_n,p) < B$ for all n, there is a point of accumulation of $\{x_n\}$ in M.

It may be seen that without requiring $x_n \leqslant q \ll p$ (or $x_n \leqslant q \ll p$) for some $q \in M$ in Definition 5.10, Minkowski space-time fails to be finitely compact (cf. Figure 5.3).

For globally hyperbolic space-times, a characterization of finite compactness more reminiscent of condition (2) above for Riemannian manifolds may be given.

LEMMA 5.11 Let $(M,g)$ be globally hyperbolic. Then $(M,g)$ is finitely compact iff for each real constant $B > 0$, the set $\{x \in M : p \ll q \leqslant x,\ d(p,x) \leqslant B\}$ is compact for any $p,q \in M$ with $q \in I^+(p)$, and the set $\{x \in M : x \leqslant q \ll p,\ d(x,p) \leqslant B\}$ is compact for any $p,q \in M$ with $p \in I^+(q)$.

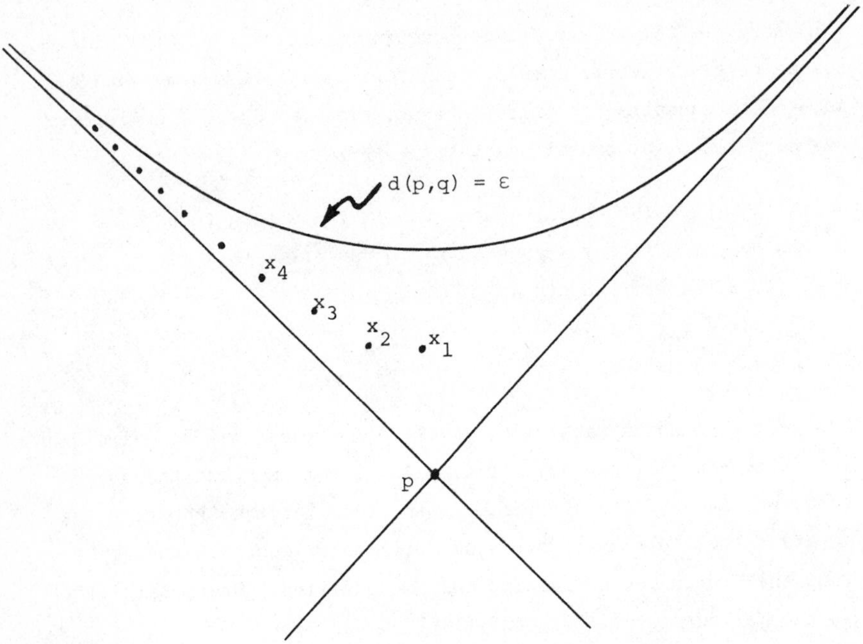


*Figure 5.3* A sequence $\{x_n\}$ in Minkowski space ($\mathbb{R}^2$ , $ds^2 = dx^2 - dy^2$) with $x_n \gg p$ for all n, $d(p,x_n) \longrightarrow 0$ as $n \longrightarrow \infty$, but such that $\{x_n\}$ has no point of accumulation, is shown.

*Proof.* This follows easily because the sets $J^+(q)$ are closed and the Lorentzian distance function is continuous since (M,g) is globally hyperbolic. □

Minkowski space-time is both timelike Cauchy complete and finitely compact. More generally, it may be shown that these concepts are equivalent for all globally hyperbolic space-times [cf. Beem (1976b, pp. 343-344)].

THEOREM 5.12 If (M,g) is globally hyperbolic, then (M,g) is finitely compact iff (M,g) is timelike Cauchy complete. Also if (M,g) is globally hyperbolic and nonspacelike geodesically complete, then (M,g) is finitely compact and timelike Cauchy complete.

REMARK 5.13 Even for the class of globally hyperbolic space-times, finite compactness, or equivalently, timelike Cauchy completeness, does *not* imply timelike geodesic completeness. Indeed, Geroch's example given in Figure 5.2 is a timelike geodesically incomplete, globally hyperbolic space-time which is finitely compact.

## 5.4 IDEAL BOUNDARIES

In this section we give brief descriptions of the b-boundary and the causal boundary for a space-time. Further details may be found in Hawking and Ellis (1973, Sections 6.8 and 8.3) or Dodson (1978).

The *b-boundary* of a space-time (M,g) will be denoted by $\partial_b M$. This boundary is formed by defining a certain positive definite metric on the bundle of linear frames L(M) over M, taking the Cauchy completion of L(M), then using the newly formed ideal points of L(M) to obtain ideal points of M. The b-boundary is particularly useful in telling whether or not some points have been removed from the space-time. Somewhat unfortunately the b-boundary often consists of just a single point [cf. Bosshard (1976), Johnson (1977)]. This boundary is not invariant under conformal changes and also is not directly related to the causal structure of (M,g). A recent discussion of the merits and demerits of the b-boundary and geodesic incompleteness may be found in the review article of Tipler, Clarke, and Ellis (1980).

Recall that a curve $\gamma : [0,a) \longrightarrow M$ is said to be *b-incomplete* if it has finite generalized affine parameter (cf. Section 5.2). Any curve $\gamma : [0,a) \longrightarrow M$ which is both b-incomplete and inextendible to t = a defines a point of $\partial_b M$ corresponding to $\gamma(a)$. In Minkowski space-time, generalized affine parameter values along a curve can be made to correspond to Euclidean arc length. Thus Minkowski space-time has an empty b-boundary and each b-incomplete curve in Minkowski space-time has an endpoint in the space-time.

The *causal boundary* of a space-time (M,g) will be denoted by $\partial_c M$. This boundary is constructed using the causal structure of the space-time. Thus it is invariant under conformal changes. We will only be interested in using this boundary for strongly causal space-times.

The causal boundary is formed using indecomposable past (resp., future) sets which do not correspond to the past (resp., future) of any point of M. A *past* (resp., *future*) *set* A is a subset of M such that $I^-(A) \subset A$ [resp., $I^+(A) \subset A$]. The open past (resp., future) sets are characterized by $I^-(A) = A$ [resp., $I^+(A) = A$]. An *indecomposable past set* (IP) is an open past set that cannot be written as a union of two proper subsets both of which are open past sets. An *indecomposable future set* (IF) is defined dually.

A *terminal indecomposable past set* (TIP) is a subset A of M such that

(1) A is an indecomposable past set.

(2) A is not the chronological past of any point $p \in M$.

A *terminal indecomposable future set* (TIF) is defined dually. The causal boundary $\partial_c M$ is formed using TIPs and TIFs after making certain identifications which will be described below [cf. Hawking and Ellis (1973, pp. 218-221)]. These identifications allow the topology of M to be extended to $M^* = M \cup \partial_c M$ in such a way that the causal completion of M is Hausdorff.

The use of TIPs and TIFs to represent ideal points of the causal boundary of (M,g) is illustrated in Figure 5.4.

We now show that a TIP may be represented as the chronological past of a future inextendible timelike curve. This result is due to Geroch, Kronheimer, and Penrose (1972, p. 551).

PROPOSITION 5.14 A subset W of the strongly causal space-time (M,g) is a TIP iff there exists a future-directed and future-inextendible timelike curve $\gamma$ such that $W = I^-(\gamma)$.

*Proof*. Assume there is a future-inextendible timelike curve $\gamma$ with $W = I^-(\gamma)$. Using the strong causality of (M,g), it follows that if W is an IP, then W is a TIP. To show W is an IP, assume that $W = U_1 \cup U_2$ for nonempty open past sets $U_1$ and $U_2$ such that

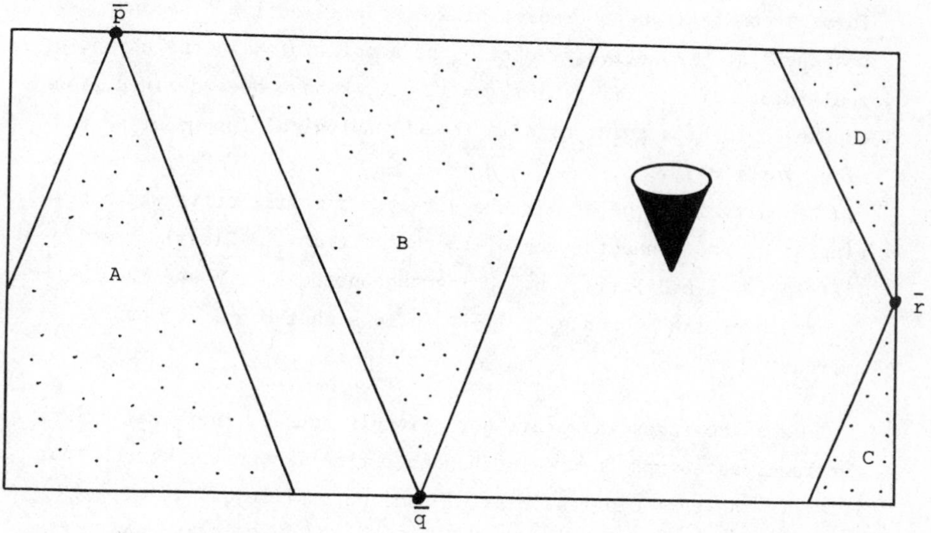


*Figure 5.4* The ideal point $\bar{p}$ in $\partial_c M$ is represented by the terminal indecomposable past set A and the ideal point $\bar{q}$ is represented by the terminal indecomposable future set B. The point $\bar{r}$ is represented by both the set C which is a TIP and the set D which is a TIF.

neither is a subset of the other. Choose $r_1 \in U_1 - U_2$ and $r_2 \in U_2 - U_1$. There must exist points $r_i' \in \gamma$ such that $r_i \in I^-(r_i')$ for $i = 1, 2$ because $U_1 \cup U_2 = I^-(\gamma)$. However, whichever $U_i$ contains the futuremost of $r_1'$ and $r_2'$ must then contain all four of $r_1$, $r_2$, $r_1'$ and $r_2'$. This contradicts either the definition of $r_1$ or of $r_2$.

On the other hand, assume that W is a TIP. If p is any point of W, then $W = [W \cap I^+(p)] \cup [W - I^+(p)]$ and thus $W = I^-(W \cap I^+(p)) \cup I^-(W - I^+(p))$ since W is a past set. Since W is an IP, either $W = I^-(W \cap I^+(p))$ or $W = I^-(W - I^+(p))$. Consequently, as $p \notin I^-(W - I^+(p))$, we have $W = I^-(W \cap I^+(p))$. Thus given any $q \neq p$ in W, there must be some point r in W which is in the chronological future of both p and q. Inductively, for each finite subset of W, there exists some point of W in the chronological future of each point of the subset. Now choose a sequence of points $p_n$ which forms a countable dense subset of W. We will define a second sequence $q_n$ inductively. Let $q_0$ be a point of W in the chronological future of $p_0$. If $q_i$ for i = 1, ..., k have been defined, then choose $q_k$ to be a point of W in the chronological future of $p_k$ and of $q_i$ for i = 1, ..., k - 1.

Finally let $\gamma$ be any future-directed timelike curve which begins at $q_0$ and connects each $q_i$ to the next $q_{i+1}$. Clearly, each $p_n$ lies in $I^-(\gamma)$ and $I^-(\gamma) \subset W$. Using the openness of W and the redenseness of the sequence $\{p_n\}$, it follows that $W = I^-(\gamma)$ as required. □

In space-times which are not strongly causal, there may exist future-directed and future-inextendible timelike curves $\gamma$ such that $I^-(\gamma)$ is the chronological past of some point [i.e., $I^-(\gamma)$ is *not* a TIP]. Consider, for example, the cylinder $\mathbb{R}^1 \times S^1$ with the flat metric $ds^2 = dt\, d\theta$ and the usual time orientation with the future corresponding to increasing t. The lower half of the cylinder $W = \{(t,\theta) : t < 0\}$ is an IP which can be represented as $I^-(\gamma)$ for a future-directed and future-inextendible timelike curve $\gamma$. However, W is not a TIP since W can be represented as the chronological past of any point on the circle t = 0. By restricting our attention to strongly causal space-times we avoid examples of this nature. For strongly causal space-times, the IPs which are not TIPs are in one-to-one correspondence with the points of M. The dual statement holds for IFs which are not TIFs.

We now define $\hat{M}$ (resp., $\check{M}$) to be the collection of all IPs (resp., IFs). Furthermore, let $M^\# = \hat{M} \cup \check{M}/\sim$ where for each $p \in M$ the element $I^-(p)$ of $\hat{M}$ is identified with the element $I^+(p)$ of $\check{M}$. The map $I^+ : M \longrightarrow M^\#$ given by $p \longrightarrow I^+(p)$ then identifies M with a subset of $M^\#$. Using this identification, the set $M^\#$ corresponds to M together with all TIPs and TIFs.

In order to define a topology on $M^\#$, first define for any $A \in \check{M}$ the sets $A^{int}$ and $A^{ext}$ by

$$A^{int} = \{V \in \hat{M} : V \cap A \neq \phi\}$$

and

$$A^{ext} = \{V \in \hat{M} : V = I^-(W) \text{ implies } I^+(W) \not\subset A\}$$

Similar definitions of $B^{int}$ and $B^{ext}$ are made for any $B \in \hat{M}$. The subbasis for a topology on $M^\#$ is then given by all sets of the form $A^{int}$, $A^{ext}$, $B^{int}$, and $B^{ext}$. The sets $A^{int}$ and $B^{int}$ are analogues of sets of the form $I^+(p)$ and $I^-(p)$, respectively. The sets $A^{ext}$ and $B^{ext}$ are analogues of $M - \overline{I^+(p)}$ and $M - \overline{I^-(p)}$, respectively.

The set $M^* = M \cup \partial_c M$ is now obtained from $M^\#$ with the above topology by identifying the smallest number of points of $M^\#$ necessary to obtain a Hausdorff space. In other words, $M^*$ is the quotient $M^\#/R_h$, where $R_h$ is the intersection of all equivalence relations R on $M^\#$ such that $M^\#/R$ is Hausdorff. The topological space $M^*$ contains a subset which may be identified with M via the map $I^+ : M \longrightarrow M^\#$ as above. Under this identification the given manifold topology on M agrees with the relative topology induced on $I^+(M)$ as a subset of $M^*$. Subtracting the subset $I^+(M)$ of $M^*$ from the space $M^*$, we obtain the causal boundary $\partial_c M$. This boundary $\partial_c M$ consists of TIPs and TIFs where identifications have been made as described above. Finally identifying M with the subset $I^+(M)$ of $M^*$, we have the desired decomposition $M^* = M \cup \partial_c M$.

## 5.5 LOCAL EXTENSIONS

In this section, extendibility and inextendibility of Lorentzian manifolds are defined. Also, two types of local extendibility are discussed. Most of the results of this section hold for Lorentzian manifolds which are not time orientable as well as for space-times.

DEFINITION 5.15 An *extension* of a Lorentzian manifold (M,g) is a Lorentzian manifold (M',g') together with an isometry $f : M \longrightarrow M'$ which maps M onto a proper open subset of M'. An analytic extension of (M,g) is an extension $f : (M,g) \longrightarrow (M',g')$ such that both Lorentzian manifolds are analytic and the map $f : M \longrightarrow M'$ is analytic. If (M,g) has no extensions, it is said to be *inextendible*.

Suppose that the Lorentzian manifold (M,g) has an extension $f : (M,g) \longrightarrow (M',g')$. Since M' is connected and f(M) is assumed to be open in M', it follows that $Bd(f(M)) = \overline{f(M)} - f(M) \neq \emptyset$, where $\overline{f(M)}$ denotes the closure of f(M) in M'. Because $Bd(f(M)) \neq \emptyset$ and the isometry $f : M \longrightarrow M'$ maps geodesics in M into geodesics in M' lying in f(M), it is easily seen that (M,g) cannot be timelike, null, or spacelike geodesically complete. Recalling that b-completeness and b.a. completeness both imply timelike geodesic completeness (cf. Section 5.2), we thus have the following criteria for Lorentzian manifolds to be inextendible.

PROPOSITION 5.16 A Lorentzian manifold (M,g) is inextendible if it is complete in any of the following ways:

(1) b-complete

(2) b.a. complete

(3) Timelike geodesically complete

(4) Null geodesically complete

(5) Spacelike geodesically complete

We now define two types of local extensions [cf. Clarke (1973, p. 207), Beem (1980), Hawking and Ellis (1973, p. 59)].

DEFINITION 5.17 Let (M,g) be a Lorentzian manifold.

(1) Suppose $\gamma : [0,a) \longrightarrow M$ is a b-incomplete curve which is not extendible to t = a in M. A *local b-boundary extension about* $\gamma$ is an open neighborhood $U \subset M$ of $\gamma$ and an extension (U',g') of $(U,g|U)$ such that the image of $\gamma$ in U' is $C^0$ extendible beyond t = a.

(2) A *local extension* of (M,g) is a connected open subset U of M having noncompact closure in M and an extension (U',g') of $(U,g|U)$ such that the image of U has compact closure in U'.

REMARK 5.18 This definition of local extension differs from the corresponding definition of local extension in Hawking and Ellis (1973, p. 59) in that U is required to be connected in Definition 5.17(2) but not in Hawking and Ellis (cf. Figures 5.5 and 5.6).

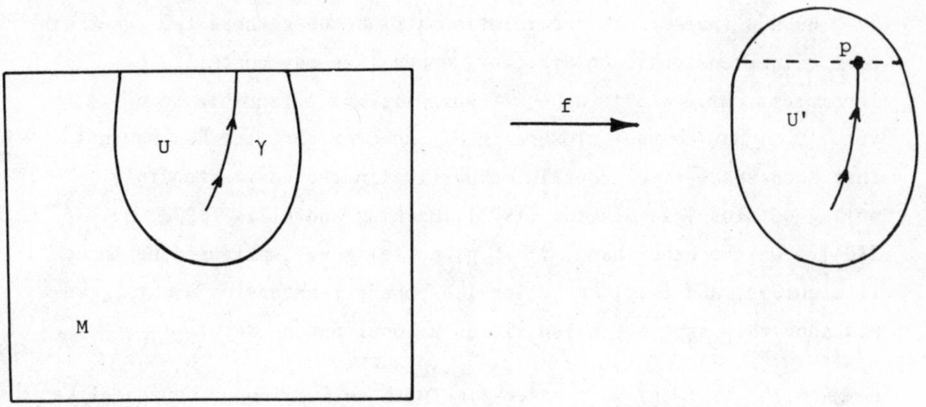


*Figure 5.5* Let $\gamma : [0,a) \longrightarrow M$ be a b-incomplete curve which is not extendible to $t = a$ in the space-time $(M,g)$. Assume that there is an isometry $f : (U,g|U) \longrightarrow (U',g')$ which takes $\gamma$ to a curve $f \circ \gamma$ having an endpoint p in U'. Then $f \circ \gamma$ may be continuously extended beyond $t = a$. Thus $(M,g)$ has a local b-boundary extension about $\gamma$.

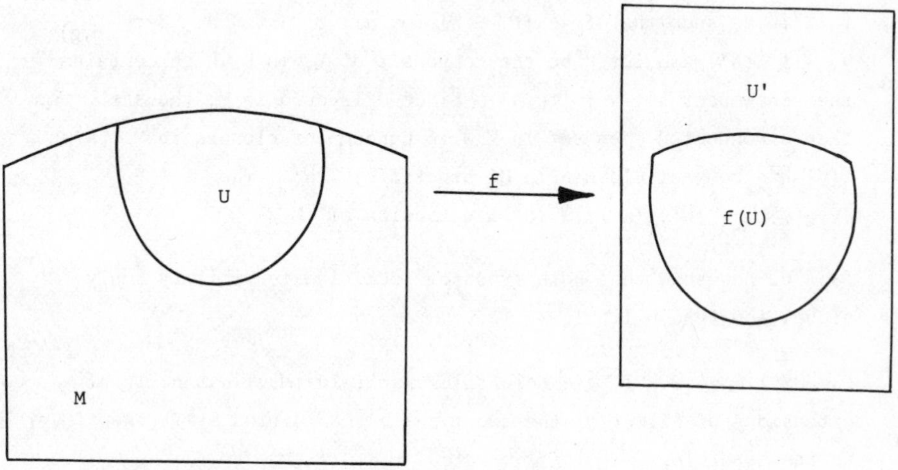


*Figure 5.6* Let U be a connected open subset of M having noncompact closure in M. A local extension is an isometry $f : (U,g|U) \longrightarrow (U',g')$ such that f(U) has compact closure in U'.

Minkowski space-time shows that even real analytic b-complete space-times may admit analytic local extensions (cf. Example 5.21). Thus a space-time may admit local extensions, but not admit local b-boundary extensions.

We now investigate the relationships between these two types of local extensions. An arbitrary space-time may contain a b-incomplete curve $\gamma : [0,a) \longrightarrow M$ which is not extendible to $t = a$, yet $\gamma[0,a)$ has compact closure in M. However, Schmidt has shown that such space-times contain compactly imprisoned inextendible null geodesics [cf. Schmidt (1973), Hawking and Ellis (1973, p. 280)]. On the other hand, if (M,g) contains no imprisoned nonspacelike curves, and (M,g) has a local b-boundary extension about $\gamma$, we now show this same extension yields a local extension.

LEMMA 5.19 If (M,g) is a space-time with no imprisoned nonspacelike curves which has a local b-boundary extension about $\gamma$, then (M,g) has a local extension.

*Proof.* Suppose that $f : (U,g|U) \longrightarrow (U',g')$ is a local b-boundary extension about $\gamma$. Then $f \circ \gamma : [0,a) \longrightarrow U'$ is extendible and $f \circ \gamma(t)$ converges to some $\overline{p} \in U'$ as $t \longrightarrow a$. Let W' be an open neighborhood of $\overline{p}$ in U' with compact closure in U'. Choose $t_0 \in [0,a)$ such that $f \circ \gamma(t) \in W'$ for all $t_0 \leqslant t < a$. Set $V_1 = f^{-1}(W')$ and let V be the component of $V_1$ in U which contains the noncompact set $\gamma \mid [t_0,a)$. Since U is open in M, the set V is then a connected open set in M with noncompact closure in M. Also f(V) has compact closure in U' since $f(V) \subset W'$. Thus $f \mid V : (V,g|V) \longrightarrow (U',g')$ is a local extension of (M,g). □

We now show that both types of local inextendibility imply global inextendibility.

PROPOSITION 5.20 If the Lorentzian manifold (M,g) has no local extensions of either of the two types of Definition 5.17, then (M,g) is inextendible.

*Proof.* Suppose (M,g) has an extension $F : (M,g) \longrightarrow (M',g')$. Let $\overline{p} \in Bd(F(M))$ and choose a geodesic $\sigma : [0,1] \longrightarrow (M',g')$ with $\sigma(0) \in F(M)$ and $\sigma(1) = \overline{p}$. Since F(M) is open in M' and $\overline{p} \notin F(M)$, there exists some $t_0 \in (0,1]$ such that $\sigma(t) \in F(M)$ for all $0 \leqslant t < t_0$ but $\sigma(t_0) \notin F(M)$. Then the curve $\gamma = F^{-1} \circ \sigma \mid [0,t_0) : [0,t_0) \longrightarrow M$

is b-incomplete, inextendible to $t = t_0$ in $M$, and has noncompact closure in $M$. Taking $U = M$, $U' = M'$, and $f = F$ in (1) of Definition 5.17, it follows that $(M,g)$ has a local b-boundary extension about $\gamma$. Taking $W$ to be any open subset about $\overline{p}$ with compact closure in $M'$, $U' = M'$, and $U$ to be the component of $F^{-1}(W)$ containing $\gamma$, we obtain a local extension $F \mid U : (U,g|U) \longrightarrow (M',g')$. □

The next example shows that Minkowski space-time has local extensions. Since Minkowski space-time is b-complete, this example shows that even though b-completeness is an obstruction to global extensions, it is *not* an obstruction to local extensions (cf. Proposition 5.16). This example is unusual in that it does *not* correspond to a local extension of $M$ over a point of either the b-boundary $\partial_b M$ or the causal boundary $\partial_c M$. It is a local extension of a set which extends to $i^0$ (cf. Figure 4.4).

EXAMPLE 5.21 Let $(M = \mathbb{R}^n, g)$ be n-dimensional Minkowski space-time and let $M' = \mathbb{R} \times T^{n-1}$ where $T^{n-1} = \{(\theta_2,\theta_3,\ldots,\theta_n) : 0 \leqslant \theta_i \leqslant 1\}$ is the $(n - 1)$-dimensional torus (using the usual identifications). We may define a Lorentzian metric $g'$ for $M'$ by $g' = (ds')^2 = -dt^2 + d\theta_2{}^2 + \cdots + d\theta_n{}^2$. Then $(M,g)$ is the universal Lorentzian covering space of $(M',g')$ with covering map $f : (M,g) \longrightarrow (M',g')$ given by

$$f(x_1,\ldots,x_n) = (x_1,x_2(\text{mod } 1),x_3(\text{mod } 1),\ldots,x_n(\text{mod } 1)).$$

Fix $\beta > 0$ and consider the curve $\gamma : [1,\infty) \longrightarrow M$ given by $\gamma(s) = (s^{-\beta},s,0,\ldots,0)$. Then $f \circ \gamma : [1,\infty) \longrightarrow M'$ is a spiral which is asymptotic to the circle $t = \theta_3 = \cdots = \theta_n = 0$ in $M'$. Let $U$ be an open tubular neighborhood about $\gamma$ in $M$ such that $U$ is contained in some open set $\{(x_1,\ldots,x_n) \in \mathbb{R}^n : 0 < x_1 < \alpha\}$ for some fixed $\alpha > 1$ and such that $f \mid U : U \longrightarrow M'$ is a homeomorphism onto its image (cf. Figure 5.7). Intuitively, the set $U$ must be chosen to be thinner as $s \longrightarrow \infty$ in order to satisfy the requirement $x_1 > 0$ for $(x_1,\ldots,x_n) \in U$. While $U$ does not have compact closure in Minkowski

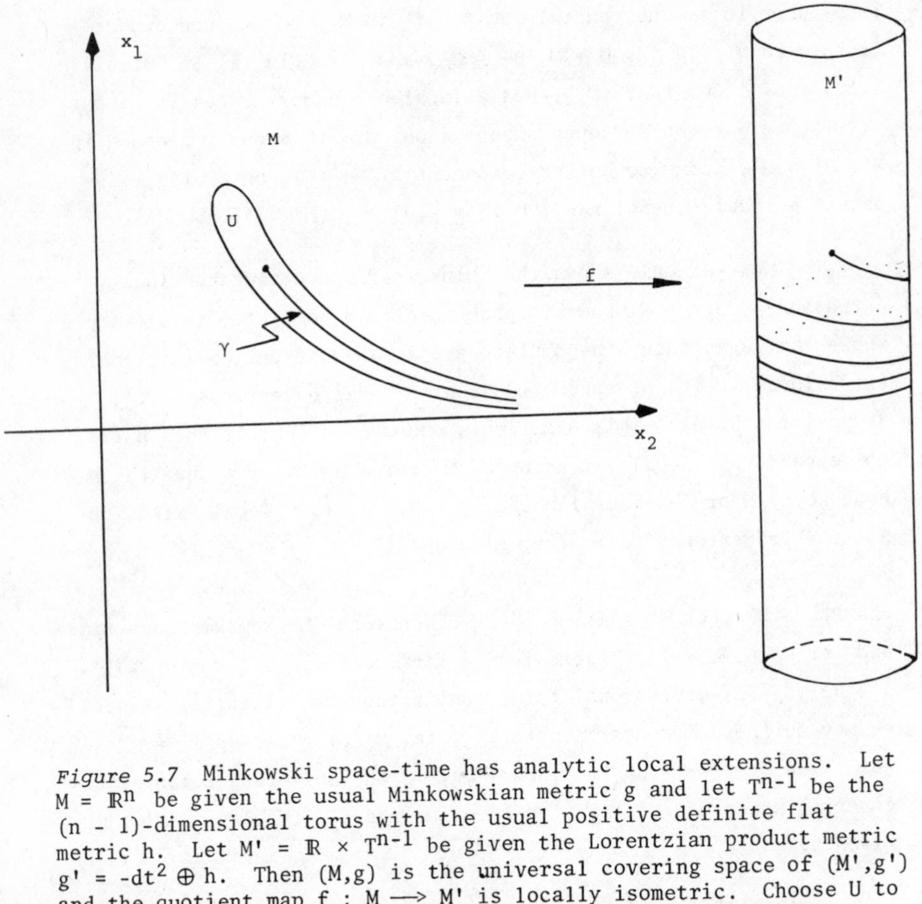


*Figure 5.7* Minkowski space-time has analytic local extensions. Let $M = \mathbb{R}^n$ be given the usual Minkowskian metric g and let $T^{n-1}$ be the (n - 1)-dimensional torus with the usual positive definite flat metric h. Let $M' = \mathbb{R} \times T^{n-1}$ be given the Lorentzian product metric $g' = -dt^2 \oplus h$. Then (M,g) is the universal covering space of (M',g') and the quotient map $f : M \longrightarrow M'$ is locally isometric. Choose U to be an open set in M about $\gamma(s) = (s^{-\beta}, s, \ldots, 0)$, $\gamma : [1,\infty) \longrightarrow M$, such that $f \mid U$ is one-to-one and f(U) has compact closure in M'. Then $f \mid U : (U, g|U) \longrightarrow (M',g')$ is an analytic local extension of Minkowski space. But this extension is *not* across points of $\partial_c M$ and *not* across points of $\partial_b M$.

space-time, f(U) does have compact closure in M' since f(U) is contained in the compact set $[0,\alpha] \times T^{n-1}$. Thus $f : (U, g|U) \longrightarrow (M',g')$ is an analytic local extension of Minkowski space-time. Notice that if $\gamma_1 : [0,a) \longrightarrow U$ is any curve with noncompact closure in M, then $f \circ \gamma_1$ cannot be extended to t = a in M'.

## 5.6 CURVATURE SINGULARITIES

Let $\partial M$ denote an ideal boundary of M (i.e., $\partial M$ represents either $\partial_b M$ or $\partial_c M$). A point $q \in \partial M$ is said to be a *regular boundary point* of M if there exists a global extension (M',g') of (M,g) such that q may be naturally identified with a point of M'. A regular boundary point may thus be regarded as being a removable singularity of M.

Let $\gamma : [0,a) \longrightarrow M$ be an inextendible curve such that $\gamma(a)$ corresponds to an ideal point of M. The curve $\gamma$ is said to define a *curvature singularity* [cf. Ellis and Schmidt (1977, p. 916)] if some component of $R_{abcd;e_1,\ldots,e_k}$ is *not* $C^0$ on [0,a] when measured in a parallelly propagated orthonormal basis along $\gamma$. A curvature singularity is an obstruction to a local b-boundary extension about $\gamma$ because if there is a local b-boundary extension about $\gamma$, then the curvature tensor and all of its derivatives measured in a parallelly propagated orthonormal basis must be continuous and hence converge to well-defined limits as $t \longrightarrow a^-$.

A b-boundary point $q \in \partial_b M$ which is neither a regular boundary point nor a curvature singularity is called a *quasi-regular singularity*. Clarke (1973, p. 208) has proven that if $\gamma : [0,a) \longrightarrow M$ is an inextendible b-incomplete curve which corresponds to a quasi-regular singularity, then there is a local b-boundary extension about $\gamma$. This shows that curvature singularities are the only real obstructions to local b-boundary extensions.

In general, it can be quite difficult to decide if a given space-time has local extensions of some type. However, for analytical local b-boundary extensions of analytic space-times, the situation is somewhat simpler (cf. Theorem 5.23).

For the proof of Theorem 5.23, it is useful to prove the following proposition about real analytic space-times and local isometries. Recall that a local isometry $F : M \longrightarrow M'$ is a map such that for each $p \in M$, there exists an open neighborhood U(p) of p on which F is an isometry. Thus local isometries are local diffeomorphisms but need not be globally one-to-one.

PROPOSITION 5.22 Let $(M,g)$ and $(M_1,g_1)$ be real analytic space-times of the same dimension and suppose that $F : M \longrightarrow M_1$ is a real analytic map. If M contains an open set U such that $F \mid U : U \longrightarrow M_1$ is an isometry, then F is a local isometry.

*Proof.* Let $W = \{m \in M : F_* v \neq 0$ for all $v \neq 0$ in $T_mM\}$ which is an open subset of M by the inverse function theorem. Since $F \mid U$ is an isometry, U is contained in W. Fix any $p \in U$ and let V be the path connected component of W containing p. We will establish the proposition by showing first that $F \mid V$ is a local isometry and second that $V = M$.

Let q be any point of V. Choose a curve $\gamma : [0,1] \longrightarrow V$ with $\gamma(0) = p$ and $\gamma(1) = q$. By the usual compactness arguments, we may cover $\gamma[0,1]$ with a finite chain of coordinate charts $(U_1,\phi_1)$, $(U_2,\phi_2)$, ..., $(U_k,\phi_k)$ such that each $U_i$ is simply connected, $F \mid U_i : U_i \longrightarrow M_1$ is an analytic diffeomorphism, and $p \in U_1 \subset U \cap V$, $q \in U_k$, and $U_i \cap U_{i+1} \neq \emptyset$ for each i with $1 \leqslant i \leqslant k - 1$. Since $U_1 \subset U \cap V$, we have $g = (F|U_1)^* g_1$ on $U_1$. Thus $g = (F|U_1)^* g_1$ on $U_1 \cap U_2$. Since $U_1 \cap U_2$ is an open subset of $U_2$ and F is a real analytic diffeomorphism of $U_2$ onto its image, it follows that $g = (F|U_2)^* g_1$ on $U_2$. Carrying on inductively, we obtain $g = (F|U_k)^* g_1$ on $U_k$; whence F is an isometry in the open neighborhood $U_k$ of q. Thus $F \mid V : V \longrightarrow M_1$ is a local isometry.

It remains to show that $V = M$. Suppose $V \neq M$. Choose any point $r_1 \in M - V$. Let $\gamma_1 : [0,1] \longrightarrow M$ be a smooth curve with $\gamma(0) = p$ and $\gamma(1) = r_1$. There is a smallest $t_0 \in [0,1]$ such that $r = \gamma(t_0) \in M - V$. Then F restricted to the neighborhood V of $\gamma_1[0,t_0)$ is a local isometry. It suffices to show that $r \in V$ to obtain the desired contradiction. Since $r \in M - V$, there exists a tangent vector $x \neq 0$ in $T_rM$ with $F_*x = 0$. Let X be the unique parallel field along $\gamma$ with $X(t_0) = x$. Then $F_* \circ X$ is a parallel field along $F \circ \gamma_1 : [0,t_0) \longrightarrow M_1$ since F is a local isometry in a neighborhood of $\gamma_1 : [0,t_0) \longrightarrow M$. But since F is smooth,

$F_*x = F_*X(t_0) = \lim_{t \to t_0^-} F_*X(t)$. Because $F_* \circ X$ is a parallel vector field for all t with $0 \leq t \leq t_0$, it follows that $\lim_{t \to t_0^-} F_*X(t) \neq 0$. Hence $F_*x \neq 0$. Thus F is nonsingular at the point r, whence $r \in V$, in contradiction. □

We are now ready to turn to the proof of Theorem 5.23 on local b-boundary extensions of real analytic space-times.

THEOREM 5.23 Suppose (M,g) is an analytic space-time with no imprisoned nonspacelike curves which has an analytic local b-boundary extension about $\gamma : [0,a) \longrightarrow M$. Then there are timelike, null, and spacelike geodesics of finite affine parameter which are inextendible in one direction and which do not correspond to curvature singularities.

The proof of Theorem 5.23 will involve two lemmas.

LEMMA 5.24 Suppose (M,g) is an analytic space-time with no imprisoned nonspacelike curves which has an analytic local b-boundary extension about $\gamma : [0,a) \longrightarrow M$. Then (M,g) has an incomplete geodesic.

*Proof*. Let $f : (U,g|U) \longrightarrow (U',g')$ be an analytic extension about $\gamma$. We may assume U contains the image of $\gamma$. Also $f \circ \gamma$ is extendible in U'. Thus $f \circ \gamma(t) \longrightarrow p \in U'$ as $t \longrightarrow a^-$. Let W' be a neighborhood of p such that W' is a convex normal neighborhood of each of its points. Then $\exp_x^{-1} : W' \longrightarrow T_xU'$ is a diffeomorphism for each fixed $x \in W'$. Assume $t_0$ is chosen with $f \circ \gamma(t) \in W'$ for all $t_0 \leq t < a$. Set $q = \gamma(t_0)$ and $r = f(q)$. Then $H = \exp_q \circ f_{*_q}^{-1} \circ \exp_r^{-1} : W' \longrightarrow M$ is analytic and is at least defined near r. The map H takes geodesics starting at r to geodesics starting at q and H preserves lengths along these geodesics. In fact, H agrees with $f^{-1}$ near r. The map H need not be one-to-one since $\exp_q$ is not necessarily one-to-one. Because the domain of $\exp_q$ is a union of line segments starting at the origin of $T_qM$, the domain V' of H must be some subset of W' which is a union of geodesic segments starting at r. Hence the set V' fails to be all of W' only when $\exp_q : T_qM \longrightarrow M$ is defined on a proper subset of $T_qM$ which does not include all of the image ${f_*}_q^{-1} \circ \exp_r^{-1}(W')$. Thus if we show $V' \neq W'$, there is some incomplete, inextendible geodesic starting at q. But the analytic maps H and $f^{-1}$ must agree on the component of $f(U) \cap V'$ which contains r. This implies $V' \neq W'$. Otherwise, H and $f^{-1}$ would agree on $f(U) \cap W'$ and hence on a neighborhood of $f \circ \gamma[t_0,a)$. This yields $H \circ f \circ \gamma = \gamma$ for $t_0 \leqslant t < a$ and implies $\gamma$ is extendible in M across the point H(p), in contradiction. □

We will continue with the same notation in the next lemma.

LEMMA 5.25 The map $H : V' \longrightarrow M$ is a local isometry.

*Proof*. The space-time (M,g) is analytic, (U',g') is analytic and H is analytic. Furthermore, H agrees with the isometry $f^{-1}$ near r and H is defined on an arcwise connected set V'. Thus Proposition 5.22 implies H is a local isometry. □

We are now ready to complete the

*Proof of Theorem 5.23* There are three cases to consider corresponding to incomplete timelike, null, and spacelike geodesics. We only give the proof for the timelike case. Let U, U', f, etc., be as in Lemmas 5.24 and 5.25. Assume without loss of generality that there is some point $x \in W'$ such that in a chronological ordering on W we have $x \ll p$ and $x \ll f \circ \gamma(t)$ for all $t_0 \leqslant t < a$ (cf. Figure 5.8).

If $x \notin V'$, let $\alpha$ be the geodesic segment in W' from r to x. Then H takes $\alpha \cap V'$ to an inextendible, incomplete timelike geodesic starting at q. Lemma 5.25 implies that this geodesic does not correspond to a curvature singularity.

If $x \in V'$, let $y = H(x)$ and define $H' = \exp_y \circ {H_*}_x \circ \exp_x^{-1} : W' \longrightarrow M$. The map H' is defined on some subset V" of W'. It is a local isometry for the same reasons that H is a local isometry and H' agrees with both H and $f^{-1}$ near r. The set V" cannot contain

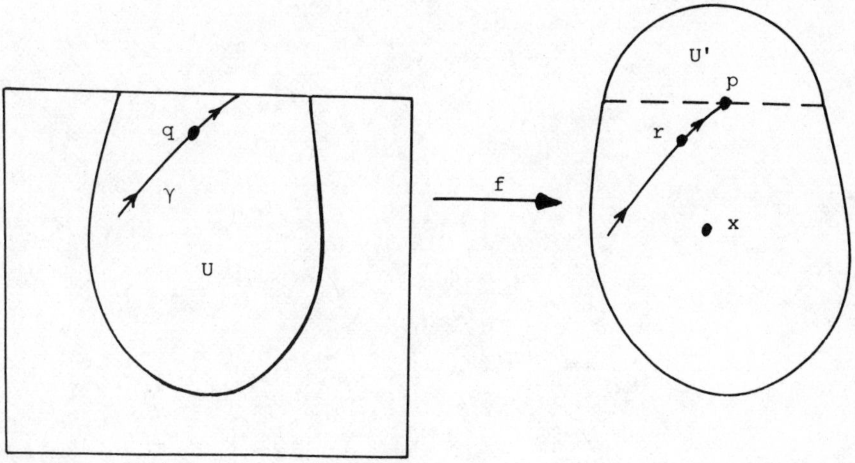


*Figure 5.8* In the proof of Theorem 5.23, the map $f : (U,g|U) \longrightarrow (U',g')$ is an isometry which is an analytic local b-boundary extension about $\gamma$. The point $p \in U'$ is the endpoint of $f \circ \gamma$ in U'. In this figure $f(q) = r$, and all points of $f \circ \gamma$ between r and p are in the chronological future of x.

all of $f \circ \gamma$ for $\gamma$ on $t_0 \leqslant t < a$ since this would yield an endpoint $H'(p)$ of $\gamma$ in M. Using $x \ll f \circ \gamma(t)$ for $t_0 \leqslant t < a$, we conclude that there is an inextendible incomplete timelike geodesic starting at y in M which does not correspond to a curvature singularity. □

REMARK 5.26 There are examples of $C^\infty$ space-times which are both geodesically complete and locally b-extendible [cf. Beem (1976c, p. 506)]. Hence the analyticity in the hypothesis of Theorem 5.23 cannot be replaced by a $C^\infty$ assumption.

COROLLARY 5.27 Let (M,g) be an analytic space-time with no imprisoned nonspacelike curves such that each timelike geodesic $\gamma : [0,a) \longrightarrow M$ which is inextendible to $t = a$ is either complete (i.e., $a = \infty$) in the indicated direction or else corresponds to a curvature singularity. Then (M,g) has no analytic local b-boundary extensions.

Chapter 6

# STABILITY OF ROBERTSON-WALKER SPACE-TIMES

In proving singularity theorems in general relativity, it is important to use hypotheses that hold not just for the given "background" Lorentzian metric $g_0$ for M, but in addition for all metrics g for M sufficiently close to $g_0$. Not only does the imprecision of astronomical measurements mean that the Lorentzian metric of the universe cannot be determined exactly, but also cosmological assumptions like the spatial homogeneity of the universe hold only approximately. Nevertheless, if an incompleteness theorem can be obtained for the idealized model $(M,g_0)$ using hypotheses valid for all metrics g for M in an open neighborhood of $g_0$, all space-times (M,g) with g sufficiently close to $g_0$ will also be incomplete. Hence if the model is believed to be sufficiently accurate, conclusions valid for the model are also valid for the actual universe.

Recall that Lor(M) denotes the space of all Lorentzian metrics for a given manifold M and that Con(M) denotes the quotient space formed by identifying all pointwise globally conformal metrics $g_1 = \Omega g_2$ for M, $\Omega : M \longrightarrow (0,\infty)$ smooth. Let $\tau : \mathrm{Lor}(M) \longrightarrow \mathrm{Con}(M)$ denote the natural projection map which assigns to each Lorentzian metric g for M the set $\tau(g) = \overline{g}$ of all Lorentzian metrics for M pointwise globally conformal to g. Given $\overline{g} \in \mathrm{Con}(M)$, set $C(M,g) = \tau^{-1}(\overline{g}) \subset \mathrm{Lor}(M)$. It is customary in general relativity to say that a curvature or causality condition for a space-time $(M,g_0)$ is *$C^r$ stable* in Lor(M) [resp., Con(M)], if the validity of the condition for $(M,g_0)$ implies the validity of the condition for all g in a $C^r$ open neighborhood of $g_0$ in Lor(M) [resp., Con(M)].

After the singularity theorems described in Chapter 8 of Hawking and Ellis (1973) were obtained, it was of interest to study the $C^r$ stability of conditions such as the existence of closed trapped surfaces, positive nonspacelike Ricci curvature, and geodesic completeness, which played such a key role in these singularity theorems. Geroch (1970) established the stability in the interval topology of global hyperbolicity in Con(M). Then Lerner (1973) made a thorough study of the stability on Lor(M) and Con(M) of causality and curvature conditions useful in general relativity. In particular, Lerner noted that the interval and quotient topologies for Con(M) coincide. Hence Geroch's stability result for global hyperbolicity holds for Con(M) in the quotient topology and thus automatically in Lor(M). Lerner also raised the following question (1973, p. 35) about the Robertson-Walker big bang models $(M,g_0)$: under small $C^2$ perturbations of the metric, does each nonspacelike geodesic remain incomplete?

The primary purpose of Chapter 6 is to answer this question affirmatively. In Section 6.1, we define the fine $C^r$ topologies and the interval topology for Con(M). Then we review stability properties of Lor(M) and Con(M) established by Geroch (1970) and Lerner (1973). In Section 6.2, using the "Euclidean norm,"

$$\|\xi - \eta\|_2 = \left\{ \sum_{i=1}^{2n} [\tilde{x}_i(\xi) - \tilde{x}_i(\eta)]^2 \right\}^{1/2}$$

induced on $(TM|_U, \tilde{x})$ by a coordinate chart $(U,x)$ for M and standard estimates from the theory of systems of ordinary differential equations in $\mathbb{R}^n$, we obtain estimates for the behavior of geodesics in $(U,x)$ under $C^1$ metric perturbations. In Section 6.3, we apply these estimates to coordinate charts adapted to the product structure $M = (a,b) \times_f H$ of a Robertson-Walker space-time (cf. Definition 4.10) to study the stability of geodesic incompleteness for such space-times. We show (Theorem 6.15) that if $((a,b) \times_f H, g_0)$ is a Robertson-Walker space-time with $a > -\infty$, then there is a fine $C^0$ neighborhood $U(g_0)$ of $g_0$ in Lor(M) such that *all* timelike geodesics of *each* space-time (M,g) are past incomplete for all $g \in U(g_0)$. If we assume $b < \infty$ as well, we may obtain (Theorem 6.16) *both* future and past incompleteness of all timelike geodesics for all $g \in U(g_0)$. A similar result (Theorem 6.19) may be established for null geodesic incompleteness using the $C^1$ topology on Lor(M). Combining these results yields the $C^1$ stability of past nonspacelike geodesic incompleteness for Robertson-Walker space-times.

At the end of Section 2.6 we have discussed the relationship between the choice of warping function $f : (a,b) \longrightarrow (0,\infty)$ and the nonspacelike geodesic incompleteness of a given Lorentzian warped product space-time $(a,b) \times_f H$.

The results in Sections 6.2 and 6.3 have been given in Beem and Ehrlich (1981).

## 6.1 STABLE PROPERTIES OF LOR(M) AND CON(M)

An equivalence relation C may be placed on the space Lor(M) of Lorentzian metrics for M by defining $g_1, g_2 \in \text{Lor}(M)$ to be equivalent if there exists a smooth conformal factor $\Omega : M \longrightarrow (0,\infty)$ such that $g_1 = \Omega g_2$. As in Chapter 1, we will denote the equivalence class of g in Lor(M) by C(M,g). The quotient space Lor(M)/C of equivalence classes will be denoted by Con(M). There is then a natural projection map $\tau : \text{Lor}(M) \longrightarrow \text{Con}(M)$ given by $\tau(g) = C(M,g)$.

The fine $C^0$ topology (cf. Section 2.2) on Lor(M) induces a *quotient topology* on Con(M) as usual. Namely, define a subset A of Con(M) to be open in this topology if the inverse image $\tau^{-1}(A)$ is open in the fine $C^0$ topology on Lor(M).

Con(M) may also be given the *interval topology* [cf. Geroch (1970, p. 447)]. Recall from our discussion of stable causality in Section 2.2 that a partial ordering may be defined on Lor(M) by $g_1 < g_2$ if $g_1(v,v) \leq 0$ implies $g_2(v,v) < 0$ for all $v \neq 0$ in TM. It may then be checked that $g_1, g_2 \in \text{Lor}(M)$ satisfy $g_1 < g_2$ iff $g_1' < g_2'$ for all $g_1' \in C(M,g_1)$ and $g_2' \in C(M,g_2)$. Thus the partial ordering $<$ for Lor(M) projects to a partial ordering on Con(M) which will also be denoted by $<$. A subbasis for the *interval topology* on Con(M) is then given by all sets of the form

$$\{C(M,g) \in Con(M) : C(M,g_1) < C(M,g) < C(M,g_2)\}$$

where $g_1$ and $g_2$ are arbitrary Lorentzian metrics for M with $g_1 < g_2$. It is known that the quotient and interval topologies agree on Con(M) [cf. Lerner (1973, p. 23)]. Thus, intuitively, two conformal classes $C(M,g_1)$ and $C(M,g_2)$ are close in either of these topologies on Con(M) iff at all points of M, the metrics $g_1$ and $g_2$ have light cones which are close in $T_pM$.

A property defined on Lor(M) which holds on a $C^r$ open subset of Lor(M) is said to be *$C^r$ stable*. Also a property defined on Lor(M) which is invariant under the conformal relation C is said to be *conformally stable* if it holds for an open set of equivalence classes in the quotient (or interval) topology on Con(M). The continuity of the projection map $\tau$ : Lor(M) $\longrightarrow$ Con(M) implies that any conformally stable property defined on Lor(M) is also $C^0$ stable on Lor(M). Furthermore, since the fine $C^r$ topology is strictly finer than the fine $C^s$ topology on Lor(M) for $r > s$, any conformally stable property defined on Lor(M) is also $C^r$ stable for all $r \geq 0$.

EXAMPLE 6.1 Stable causality is conformally stable and hence also $C^r$ stable for all $r \geq 0$. Indeed, a metric $g_0 \in$ Lor(M) may be defined to be stably causal if the property of causality is $C^0$ stable in Lor(M) at $g_0$.

A second example of a conformally stable property is furnished by a result of Geroch (1970, p. 448).

THEOREM 6.2 Global hyperbolicity is conformally stable and hence $C^r$ stable in Lor(M) for all $r \geq 0$.

It may also be shown that if S is a smooth Cauchy surface for (M,g), there exists a $C^0$ neighborhood U of g in Lor(M) such that if $g_1 \in U$, then S is a Cauchy hypersurface for $(M,g_1)$ [cf. Geroch (1970, p. 448)].

The next two results have been obtained by Lerner (1973).

PROPOSITION 6.3 If (M,g) is a Lorentzian manifold such that $g(v,v) \leqslant 0$ and $v \neq 0$ in TM imply $\mathrm{Ric}(g)(v,v) > 0$, then there is a fine $C^2$ neighborhood U(g) of g in Lor(M) such that for all $g_1 \in U(g)$, the relations $g_1(v,v) \leqslant 0$ and $v \neq 0$ in TM imply $\mathrm{Ric}(g_1)(v,v) > 0$.

PROPOSITION 6.4 Geodesic completeness if a $C^r$ stable property in Lor(M) for all $r \geqslant 2$.

By Proposition 5.16, a geodesically complete Lorentzian manifold is globally inextendible. Thus Propositions 5.16 and 6.4 imply that if (M,g) is geodesically complete, there exists a fine $C^2$ neighborhood U(g) of g in Lor(M) such that $(M,g_1)$ is inextendible for each $g_1 \in U(g)$. Also, if (M,g) is an extendible Lorentzian manifold and U(g) is any fine $C^r$ neighborhood of g in Lor(M), there exists $g_1 \in U(g) - \{g\}$ such that $(M,g_2)$ is extendible.

In this chapter, the manifold M is always fixed. However, one-parameter families $(M_\lambda, g_\lambda)$ of manifolds and Lorentzian metrics have been considered in general relativity [cf. Geroch (1969)].

## 6.2 THE $C^1$ TOPOLOGY AND GEODESIC SYSTEMS

If (M,g) is an arbitrary Lorentzian manifold, then metrics in Lor(M) which are close to g in the fine $C^1$ topology have geodesic systems which are close to the geodesic system of g. The purpose of this section is to give a more analytic formulation of this concept needed for our investigation of the $C^1$ stability of null geodesic incompleteness for Robertson-Walker space-times in Section 6.3.

We begin by recalling a well-known estimate from the theory of ordinary differential equations [cf. Birkhoff and Rota (1969, p. 155)]. We will always use $\|x\|_2$ to denote the Euclidean norm $[\Sigma_{i=1}^m x_i^2]^{1/2}$ of the point $x = (x_1, x_2, \ldots, x_m) \in \mathbb{R}^m$.

PROPOSITION 6.5 Suppose that $f = (f_1,\dots,f_m)$ and $h = (h_1,\dots,h_m)$ are continuous functions defined on a common domain $D \subset \mathbb{R} \times \mathbb{R}^m$ and suppose that f satisfies the Lipschitz condition

$$\|f(s,x) - f(s,\bar{x})\|_2 \leq L\|x - \bar{x}\|_2$$

for all $(s,x),(s,\bar{x}) \in D$. Let $x(s) = (x_1(s),x_2(s),\dots,x_m(s))$ and $y(s) = (y_1(s),\dots,y_m(s))$ be solutions for $0 \leq s \leq b$ of the differential equations

$$\frac{dx}{ds} = f(s,x) \qquad \frac{dy}{ds} = h(s,y)$$

respectively. Then if $\|f(s,x) - h(s,x)\|_2 \leq \varepsilon$ for all $(s,x) \in D$ with $0 \leq s \leq b$, we have

$$\|x(s) - y(s)\|_2 \leq \|x(0) - y(0)\|_2 e^{Ls} + \frac{\varepsilon}{L}(e^{Ls} - 1)$$

for all $0 \leq s \leq b$.

Now let M be a smooth manifold and let (U,x) be any coordinate chart for M. We may obtain an associated coordinate chart $\tilde{x} = (x_1,x_2,\dots,x_n,x_{n+1},\dots,x_{2n})$ for $TM|_U$ as follows. Let $\partial/\partial x_1, \dots, \partial/\partial x_n$ be the basis vector fields defined on U by the local coordinates $x = (x_1,\dots,x_n)$. Given $v \in T_qM$ for $q \in U$, we may write $v = \Sigma_{i=1}^n a_i \, \partial/\partial x_i|_q$. Then $\tilde{x}(v)$ is defined to be $\tilde{x}(v) = (x_1(v),x_2(v),\dots,x_n(v),a_1,a_2,\dots,a_n)$. These coordinate charts may then be used to define Euclidean coordinate distances on U and $TM|_U$. Explicitly, given $p,q \in U$ and $v,w \in TM|_U$, set

$$\|p - q\|_2 = \left\{\sum_{i=1}^{n} [x_i(p) - x_i(q)]^2\right\}^{1/2}$$

and

$$\|v - w\|_2 = \left\{\sum_{i=1}^{2n} [x_i(v) - x_i(w)]^2\right\}^{1/2}$$

respectively. Also if $r \geq 0$ is given, we will use the notation

$\|g_1 - g_2\|_{r,U} < \delta$ for $g_1, g_2 \in \text{Lor}(M)$ and a positive constant $\delta > 0$ to mean that calculating with the local coordinates $(U,x)$, all the corresponding entries of the two metric tensors and all their corresponding partial derivatives up to order r are $\delta$-close at each point of U.

We will denote the Christoffel symbols of the second kind for $g_1, g_2 \in \text{Lor}(M)$ by $\Gamma^i_{jk}(g_1)$ and $\Gamma^i_{jk}(g_2)$, respectively. Then for $a = 1, 2$, the geodesic equations in the coordinate chart $(U,x)$ for $(M,g_a)$ are given by

$$\begin{aligned} \frac{dx_i}{ds} &= x_{i+n} \\ \frac{dx_{i+n}}{ds} &= -\Gamma^i_{jk}(g_a)x_{j+n}x_{k+n} \end{aligned} \tag{6.1}$$

for $1 \leqslant i,j,k \leqslant n$ where we employ the Einstein summation convention throughout this chapter.

We will use the notation $\exp_q[g_a]$ for the exponential map at $q \in (M,g_a)$, $a = 1, 2$. If $v \in TM|_U$, then $s \longrightarrow \exp_q[g_a](sv)$ is the solution of (6.1) in U with initial conditions $(q,v)$ for $(M,g_a)$. In order to apply Proposition 6.5 to these exponential maps, we identify $TM|_U$ with a subset of $\mathbb{R}^{2n}$ using the coordinate chart $(TM|_U, \tilde{x})$ and define $f(s,X) = f(X)$ and $h(s,X) = h(X)$ by

$$f_i(X) = h_i(X) = x_{i+n}$$

$$f_{i+n}(X) = -\Gamma^i_{jk}(g_1)x_{j+n}x_{k+n}$$

and $$h_{i+n}(X) = -\Gamma^i_{jk}(g_2)x_{j+n}x_{k+n}$$

for $1 \leqslant i,j,k \leqslant n$ and $X = (x_1,\ldots,x_{2n}) \in \mathbb{R}^{2n}$.

LEMMA 6.6 Let $(U,x)$ be a local coordinate chart for the n-manifold M. Let $(p,v) \in TM|_U$ and assume that $c_1(s) = \exp_p[g_1](sv)$ lies in U for all $0 \leqslant s \leqslant b$. Given $\varepsilon > 0$, there exists a constant $\delta > 0$ such that $\|v - w\|_2 < \delta$ and $\|g_1 - g_2\|_{1,U} < \delta$ imply that $c_2(s) = \exp_q[g_2](sw)$ lies in U for all $0 \leqslant s \leqslant b$ and, moreover,

$$|(x_j \circ c_1)(s) - (x_j \circ c_2)(s)| < \varepsilon$$

and

$$|(x_j \circ c_1)'(s) - (x_j \circ c_2)'(s)| < \varepsilon$$

for all $1 \leqslant j \leqslant n$, $0 \leqslant s \leqslant b$.

*Proof.* Let $f(X)$ and $h(X)$ be defined as above. Then $X(s) = (c_1(s), c_1'(s))$ and $Y(s) = (c_2(s), c_2'(s))$ are solutions to the differential equations $dX/ds = f(X)$ and $dY/ds = h(Y)$, respectively. Choose $D_0$ to be an open set in $TM|_U$ about the image of the curve $X(s)$ such that $\overline{D}_0$ is compact. Then there exists a constant $L$ such that $f$ satisfies a Lipschitz condition $\| f(X) - f(\overline{X})\|_2 \leqslant L\| X - \overline{X}\|_2$ on $D_0$.

We may make the term $\| X(0) - Y(0)\|_2$ in the estimate of Proposition 6.5 as small as required by making $\| v - w\|_2$ small. Furthermore, since the Christoffel symbols depend only on the coefficients of the metric tensor and on their first partial derivatives, we may make $\| f(X) - h(X)\|_2$ as small as we wish on $D_0$ by requiring that $\| g_1 - g_2\|_{1,U}$ be small. Hence for a sufficiently small $\delta > 0$, Proposition 6.5 may be applied to guarantee that $c_2(s) \in U$ for all $0 \leqslant s \leqslant b$ and also to yield the estimate $\| X(s) - Y(s)\|_2 < \varepsilon$ for all $0 \leqslant s \leqslant b$. Consequently, $|X_i(s) - Y_i(s)| < \varepsilon$ for all $1 \leqslant i \leqslant 2n$ and $0 \leqslant s \leqslant b$. In view of (6.1), this establishes the desired inequalities. □

The following slightly more technical lemma needed in Section 6.3 follows directly from Lemma 6.6 by using the triangle inequality and the continuity of the geodesic solution $X(s) = (c_1(s), c_1'(s))$.

LEMMA 6.7 Let $(U,x)$ be a local coordinate chart on the $n$ manifold $M$. Suppose that $c_1(v) = \exp_p[g_1](sv)$ lies in $U$ for all $0 \leqslant s \leqslant b$. Let $\varepsilon > 0$ and $s_1$ with $0 < s_1 < b$ be given. Then there exists a constant $\delta > 0$ such that if $\| v - w\|_2 < \delta$, $\| g_1 - g_2\|_{1,U} < \delta$ and $|s_0 - s_1| < \delta$, the geodesic $c_2(s) = \exp_q[g_2](sw)$ lies in $U$ for all $0 \leqslant s \leqslant b$ and, moreover,

$$|(x_j \circ c_1)(s_1) - (x_j \circ c_2)(s_0)| < \varepsilon \tag{6.2}$$

and

$$|(x_j \circ c_1)'(s_1) - (x_j \circ c_2)'(s_0)| < \varepsilon \tag{6.3}$$

for all $1 \leqslant j \leqslant n$.

Furthermore, if $(x_1 \circ c_1)'(s_1) \neq 0$, then the constant $\delta > 0$ may be chosen such that

$$1 - \varepsilon < \left| \frac{(x_1 \circ c_2)'(s_0)}{(x_1 \circ c_1)'(s_1)} \right| < 1 + \varepsilon$$

## 6.3 STABILITY OF GEODESIC INCOMPLETENESS FOR ROBERTSON-WALKER SPACE-TIMES

In this section, we investigate the stability in the space of Lorentzian metrics of the nonspacelike geodesic incompleteness of Robertson-Walker space-times $M = (a,b) \times_f H$ (cf. Definition 4.10). It turns out, however, that the proof of the $C^0$ stability of timelike geodesic incompleteness only uses the homogeneity of the Riemannian factor $(H,h)$ and not the isotropy of $(H,h)$. Accordingly, we will formulate the results in the first portion of this section for the larger class of Lorentzian warped products $M = (a,b) \times_f H$ with $-\infty \leqslant a < b \leqslant +\infty$ and $(H,h)$ a homogeneous Riemannian manifold. We will let $x_1 = t$ denote the usual coordinate on $(a,b)$ throughout this section.

In order to study the geodesic incompleteness of such space-times under metric perturbations, it is helpful to use coordinates adapted to the product structure. Fix $p = (t_1,h_1) \in (a,b) \times H$. Since the submanifold $\{t_1\} \times H$ of $M$ is spacelike, the Lorentzian metric $g$ for $M$ restricts to a positive definite inner product on the tangent space to this submanifold at $p$. Identifying $\{t_1\} \times H$ and $H$, we may use the orthonormal basis for the tangent space to $\{t_1\} \times H$ at $p$ to define Riemann normal coordinates $x_2, \ldots, x_n$ for $H$ in a neighborhood V of $h_1$. Then $(x_1, x_2, \ldots, x_n)$ defines a coordinate system for M on $(a,b) \times V$. By construction, g has the form $\mathrm{diag}\{-1,+1,\ldots,+1\}$ at p in these coordinates. Because the submanifold $\{t_1\} \times H$ is not necessarily totally geodesic if f is nonconstant, these coordinates are not necessarily normal coordinates. Nonetheless, the coordinates $(x_1, x_2, \ldots, x_n)$ are well adapted to the product structure since the level sets $x_1(t) = \lambda$ are just $\{\lambda\} \times V$. We will say that coordinates $(x_1, x_2, \ldots, x_n)$ constructed as above are *adapted at* $p \in M$ and call such coordinates *adapted coordinates*. It will also be useful to define adapted normal neighborhoods.

DEFINITION 6.8 A convex normal neighborhood U of (M,g) with compact closure $\overline{U}$ is said to be an *adapted normal neighborhood* if $\overline{U}$ is covered by adapted coordinates $(x_1, x_2, \ldots, x_n)$ which are adapted at some point of U such that the following hold:

(1) At every point of U, the components $g_{ij}$ of the metric tensor g expressed in the given coordinates $(x_1, x_2, \ldots, x_n)$ differ from the matrix $\mathrm{diag}\{-1,+1,\ldots,+1\}$ by at most 1/2.

(2) The metric g satisfies $g <_U g_0$, where $g_0$ is the Minkowskian metric $ds^2 = -2\, dx_1^2 + \cdots + dx_n^2$ for U (cf. the definition of stably causal in Section 2.2 for the notation $g <_U g_0$).

Thus on the neighborhood U of Definition 6.8, the Lorentzian metric g may be expressed as

$$g \mid U = -dx_1^2 + dx_2^2 + \cdots + k_{ij}\, dx_i\, dx_j$$

where the functions $k_{ij} : U \longrightarrow \mathbb{R}$ satisfy $|k_{ij}| \leq 1/2$ for all $1 \leq i,j \leq n$.

For use in the sequel, we need to establish the existence of countable chains $\{U_k\}$ of adapted normal neighborhoods covering future-directed, past-inextendible timelike geodesics of the form $c(t) = (t, y_0)$. Since in Definition 6.10 and in Lemma 6.11, it is possible that $a = -\infty$, we adopt the following convention throughout this section.

CONVENTION 6.9 Let $\omega_0$ denote any fixed interior point of the interval $(a,b)$.

Now we make the following definition.

DEFINITION 6.10 Let $M = (a,b) \times_f H$ be a Lorentzian warped product with metric $\bar{g} = -dt^2 \oplus fh$. Fix any $y_0 \in H$ and let $c : (a,\omega_0] \longrightarrow (M,\bar{g})$ be the future-directed, past-inextendible timelike geodesic given by $c(t) = (t,y_0)$. A countable covering $\{U_k\}_{k=1}^{\infty}$ of c by open sets and a strictly monotone decreasing sequence $\{t_k\}_{k=1}^{\infty}$ with $t_1 = \omega_0$ and $t_k \longrightarrow a^+$ as $k \longrightarrow \infty$ is said to be an *admissible chain* for $c : (a,\omega_0) \longrightarrow (M,\bar{g})$ if

(1) Each $U_k$ is an adapted normal neighborhood containing $c(t_k) = (t_k,y_0)$ which is adapted at some point of c.

(2) For each k, every future-directed and past-inextendible nonspacelike curve $\sigma(t) = (t,\sigma_2(t))$ with $\sigma(t_k) = (t_k,y_0)$ remains in $U_k$ for all t with $t_{k+1} \leqslant t \leqslant t_k$.

Any future-directed nonspacelike curve $\sigma$ in $(M,\bar{g})$ may be given a parameterization of the form $\sigma(t) = (t,\sigma_1(t))$. Thus condition (2) applies to all future-directed nonspacelike curves issuing from $(t_k,y_0)$. We now show that admissible chains exist.

LEMMA 6.11 Let $M = (a,b) \times_f H$ with $a \geqslant -\infty$ and $\bar{g} = -dt^2 \oplus fh$ be a Lorentzian warped product. For any $y_0 \in H$, the timelike geodesic $c : (a,\omega_0] \longrightarrow (M,\bar{g})$ given by $c(t) = (t,y_0)$ has a covering by an admissible chain.

*Proof.* We will say that $\{U_k\}$, $\{t_k\}$ is an admissible chain for $c \mid (\theta,\omega_0]$, $\theta \geqslant a$, if $\{U_k\}$, $\{t_k\}$ satisfy the properties of Definition 6.10 except that $t_k \longrightarrow \theta^+$ as $k \longrightarrow \infty$ instead of $t_k \longrightarrow a^+$ as $k \longrightarrow \infty$. Set

$$\tau = \inf\{\theta \in [a,\omega_0] : \text{there is an admissible chain } \{U_k\}, \{t_k\} \text{ for } c \mid (\theta',\omega_0] \text{ for all } \theta' \geqslant \theta\}$$

We must show that $\tau = a$.

By taking an adapted normal neighborhood centered at $c(\omega_0)$, it is easily seen that $\tau < \omega_0$. Suppose that $\tau > a$. Let U be any adapted normal neighborhood adapted at the point $(\tau, y_0) \in M$. Choose $r > \tau$ such that all future-directed nonspacelike curves $\sigma(t) = (t, \sigma_1(t))$ originating at $(r, y_0)$ lie in U for all $\tau - \varepsilon \leqslant t \leqslant r$ where $\varepsilon > 0$. There exists an admissible chain $\{U_n\}$, $\{t_n\}$ for $c((r + \tau)/2, \omega_0]$ with $t_m < r$ for some m. Define $\tilde{U}_m = U_{m+1} = U$. Extending the finite chain $\{U_1, U_2, \ldots, U_{m-1}, \tilde{U}_m, U_{m+1}\}$, $\{t_1, t_2, \ldots, t_{m-1}, t_m, \tau - \varepsilon\}$ to an infinite admissible chain yields the required contradiction. □

We now show that the subset of $U_k$ for which property (2) of Definition 6.10 holds may be extended from the point $(t_k, y_0)$ to a neighborhood $\{t_k\} \times V_k(y_0)$ in $\{t_k\} \times H$. The notation $\| g - g_1 \|_{0,U_k} < \delta$ has been defined in Section 6.2.

LEMMA 6.12 Let $\{U_k\}$, $\{t_k\}$ be an admissible chain for the timelike geodesic $c(t) = (t, y_0)$, $c : (a, \omega_0] \longrightarrow (M, g)$. For each k, there is a neighborhood $V_k(y_0)$ of $y_0$ in H such that any future-directed nonspacelike curve $\sigma(t) = (t, \sigma_1(t))$ with $\sigma(t_k) \in \{t_k\} \times V_k(y_0)$ remains in $U_k$ for all t with $t_{k+1} \leqslant t \leqslant t_k$. Furthermore, $V_k(y_0)$ and $\delta > 0$ may be chosen such that if $g_1 \in \mathrm{Lor}(M)$ and $\| g - g_1 \|_{0,U_k} < \delta$, then the following two conditions are satisfied: If $\gamma(t) = (t, \gamma_1(t))$ is any nonspacelike curve of $(M, g_1)$ with $\gamma(t_k) \in \{t_k\} \times V_k(y_0)$, then

(1) $\gamma$ remains in $U_k$ for $t_{k+1} \leqslant t \leqslant t_k$.

(2) The $g_1$-length of $\gamma \mid [t_{k+1}, t_k]$ is at most $\sqrt{6}n(t_k - t_{k+1})$.

*Proof.* First recall that $\pi : M = (a,b) \times H \longrightarrow \mathbb{R}$ given by $\pi(t,h) = t$ serves as a global time function for M. In particular, the vector field $\nabla\pi$ satisfies $g(\nabla\pi, \nabla\pi) < 0$ at all points of M. Define $\tilde{g} \in \mathrm{Lor}(M)$ by

$$\tilde{g}(x,y) = g(x,y) - g(x, \nabla\pi) g(y, \nabla\pi)$$

It follows that $g < \tilde{g}$ on M so that $U_2 = \{\bar{g}_2 \in \mathrm{Con}(M) : \bar{g}_2 < \tau(\tilde{g})\}$

is an open neighborhood of $C(M,g)$ in $\mathrm{Con}(M)$. Let $U_1 = \tau^{-1}(U_2)$. Then $U_1$ is a $C^0$-open neighborhood of $g$ in $\mathrm{Lor}(M)$ such that if $g_1 \in U_1$, the projection map $\pi : M \longrightarrow \mathbb{R}$ is a global time function for $(M,g_1)$. Hence the hypersurfaces $\{t\} \times H$, $t \in (a,b)$, remain spacelike in $(M,g_1)$. Thus any nonspacelike curve $\gamma$ of $(M,g_1)$, $g_1 \in U_1$, may be parameterized as $\gamma(t) = (t,\gamma_1(t))$. Thus the lemma will apply to any nonspacelike curve of $(M,g_1)$ originating at any point of $\{t_k\} \times V_k(y_0)$ provided $g_1 \in U_1$ is sufficiently close to $g$ on $U_k$.

Let $(x_1,\ldots,x_n)$ denote the given adapted coordinate system for the adapted normal neighborhood $U_k$. In view of condition (2) of Definition 6.8, we may find $\delta_1 > 0$ such that $\|g - g_1\|_{0,U_k} < \delta_1$ implies $g_1 < g_0$ on $U_k$ where $g_0$ is the Lorentzian metric on $U_k$ given in the adapted local coordinates by $g_0 = -3dx_1^{\,2} + dx_2^{\,2} + \cdots + dx_n^{\,2}$. Secondly, since $C^0$-close metrics have close light cones, it follows by a compactness argument that there exists a neighborhood $V_k(y_0)$ of $y_0$ in $H$ and a constant $\delta_2 > 0$ such that if $g_1 \in \mathrm{Lor}(M)$ satisfies $\|g - g_1\|_{0,U_k} < \delta_2$ and $\gamma(t) = (t,\gamma_1(t))$ is any future-directed nonspacelike curve of $(M,g_1)$ with $\gamma(t_k) \in \{t_k\} \times V_k(y_0)$, then $\gamma(t) \in U_k$ for $t_{k+1} \leq t \leq t_k$.

It remains to establish the length estimate (2). Set $\delta = \min(\delta_1,\delta_2,1/2)$. Suppose that $g' \in \mathrm{Lor}(M)$ satisfies $\|g' - g\|_{0,U_k} \leq \delta$ and let $\gamma(t) = (t,\gamma_1(t))$ be any nonspacelike curve of $(M,g')$ with $\gamma(t_k) \in \{t_k\} \times V_k(y_0)$, $\gamma : [t_{k+1},t_k] \longrightarrow M$. Let $L(\gamma)$ denote the length of $\gamma$ in $(M,g')$. Thus

$$L(\gamma) = \int_{t_{k+1}}^{t_k} \sqrt{\sum_{i,j} - g'_{ij}(\gamma(t))\gamma'_i(t)\gamma'_j(t)}\; dt$$

From Definition 6.8 and the choice of the $\delta$'s, we have $|g'_{ij}| \leq (1 + 1/2) + 1/2 = 2$ and $|\gamma'_i(t)| \leq \sqrt{3}$ for all $1 \leq i,j \leq n$. Thus Thus $L(\gamma) \leq \int_{t_{k+1}}^{t_k} \sqrt{2n^2(\sqrt{3})^2}\, dt = \sqrt{6}n(t_k - t_{k+1})$ as required. □

Assuming now that the Riemannian factor (H,h) of the Lorentzian warped product is homogeneous, we may extend Lemma 6.12 from $U_k$ to $[t_{k+1},t_k] \times H$. We will use the notation $|g_1 - g|_0 < \delta$ as defined in Section 2.2.

LEMMA 6.13 Let (M,g) be a Lorentzian warped product with (H,h) homogeneous and let $\{U_k\}$, $\{t_k\}$ be an admissible chain for $c(t) = (t,y_0)$, $c : (a,\omega_0] \longrightarrow M$. For each k, there is a continuous function $\delta_k : [t_{k+1},t_k] \times H \longrightarrow (0,\infty)$ such that if $g_1 \in \mathrm{Lor}(M)$ and $|g - g_1|_0 < \delta_k$ on $[t_{k+1},t_k] \times H$, then any nonspacelike curve $\gamma(t) = (t,\gamma_1(t))$, $\gamma : [t_{k+1},t_k] \longrightarrow (M,g_1)$, joining any point of $\{t_{k+1}\} \times H$ to any point of $\{t_k\} \times H$ has length at most $\sqrt{6}n(t_k - t_{k+1})$.

*Proof.* Fix any $k > 0$. Let $\delta > 0$ be the constant given by Lemma 6.12 such that if $g_1 \in \mathrm{Lor}(M)$ satisfies $\|g_1 - g\|_{0,U_k} < \delta$, then any nonspacelike curve $\gamma(t) = (t,\gamma_1(t))$ in $(M,g_1)$ with $\gamma(t_k) \in \{t_k\} \times V_k(y_0)$ remains in $U_k$ for $t_{k+1} \leqslant t \leqslant t_k$ and has length at most $\sqrt{6}n(t_k - t_{k+1})$. Also let $(x_1,\ldots,x_n)$ denote the given adapted coordinates for $U_k$.

We may find isometries $\{\phi_i\}_{i=1}^{\infty}$ in I(H) such that if $y_i = \phi_i(y_0)$ and $V_k(y_i) = \phi_i(V_k(y_0))$, then the sets $\{V_k(y_i)\}_{i=1}^{\infty}$ together with $V_k(y_0)$ form a locally finite covering of H. Let $\Phi_i : M \longrightarrow M$ be the isometry given by $\Phi_i(t,h) = (t,\phi_i(h))$ and set $\tilde{U}_i = \Phi_i(U_k)$ for each i. Then the sets $\{\tilde{U}_i\}$ cover $[t_{k+1},t_k] \times H$ and $(x_1, x_2 \circ \Phi_i^{-1}, \ldots, x_n \circ \Phi_i^{-1})$ form adapted local coordinates for $\tilde{U}_i$ for each i. Since everything is constructed with isometries, the constant $\delta > 0$ that works in Lemma 6.12 for $U_k$ and $c(t) = (t,y_0)$ works equally well for each $\tilde{U}_i$ and $\Phi_i \circ c$ provided that the adapted coordinates $(x_1, x_2 \circ \Phi_i^{-1}, \ldots, x_n \circ \Phi_i^{-1})$ are used for $\tilde{U}_i$. If we let $\delta_k : [t_{k+1},t_k] \times H \longrightarrow M$ be any continuous function such that $\|g_1 - g\|_0 < \delta_k$ on $[t_{k+1},t_k] \times H$ implies that $\|g_1 - g\|_{0,\tilde{U}_i} < \delta$ for each i, then the lemma is immediate from Lemma 6.12. □

We are now ready to prove the $C^0$ stability of timelike geodesic incompleteness for Lorentzian warped products $M = (a,b) \times_f H$ with $a > -\infty$ and (H,h) homogeneous.

THEOREM 6.14 Let $(M,g)$ be a warped product space-time $M = (a,b) \times_f H$ with $a > -\infty$, $g = -dt^2 \oplus fh$, and $(H,h)$ a homogeneous Riemannian manifold. Then there exists a fine $C^0$ neighborhood $U(g)$ of $g$ in $\mathrm{Lor}(M)$ of globally hyperbolic metrics such that all timelike geodesics of $(M,g_1)$ are past incomplete for each $g_1 \in U(g)$.

*Proof.* Fix any $y_0 \in M$ and let $c : (a,\omega_0] \longrightarrow M$ be the past-inextendible future-directed geodesic given by $c(t) = (t,y_0)$. Let $\{U_k\}$, $\{t_k\}$ be an admissible chain for $c$, guaranteed by Lemma 6.11. Also choose $\delta_k : [t_{k+1},t_k] \times H \longrightarrow (0,\infty)$ for each $t_k$ according to Lemma 6.13. Let $\delta : M \longrightarrow (0,\infty)$ be a continuous function such that $\delta(q) \leqslant \delta_k(q)$ for all $q \in [t_{k+1},t_k] \times H$, each $k > 0$. Set $V_1(g) = \{g_1 \in \mathrm{Lor}(M) : |g_1 - g|_0 < \delta\}$. Since global hyperbolicity is a $C^0$-open condition, we may also assume that all metrics in $V_1(g)$ are globally hyperbolic.

By the first paragraph of the proof of Lemma 6.12 we may choose a $C^0$ neighborhood $V_2(g)$ of $g$ in $\mathrm{Lor}(M)$ such that for all $g_1 \in V_2(g)$, each hypersurface $\{t\} \times H$, $t \in (a,b)$, is spacelike in $(M,g_1)$. Then every nonspacelike curve $\gamma : (\alpha,\beta) \longrightarrow (M,g_1)$ may be parameterized in the form $\gamma(t) = (t,\gamma_1(t))$. Hence Lemma 6.12 may be applied to all inextendible nonspacelike geodesics in $(M,g_1)$, $g_1 \in V_2(g)$.

Now $U(g) = V_1(g) \cap V_2(g)$ is a fine $C^0$ neighborhood of $g$ in the $C^0$ topology. Let $g_1 \in U(g)$ and let $\gamma : (\alpha,\beta) \longrightarrow (M,g_1)$ be any future-directed inextendible timelike geodesic. We may assume that $\{t_1\} \times H$ is a Cauchy surface for $(M,g_1)$ by the arguments of Geroch (1970, p. 448) and hence there exists an $s_0 \in (\alpha,\beta)$ such that $\gamma(s_0) \in \{t_1\} \times H$. In passing from $\{t_{k+1}\} \times H$ to $\{t_k\} \times H$, the $g_1$ length of $\gamma$ is at most $\sqrt{6}n(t_k - t_{k+1})$ by Lemma 6.13 for each $k$. Summing up these estimates, it follows that the $g_1$ length of $\gamma \mid (\alpha,s_0]$ is at most $\sqrt{6}n(t_1 - a)$. Since $\gamma \mid [\alpha,s_0]$ is a past-inextendible timelike geodesic of finite $g_1$ length, it follows that $\gamma$ is past incomplete in $(M,g_1)$. $\square$

Since the Riemannian factor $(H,h)$ of a Robertson-Walker space-time is homogeneous, we obtain the following corollary to Theorem

6.14 which settles affirmatively for timelike geodesics the question raised by Lerner (1973, p. 35).

THEOREM 6.15 Let $(M,g)$ be a Robertson-Walker space-time $M = (a,b) \times_f H$ with $a > -\infty$. Then there exists a fine $C^0$ neighborhood $U(g)$ of $g$ in $\mathrm{Lor}(M)$ of globally hyperbolic metrics such that all timelike geodesics of $(M,g_1)$ are past incomplete for each $g_1 \in U(g)$.

If we change the time function on $(M,g)$ to $\pi_1(t,h) = -t$, $\pi_1 : M \longrightarrow \mathbb{R}$, and apply Lemmas 6.12 and 6.13 to the resulting space-time, we obtain the exact analogue of these lemmas for the future-directed timelike geodesic $c : [\omega_0,b) \longrightarrow (M,g)$ given by $c(t) = (t,y_0)$ in the given space-time. Hence if $(M,g)$ is a Lorentzian warped product $M = (a,b) \times_f H$ with $(H,h)$ homogeneous and $b < \infty$, the same proof as for Theorem 6.14 yields the $C^0$ stability of the future timelike geodesic incompleteness. Combining this remark with Theorem 6.14 then yields the following result.

THEOREM 6.16 Let $(M,g)$ be a Lorentzian warped product $M = (a,b) \times_f H$, $g = -dt^2 \oplus fh$ with both $a$ and $b$ finite and $(H,h)$ homogeneous. Then there is a fine $C^0$ neighborhood $U(g)$ of $g$ in $\mathrm{Lor}(M)$ of globally hyperbolic metrics such that all timelike geodesics of $(M,g_1)$ for each $g_1 \in U(g)$ are both past incomplete and future incomplete.

It is interesting to note that while the finiteness of $a$ and $b$ is essential to the proof of Theorem 6.16, the proof is independent of the particular choice of warping function $f : (a,b) \longrightarrow (0,\infty)$. While the homogeneity of the Riemannian factor $(H,h)$ is also used in the proof of Theorem 6.16, no other geometric or topological property of $(H,h)$ is needed.

In general relativity and cosmology, closed big bang models for the universe are considered [cf. Hawking and Ellis (1973, Section 5.3)]. These models are Robertson-Walker space-times for which $b - a < \infty$ and $H$ is compact. Hence Theorem 6.16 implies, in particular, the $C^0$ stability of timelike geodesic incompleteness for these models.

We now turn to the proof of the $C^1$ stability of null geodesic incompleteness for Robertson-Walker space-times. Taking $M = (0,1) \times_f \mathbb{R}$ with $f(t) = (2t)^{-2}$ and $\bar{g} = -dt^2 \oplus f\, dx^2$, it may be checked using the results of Section 2.6 that the curve $\gamma : (-\infty,0) \longrightarrow (M,\bar{g})$ given by $\gamma(t) = (e^t, e^{2t})$ is a past complete null geodesic. Thus by choosing the warping function suitably, it is possible to construct Robertson-Walker space-times with $a > -\infty$ which are past null geodesically complete. Thus unlike the proof of stability for timelike geodesic incompleteness, it is necessary to assume that $(M,g)$ contains a past-incomplete (resp., past- and future-incomplete) null geodesic to obtain the null analogue of Theorem 6.15 (resp., Theorem 6.16). Not surprisingly, the proof of the $C^1$ stability of null geodesic incompleteness is more complicated than for the timelike case since affine parameters must be used instead of Lorentzian arc length to establish null incompleteness. Also for the proof of Lemma 6.18, we need the isotropy as well as the homogeneity of $(H,h)$. Thus we will assume that $M = (a,b) \times_f H$ is a Robertson-Walker space-time in the rest of this section.

Let $(V, x_1, \ldots, x_n)$ be an adapted normal neighborhood of $(M,g)$ with adapted coordinates $(x_1, \ldots, x_n)$. For the proof of Lemma 6.18, it is necessary to define a distance between compact subsets of vectors that are null for different Lorentzian metrics for M and are attached at different points of V. Recall from Section 6.2 that local coordinates $(x_1, \ldots, x_n)$ for V give rise to local coordinates $\tilde{x} = (x_1, \ldots, x_n, x_{n+1}, \ldots, x_{2n})$ for $TV = TM|_V$. Thus given any $q \in V$, $g_1 \in \mathrm{Lor}(M)$, and $\alpha > 0$, we may define

$$S(q,\alpha,g_1) = \{v \in T_qM : g_1(v,v) = 0 \text{ and } x_{n+1}(v) = -\alpha\}$$

Then $S(q,\alpha,g_1)$ is a compact subset of $T_qM$ for any $\alpha > 0$ and $g_1 \in \mathrm{Lor}(M)$. Given $p,q \in V$, $g_1, g_2 \in \mathrm{Lor}(M)$, and $\alpha_1, \alpha_2 > 0$, define the *Hausdorff distance* between $S(p,\alpha_1,g_1)$ and $S(q,\alpha_2,g_2)$ by

$$\operatorname{dist}(S(p,\alpha_1,g_1),S(q,\alpha_2,g_2)) = \sup_w \inf_v \left\{ \sum_{i=1}^{2n} [x_i(v) - x_i(w))^2]^{1/2} : \right.$$
$$\left. w \in S(q,\alpha_2,g_2),\ v \in S(p,\alpha_1,g_1) \right\}$$

The continuity of the components of the metric tensor g as functions $g_{ij} : V \times V \longrightarrow \mathbb{R}$ and the closeness of light cones for Lorentzian metrics close in the $C^0$ topology imply the continuity of this distance in p, $\alpha$, and g [cf. Busemann (1955, pp. 11-12)].

LEMMA 6.17 Let V be an adapted normal neighborhood adapted at $p \in (M,g)$. Given $\alpha > 0$ and $\varepsilon > 0$, there exists $\delta > 0$ such that $\|p - q\|_2 < \delta$, $g_1 \in \mathrm{Lor}(M)$ with $\|g - g_1\|_{0,V} < \delta$, and $|\alpha_1 - \alpha| < \delta$ imply that $\operatorname{dist}(S(q,\alpha_1,g_1),S(p,\alpha,g)) < \varepsilon$.

Now let (M,g) be a Robertson-Walker space-time $(a,b) \times_f H$ which is past null incomplete. Thus some past-directed, past-inextendible null geodesic $c : [0,A) \longrightarrow (M,g)$ is past incomplete, (i.e., $A < \infty$). Since (H,h) is isotropic and spatially homogeneous and since isometries map geodesics to geodesics, it follows that all null geodesics are past incomplete. We now fix through the proof of Theorem 6.19 this past-inextendible, past-incomplete null geodesic $c : [0,A) \longrightarrow (M,g)$ with the given parameterization.

Let $(\omega_0,y_0) = c(0) \in M = (a,b) \times H$. With this choice of $\omega_0$, apply Lemma 6.11 to the future-directed timelike geodesic $t \longrightarrow (t,y_0)$, $t \leq \omega_0$, to get an admissible chain $\{U_k\}$, $\{t_k\}$ for this timelike geodesic. Using this choice of $\{t_k\}$, we may find $s_k$ with $0 = s_1 < s_2 < \cdots < s_k < \cdots < A$ such that $c(s_k) \in \{t_k\} \times H$ for each k. Set $\Delta s_k = s_{k+1} - s_k$. As above, let $x_1 : M \longrightarrow \mathbb{R}$ denote the projection map $x_1(t,h) = t$ on the first factor of $M = (a,b) \times_f H$. Notice that if $(V,x_1,x_2,\ldots,x_n)$ is any adapted coordinate chart, then the coordinate function $x_1 : V \longrightarrow \mathbb{R}$ coincides with this projection map. If $\gamma$ is any smooth curve of M which intersects each hypersurface $\{t\} \times H$ of M exactly once and $\gamma(s) \in \{t\} \times H$, we will say that $|(x_1 \circ \gamma)'(s)|$ is the $x_1$ speed of $\gamma$ at $\{t\} \times H$. In particular, we will denote by $\alpha_k = |(x_1 \circ c)'(s_k)|$

the $x_1$ speed of the fixed null geodesic $c : [0,A) \longrightarrow (M,g)$ at $\{t_k\} \times H$ for each k.

LEMMA 6.18 Let $\varepsilon > 0$ be given. Then for each $k > 0$, there exists a continuous function $\delta_k : [t_{k+1} t_k] \times H \longrightarrow (0,\infty)$ with the following properties. Let $g_1 \in \mathrm{Lor}(M)$ with $|g - g_1|_1 < \delta_k$ on $[t_{k+1}, t_k] \times H$ and let $\gamma : [0,B) \longrightarrow M$ be any past-directed, past-inextendible null geodesic with $\gamma(0) \in \{t_k\} \times H$ and with $x_1$ speed of $\alpha_k$ at $\{t_k\} \times H$. Then $\gamma$ reaches $\{t_{k+1}\} \times H$ with an increase in affine parameter of at most $2\Delta s_k$ and moreover, the $x_1$ speed $\theta$ of $\gamma$ at $\{t_{k+1}\} \times H$ satisfies the estimate

$$1 - \varepsilon < \left|\frac{\theta}{\alpha_{k+1}}\right| < 1 + \varepsilon$$

*Proof.* Let $c : [0,A) \longrightarrow (M,g)$ be the given past-incomplete null geodesic as above. Fix $k > 0$. By the spatial homogeneity of Robertson-Walker space-times, we may find an isometry $\phi \in I(H)$ such that $\psi = \mathrm{id} \times \phi \in I(M,g)$ satisfies $\psi(c(s_k)) = (t_k, y_0)$, with $y_0$ as above. Since k is fixed during the course of this proof, we may set $p = (t_k, y_0)$ without danger of confusion. Put $c_1(s) = \psi \circ c(s + s_k)$. Then $c_1$ is a past-inextendible, past-incomplete null geodesic of $(M,g)$ with $c_1(0) \in \{t_k\} \times H$, $c_1(\Delta s_k) \in \{t_{k+1}\} \times H$, and $c_1(s) = \exp_p[g](sv)$ for $v = \psi_*(c'(s_k))$. Choose $b > 0$ with $\Delta s_k \leqslant b \leqslant 2\Delta s_k$ such that $c_1(s) \in U_k$ for all s with $0 \leqslant s \leqslant b$. Since $b > \Delta s_k$, we have $c_1(b) \in \{t\} \times H$ for some $t < t_{k+1}$. Hence $(x_1 \circ c_1)(\Delta s_k) - (x_1 \circ c_1)(b) = t_{k+1} - t > 0$. Set $\varepsilon_1 = \min\{\varepsilon, t_{k+1} - (x_1 \circ c_1)(b)\} > 0$.

Now let $g_1 \in \mathrm{Lor}(M)$ and let $q \in U_k \cap (\{t_k\} \times H)$. Suppose that $\gamma : [0,B) \longrightarrow (M,g_1)$ is any past-directed, past-inextendible null $g_1$ geodesic with $\gamma(0) = q$ and with $x_1$ speed $\alpha_k$ at q. Then $w = \gamma'(0) \in T_qM$ satisfies $g_1(w,w) = 0$ and $x_{n+1}(w) < 0$. Moreover, $\gamma(s) = \exp_q[g_1](sw)$. Applying Lemmas 6.6 and 6.7 to $c_1$ and $c_2 = \gamma$ with the constant $\varepsilon_1$ as above, we may find a constant $\delta_0 > 0$ with $0 < \delta_0 < \Delta s_k$ such that $\|v - w\|_2 < \delta_0$, $\|g - g_1\|_{1,U_k} < \delta_0$, and $|s_0 - \Delta s_k| < \delta_0$ imply that

$$|(x_1 \circ c_1)(s) - (x_1 \circ c_2)(s)| < \varepsilon_1 \leq t_{k+1} - (x_1 \circ c_1)(b) \quad (1)$$

for all s with $0 \leq s \leq b$.

$$1 - \varepsilon_1 < \left| \frac{(x_1 \circ c_2)'(s_0)}{(x_1 \circ c_1)'(\Delta s_k)} \right| < 1 + \varepsilon_1 \quad (2)$$

Setting s = b in (1), we obtain $|(x_1 \circ c_1)(b) - (x_1 \circ c_2)(b)| < t_{k+1} - (x_1 \circ c_1)(b)$ from which $(x_1 \circ c_2)(b) < t_{k+1}$. Hence there exists an s' with $0 < s' < b$ such that $(x_1 \circ c_2)(s') = t_{k+1}$. But then $s' < b < 2\Delta s_k$, which shows that the increase in affine parameter of $c_2$ in passing from $\{t_k\} \times H$ to $\{t_{k+1}\} \times H$ is less than $2\Delta s_k$ provided that $\delta_0$ is chosen as above.

For any geodesic $c_2(s) = \exp[g_1](sw)$ with $g_1$ and w $\delta_0$-close to g and v as above, let s' denote the value of the affine parameter s of $c_2$ such that $x_1 \circ c_2(s') = t_{k+1}$. As $\delta_0 \longrightarrow 0$, the corresponding value of s' must approach $\Delta s_k$ by Lemma 6.6. Thus by continuity, we may choose $\delta_1$ with $0 < \delta_1 < \delta_0$ such that for any geodesic $c_2(s) = \exp[g_1](sw)$ with $g_1$ and w both $\delta_1$-close to g and v, we have $x_1 \circ c_2(s') = t_{k+1}$ for some $s' \in [\Delta s_k - \delta_0, \Delta s_k + \delta_0]$. Hence as $\delta_1 < \delta_0$, we may apply estimate (2) above with $s_0 = s'$ to obtain

$$1 - \varepsilon < \left| \frac{(x_1 \circ c_2)'(s')}{(x_1 \circ c_1)'(\Delta s_k)} \right| = \left| \frac{\theta}{\alpha_{k+1}} \right| < 1 + \varepsilon \quad (6.4)$$

We now need to extend these estimates from a neighborhood of $v \in T_pM$ to a neighborhood of $S(p,\alpha_k,g)$. To this end, note that since $M = (a,b) \times_f H$ is a warped product with $f : (a,b) \longrightarrow \mathbb{R}$ of H and the one-dimensional factor (a,b), then I(H) acts transitively on $S(p,\alpha_k,g)$. Thus given any $z \in S(p,\alpha_k,g)$, we may apply the previous arguments using the *same* admissible chain $\{U_k\}$, $\{t_k\}$ to find a constant $\delta_1(z) > 0$ such that if $w \in TM$ satisfies $\|w - z\|_2 < \delta_1(z)$, $\gamma(w) \in U_k \cap (\{t_k\} \times H)$, $\|g_1 - g\|_{1,U_k} < \delta_1(z)$, and $c_2(s) = \exp[g_1](sw)$ has $x_1$ speed $\alpha_k$ at $\{t_k\} \times H$, then $c_2$ has an increase in affine parameter of at most $2\Delta s_k$ in passing from $\{t_k\} \times H$ to $\{t_{k+1}\} \times H$ and satisfies estimate (6.4). Using the compactness of $S(p,\alpha_k,g)$, we may choose null vectors $v_1, v_2, \ldots, v_j \in S(p,\alpha_k,g)$ such that $S(p,\alpha_k,g)$ is covered by the sets $\{w \in S(p,\alpha_k,g) : \|w - v_m\|_2 < \delta_1(v_m)\}$ for $m = 1, 2, \ldots, j$. Set $\delta_2 = \min\{\delta_1(v_m) : 1 \leq m \leq j\}$. By Lemma 6.17 we may find a constant $\delta_3$ with $0 < \delta_3 < \delta_2$ such that if $\|p - q\|_2 < \delta_3$, $\|g_1 - g\|_{1,U_k} < \delta_3$, and $w \in S(p,\alpha_k,g_1)$, then $\|w - v_m\|_2 < \delta_1(v_m)$ for some m. Hence $\delta_3$ has the following properties. If $\gamma : [0,B) \longrightarrow (M,g_1)$ is any past-inextendible, past-directed null geodesic of $(M,g_1)$ such that $\|g_1 - g\|_{1,U_k} < \delta_3$, $\gamma(0) \in (\{t_k\} \times H) \cap \{q \in U_k : \|p - q\|_2 < \delta_3\}$ where $p = (t_k, y_0)$, and $\gamma$ has $x_1$ speed $\alpha_k$ at $\gamma(0)$, then the conclusions of the theorem apply to $\gamma$. Since I(H) acts transitively on H, we may extend this result from $(\{t_k\} \times H) \cap \{q \in U_k : \|p - q\|_2 < \delta_3\}$ to all of $\{t_k\} \times H$ just as in the proof of Lemma 6.13. The function $\delta_k : [t_{k+1}, t_k] \times H \longrightarrow (0,\infty)$ may be constructed exactly as in Lemma 6.13. □

With Lemma 6.18 in hand, we are now ready to prove the $C^1$ stability of past null geodesic incompleteness for Robertson-Walker space-times. Since Robertson-Walker space-times are isotropic and spatially homogeneous, past incompleteness of one inextendible null geodesic implies past incompleteness of all null geodesics. Thus the stability theorem may be formulated as follows.

THEOREM 6.19 Let (M,g) be a Robertson-Walker space-time containing an inextendible null geodesic which is past incomplete. Then there is a fine $C^1$ neighborhood U(g) of g in Lor(M) of globally hyperbolic metrics such that *all* null geodesics of $(M,g_1)$ are past incomplete for each $g_1 \in U(g)$.

*Proof.* Let $M = (a,b) \times_f H$ and let $c : [0,A) \longrightarrow (M,g)$ be the given inextendible past-incomplete null geodesic. With $\omega_0 = x_1(c(0))$, let $\{U_k\}$, $\{t_k\}$, $\{s_k\}$, and $\{\alpha_k\}$ be chosen as in the paragraph preceding Lemma 6.18. Let $\{\beta_k\}$ be a sequence of real numbers with $0 < \beta_k < 1$ for each k such that $1/2 < \Pi_{k=1}^{\infty} (1 - \beta_k) < 1$. Thus for each $m \geq 1$, we have

$$1 < \prod_{k=1}^{m} (1 - \beta_k)^{-1} < 2 \tag{6.5}$$

For each $k \geq 1$, we apply Lemma 6.18 with $\varepsilon = \beta_k$ to obtain a continuous function $\delta_k : [t_{k+1}, t_k] \times H \longrightarrow (0,\infty)$ with the properties of Lemma 6.18. Choose a continuous function $\delta : M \longrightarrow (0,\infty)$ such that $\delta(q) < \delta_k(q)$ for all k with q in the domain of $\delta_k$ for each $q \in M$. Let $U_1(g) = \{g_1 \in \mathrm{Lor}(M) : |g_1 - g|_1 < \delta\}$. Also choose a $C^1$-open neighborhood $U_2(g)$ of g in Lor(M) such that all metrics in $U_2(g)$ are globally hyperbolic and such that each hypersurface $\{t\} \times H$ is spacelike in $(M,g_1)$ for all $t \in (a,b)$ and any $g_1 \in U_2(g)$ (cf. Lemma 6.12). Set $U(g) = U_1(g) \cap U_2(g)$.

Now suppose that $g_1 \in U(g)$ and that $\gamma : [0,B) \longrightarrow M$ is any past-directed and past-inextendible null geodesic of $(M,g_1)$. Reparameterizing $\gamma$ if necessary, we may assume that $x_1(\gamma(0)) = t_k$ for some $k \geq 1$ and that $\gamma$ has $x_1$ speed $\alpha_k$ at $\{t_k\} \times H$. By Lemma 6.18, $\gamma$ changes in affine parameter by at most $2\Delta s_k$ in passing from $\{t_k\} \times H$ to $\{t_{k+1}\} \times H$. In order to apply Lemma 6.18 to $\gamma$ as $\gamma$ passes from $\{t_{k+1}\} \times H$ to $\{t_{k+2}\} \times H$, it may be necessary to reparameterize $\gamma$ at $\{t_{k+1}\} \times H$ to have $x_1$ speed $\alpha_{k+1}$ at $\{t_{k+1}\} \times H$. Nonetheless, if $\theta_{k+1}$ denotes the $x_1$ speed of $\gamma$ at $\{t_{k+1}\} \times H$, we have $1 - \beta_k < |\theta_{k+1}/\alpha_{k+1}| < 1 + \beta_k$ from Lemma 6.18. Thus the $x_1$ speed of $\gamma$ at $\{t_{k+1}\} \times H$ cannot be less than $(1 - \beta_k)\alpha_{k+1}$. Hence the affine parameter of $\gamma$ increases in passing from $\{t_{k+1}\} \times H$ to $\{t_{k+2}\} \times H$ by at most $2(1 - \beta_k)^{-1}\, \Delta s_{k+1}$. Arguing inductively, it may be seen that $\gamma$ increases in affine parameter by at most $2\Delta s_{k+\ell} \prod_{i=0}^{\ell-1} (1 - \beta_{k+i})^{-1}$ in passing from $\{t_{k+\ell}\} \times H$ to $\{t_{k+\ell+1}\} \times H$. Using inequality (6.5), we thus have that $\gamma$ increases in affine parameter by at most $4\Delta s_{k+\ell}$ in passing from $\{t_{k+\ell}\} \times H$ to $\{t_{k+\ell+1}\} \times H$. Since $\Sigma_{k=1}^{\infty} \Delta s_k = A$, it follows that the total affine length B of $\gamma$ is less than 4A. Since $4A < \infty$, it follows that $\gamma$ is past incomplete as required. □

By reversing the time orientation, we may obtain the analogue of Theorem 6.19 for Robertson-Walker space-times having future-incomplete null geodesics. Thus Theorem 6.19 implies the following result.

THEOREM 6.20 Let (M,g) be a Robertson-Walker space-time containing an inextendible null geodesic which is both past and future incomplete. Then there is a fine $C^1$ neighborhood U(g) of g in Lor(M) of globally hyperbolic metrics such that all null geodesics of $(M,g_1)$ are past and future incomplete for each $g_1 \in U(g)$.

We now obtain two stability theorems for nonspacelike geodesic incompleteness by combining Theorems 6.15 and 6.19 and by combining Theorems 6.16 and 6.20. The first of these theorems applies to all big bang models and the second theorem applies to the closed big bang models.

THEOREM 6.21 Let (M,g) be a Robertson-Walker space-time of the form $(a,b) \times_f H$, where $a > -\infty$. Assume that (M,g) contains a past-incomplete and past-inextendible null geodesic. Then there is a fine $C^1$ neighborhood U(g) of g in Lor(M) of globally hyperbolic metrics such that *all* nonspacelike geodesics of $(M,g_1)$ are past incomplete for *each* $g_1 \in U(g)$.

THEOREM 6.22 Let (M,g) be a Robertson-Walker space-time of the form $(a,b) \times_f H$ where both a and b are finite. Assume that (M,g) contains an inextendible null geodesic which is both past and future incomplete. Then there is a fine $C^1$ neighborhood U(g) of g in Lor(M) of globally hyperbolic metrics such that *all* nonspacelike geodesics of $(M,g_1)$ are *both* past and future incomplete for *each* $g_1 \in U(g)$.

Chapter 7

# MAXIMAL GEODESICS AND CAUSALLY DISCONNECTED SPACE-TIMES

Many basic properties of complete, noncompact Riemannian manifolds stem from the principle that a limit curve of a sequence of minimal geodesics is itself a minimal geodesic. After the correct formulation of completeness had been given by Hopf and Rinow (1931), Rinow (1932), and Myers (1935) were able to establish the existence of a geodesic ray issuing from every point of a complete noncompact Riemannian manifold using this principle. Here a geodesic $\gamma : [0,\infty) \longrightarrow (N,g_0)$ is said to be a *ray* if $\gamma$ realizes the Riemannian distance between every pair of its points. Rinow and Myers constructed the desired geodesic ray as follows. Since $(N,g_0)$ is complete and noncompact, there exists an infinite sequence $\{p_n\}$ of points in N such that $d_0(p,p_n) \longrightarrow \infty$ as $n \longrightarrow \infty$ for all points $p \in N$. Let $\gamma_n$ be a minimal (i.e., distance realizing) unit speed geodesic segment from $p = \gamma_n(0)$ to $p_n$. This segment exists by the completeness of $(N,g_0)$. If $v \in T_pN$ is any accumulation point of the sequence $\{\gamma_n'(0)\}$ of unit tangent vectors in $T_pN$, then $\gamma(t) = \exp_p tv$ is the required geodesic ray. Intuitively, $\gamma$ is a ray since it is a limit curve of some subsequence of the minimal geodesic segments $\{\gamma_n\}$. The existence of geodesic rays through every point has been an essential tool in the recent structure theory of both positively curved [cf. Cheeger and Gromoll (1971, 1972)] and negatively curved [cf. Eberlein and O'Neill (1973)] complete noncompact Riemannian manifolds.

A second application of this basic principle of constructing geodesics as limits of minimal geodesic segments is a concrete geometric realization for complete Riemannian manifolds of the theory of ends for noncompact Hausdorff topological spaces [cf. Cohn-Vossen (1936)]. An infinite sequence $\{p_n\}$ of points in a manifold is said to *diverge to infinity* if, given any compact subset K, only finitely many members of the sequence are contained in K. If a complete Riemannian manifold $(N, g_0)$ has more than one end, there exists a compact subset K of N and sequences $\{p_n\}$ and $\{q_n\}$ which diverge to infinity such that $0 < d_0(p_n, q_n) \longrightarrow \infty$ and every curve from $p_n$ to $q_n$ meets K for each n. Let $\gamma_n$ be a minimal (i.e., distance-realizing) geodesic segment from $p_n$ to $q_n$. Since each $\gamma_n$ meets K, a limit geodesic $\gamma : \mathbb{R} \longrightarrow M$ may be constructed. Moreover, $\gamma$ is minimal as a limit of a sequence of minimal curves. Then "$\gamma(-\infty)$" corresponds to the end of N represented by $\{p_n\}$ and "$\gamma(+\infty)$" to the end represented by $\{q_n\}$. In particular, a complete Riemannian manifold with more than one end contains a line, i.e., a geodesic $\gamma : (-\infty, \infty) \longrightarrow N$ that is distance realizing between any two of its points.

Motivated by these Riemannian constructions, we study similar existence theorems for geodesic rays and lines in strongly causal space-times. From the viewpoint of general relativity, it is desirable to have constructions that are valid not only for globally hyperbolic subsets of space-times, but also for strongly causal space-times. However, if we only assume strong causality, it is not true in general that causally related points may be joined by maximal geodesic segments. Thus a slightly weaker principle for construction of maximal geodesics is needed for Lorentzian manifolds than for complete Riemannian manifolds. Namely, in strongly causal space-times, limit curves of sequences of "almost maximal" curves are maximal and hence also geodesics. In Section 7.1, we give two methods for constructing families of almost maximal curves whose limit curves in strongly causal space-times are maximal geodesics. The strong causality is needed to insure the upper semicontinuity of arc length in the $C^0$ topology on curves and also so that Proposition 2.21

may be applied. In Section 7.2, we apply this construction to prove the existence of past- and future-directed nonspacelike geodesic rays issuing from every point of a strongly causal space-time.

In Section 7.3, we study the class of causally disconnected space-times. Here a space-time is said to be *causally disconnected by a compact set* K if there are two infinite sequences $\{p_n\}$ and $\{q_n\}$, both diverging to infinity, such that $p_n \leqslant q_n$, $p_n \neq q_n$, and all nonspacelike curves from $p_n$ to $q_n$ meet K for each n. A space-time (M,g) admitting such a compact K causally disconnecting two divergent sequences is said to be *causally disconnected*. It is evident from the definition that causal disconnection is a global conformal invariant of C(M,g). Applying the principle of Section 7.1, we show that if the strongly causal space-time (M,g) is causally disconnected by the compact set K, then (M,g) contains a nonspacelike geodesic line $\gamma : (a,b) \longrightarrow M$ which intersects K. That is, $d(\gamma(s),\gamma(t)) = L(\gamma|[s,t])$ for all s, t with $a < s \leqslant t < b$. This result is essential to the proof of the singularity theorem 6.3 in Beem and Ehrlich (1979a) as will be seen in Chapter 11. We conclude this chapter by studying conditions on the global geodesic structure of a given space-time (M,g) which imply that (M,g) is causally disconnected. In particular, we show that all two-dimensional globally hyperbolic space-times are causally disconnected. Also, it follows from one of these conditions and the existence of nonspacelike geodesic lines in strongly causal, causally disconnected space-times that a strongly causal space-time containing no future-directed null geodesic rays contains a timelike geodesic line.

## 7.1 ALMOST MAXIMAL CURVES AND MAXIMAL GEODESICS

The purpose of this section is to show how geodesics may be constructed as limits of "almost maximal" curves in strongly causal space-times. In both constructions, the upper semicontinuity of Lorentzian arc length in the $C^0$ topology on curves for strongly causal space-times and the lower semicontinuity of Lorentzian distance play an important role. The strong causality of $(M,g)$ is also essential so that convergence in the limit curve sense and in the $C^0$ topology on curves are closely related, cf. Proposition 2.21. The first construction may be applied to pairs of chronologically related points $p$, $q$ with $d(p,q) < \infty$. While this approach is thus sufficient to show the existence of nonspacelike geodesic rays in globally hyperbolic space-times [cf. Beem and Ehrlich (1979c, Theorem 4.2)], it is not valid for points at infinite distance. Accordingly, for use in Sections 7.2 and 7.3, we give a second construction which may be used in arbitrary strongly causal space-times.

Let $(M,g)$ be an arbitrary space-time and suppose that $p$ and $q$ are distinct points of $M$ with $p \leqslant q$. If $d(p,q) = 0$, then letting $\gamma$ be any future-directed nonspacelike curve from $p$ to $q$, we have $L(\gamma) \leqslant d(p,q) = 0$. Hence $L(\gamma) = d(p,q)$ and $\gamma$ may be reparameterized to a maximal null geodesic segment from $p$ to $q$ by Theorem 3.13. Thus suppose that $p \ll q$ or, equivalently, that $d(p,q) > 0$. If $d(p,q) < \infty$ as well, then by Definition 3.1 there exists a future-directed nonspacelike curve $\gamma$ from $p$ to $q$ with

$$d(p,q) \geqslant L(\gamma) \geqslant d(p,q) - \varepsilon \tag{7.1}$$

for any $\varepsilon > 0$. Of course, inequality (7.1) is only a restriction on $L(\gamma)$ provided $\varepsilon < d(p,q)$. In this case, we will call $\gamma$ an *"almost maximal" curve*.

We note the following elementary consequence of the reverse triangle inequality.

REMARK 7.1 Let $\gamma : [0,1] \longrightarrow M$ be a future-directed nonspacelike curve from $p$ to $q$, $p \neq q$, with

$$d(p,q) - \varepsilon \leqslant L(\gamma) < \infty$$

Then for any $s < t$ in $[0,1]$, we have

$$L(\gamma|[s,t]) \geqslant d(\gamma(s),\gamma(t)) - \varepsilon$$

*Proof.* Assume that $L(\gamma|[s,t]) < d(\gamma(s),\gamma(t)) - \varepsilon$ for some $s < t$ in $[0,1]$. Then

$$\begin{aligned}L(\gamma) &= L(\gamma|[0,s]) + L(\gamma|[s,t]) + L(\gamma|[t,1])\\ &\leqslant d(\gamma(0),\gamma(s)) + L(\gamma|[s,t]) + d(\gamma(t),\gamma(1))\\ &< d(p,\gamma(s)) + d(\gamma(s),\gamma(t)) - \varepsilon + d(\gamma(t),q)\\ &\leqslant d(p,q) - \varepsilon\end{aligned}$$

in contradiction. □

We are now ready to give an example of the principle that for strongly causal space-times, limits of almost maximal curves are maximal geodesics. Strong causality is essential here since convergence in the limit curve sense and in the $C^0$ topology are closely related for strongly causal space-times, but not for arbitrary space-times.

PROPOSITION 7.2 Let $(M,g)$ be a strongly causal space-time. Suppose that $p_n \longrightarrow p$ and $q_n \longrightarrow q$ where $p_n \leqslant q_n$ for each n and $0 < d(p,q) < \infty$. Let $\gamma_n : [a,b) \longrightarrow M$ be a future-directed nonspacelike curve from $p_n$ to $q_n$ with

$$d(p_n,q_n) \geqslant L(\gamma_n) \geqslant d(p_n,q_n) - \varepsilon_n > 0 \qquad (7.2)$$

where $\varepsilon_n \longrightarrow 0$ as $n \longrightarrow \infty$. If $\gamma : [a,b] \longrightarrow M$ is a limit curve of the sequence $\{\gamma_n\}$ with $\gamma(a) = p$ and $\gamma(b) = q$, then $L(\gamma) = d(p,q)$. Thus $\gamma$ may be reparameterized to be a smooth maximal geodesic from p to q.

*Proof*. First, $\gamma$ is nonspacelike by Lemma 2.16. Second, by Proposition 2.21, a subsequence $\{\gamma_m\}$ of $\{\gamma_n\}$ converges to $\gamma$ in the $C^0$ topology on curves. By the upper semicontinuity of arc length in this topology for strongly causal space-times (cf. Remark 2.22), we then have

$$\begin{aligned}L(\gamma) &\geqslant \limsup L(\gamma_m)\\ &\geqslant \limsup\, [d(p_m,q_m) - \varepsilon_m] \qquad \text{by (7.2)}\\ &\geqslant d(p,q)\end{aligned}$$

using the lower semicontinuity of Lorentzian distance (Lemma 3.4). But by definition of distance, $d(p,q) \geqslant L(\gamma)$. Thus $L(\gamma) = d(p,q)$ and the last statement follows from Theorem 3.13. □

We now consider a second method for constructing maximal geodesics in strongly causal space-times $(M,g)$ which may be applied to points at infinite Lorentzian distance. For this purpose, we fix throughout the rest of Chapter 7 an arbitrary point $p_0 \in M$ and a complete (positive definite) Riemannian metric $h$ for the paracompact manifold $M$. Let $d_0 : M \times M \longrightarrow \mathbb{R}$ denote the Riemannian distance function induced on $M$ by $h$. For all positive integers $n$, the sets

$$\overline{B}_n = \{m \in M : d_0(p_0,m) \leqslant n\}$$

are compact by the Hopf-Rinow theorem. Thus the sets $\{\overline{B}_n : n > 0\}$ form a compact exhaustion of $M$ by connected sets. For each $n$, let

$$d[\overline{B}_n] : \overline{B}_n \times \overline{B}_n \longrightarrow \mathbb{R} \cup \{\infty\}$$

denote the Lorentzian distance function induced on $\overline{B}_n$ by the inclusion $\overline{B}_n \subset (M,g)$. That is, given $p \in \overline{B}_n$ set $d[\overline{B}_n](p,q) = 0$ if $q \notin J^+(p,\overline{B}_n)$ and for $q \in J^+(p,\overline{B}_n)$ let $d[\overline{B}_n](p,q)$ be the supremum of lengths of future-directed nonspacelike curves from $p$ to $q$ which are contained in $\overline{B}_n$. It is then immediate that $d[\overline{B}_n](p,q) \leqslant d(p,q)$ for all $p,q \in \overline{B}_n$. However, $d[\overline{B}_n]$ is *not* in general the restriction of the given Lorentzian distance function $d$ of $(M,g)$ to the set $\overline{B}_n \times \overline{B}_n$. Nonetheless, for strongly causal space-times, these two distances coincide "in the limit."

LEMMA 7.3 Let $(M,g)$ be strongly causal. Then for all $p,q \in M$, we have $d(p,q) = \lim d[\overline{B}_n](p,q)$.

*Proof.* Since $d[\overline{B}_n](p,q) \leqslant d(p,q)$, the desired equality is obvious if $d(p,q) = 0$. Thus suppose that $d(p,q) > 0$. By definition of Lorentzian distance, we may find a sequence $\{\gamma_k\}$ of future-directed nonspacelike curves from $p$ to $q$ such that $L(\gamma_k) \longrightarrow d(p,q)$ as $k \longrightarrow \infty$. [If $d(p,q) = \infty$, choose $\{\gamma_k\}$ such that $L(\gamma_k) \geqslant k$ for each $k$.] Since the image of $\gamma_k$ in $M$ is compact and the Riemannian distance function $d_0 : M \times M \longrightarrow \mathbb{R}$ is continuous and finite valued, there exists an $n(k) > 0$ for each $k$ such that $\gamma_k \subset \overline{B}_j$ for all $j \geq n(k)$. Thus $d(p,q) = \lim L(\gamma_k) \leq \lim d[\overline{B}_n](p,q)$. Hence as $d[\overline{B}_n](p,q) \leq d(p,q)$ for each $n$, the lemma is established. □

It will be convenient to introduce the following notational convention for use during the rest of this chapter.

NOTATIONAL CONVENTION 7.4 Let $\gamma$ be a future-directed nonspacelike curve in a causal space-time. Suppose $p = \gamma(s)$ and $q = \gamma(t)$ with $s < t$ and $p \neq q$. We will let $\gamma[p,q]$ denote the restriction of $\gamma$ to the interval $[s,t]$.

For strongly causal space-times, the Lorentzian distance function $d$ and the $d[\overline{B}_n]$'s are related by the following lower semicontinuity.

LEMMA 7.5 Let $(M,g)$ be strongly causal. If $p_n \longrightarrow p$ and $q_n \longrightarrow q$, then $d(p,q) \leq \liminf d[\overline{B}_n](p_n,q_n)$.

*Proof*. If $d(p,q) = 0$, there is nothing to prove. Thus we first assume that $0 < d(p,q) < \infty$. Let $\varepsilon > 0$ be given. By definition of Lorentzian distance and standard results from elementary causality theory [cf. Penrose (1972, pp. 15-16)], a timelike curve $\gamma$ from $p$ to $q$ may be found with $d(p,q) - \varepsilon < L(\gamma) \leq d(p,q)$. Since $\gamma$ is timelike and $L(\gamma) > d(p,q) - \varepsilon$, we may find $r_1, r_2 \in \gamma$ with $d(p,q) - \varepsilon < L(\gamma[r_1,r_2])$ and $p \ll r_1 \ll r_2 \ll q$. Since $I^-(r_1)$ and $I^+(r_2)$ are open and $p_n \longrightarrow p$, $q_n \longrightarrow q$, we have $p_n \ll r_1 \ll r_2 \ll q_n$ for all $n$ sufficiently large. Also $\gamma \subset \overline{B}_n$, $p_n \in J^-(r_1,\overline{B}_n)$, and $q_n \in J^+(r_2,\overline{B}_n)$ for all $n$ sufficiently large. Consequently, $d(p,q) - \varepsilon < L(\gamma[r_1,r_2]) \leq d[\overline{B}_n](p_n,q_n)$ for all large $n$. Since $\varepsilon > 0$ was arbitrary, we thus have $d(p,q) \leq \liminf d[\overline{B}_n](p_n,q_n)$ in the case that $0 < d(p,q) < \infty$. Assume finally that $d(p,q) = \infty$. Choosing timelike curves $\gamma_k$ from $p$ to $q$ with $L(\gamma_k) \geq k$ for each $k$, we have for each $k$ that $d[\overline{B}_n](p_n,q_n) \geq k - \varepsilon$ for all $n$ sufficiently large as above. Hence $\lim d[\overline{B}_n](p_n,q_n) = \infty$ as required. □

Since we are assuming that $(M,g)$ is strongly causal but not necessarily globally hyperbolic, it is possible that the Lorentzian distance function $d : M \times M \longrightarrow \mathbb{R} \cup \{\infty\}$ assumes the value $+\infty$. Nonetheless, for any given $\overline{B}_n$, the distance function $d[\overline{B}_n]$ : $\overline{B}_n \times \overline{B}_n \longrightarrow \mathbb{R} \cup \{\infty\}$ is finite valued. This is a consequence of the compactness of the $\overline{B}_n$'s and the compactness of certain subspaces of nonspacelike curves in the $C^0$ topology on curves [cf. Penrose (1972, p. 50, Theorem 6.5)]. Moreover, this compactness also implies the existence of curves realizing the $d[\overline{B}_n]$ distance for points $p,q \in \overline{B}_n$ with $q \in J^+(p,\overline{B}_n)$.

LEMMA 7.6 Let $(M,g)$ be a strongly causal space-time and let $n > 0$ be arbitrary. If $q \in J^+(p,\overline{B}_n)$, then $d[\overline{B}_n](p,q) < \infty$ and there exists a future-directed nonspacelike curve $\gamma$ in $\overline{B}_n$ joining $p$ to $q$ which satisfies $L(\gamma) = d[\overline{B}_n](p,q)$.

*Proof.* By definition of the distance $d[\overline{B}_n]$, if $d[\overline{B}_n](p,q) = 0$ and $q \in J^+(p,\overline{B}_n)$, there exists a future-directed nonspacelike curve $\gamma$ in $\overline{B}_n$ from $p$ to $q$ with $L(\gamma) \leqslant d[\overline{B}_n](p,q) = 0$. Hence $L(\gamma) = d[\overline{B}_n](p,q)$ as required. Thus we may suppose that $d[\overline{B}_n](p,q) > 0$. Again by definition of $d[\overline{B}_n]$, we may find a sequence $\{\gamma_k\}$ of future-directed nonspacelike curves from $p$ to $q$ with $L(\gamma_k) \longrightarrow d[\overline{B}_n](p,q)$. (If $d[\overline{B}_n](p,q) = \infty$, choose $\gamma_k$ with $L(\gamma_k) \geqslant k$ for each $k$.) Since $\overline{B}_n$ is compact and $(M,g)$ is strongly causal, there exists a future-directed nonspacelike curve $\gamma$ in $\overline{B}_n$ joining $p$ to $q$ with the property that a subsequence $\{\gamma_m\}$ of $\{\gamma_k\}$ converges to $\gamma$ in the $C^0$ topology on curves by Theorem 6.5 of Penrose (1972, pp. 50-51). But then using the upper semicontinuity of arc length in the $C^0$ topology on curves, we have $d[\overline{B}_n](p,q) = \lim L(\gamma_m) \leqslant L(\gamma)$ which implies the finiteness of $d[\overline{B}_n](p,q)$. Since $L(\gamma) < d[\overline{B}_n](p,q)$ from the definition, we also have $d[\overline{B}_n](p,q) = L(\gamma)$ as required. □

Now let $p,q \in M$ with $p \leqslant q$, $p \neq q$ be arbitrary. Choose any nonspacelike curve $\gamma_0$ from $p$ to $q$. Since the image of $\gamma_0$ is compact in $M$ and the Riemannian distance function is continuous, we may find an $N > 0$ such that $\gamma_0$ is contained in $\overline{B}_N$. Hence $q \in J^+(p,\overline{B}_n)$ for all $n \geq N$. Thus using Lemma 7.6, we may find a future-directed nonspacelike curve $\gamma_n$ from p to q with $L(\gamma_n) = d[\overline{B}_n](p,q)$ for each $n \geq N$. For $C^0$ limit curves of the sequence $\{\gamma_n\}$, we then have the following analogue of Proposition 7.2.

PROPOSITION 7.7 Let (M,g) be strongly causal and let $p,q \in M$ be distinct points with $p \leq q$. For all $n > 0$ sufficiently large, let $\gamma_n$ be a future-directed nonspacelike curve from p to q in $\overline{B}_n$ with $L(\gamma_n) = d[\overline{B}_n](p,q)$. If $\gamma$ is a nonspacelike curve from p to q such that $\{\gamma_n\}$ converges to $\gamma$ in the $C^0$ topology on curves, then $L(\gamma) = d(p,q)$ and hence $\gamma$ may be reparameterized to a maximal geodesic segment from p to q.

*Proof.* Using Lemma 7.5 and the upper semicontinuity of arc length in the $C^0$ topology on curves in strongly causal space-times, we have

$$d(p,q) \leq \liminf d[\overline{B}_n](p,q) = \liminf L(\gamma_n)$$
$$\leq \limsup L(\gamma_n) \leq L(\gamma) \leq d(p,q)$$

as required. □

Now let p, q be distinct points of an arbitrary strongly causal space-time with $p \leq q$ and let a sequence $\{\gamma_n\}$ of nonspacelike curves from p to q be chosen as in Proposition 7.7. While a limit curve $\gamma$ for the sequence $\{\gamma_n\}$ with $\gamma(0) = p$ may always be extracted by Proposition 2.18, we have no guarantee that $\gamma$ reaches q unless (M,g) is globally hyperbolic. Indeed, if $d(p,q) = \infty$, there is no maximal geodesic from p to q and hence no limit curve $\gamma$ of the sequence $\{\gamma_n\}$ with $\gamma(0) = p$ passes through q. Thus the hypothesis that $\gamma$ joins p to q in Proposition 7.7 together with the conclusion of Proposition 7.7 implies that $d(p,q) < \infty$. On the other hand, the condition that $d(p,q) < \infty$ does *not* imply that any limit curve $\gamma$ of $\{\gamma_n\}$ with $\gamma(0) = p$ reaches q when (M,g) is not globally hyperbolic. Examples may easily be constructed by deleting points from Minkowski space-time.

## 7.2 NONSPACELIKE GEODESIC RAYS IN STRONGLY CAUSAL SPACE-TIMES

The purpose of this section is to establish the existence of past- and future-directed nonspacelike geodesic rays issuing from every point of a strongly causal space-time (M,g).

DEFINITION 7.8 A *future-directed* (resp., *past-directed*) *nonspacelike geodesic ray* $\gamma : [0,a) \longrightarrow (M,g)$ is a future- (resp., past-) directed, future- (resp., past-) inextendible nonspacelike geodesic with the property that $d(\gamma(0),\gamma(t)) = L(\gamma|[0,t])$ (resp., $d(\gamma(t),\gamma(0)) = L(\gamma|[0,t])$) for all t with $0 \leq t < a$.

The reverse triangle inequality then implies that a nonspacelike geodesic ray is maximal between any pair of its points.

Using Lemmas 7.5 and 7.6 we first prove a proposition that will be needed not only for the proof of the existence of nonspacelike geodesic rays, but also for the proof of the existence of nonspacelike geodesic lines in strongly causal, causally disconnected space-times in Section 7.3. Let $\overline{B}_n$ and $d[\overline{B}_n] : \overline{B}_n \times \overline{B}_n \longrightarrow \mathbb{R}$ be constructed as in Section 7.1.

PROPOSITION 7.9 Let (M,g) be a strongly causal space-time and let K be any compact subset of M. Suppose that p and q are distinct points of M such that $p \leq q$ and every future-directed nonspacelike curve from p to q meets K. Then at least one of the following holds:

(1) There exists a future-directed maximal nonspacelike geodesic segment from p to q which intersects K.

(2) There exists a future-directed maximal nonspacelike geodesic which starts at p, intersects K, and is future inextendible.

(3) There exists a future-directed maximal nonspacelike geodesic which ends at q, intersects K, and is past inextendible.

(4) There exists a maximal nonspacelike geodesic with intersects K and is both past and future inextendible.

*Proof*. Let $\gamma_0$ be any future-directed nonspacelike curve in M from p to q. Since $K \cup \{\gamma_0\}$ is compact, there exists an $N > 0$ such that $K \cup \{\gamma_0\}$ is contained in $\overline{B}_n$ for all $n \geq N$. Hence $q \in J^+(p,\overline{B}_n)$ for all $n \geq N$. Thus by Lemma 7.6, for each $n \geq N$ there exists a future-directed nonspacelike curve $\gamma_n$ in $\overline{B}_n$ joining p to q with $L(\gamma_n) = d[\overline{B}_n](p,q)$. By hypothesis, each $\gamma_n$ intersects K in some point $r_n$. Since K is compact, there exists a point $r \in K$ and a subsequence $\{r_m\}$ of $\{r_n\}$ such that $r_m \longrightarrow r$ as $m \longrightarrow \infty$. Extend each curve $\gamma_m$ to a past- and future-extendible nonspacelike curve which we will still denote by $\gamma_m$. By Proposition 2.18, there exists an inextendible nonspacelike limit curve $\gamma$ for the subsequence $\{\gamma_m\}$ such that $\gamma$ contains r. Relabeling if necessary, we may assume that $\{\gamma_m\}$ distinguishes $\gamma$.

Now the limit curve $\gamma$ may contain both p and q, only p, only q, or neither p nor q. These four cases give rise respectively to the four cases (1) to (4) of the proposition. Since the proofs are similar, we will only give the proof for the second case. Thus we assume that $\gamma : (a,b) \longrightarrow M$ contains $p = \gamma(t_0)$ but not q. We must show that $\gamma \mid [t_0,b)$ is maximal. To this end, let x be an arbitrary point of $\gamma \mid [t_0,b)$. Since $\{\gamma_m\}$ distinguishes $\gamma$, we may find points $x_m \in \gamma_m$ with $x_m \longrightarrow x$ as $m \longrightarrow \infty$. Passing to a subsequence $\{\gamma_k\}$ of $\{\gamma_m\}$ if necessary, we may assume by Proposition 2.21 that $\gamma_k[p,x_k]$ converges to $\gamma[p,x]$ in the $C^0$ topology on curves (recall Notational Convention 7.4). Since $\gamma[p,x]$ is closed in M and $q \notin \gamma$, there exists an open set V containing $\gamma[p,x]$ with $q \notin V$. Since $\gamma_k[p,x_k] \longrightarrow \gamma[p,x]$ in the $C^0$ topology on curves, there exists an $N_1 > 0$ such that $\gamma_k[p,x_k] \subset V$ for all $k \geq N_1$. Hence $q \notin \gamma_k[p,x_k]$ for all $k \geq N_1$. Thus $\gamma_k[p,x_k] \subset \gamma_k[p,q]$ for all $k \geq N_1$ which implies that $L(\gamma_k[p,x_k]) = d[\overline{B}_k](p,x_k)$ for all $k \geq N_1$. By Lemma 7.5 and the upper semicontinuity of arc length in the $C^0$ topology on curves for strongly causal space-times, we have

$$d(p,x) \leq \liminf d[\overline{B}_k](p,x_k) = \liminf L(\gamma_k[p,x_k])$$
$$\leq \limsup L(\gamma_k[p,x_k]) \leq L(\gamma[p,x])$$

Since $L(\gamma[p,x]) \leq d(p,x)$ by definition of Lorentzian distance, we thus have $d(p,x) = L(\gamma[p,x])$ as required. □

For globally hyperbolic space-times, case (1) of Proposition 7.9 always applies because $J^+(p) \cap J^-(q)$ is compact and no inextendible nonspacelike curve is past or future imprisoned in a compact set. However, space-times which are strongly causal but not globally hyperbolic and which have chronologically related points $p \ll q$ to which exactly one of cases (2) to (4) applies, may be constructed by deleting points from Minkowski space-time.

With Proposition 7.9 in hand, we are now ready to prove the existence of past and future-directed nonspacelike geodesic rays issuing from each point of a strongly causal space-time. By the usual duality, it suffices to show the existence of a future-directed ray at each point.

THEOREM 7.10 Let $(M,g)$ be a strongly causal space-time and let $p \in M$ be arbitrary. Then there exists a future-directed nonspacelike geodesic ray $\gamma : [0,a) \longrightarrow M$ with $\gamma(0) = p$, i.e., $d(p,\gamma(t)) = L(\gamma|[0,t])$ for all $t$ with $0 \leqslant t < a$.

*Proof.* Let $c : [0,b) \longrightarrow M$ be a future-directed, future-inextendible timelike curve with $c(0) = p$. Since $(M,g)$ is strongly causal, $c$ cannot be future imprisoned in any compact set (cf. Proposition 2.9). Thus there is a sequence $\{t_n\}$ with $t_n \longrightarrow b$ such that $d_0(p,c(t_n)) \longrightarrow \infty$ as $n \longrightarrow \infty$. Set $q_n = c(t_n)$ for each $n$.

We now apply Proposition 7.9 to each pair $p$, $q_n$ with $K = \{p\}$. Thus for each $n$, either (1) there is a maximal future-directed nonspacelike geodesic segment from $p$ to $q_n$, or (2) there is a future-directed, future-inextendible nonspacelike geodesic ray starting at $p$. If case (2) occurs for some $n$, we are done. Thus assume that for each $n$, there is a maximal future-directed nonspacelike geodesic segment $\gamma_n$ from $p$ to $q_n$. Extend each $\gamma_n$ to a future-directed, future-inextendible nonspacelike curve, still denoted by $\gamma_n$. By Proposition 2.18, the sequence $\{\gamma_n\}$ has a future-directed, future-inextendible nonspacelike limit curve $\gamma : [0,a) \longrightarrow M$ with $\gamma(0) = p$. Relabeling the $q_n$'s if necessary, we may suppose that the sequence $\{\gamma_n\}$ itself distinguishes $\gamma$.

It remains to show that if $x \in \gamma$ with $p \neq x$ is arbitrary, then $L(\gamma[p,x]) = d(p,x)$. Since $\{\gamma_n\}$ distinguishes $\gamma$, we may choose $x_n \in \gamma_n$ for each n such that $x_n \longrightarrow x$ as $n \longrightarrow \infty$. Choose a subsequence $\{\gamma_m\}$ of $\{\gamma_n\}$ by Proposition 2.21 such that $\{\gamma_m[p,x_m]\}$ converges to $\gamma[p,x]$ in the $C^0$ topology on curves. Hence there exists an $N > 0$ such that $\gamma_m[p,x_m] \subset \overline{B}_N$ for all $m \geq N$. Since $\overline{B}_N$ is compact and $d_0(p,q_n) \longrightarrow \infty$, there exists an $N_1 \geq N$ such that $q_m \notin \overline{B}_N$ for all $m \geq N_1$. Hence $L(\gamma_m[p,x_m]) = d(p,x_m) = d[\overline{B}_m](p,x_m)$ for all $m \geq N_1$. Using Lemma 7.5 and the upper semicontinuity of Lorentzian arc length in the $C^0$ topology on curves, we then obtain

$$\begin{aligned} d(p,x) &\leq \liminf d[\overline{B}_m](p,x_m) = \liminf L(\gamma_m[p,x_m]) \\ &\leq \limsup L(\gamma_m[p,x_m]) \leq L(\gamma[p,x]) \end{aligned}$$

whence $d(p,x) = L(\gamma[p,x])$ as in Proposition 7.9. □

## 7.3 CAUSALLY DISCONNECTED SPACE-TIMES AND NONSPACELIKE GEODESIC LINES

In this section we define and study the class of causally disconnected space-times. Our definition of this class of space-times is motivated by the geometric realization discussed in the introduction to this chapter of the ends of a noncompact complete Riemannian manifold by geodesic lines [cf. Freudenthal (1931) for the original definition of ends of a noncompact Hausdorff topological space]. Recall that an infinite sequence in a noncompact topological space is said to *diverge to infinity* if given any compact subset C, only finitely many elements of the sequence are contained in C.

DEFINITION 7.11 A space-time (M,g) is said to be *causally disconnected by a compact set* K if there exist two infinite sequences $\{p_n\}$ and $\{q_n\}$ diverging to infinity such that for each n, $p_n \leq q_n$, $p_n \neq q_n$, and all future-directed nonspacelike curves from $p_n$ to $q_n$ meet K. A space-time (M,g) that is causally disconnected by some compact K is said to be *causally disconnected.*

Note first that if $k \neq n$, then $p_k$ is not necessarily causally related to $q_n$ or $p_n$. Also the compact set K may be quite different from a Cauchy surface (cf. Theorem 2.13) and nonglobally hyperbolic, strongly causal space-times may be causally disconnected even though they contain no Cauchy surfaces. An example is provided by a Reissner-Nordström space-time with $e^2 = m^2$ (cf. Figure 7.1).

It is immediate from Definition 7.11 that if (M,g) is causally disconnected and $g_1 \in C(M,g)$ is arbitrary, then $(M,g_1)$ is causally disconnected. Thus causal disconnection is a global conformal invariant.

We have previously used a more restrictive version of the concept of causal disconnection in which we assumed in addition to the conditions of Definition 7.11 that $0 < d(p_n,q_n) < \infty$ for each n, [cf. Beem and Ehrlich (1979a, p. 171), (1979c)]. With this additional condition, our previous definition was, in general, conformally invariant only for the class of globally hyperbolic space-times.

We now give the following definition.

DEFINITION 7.12 Let (M,g) be an arbitrary space-time. A past- and future- inextendible, future-directed nonspacelike geodesic $\gamma$ : $(a,b) \longrightarrow M$ is said to be a *nonspacelike geodesic line* if $L(\gamma|[s,t]) = d(\gamma(s),\gamma(t))$ for all s, t with $a < s \leq t < b$.

We now establish the existence of nonspacelike geodesic lines for strongly causal, causally disconnected space-times. This result will be an important ingredient in the proof of singularity theorems for causally disconnected space-times in Section 11.4.

THEOREM 7.13 Let (M,g) be a strongly causal space-time which is causally disconnected by a compact set K. Then M contains a nonspacelike geodesic line $\gamma : (a,b) \longrightarrow M$ which intersects K.

*Proof.* Let K, $\{p_n\}$, and $\{q_n\}$ be as in Definition 7.11. Applying Proposition 7.9 to K, $p_n$, and $q_n$ for each n, we obtain a future-directed nonspacelike geodesic $\gamma_n$ intersecting K at some point $r_n$ and satisfying at least one of the cases (1) to (4) of Proposition

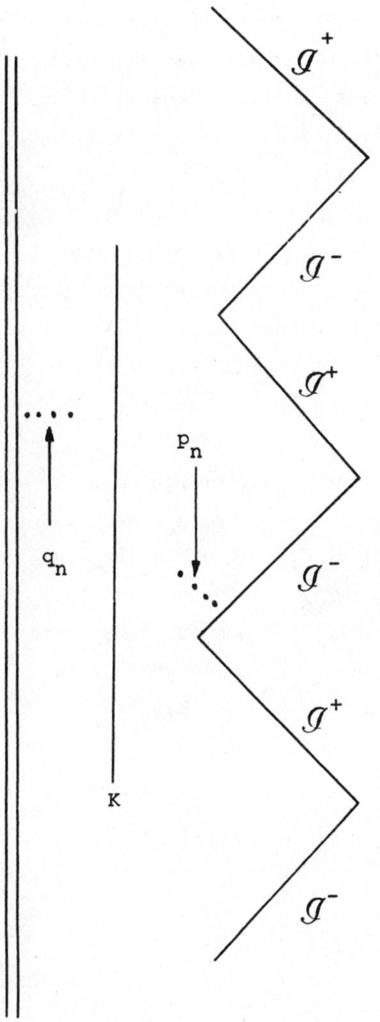


*Figure 7.1* The Penrose diagram for a Reissner-Nordström space-time with $e^2 = m^2$ containing a causally disconnected set K and associated divergent sequences $\{p_n\}$ and $\{q_n\}$ is shown. This space-time contains no Cauchy surfaces because it is not globally hyperbolic.

7.9. If case (4) holds for any $\gamma_n$, then we are done. Thus assume that no $\gamma_n$ satisfies case (4). Hence at least one of cases (1), (2), or (3) holds for infinitely many n. Since the proofs are similar, we will only give the proof assuming that case (2) holds for infinitely many n. Passing to a subsequence if necessary, we may suppose that condition (2) holds for all n. Since K is compact, there exists a subsequence $\{r_m\}$ of $\{r_n\}$ such that $r_m \longrightarrow r$ as $m \longrightarrow \infty$. Extend each $\gamma_m$ past $p_m$ to get a nonspacelike curve, still denoted by $\gamma_m$, that is past as well as future inextendible for each m. By Proposition 2.18, the sequence $\{\gamma_m\}$ has a future-directed, past- and future-inextendible nonspacelike limit curve $\gamma$ with $r \in \gamma$. Relabeling if necessary, we may assume that $\{\gamma_m\}$ distinguishes $\gamma$. We will prove that $\gamma$ is the required nonspacelike line. To show this, it suffices to show that if $x,y \in \gamma$ are distinct points with $x \leqslant r \leqslant y$, then $L(\gamma[x,y]) = d(x,y)$. Since $\{\gamma_m\}$ distinguishes $\gamma$, we may find points $x_m, y_m \in \gamma_m$ such that $x_m \longrightarrow x$ and $y_m \longrightarrow y$ as $m \longrightarrow \infty$. Passing to a subsequence $\{\gamma_k\}$ of $\{\gamma_m\}$ if necessary, we may suppose by Proposition 2.21 that $\gamma_k[x_k,y_k]$ converges to $\gamma[x,y]$ in the $C^0$ topology on curves. Since $\gamma[x,y]$ is compact in M, there exists an $N > 0$ such that $\gamma[x,y] \subset \mathrm{Int}(\overline{B}_N)$. By definition of the $C^0$ topology on curves, there is then an $N_1 \geqslant N$ such that $\gamma_k[x_k,y_k] \subset \mathrm{Int}(\overline{B}_N)$ for all $k \geqslant N_1$. Since $\{p_k\}$ diverges to infinity and $\overline{B}_N$ is compact, there is an $N_2 \geqslant N_1$ such that $p_k \notin \overline{B}_N$ for all $k \geqslant N_2$. Consequently, $x_k$ comes after $p_k$ on $\gamma_k$ for all $k \geqslant N_2$ so that $\gamma_k[x_k,y_k]$ is maximal for all $k \geqslant N_2$. We thus have

$$\begin{aligned} d(x,y) &\leqslant \liminf d(x_k,y_k) = \liminf L(\gamma_k[x_k,y_k]) \\ &\leqslant \limsup L(\gamma_k[x_k,y_k]) \leqslant L(\gamma[x,y]) \leqslant d(x,y) \end{aligned}$$

Hence $d(x,y) = L(\gamma[x,y])$ as required. □

We now give several criteria in terms of the global geodesic structure for globally hyperbolic space-times and for strongly causal space-times to be causally disconnected. In particular, we are able to show that all two-dimensional globally hyperbolic space-times are causally disconnected. Also one of our criteria (Proposition 7.18) together with Theorem 7.13 implies that if a strongly causal space-time $(M,g)$ has no null geodesic rays, then $(M,g)$ contains a timelike geodesic line.

Recall that an inextendible null geodesic $\gamma : (a,b) \longrightarrow (M,g)$ is said to be a *null geodesic line* if $d(\gamma(s),\gamma(t)) = 0$ for all s, t with $a < s \leqslant t < b$.

PROPOSITION 7.14 Let $(M,g)$ be globally hyperbolic. If $(M,g)$ contains a null geodesic line, then $(M,g)$ is causally disconnected.

*Proof.* Let $c : (a,b) \longrightarrow M$ be the given null geodesic line. Let r be any point of c and choose K to be any compact subset of M with $r \in \mathrm{Int}(K)$. Choose sequences $t_n \longrightarrow a^+$ and $t_n' \longrightarrow b^-$ such that $c(t_n) \leqslant r \leqslant c(t_n')$ for each n. Let $p_n = c(t_n)$. It suffices to show that for each n there is some $q_n$ with $c(t_n') \ll q_n$ such that all nonspacelike curves from $p_n$ to $q_n$ meet K.

If no such $q_n$ existed for some fixed n, then there would be a sequence of points $\{x_k\}$ converging to $c(t_n')$ with $c(t_n') \ll x_k$ for each k and a sequence of future-directed nonspacelike curves $\gamma_k$ from $c(t_n)$ to $x_k$ such that $K \cap \gamma_k = \emptyset$ for each k. The sequence $\{\gamma_k\}$ would have a future-directed nonspacelike limit curve $\gamma$ starting at $c(t_n)$. Since $(M,g)$ is globally hyperbolic, $\gamma$ would join $c(t_n)$ to $c(t_n')$. On the other hand, because c is a maximal null geodesic, the only (up to parameterization) nonspacelike curve from $c(t_n)$ to $c(t_n')$ is the null geodesic $c \mid [t_n,t_n']$, cf. Lemma 8.13. Thus $\gamma \subset c$ and the sequence $\{\gamma_k\}$ must intersect K for some large k. □

Proposition 7.14 implies that Minkowski space-time, de Sitter space-time, and the Friedman cosmological models are all causally disconnected. Also, the Einstein static universe (cf. Example 4.11) shows that there are globally hyperbolic, causally disconnected space-times which do not have null geodesic lines. Thus the existence of a null geodesic line is *not* a necessary condition for a globally hyperbolic space-time to be causally disconnected.

In the next proposition, we will give a sufficient condition for a strongly causal space-time (M,g) to be causally disconnected. For the proof of this result (Proposition 7.18), it is necessary to recall some additional concepts from elementary causality theory. A subset S of (M,g) is said to be *achronal* if no two points of S are chronologically related. Given a closed subset S of (M,g), the *future Cauchy development* or domain of dependence $D^+(S)$ of S is defined as the set of all points q such that every past-inextendible nonspacelike curve from q intersects S. The *future Cauchy horizon* $H^+(S)$ is given by $H^+(S) = \mathrm{cl}(D^+(S)) - I^-(D^+(S))$. The *future horismos* $E^+(S)$ of S is defined to be $E^+(S) = J^+(S) - I^+(S)$. An achronal set S is said to be *future trapped* if $E^+(S)$ is compact. Details about these concepts may be found in Hawking and Ellis (1973, pp. 102, 202, 183, and 267, respectively).

For the proof of Proposition 7.18, we also need to use a result first obtained in Hawking and Penrose (1970, p. 537, Lemma 2.12). In the text of Hawking and Ellis (1973), this result is presented slightly differently during the course of the proof of Theorem 2 of Hawking and Ellis (1973, p. 266). In the proof of this theorem, it is assumed that $\dim M \geqslant 3$ and that (M,g) has everywhere positive nonspacelike Ricci curvatures and satisfies the generic condition [conditions (1) and (2) of Theorem 2]. However, it may be seen that in the proof of Lemma 8.2.1 and the following corollary in Hawking and Ellis (1973, pp. 267-269), it is only necessary to assume that (M,g) is strongly causal to obtain our Lemma 7.15 and Corollary 7.16. We now state these two results for completeness.

LEMMA 7.15 Let A be a closed subset of the strongly causal space-time (M,g). Then $H^+(\mathrm{cl}(E^+(A)))$ is noncompact or empty.

From this lemma, one obtains as in Hawking and Penrose (1970, p. 537) or Hawking and Ellis (1973, pp. 268-269) the following corollary.

COROLLARY 7.16 Let (M,g) be strongly causal. If S is future trapped in (M,g), i.e., $E^+(S)$ is compact, then there is a future-inextendible timelike curve $\gamma$ contained in $D^+(E^+(S))$.

It will also be convenient to prove the following lemma for the proof of Proposition 7.18

LEMMA 7.17 Let (M,g) be strongly causal. If $E^+(p)$ is noncompact, then $E^+(p)$ contains an infinite sequence $\{q_n\}$ which diverges to infinity.

*Proof.* If $E^+(p)$ is closed, this is immediate since a closed and noncompact subset of M must be unbounded with respect to $d_0$. Thus assume that $E^+(p)$ is not closed. Then there exists an infinite sequence $\{x_n\} \subset E^+(p)$ such that $x_n \longrightarrow x \notin E^+(p)$ as $n \longrightarrow \infty$. Since $x_n \in E^+(p)$, we have $d(p,x_n) = 0$ and hence as $x_n \in J^+(p)$, there exists a maximal future-directed null geodesic segment $\gamma_n$ from p to $x_n$ for each n. Extend each $\gamma_n$ beyond $x_n$ to a future-inextendible nonspacelike curve still denoted by $\gamma_n$. By Proposition 2.18, the sequence $\{\gamma_n\}$ has a future-inextendible, future-directed nonspacelike limit curve $\gamma : [0,a) \longrightarrow M$ with $\gamma(0) = p$. We may assume that the sequence $\{\gamma_n\}$ itself distinguishes $\gamma$. If $x \in \gamma$, then $x \in J^+(p)$. Since $d(p,x) \leqslant \liminf d(p,x_n) = 0$, we then have $x \in J^+(p) - I^+(p) = E^+(p)$, in contradiction to the assumption that $x \notin E^+(p)$. Thus $x \notin \gamma$. We now show that $\gamma[0,a)$ is contained in $E^+(p)$. To this end, let $z \in \gamma$ be arbitrary. Since $\{\gamma_n\}$ distinguishes $\gamma$, we may find $z_n \in \gamma_n$ such that $z_n \longrightarrow z$ as $n \longrightarrow \infty$. By Proposition 2.21, there is a subsequence $\{\gamma_k\}$ of $\{\gamma_n\}$ such that $\gamma_k[p,z_k]$ converges to $\gamma[p,z]$ in the $C^0$ topology on curves. Since $x \notin \gamma$, we may find an open set U containing $\gamma[p,z]$ such that $x \notin \overline{U}$. Since $x_k \longrightarrow x$, it follows that $z_k$ comes before $x_k$ on $\gamma_k$ for all k sufficiently large. Thus $\gamma[p,z_k]$ is maximal and $d(p,z_k) = 0$ for all k sufficiently large. Hence $d(p,z) \leqslant \liminf d(p,z_k) = 0$. Since z was arbitrary, we have thus shown that $d(p,z) = 0$ for all $z \in \gamma$. Since $\gamma$ is a nonspacelike curve, $\gamma$ is then a maximal future-directed, future-inextendible null geodesic ray. Letting $\{t_n\}$ be any infinite sequence with $t_n \longrightarrow a^-$ and setting $q_n = \gamma(t_n)$, gives the required divergent sequence. □

With these preliminaries completed, we may now obtain a sufficient condition for strongly causal space-times to be causally disconnected. Minkowski space-time shows that this condition is not a necessary condition. Recall that a future-directed future-inextendible null geodesic $\gamma : [0,a) \longrightarrow M$ is said to be a *null geodesic ray* if $d(\gamma(0),\gamma(t)) = 0$ for all t with $0 \leqslant t < a$.

PROPOSITION 7.18 Let (M,g) be strongly causal. If $p \in M$ is *not* the origin of any future- (resp., past-) directed null geodesic ray, then (M,g) is causally disconnected by the future (resp., past) horismos $E^+(p) = J^+(p) - I^+(p)$ [resp., $E^-(p) = J^-(p) - I^-(p)$] of p.

*Proof.* We first show that the assumption that p is not the origin of any future-directed null geodesic ray implies that $E^+(p)$ is compact. For suppose that $E^+(p)$ is noncompact. Then there exists an infinite sequence $\{q_n\} \subset E^+(p)$ which diverges to infinity by Lemma 7.17. Since $q_n \in E^+(p)$, we have $d(p,q_n) = 0$ for all n. As $q_n \in J^+(p)$, there exists a future-directed null geodesic $\gamma_n$ from p to $q_n$ by Corollary 3.14. Extend each $\gamma_n$ beyond $q_n$ to a future-inextendible nonspacelike curve, still denoted by $\gamma_n$. Let $\gamma$ be a future-inextendible nonspacelike limit curve of the sequence $\{\gamma_n\}$ with $\gamma(0) = p$ guaranteed by Proposition 2.18. Using Proposition 2.21 and the fact that the $q_n$'s diverge to infinity, it may be shown along the lines of the proof of Theorem 7.10 that if q is any point on $\gamma$ with $q \neq p$, $q \geqslant p$, then $L(\gamma[p,q]) = d(p,q)$. Thus $\gamma$ may be reparameterized to a null geodesic ray issuing from p, in contradiction. Hence $E^+(p)$ is compact.

We now show that $E^+(p)$ causally disconnects (M,g). Since $E^+(p)$ is compact, the set {p} is future trapped in M. Thus by Corollary 7.16, there is a future-inextendible timelike curve $\gamma$ contained in $D^+(E^+(p))$. Extend $\gamma$ to a past- as well as future-inextendible timelike curve, still denoted by $\gamma$. From the definition of $D^+(E^+(p))$, the curve $\gamma$ must meet $E^+(p)$ at some point r. Since $E^+(p)$

is achronal and $\gamma$ is timelike, $\gamma$ meets $E^+(p)$ at no other point than $r$. Now let $\{p_n\}$ and $\{q_n\}$ be two sequences on $\gamma$ both of which diverge to infinity and which satisfy $p_n \ll r \ll q_n$ for each $n$ (cf. Proposition 2.9). To show that $\{p_n\}$, $\{q_n\}$, and $E^+(p)$ causally disconnect $(M,g)$, we must show that for each $n$, every nonspacelike curve $\lambda : [0,1] \longrightarrow M$ with $\lambda(0) = p_n$ and $\lambda(1) = q_n$ meets $E^+(p)$. Given $\lambda$, extend $\lambda$ to a past-inextendible curve $\tilde{\lambda}$ by traversing $\gamma$ up to $p_n$ and then traversing $\lambda$ from $p_n$ to $q_n$ (cf. Figure 7.2). As $q_n \in D^+(E^+(p))$, the curve $\tilde{\lambda}$ must intersect $E^+(p)$. Since $\gamma$ meets $E^+(p)$ only at $r$, it follows $\lambda$ intersects $E^+(p)$. Thus $\{p_n\}$, $\{q_n\}$, and $K = E^+(p)$ causally disconnect $(M,g)$ as required. □

Combining Theorem 7.13 and Proposition 7.18, we obtain the following result on the geodesic structure of strongly causal space-times with no null geodesic rays. Examples of such space-times are the Einstein static universes (Example 4.11).

THEOREM 7.19 Let $(M,g)$ be a strongly causal space-time such that no point is the origin of any future-directed null geodesic ray. Then $(M,g)$ contains a timelike geodesic line.

*Proof.* By Proposition 7.18, $(M,g)$ is causally disconnected. Thus $(M,g)$ contains a nonspacelike line by Theorem 7.13. By hypothesis, the line must be timelike rather than null. □

An equivalent result may be formulated using the hypothesis that no point is the origin of any past-directed null geodesic ray.

Using Propositions 7.14 and 7.18, we are now able to show that all two-dimensional globally hyperbolic space-times are causally disconnected. We first establish the following lemma.

LEMMA 7.20 Let $(M,g)$ be a two-dimensional globally hyperbolic space-time. If $E^+(p)$ [resp., $E^-(p)$] is not compact, then both of the future- (resp., past-) directed and future- (resp., past-) inextendible null geodesics starting at $p$ are maximal.

*Proof.* Assume $E^+(p)$ is noncompact for some $p \in M$. Let $c_1 : [0,a) \longrightarrow M$ and $c_2 : [0,b) \longrightarrow M$ be the two future-directed,

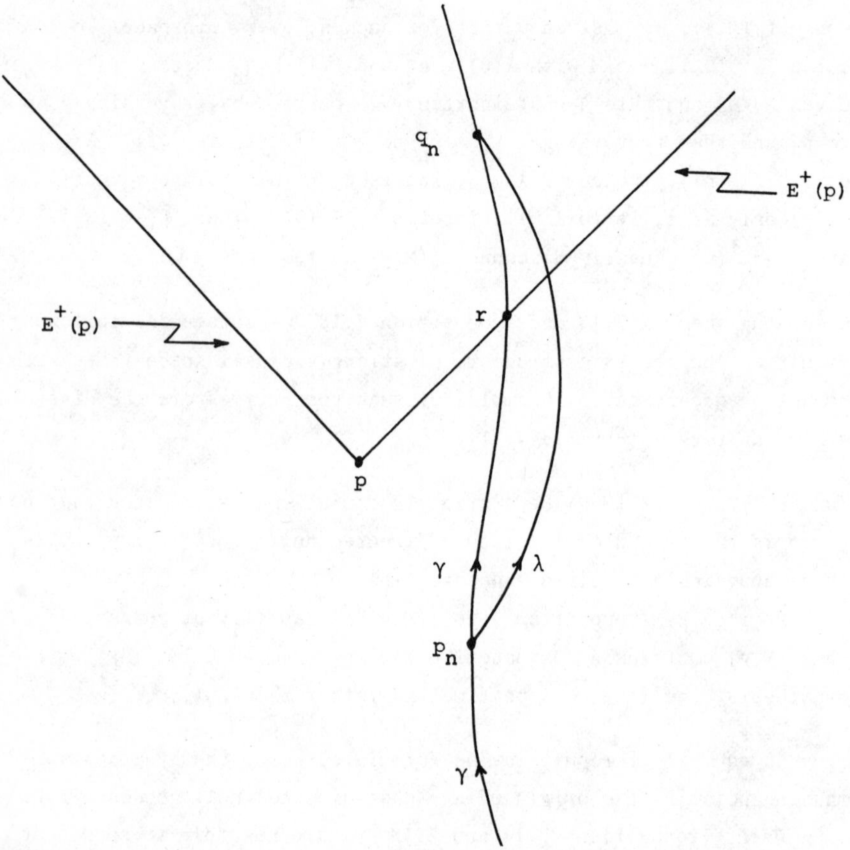


*Figure 7.2* In the proof of Proposition 7.18 the set $E^+(p)$ is shown to causally disconnect (M,g) if p is not the origin of any future directed null geodesic ray. The timelike curve $\gamma$ intersects $E^+(p)$ in the single point r and any nonspacelike curve $\lambda$ from $p_n$ to $q_n$ must meet $E^+(p)$.

future-inextendible null geodesics with $c_1(0) = c_2(0) = p$. Suppose that $c_1$ is not maximal. Setting $t_0 = \sup\{t \in [0,a) : d(p,c_1(t)) = 0\}$, we have $0 < t_0 < a$ since $c_1$ is not maximal and $(M,g)$ is strongly causal (cf. Section 8.2). Let $q = c_1(t_0)$ and choose $t_n \longrightarrow t_0^+$ with $t_n < a$ for each n. By the lower semicontinuity of Lorentzian distance, we have $d(p,q) = 0$. Since $c_1(t_n) \in I^+(p)$ and $(M,g)$ is globally hyperbolic, there is a maximal future-directed timelike geodesic $\gamma_n$ from p to $c_1(t_n)$ for each n. The sequence $\{\gamma_n\}$ has a limit curve $\gamma$ which is a nonspacelike geodesic, and $\gamma$ joins p to q by Corollary 2.19. Furthermore, $d(p,q) = 0$ implies that $\gamma$ is a null geodesic. Since $\dim M = 2$, $\gamma$ is either a geodesic subsegment of $c_1$ or a subsegment of $c_2$. If $\gamma \subset c_2$, then $c_2$ passes through q at some parameter value $t_0'$. In this case $E^+(p) = c_1 \mid [0,t_0] \cup c_2 \mid [0,t_0']$ is compact and we have a contradiction.

Assume $\gamma \subset c_1$ and let $U(q)$ be a convex normal neighborhood about q. Let $U(q)$ be chosen so small that no nonspacelike curve which leaves $U(q)$ ever returns. The inextendible null geodesics of $U(q)$ may be divided into two disjoint families $F_1$ and $F_2$ each of which cover $U(q)$ simply (cf. Section 2.4). Let $F_1$ be the class which contains the null geodesic $c_1 \cap U(q)$. Let $c_3$ be the unique null geodesic of $F_2$ which contains q. For some fixed large n the timelike geodesic $\gamma_n$ must intersect $c_3$ at some point r. We must have $q \leqslant r$ since if $r \leqslant q$ then $p \ll r$ would yield $p \ll q$ which is false. However, $q \leqslant r$, $q = c_1 \cap c_3$, $q \leqslant c_1(t_n)$, and $r \in c_3$ imply $r \nleqslant c_1(t_n)$, [cf. Busemann and Beem (1966, p. 245)]. This contradicts the fact that $r \ll c_1(t_n)$ because r comes before $c_1(t_n)$ on the timelike geodesic $\gamma_n$. This establishes the lemma. □

THEOREM 7.21 All two-dimensional globally hyperbolic space-times are causally disconnected.

*Proof.* Let $(M,g)$ be a two-dimensional globally hyperbolic space-time. If $E^+(p_0)$ is compact for some $p_0 \in M$, then each future-directed null geodesic starting at $p_0$ has a cut point and thus fails to be globally maximal (cf. Section 8.2). Thus (M,g) is causally disconnected by Proposition 7.18.

Now suppose $E^+(p)$ is noncompact for each $p \in M$. Assume $c : (a,b) \longrightarrow M$ is a future-directed, inextendible null geodesic. Let s, t with $a < s \leq t < b$ be arbitrary. Applying Lemma 7.20 at the point $p = c(s)$, we find that $c \mid [s,b)$ is maximal and thus $d(c(s),c(t)) = L(c|[s,t])$. This implies that c is maximal and hence c is a null line. Using Proposition 7.14, it follows that (M,g) is causally disconnected. □

We obtain the following corollary to Theorems 7.13 and 7.21.

COROLLARY 7.22 Let (M,g) be any globally hyperbolic space-time of dimension 2. Then (M,g) contains a nonspacelike line.

Chapter 8

# THE LORENTZIAN CUT LOCUS

Let $c : [0,\infty) \longrightarrow N$ be a geodesic in a complete Riemannian manifold starting at $p = c(0)$. Consider the set of all points q on c such that the portion of c from p to q is the unique shortest curve in all of N joining p to q. If this set has a farthest limit point, this limit point is called the *cut point* of p along the ray c. The cut locus C(p) is then defined to be the set of cut points along all geodesic rays starting at p. Since nonhomothetic conformal changes do not preserve pregeodesics, the cut locus of a point in a manifold is *not* a conformal invariant.

The cut locus has played a key role in modern global Riemannian geometry, notably in connection with the Sphere Theorem of Rauch (1951), Klingenberg (1959, 1961), and Berger (1960). The notion of cut point, as opposed to the related but different concept of conjugate point, was first defined by Poincaré (1905). An observation of Poincaré (1905) important in the later work of Rauch, Klingenberg, and Berger was that, for a complete Riemannian manifold, if q is on the cut locus of p, then either q is conjugate to p, or else there exist at least two geodesic segments of the same shortest length joining p to q [cf. Whitehead (1935)]. Klingenberg (1959, p. 657) then showed that if q is a closest cut point to p and q is not conjugate to p, there is a geodesic loop at p containing q. Klingenberg used this result to obtain an upper bound for the injectivity radius of a positively curved, complete Riemannian manifold in terms of a lower bound for the sectional curvature and the length of the shortest nontrivial smooth closed geodesic on N.

The importance of the cut locus in Riemannian geometry suggests investigating the analogous concepts and results for timelike and null geodesics in space-times. The central role that conjugate points, which are closely related to cut points, have played in singularity theory in general relativity (cf. Chapter 11) supports this idea. While there are many similarities between the Riemannian cut locus and the locus of timelike cut points in a space-time, there are also striking differences between the Riemannian and Lorentzian cut loci. Most notably, null cut points are invariant under conformal changes. Thus the null cut locus is an invariant of the causal structure of the space-time $(M,g)$. This may be used [cf. Beem and Ehrlich (1979a, Corollary 5.3)] to show that if $(M,g)$ is a Friedmann cosmological model, then there is a $C^2$ neighborhood $U(g)$ in the space $C(M,g)$ of Lorentzian metrics for M globally conformal to g such that every null geodesic in $(M,g_1)$ is incomplete for all metrics $g_1 \in U(g)$.

Because there are intrinsic differences between null and timelike cut points, we prefer to treat these cases separately. One such difference is that unlike null cut points, timelike cut points are not invariant under global conformal changes of Lorentzian metric.

In Section 8.1, we consider the analogue of Riemannian cut points for timelike geodesics. In the Lorentzian setting, timelike geodesics locally maximize the arc length between any two of their points. Thus the appropriate question to ask in defining timelike cut points is whether the portion of the given timelike geodesic segment from p to q is the *longest* nonspacelike curve in *all* of M joining p to q. This may be conveniently formulated using the Lorentzian distance function. Let $\gamma : [0,a) \longrightarrow M$ be a future-directed, future-inextendible timelike geodesic in an arbitrary space-time. Set $t_0 = \sup\{t \in [0,a) : d(\gamma(0),\gamma(t)) = L(\gamma|[0,t])\}$. If $0 < t_0 < a$, then $\gamma(t_0)$ is said to be the *future timelike cut point* of $\gamma(0)$ along $\gamma$. The future timelike cut point $\gamma(t_0)$ then has the following desired properties: (i) if $t < t_0$, then $\gamma \mid [0,t]$

is the only maximal timelike curve from $\gamma(0)$ to $\gamma(t)$ up to reparameterization; (ii) $\gamma \mid [0,t]$ is maximal for any $t \leq t_0$; (iii) if $t > t_0$, there exists a future-directed nonspacelike curve $\sigma$ from $\gamma(0)$ to $\gamma(t)$ with $L(\sigma) > L(\gamma|[0,t])$; and (iv) the future cut point $\gamma(t_0)$ comes at or before the first future conjugate point of $\gamma(0)$ along $\gamma$.

Since many of the theorems for Riemannian cut points are true only for *complete* Riemannian manifolds, it is not surprising that the more "global" results given in the rest of Section 8.1 often require global hyperbolicity. What is essential here is to know that chronologically related points at arbitrarily large distances may be joined by maximal timelike geodesic segments. Even so, the proofs are more technical for space-times than for Riemannian manifolds. Instead of using the exponential map directly as in Riemannian geometry it is necessary to regard a sequence of maximal timelike geodesics as a sequence of nonspacelike curves, extract a limit curve, take a subsequence converging to the limit in the $C^0$ topology (by Section 2.3), and finally using the upper semicontinuity of arc length, prove that the limit curve is maximal and hence a geodesic. This technical argument, isolated in Lemma 8.6, yields the following analogue of Poincaré's theorem for complete Riemannian manifolds. If $(M,g)$ is a globally hyperbolic space-time and $q$ is the future cut point of $p$ along the timelike geodesic segment $c$ from $p$ to $q$, then either one or possibly both of the following holds: (a) $q$ is the first future conjugate point to $p$ or (b) there exist at least two maximal geodesic segments from $p$ to $q$.

In Section 8.2, we study null cut points. Even though null geodesics have zero arc length, null cut points may still be defined using the Lorentzian distance function. Let $\gamma : [0,a) \longrightarrow M$ be a future-directed, future-inextendible null geodesic with $p = \gamma(0)$. Set $t_0 = \sup\{t \in [0,a) : d(p,\gamma(t)) = 0\}$. If $0 < t_0 < a$, then $\gamma(t_0)$ is called the future null cut point of $\gamma(0)$ along $\gamma$. The null cut point, if it exists, has the properties that (i) $\gamma$ is maximizing up to and including the null cut point; (ii) there is no timelike curve joining p to $\gamma(t)$ for any $t \leq t_0$; (iii) if $t_0 < t < a$, there is a timelike curve from $\gamma(0)$ to $\gamma(t)$; and (iv) the future null cut point comes at or before the first future conjugate point of $\gamma(0)$ along $\gamma$. For globally hyperbolic space-times, it is true for null as well as timelike cut points that the analogue of Poincaré's theorem for complete Riemannian manifolds is valid. Thus if (M,g) is globally hyperbolic and q is the future null cut point of $p = \gamma(0)$ along the null geodesic $\gamma$, then either one or possibly both of the following holds: (a) q is the first future conjugate point of p along $\gamma$ or (b) there exist at least two maximal null geodesic segments from p to q. We conclude Section 8.2 by using null cut points to prove singularity theorems for null geodesics following Beem and Ehrlich (1979a, Section 5).

The nonspacelike cut locus, the union of the null and timelike cut loci of a given point, is studied in Section 8.3. For complete Riemannian manifolds, if q is a closest cut point to p, then either q is conjugate to p or there is a geodesic loop based at p passing through q. The globally hyperbolic analogue (Theorem 8.24) of this result has a slightly different flavor, however. If (M,g) is a globally hyperbolic space-time and $q \in M$ is a closest (nonspacelike) cut point to p, then q is either conjugate to p, or else q is a *null* cut point to p. Thus there is no closest nonconjugate timelike cut point to p. We also show that for globally hyperbolic space-times, the nonspacelike and null cut loci are closed (Proposition 8.29). It can be seen (Example 8.28) that the hypothesis of global hyperbolicity is necessary here.

## 8.1 THE TIMELIKE CUT LOCUS

Recall that a future-directed nonspacelike curve $\gamma$ from p to q is said to be maximal if $d(p,q) = L(\gamma)$. We saw above (Theorem 3.13) that a maximal future-directed nonspacelike curve may be reparameterized to be a geodesic. We also recall the following analogue of a classical result from Riemannian geometry. The proof may be given along the lines of Kobayashi (1967, p. 99), using in place of the minimal geodesic segment from $p_1$ to $p_2$ in the Riemannian proof the fact that if $p \ll q$ and p and q are contained in a convex normal neighborhood, then p and q may be joined by a maximal timelike geodesic segment which lies in this neighborhood.

LEMMA 8.1 Let $c : [0,a] \longrightarrow M$ be a maximal timelike geodesic segment. Then for any s, t with $0 \leqslant s < t < a$, the curve $c \mid [s,t]$ is the *unique* maximal geodesic segment (up to parameterization) from c(s) to c(t).

Before commencing our study of the timelike cut locus, we need to define the unit future observer bundle $T_{-1}M$ [cf. Thorpe (1977a, 1977b)].

DEFINITION 8.2 Let $T_{-1}M = \{v \in TM : g(v,v) = -1$ and v is future directed$\}$. Given $p \in M$, let $T_{-1}M|_p$ denote the fiber of $T_{-1}M$ at p. Also given $v \in T_{-1}M$, let $c_v$ denote the unique timelike geodesic with $c_v'(0) = v$.

It is immediate from the triangle inequality that if $\gamma : [0,a) \longrightarrow M$ is a future-directed nonspacelike geodesic and $d(\gamma(0),\gamma(s)) = L(\gamma|[0,s])$, then $d(\gamma(0),\gamma(t)) = L(\gamma|[0,t])$ for all t with $0 \leqslant t \leqslant s$. Also, the triangle inequality implies that if $d(\gamma(0),\gamma(s)) > L(\gamma|[0,s])$, then $d(\gamma(0),\gamma(t)) > L(\gamma|[0,t])$ for all t with $s \leqslant t < a$. Hence the following definition makes sense.

DEFINITION 8.3 Define the function $s : T_{-1}M \longrightarrow \mathbb{R} \cup \{\infty\}$ by $s(v) = \sup\{t \geqslant 0 : d(\pi(v),c_v(t)) = t\}$.

We may first note that if $d(p,p) = \infty$, then $s(v) = 0$ for all $v \in T_{-1}M$ with $\pi(v) = p$. Also $s(v) > 0$ for all $v \in T_{-1}M$ if (M,g) is strongly causal. The number s(v) may be interpreted as the "largest" parameter value t such that $c_v$ is a maximal geodesic between $c_v(0)$ and $c_v(t)$. Indeed from Lemma 8.1 we know that the following result holds.

COROLLARY 8.4 For $0 < t < s(v)$, the geodesic $c_v : [0,t] \longrightarrow M$ is the *unique* maximal timelike curve (up to reparameterization) from $c_v(0)$ to $c_v(t)$.

The function s fails to be upper semicontinuous for arbitrary space-times as may easily be seen by deleting a point from Minkowski space. But for timelike geodesically complete space-times, we have the following proposition.

PROPOSITION 8.5 If $(M,g)$ is timelike geodesically complete, then $s : T_{-1}M \longrightarrow \mathbb{R} \cup \{\infty\}$ is upper semicontinuous.

*Proof.* It suffices to show the following. Let $v_n \longrightarrow v$ in $T_{-1}M$ with $\{s(v_n)\}$ converging in $\mathbb{R} \cup \{\infty\}$. Then $s(v) \geqslant \lim s(v_n)$. If $s(v) = \infty$, there is nothing to prove. Hence we assume that $s(v) < \lim s(v_n) = A$ and $s(v) < \infty$ and derive a contradiction.

We may choose $\delta > 0$ such that $s(v) + \delta < A$ and assume that $s(v_n) \geqslant s(v) + \delta = b$ for all n. Let $c_n = c_{v_n}$. Since $b \leqslant s(v_n)$, we have $d(\pi(v_n), c_n(b)) = b$ for all n. Since $v_n \longrightarrow v$, we have by lower semicontinuity of distance that $d(\pi(v), c_v(b)) \leqslant \liminf d(\pi(v_n), c_n(b)) = b$. Thus $d(\pi(v), c_v(b)) \leqslant b = L(c_v|[0,b])$, this last equality by definition of arc length. On the other hand, $d(\pi(v), c_v(b)) \geqslant L(c_v|[0,b])$ so that $d(\pi(v), c_v(b)) = L(c_v|[0,b]) = b$. Hence $s(v) \geqslant b = s(v) + \delta$, in contradiction. □

In order to prove the lower semicontinuity of s for globally hyperbolic space-times, it will first be useful to establish the following lemma.

LEMMA 8.6 Let $(M,g)$ be a globally hyperbolic space-time and let $\{p_n\}$ and $\{q_n\}$ be two infinite sequences of points with $p_n \longrightarrow p$ and $q_n \longrightarrow q$ where $p \ll q$. Assume $c_n : [0, d(p_n, q_n)] \longrightarrow M$ is a unit speed maximal geodesic segment from $p_n$ to $q_n$ and set $v_n = c_n'(0) \in T_{-1}M$. Then the sequence $\{v_n\}$ has a timelike limit vector $w \in T_{-1}M$. Moreover, $c_w : [0, d(p,q)] \longrightarrow M$ is a maximal geodesic segment from p to q.

*Proof.* By Corollary 2.19, there is a nonspacelike future-directed limit curve c of $c_n$ from p to q. By Proposition 2.21 and Remark 2.22, we have $L(c) \geq \limsup L(c_n) = \lim d(p_n,q_n) = d(p,q) > 0$. Thus as $d(p,q) \geq L(c)$, it follows that $L(c) = d(p,q) > 0$. Hence Theorem 3.13 implies that the curve c may be reparameterized to be a maximal timelike geodesic segment from p to q. Finally, $w = c'(0)/[-g(c'(0),c'(0))]^{1/2}$ is the required tangent vector. □

We are now ready to prove the lower semicontinuity of s for globally hyperbolic space-times.

PROPOSITION 8.7 If (M,g) is globally hyperbolic, then $s : T_{-1}M \to \mathbb{R} \cup \{\infty\}$ is lower semicontinuous.

*Proof.* It suffices to prove that if $v_n \to v$ in $T_{-1}M$ and $s(v_n) \to A$ in $\mathbb{R} \cup \{\infty\}$, then $s(v) \leq A$. If $A = \infty$, there is nothing to prove. Thus suppose $A < \infty$. Assuming $s(v) > A$, we will derive a contradiction.

Choose $\delta > 0$ such that $A + \delta < s(v)$. Define $b_n = s(v_n) + \delta$ and let $N_0$ be such that $b_n < s(v)$ for all $n \geq N_0$. Put $c_n = c_{v_n}$, $p_n = c_n(0)$, and $q = c_v(A + \delta)$. Since $v_n \to v$ and $c_v$ is defined for some parameter values beyond $A + \delta$, the geodesics $c_n$ must be defined for some parameter values past $b_n$ whenever n is larger than some $N \geq N_0$. Let $q_n = c_n(b_n)$ for $n \geq N$. Now $c_n \mid [0,b_n]$ cannot be maximal since $b_n > s(v_n)$. Because M is globally hyperbolic and $c_n(0) \ll c_n(b_n)$, we may find maximal unit speed timelike geodesic segments $\gamma_n : [0,d(p_n,q_n)] \to M$ from $p_n$ to $q_n$. Set $w_n = \gamma_n'(0)$. Since $c_v \mid [0,s(v))$ is a maximal geodesic and thus has no conjugate points, it is impossible for v to be a limit direction of $\{w_n : n \geq N\}$. Thus the maximal geodesic $c_w$ joining p to q given by Lemma 8.6 applied to $\{w_n\}$ is different from $c_v$. This then implies that $s(v) \leq A + \delta$ which contradicts $A + \delta < s(v)$. □

The following example of a strongly causal space-time which is not globally hyperbolic shows that the hypothesis of global hyperbolicity is necessary for the lower semicontinuity of s in Proposition 8.7. Let $\mathbb{R}^2$ be given the usual Minkowski metric $ds^2 = dx^2 - dy^2$ and let (M,g) be the space-time formed by equipping $M = \mathbb{R}^2 - \{(1,y) \in \mathbb{R}^2 : 1 \leq y \leq 2\}$ with the induced metric (cf. Figure 8.1). Let $p = (0,0)$, $p_n = (1/n,0)$, and let $v = \partial/\partial y|_p$, $v_n = \partial/\partial y|_{p_n}$ for all $n \geq 1$. Then $v_n \longrightarrow v$ as $n \longrightarrow \infty$. Also let $\gamma(t) = (0,t)$ and $\gamma_n(t) = (p_n,t)$ for all $t \geq 0$. Conformally changing g on the compact set C shown in Figure 8.1 which is blocked from $I^+(p)$ by the slit $\{(1,y) \in \mathbb{R}^2 : 1 \leq y \leq 2\}$, we obtain a metric $\tilde{g}$ for M with the following properties. First $\gamma$ is still a maximal geodesic in $(M,\tilde{g})$ so that $s(v) = +\infty$. But for each n, there exists a timelike curve $\sigma_n$ passing through the set C and joining $p_n$ to a point $q_n = (1/n,y_n)$ on $\gamma_n$ with $y_n \leq 4$ such that $L(\sigma_n) > L(\gamma[p_n,q_n])$. Hence $\gamma[p_n,q_n]$ fails to be maximal for all n so that $s(v_n) \leq 4$ for all n. Thus s is not lower semicontinuous. Note that $(M,\tilde{g})$ is strongly causal but $(M,\tilde{g})$ is not globally hyperbolic since $J^+((1,0)) \cap J^-((1,3))$ is not compact. The analogous construction may be applied to n-dimensional Minkowski space to produce a strongly causal n dimensional space-time for which the function s fails to be lower semicontinuous.

Globally hyperbolic examples also may be constructed for which s is not upper semicontinuous.

Combining Propositions 8.5 and 8.7 we obtain the following result.

THEOREM 8.8 If (M,g) is globally hyperbolic and timelike geodesically complete, then $s : T_{-1}M \longrightarrow \mathbb{R} \cup \{\infty\}$ is continuous.

We are now ready to define the timelike cut locus.

DEFINITION 8.9 The *future timelike cut locus* $\Gamma^+(p)$ *in* $T_pM$ is defined to be $\Gamma^+(p) = \{s(v)v : v \in T_{-1}M|_p \text{ and } 0 < s(v) < \infty\}$. The *future timelike cut locus* $C_t^+(p)$ *of p in* M is defined to be $C_t^+(p) = \exp_p(\Gamma^+(p))$. If $0 < s(v) < \infty$ and $c_v(s(v))$ exists, then the point

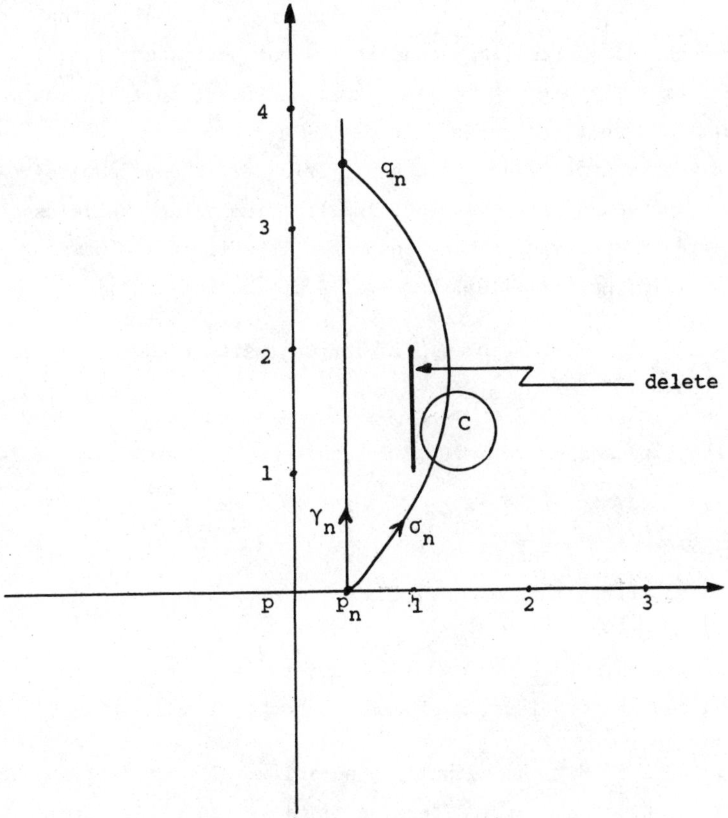


*Figure 8.1* A strongly causal space-time $(M,\tilde{g})$ is shown with unit timelike tangent vectors $v_n$ which converges to v, but with $s(v) = +\infty$ and $s(v_n) \leqslant 4$ for all $n \geqslant 1$. Hence $s : T_{-1}M \longrightarrow \mathbb{R} \cup \{\infty\}$ is not lower semicontinuous.

$c_v(s(v))$ is called the *future cut point* of $p = c_v(0)$ *along* $c_v$. The *past timelike cut locus* $C_t^-(p)$ and past cut points may be defined dually.

We may then interpret $s(v)$ as measuring the distance from p up to the future cut point along $c_v$. Thus Theorem 8.8 implies that for globally hyperbolic, timelike geodesically complete space-times, the distance from a fixed $p \in M$ to its future cut point in the direction $v \in T_{-1}M|_p$ is a continuous function of v.

We now give Lorentzian analogues of two well-known results relating cut and conjugate points on complete Riemannian manifolds. The following property of conjugate points is well known {Hawking and Ellis (1973, pp. 111-116), Lerner [1972, Theorem 4(6)]}.

THEOREM 8.10 A timelike geodesic is not maximal beyond the first conjugate point.

In the language of Definition 8.9 this may be restated as follows.

COROLLARY 8.11 The future cut point of $p = c_v(0)$ along $c_v$ comes no later than the first future conjugate point of p along $c_v$.

Utilizing this fact, we may now prove the second basic result on cut and conjugate points in globally hyperbolic space-times.

THEOREM 8.12 Let (M,g) be globally hyperbolic. If $q = c(t)$ is the future cut point of $p = c(0)$ along the timelike geodesic c from p to q, then either one or possibly both of the following holds:

(1) q is the first future conjugate point of p along c.

(2) There exist at least two maximal timelike geodesic segments from p to q.

*Proof*. Without loss of generality we may suppose that $c = c_v$ for some $v \in T_{-1}M$ and thus that $t = d(p,q) = s(v)$. Let $\{t_n\}$ be a monotone decreasing sequence of real numbers converging to t. Since $c(t) \in M$, the points $c(t_n)$ exist for n sufficiently large.

By global hyperbolicity, we may join $c(0)$ to $c(t_n)$ by a maximal timelike geodesic $c_n = c_{v_n}$ with $v_n \in T_{-1}M|_p$. Since $t_n > t = s(v)$, we have $v \neq v_n$ for all n. Let $w \in T_{-1}M$ be the timelike limiting vector for $\{v_n\}$ given by Lemma 8.6. If $v \neq w$, then c and $c_w$ are two maximal timelike geodesic segments from p to q.

It remains to show that if $v = w$, then q is the first future conjugate point of p along c. If $v = w$, then there is a subsequence $\{v_m\}$ of $\{v_n\}$ with $v_m \longrightarrow v$. If v were not a conjugate point, there would be a neighborhood U of v in $T_{-1}M|_p$ such that $\exp_p : U \longrightarrow M$ is injective. On the other hand, since $c_n$ and $c \mid [0,t_n]$ join $c(0)$ to $c(t_n)$ and $v_m \longrightarrow v$, no such neighborhood U can exist. Thus q is a future conjugate point of p along c. By Corollary 8.11, q must be the first future conjugate point of p along c. □

Theorem 8.12 has the immediate implication that for globally hyperbolic space-times, $q \in C_t^+(p)$ iff $p \in C_t^-(q)$.

The timelike cut locus of a timelike geodesically complete, globally hyperbolic space-time has the following structural property which refines Theorem 8.12. We know from this theorem that if $q \in C_t^+(p)$ and q is not conjugate to p, then there exist at least two maximal geodesic segments from p to q. Accordingly, it makes sense to consider the set

$$\mathrm{Seg}(p) = \{q \in C_t^+(p) : \text{there exist at least two future-directed maximal geodesic segments from p to q}\}$$

Since $s : T_{-1}M \longrightarrow \mathbb{R} \cup \{\infty\}$ is continuous by Theorem 8.8 and maximal geodesics joining any pair of causally related points exist in globally hyperbolic space-times, it may be shown that Seg(p) is dense in $C_t^+(p)$ for all $p \in M$. The proof may be given along the lines of Wolter's proof of Lemma 2 for complete Riemannian manifolds, [cf. Wolter (1979, p. 93)].

The dual result also holds for the past timelike cut locus $C_t^-(p)$.

## 8.2 THE NULL CUT LOCUS

The definition of null cut point has been given in Beem and Ehrlich (1979a, Section 5) where this concept was used to prove null geodesic incompleteness for certain classes of space-times. Let $\gamma : [0,a) \longrightarrow (M,g)$ be a future-directed null geodesic with endpoint $p = \gamma(0)$. Set $t_0 = \sup\{t \in [0,a) : d(p,\gamma(t)) = 0\}$. If $0 < t_0 < a$, we will say $\gamma(t_0)$ is the *future null cut point* of p on $\gamma$. Past null cut points are defined dually. Let $C_N^+(p)$ [resp., $C_N^-(p)$] denote the *future* (resp., *past*) *null cut locus* of p consisting of all future (resp., past) null cut points of p. The definition of $C_N^+(p)$ together with the lower semicontinuity of distance yields $d(p,q) = 0$ for all $q \in C_N^+(p)$. We then define the *future nonspacelike cut locus* to be $C^+(p) = C_t^+(p) \cup C_N^+(p)$. The past nonspacelike cut locus is defined dually. For a subclass of globally hyperbolic space-times, Budic and Sachs (1976) have given a different but equivalent definition of null cut point using null generators for the boundary of $I^+(p)$.

The geometric significance of null cut points is similar to that of timelike cut points. The geodesic $\gamma$ is maximizing from p up to and including the cut point $\gamma(t_0)$. That is, $L(\gamma|[0,t]) = d(p,\gamma(t)) = 0$ for all $t \leqslant t_0$. Thus there is no timelike curve joining p to $\gamma(t)$ for any t with $t \leqslant t_0$. In contrast, the geodesic $\gamma$ is no longer maximizing beyond the cut point $\gamma(t_0)$. In fact each point $\gamma(t)$ for $t_0 < t < a$ may be joined to p by a timelike curve.

Utilizing Proposition 2.19 of Penrose (1972, p. 15) and the definition of maximality, the following lemma is easily established.

LEMMA 8.13 Let (M,g) be a causal space-time. If there are two null geodesic segments from p to q, then q comes on or after the null cut point of p on each of the two segments.

The cylinder $S^1 \times \mathbb{R}$ with Lorentzian metric $ds^2 = d\theta\, dt$ shows that the assumption that (M,g) is causal is needed in Lemma 8.13.

We next prove the null analogue of Lemma 8.6.

LEMMA 8.14 Let $(M,g)$ be globally hyperbolic and let $c : [0,t] \longrightarrow (M,g)$ be a future-directed null geodesic from $p = c(0)$ to $q = c(t)$ with $d(p,q) = 0$. Assume that $p_n \longrightarrow p$, $q_n \longrightarrow q$ and $p_n \leqslant q_n$. Let $c_n$ be a maximal geodesic joining $p_n$ to $q_n$ with initial direction $v_n$. Then the set of directions $\{v_n\}$ has a limit direction $w$ and $c_w$ is a maximal null geodesic from $p$ to $q$.

*Proof.* Using Corollary 2.19, we obtain a future-directed nonspacelike limit curve $\lambda$ from $p$ to $q$. Since $d(p,q) = 0$, the curve $\lambda$ must satisfy $L(\lambda) = 0$. It follows that $\lambda$ may be reparameterized to a maximal null geodesic. □

We may now obtain the null analogue of Theorem 8.12.

THEOREM 8.15 Let $(M,g)$ be globally hyperbolic and let $q = c(t)$ be the future null cut point of $p = c(0)$ along the null geodesic $c$. Then either one or possibly both of the following holds:

(1) $q$ is the first future conjugate point of $p$ along $c$.

(2) There exist at least two maximal null geodesic segments from $p$ to $q$.

*Proof.* Let $v = c'(0)$ and let $t_n$ be a monotone decreasing sequence of real numbers with $t_n \longrightarrow t$. Since $q \in M$, we know that $c(t_n)$ exists for all sufficiently large $n$. Since $(M,g)$ is globally hyperbolic, we may find maximal nonspacelike geodesics $c_n$ with initial directions $v_n$ joining $p$ to $c(t_n)$. By Lemma 8.14 the set of directions $\{v_n\}$ has a limit direction $w$. If $v \neq w$, then the geodesic $c_w$ is a second maximal null geodesic joining $p$ to $q$ as $d(p,q) = 0$. If $v = w$, then $q$ is conjugate to $p$ along $c$. Since $d(p,q) = 0$, $q$ must be the first conjugate point of $c$ (cf. Theorem 9.72). □

We now show how the null cut locus may be applied to prove theorems on the stability of null geodesic incompleteness. The key ideas needed for this application are first, that many physically interesting space-times may be conformally imbedded in a portion of the Einstein static universe (cf. Example 4.11) that is free of null cut points, and second, that null cut points are invariant under conformal changes of metric. A discussion of global conformal imbeddings of anti-de Sitter space-time and the Friedmann cosmological models in the Einstein static universe is given in Penrose (1968) [cf. Hawking and Ellis (1973, pp. 132, 141].

We now digress briefly to give an explicit computational discussion of the well-known fact that prenull geodesics are invariant under global conformal changes. An indirect proof of the conformal invariance of null cut points may also be given using the Lorentzian distance function.

Recall that a smooth curve $\gamma : J \longrightarrow M$ is said to be a *pregeodesic* if $\gamma$ can be reparameterized to a smooth curve c which satisfies the geodesic differential equation $\nabla_{c'}c'(t) = 0$. Also recall the following definition.

DEFINITION 8.16 A diffeomorphism $f : (M_1,g_1) \longrightarrow (M_2,g_2)$ of $M_1$ onto $M_2$ is said to be a *global conformal diffeomorphism* if there exists a smooth function $\Omega : M_1 \longrightarrow \mathbb{R}$ such that

$$f^*g_2 = e^{2\Omega}g_1$$

The space-time $(M_1,g_1)$ is said to be *globally conformally diffeomorphic to an open subset* U of $(M_2,g_2)$ if there exists a diffeomorphism $f : M_1 \longrightarrow U$ and a smooth function $\Omega : M_1 \longrightarrow \mathbb{R}$ such that

$$f^*(g_2|U) = e^{2\Omega}g_1$$

The purpose of using the factor $e^{2\Omega}$ rather than just a positive-valued smooth function in Definition 8.16 is to give the simplest possible formula relating the connections $\nabla$ for $(M_1,g_1)$ and $\tilde{\nabla}$ for $(M_2,g_2)$. Explicitly, it may be shown that if $f^*g_2 = e^{2\Omega}g_1$, where $f : (M_1,g_1) \longrightarrow (M_2,g_2)$ is any smooth function, then

$$\begin{aligned}\tilde{\nabla}_{f_*X}f_*Y\Big|_{f(p)} = {} & f_*\nabla_X Y\Big|_p + X_p(\Omega)f_*Y(p) + Y_p(\Omega)f_*X(p) \\ & - g_1(X(p),Y(p))f_*(\text{grad }\Omega(p))\end{aligned} \tag{8.1}$$

for any pair X, Y of vector fields on $M_1$. Here "grad $\Omega$" denotes the gradient vector field of $\Omega$ with respect to the metric $g_1$ for $M_1$.

Using formula (8.1), it is now possible to show that if $\gamma : J \longrightarrow M_1$ is a null geodesic in $(M_1,g_1)$, then $\sigma = f \circ \gamma : J \longrightarrow M_2$ is a null pregeodesic in $(M_2,g_2)$. The crux of the matter is that because $\gamma'(t)$ is null and $\nabla_{\gamma'}\gamma' = 0$, formula (8.1) simplifies to

$$\tilde{\nabla}_{\sigma'}\sigma'(t) = 2\gamma'(t)(\Omega)\sigma'(t)$$

for all $t \in J$. Note, however, that if $\gamma$ were timelike, the factor $g_1(\gamma',\gamma')f_*(\text{grad } \Omega)$ in formula (8.1) would prevent $f \circ \gamma$ from being a timelike pregeodesic in $(M_2,g_2)$.

LEMMA 8.17 Let $f : (M_1,g_1) \longrightarrow (M_2,g_2)$ be a global conformal diffeomorphism of $M_1$ onto $M_2$. Then $\gamma : J \longrightarrow (M_1,g_1)$ is a null pregeodesic for $(M_1,g_1)$ iff $f \circ \gamma$ is a null pregeodesic for $(M_2,g_2)$.

*Proof.* Since $f^{-1} : (M_2,g_2) \longrightarrow (M_1,g_1)$ is also a conformal diffeomorphism, it is enough to show that if $\gamma : J \longrightarrow M_1$ is a null geodesic, then $\sigma = f \circ \gamma$ is a null pregeodesic of $(M_2,g_2)$. That is, we must show that $\sigma$ can be reparameterized to be a null geodesic of $(M_2,g_2)$. But it has already been shown that

$$\tilde{\nabla}_{\sigma'}\sigma'(t) = f(t)\sigma'(t) \tag{8.2}$$

for some smooth function $f : J \longrightarrow \mathbb{R}$. Just as in the classical theory of projectively equivalent connections in Riemannian geometry, however, formula (8.2) implies that $\sigma$ is a pregeodesic [cf. Spivak (1970, pp. 6-35 and 6-37)]. □

We are now ready to prove the conformal invariance of null cut points under global conformal diffeomorphisms $f : (M_1,g_1) \longrightarrow (M_2,g_2)$. Notice that if $(M_1,g_1)$ is time oriented by the vector field $X_1$, then $(M_2,g_2)$ is time oriented either by $X_2 = f_*X_1$ or $-X_2$. If $M_2$ is time oriented by $X_2$, then f maps future-directed curves to future-directed curves and thus future (resp., past) null cut points to future (resp., past) null cut points in Proposition 8.18. On the other hand, if $M_2$ is time oriented by $-X_2$, then f maps future (resp., past) null cut points to past (resp., future) null cut points since f maps future-directed curves to past-directed curves.

PROPOSITION 8.18 Let $f : (M_1,g_1) \longrightarrow (M_2,g_2)$ be a global conformal diffeomorphism of $(M_1,g_1)$ onto $(M_2,g_2)$. Let $\gamma : [0,a) \longrightarrow (M_1,g_1)$ be a null geodesic of $M_1$. If $q = \gamma(t_0)$ is a null cut point of $p = \gamma(0)$ along the geodesic $\gamma$, then f(q) is a null cut point of f(p) along the null pregeodesic $f \circ \gamma : [0,a) \longrightarrow (M_2,g_2)$.

*Proof.* It suffices to assume that q is the future null cut point of p along $\gamma$ and that $(M_2,g_2)$ is time oriented so that $f \circ \gamma$ is a future-directed null curve in $M_2$. Let $\sigma : [0,b) \longrightarrow (M_2,g_2)$ be a reparameterization of $f \circ \gamma$ to a future-directed null geodesic guaranteed by Lemma 8.17 with $f(p) = \sigma(0)$. Then $f(q) = \sigma(t_1)$ for some $t_1 \in (0,b)$. Let $d_i$ denote the Lorentzian distance function of $(M_i,g_i)$ for $i = 1, 2$.

We show first that $d_2(\sigma(0),\sigma(t)) = 0$ for any t with $0 \leqslant t \leqslant t_1$. For suppose $d_2(\sigma(0),\sigma(t)) \neq 0$ for some t with $0 \leqslant t \leqslant t_1$. Then we may find a future-directed nonspacelike curve $\beta$ in $M_2$ from $\sigma(0)$ to $\sigma(t)$ with $L_{g_2}(\beta) > 0$. Thus $f^{-1} \circ \beta$ is a future-directed nonspacelike curve in $M_1$ from p to $f^{-1}(\sigma(t))$ with $L_{g_1}(f^{-1} \circ \beta) > 0$. Since $f(q) = \sigma(t_1)$, we have $f^{-1}(\sigma(t)) = \gamma(t_2)$ for some $t_2 \leqslant t_0$. Hence $d_1(p,\gamma(t_2)) \geqslant L_{g_1}(f^{-1} \circ \beta) > 0$, which contradicts the fact that $d_1(p,\gamma(t_2)) = 0$ since $t_2 \leqslant t_0$ and $\gamma(t_0) = q$ was the future null cut point to p along $\gamma$.

We show now that $d_2(\sigma(0),\sigma(t)) \neq 0$ for any $t > t_1$. This then makes $f(q) = \sigma(t_1)$ the future null cut point of f(p) along $\sigma$ as required. To this end, fix $t > t_1$. There is then a $t_2 > t_0$ so that $f^{-1}(\sigma(t)) = \gamma(t_2)$. Since $\gamma(t_0) = q$ is the future null cut of p along $\gamma$, we have $d_1(p,\gamma(t_2)) > 0$. Hence there is a future-directed nonspacelike curve $\alpha$ from p to $\gamma(t_2)$ with $L_{g_1}(\alpha) > 0$. Then $f \circ \alpha$

is a future-directed nonspacelike curve from f(p) to $\sigma(t)$. Hence $d_2(f(p),\sigma(t)) \geq L_{g_2}(f \circ \alpha) > 0$ as required. □

With Proposition 8.18 in hand, we may now apply the null cut locus to study null geodesic incompleteness. Recall that a geodesic is said to be incomplete if it cannot be extended to all values of an affine parameter (cf. Definition 5.2).

Let R and Ric denote the curvature tensor and Ricci curvature tensor of (M,g), respectively. An inextendible null geodesic $\gamma$ is said to satisfy the generic condition if for some parameter value t, there exists a nonzero tangent vector $v \in T_{\gamma(t)}M$ with $g(v,\gamma'(t)) = 0$ such that $R(v,\gamma'(t))\gamma'(t)$ is nonzero and is *not* proportional to $\gamma'(t)$ (cf. Definition 11.7, Section 11.2). In particular, if $Ric(v,v) > 0$ for all null vectors $v \in TM$, then every inextendible null geodesic of (M,g) satisfies the generic condition. In Section 11.2, it will be seen that $\dim M \geq 3$ is necessary for the null generic condition to be satisfied. Thus we will assume that $\dim M \geq 3$ in the following proposition.

PROPOSITION 8.19 Let (M,g) be a space-time of dimension $\geq 3$ such that all inextendible null geodesics satisfy the generic condition and such that $Ric(v,v) \geq 0$ for all null vectors. If (M,g) is globally conformally diffeomorphic to an open subset of a space-time (M',g') which has no null cut points, then *all* null geodesics of (M,g) are incomplete.

*Proof.* Proposition 4.4.5 of Hawking and Ellis (1973, p. 101) shows that if the Ricci curvature is nonnegative on all null vectors, then each complete null geodesic which satisfies the generic condition has conjugate points (cf. Proposition 11.17). Since maximal geodesics do not contain conjugate points, we need only show that all null geodesics of (M,g) are maximal to establish their incompleteness. Assume that $\gamma$ is a future-directed null geodesic from p to q which is not maximal. Then there is a timelike curve from p to q. Since conformal diffeomorphisms take null geodesics to null pregeodesics and timelike curves to timelike curves, the image of $\gamma$ must be a nonmaximal null geodesic in $(M',g')$. This implies that $(M',g')$ has a null cut point and hence yields a contradiction. □

Recall that the four-dimensional Einstein static universe (Example 4.11) is the cylinder $x^2 + y^2 + z^2 + w^2 = 1$ imbedded in $\mathbb{R}^5$ with the metric induced from the Minkowski metric $ds^2 = -dt^2 + dx^2 + dy^2 + dz^2 + dw^2$. The Einstein static universe is thus $\mathbb{R} \times S^3$ with a Lorentzian product metric. The geodesics and null cut points are easy to determine in this space-time. The null cut locus of the point $(t,x,y,z,w)$ merely consists of the two points $(t \pm \pi, -x, -y, -z, -w)$. Consequently, the subset $M' = \{(t,x,y,z,w) : 0 < t < \pi,\ x^2 + y^2 + z^2 + w^2 = 1\}$ has the property that $C_N(p) \cap M' = \emptyset$ for all $p \in M'$. Proposition 8.19 then has the following consequence.

COROLLARY 8.20 Let $(M,g)$ be a space-time such that all null geodesics satisfy the generic condition and such that $\mathrm{Ric}(v,v) \geqslant 0$ for all null vectors. If $(M,g)$ is globally conformally diffeomorphic to some portion of the subset $M'$ of the Einstein static universe, then *all* null geodesics of $(M,g)$ are incomplete.

Since Minkowski space-time is free of null cut points, the above result remains valid if we replace $M'$ with Minkowski space-time.

Friedmann space-times are used in cosmology as models of the universe. In these spaces it is assumed that the universe is filled with a perfect fluid having zero pressure. We will also assume that the cosmological constant $\Lambda$ is zero for these models. These space-times are then Robertson-Walker spaces (cf. Section 4.4) with $p = \Lambda = 0$. These space-times may be conformally imbedded in the subset $M'$ of the Einstein static universe defined above, [cf. Hawking and Ellis (1973, pp. 139-141)]. Furthermore,

$\text{Ric}(g)(v,v) > 0$ for all nonspacelike vectors in a Friedmann cosmological model $(M,g)$. By Proposition 6.3, there is a $C^2$ neighborhood $U(g)$ of $g$ in $C(M,g)$ such that $\text{Ric}(g_1)(v,v) > 0$ for all $g_1 \in U(g)$. Since $\text{Ric}(g_1)(v,v) > 0$ implies that the generic condition is satisfied by all null geodesics in $(M,g_1)$, Corollary 8.20 yields the following result.

COROLLARY 8.21 Let $(M,g)$ be a Friedmann cosmological model. There is a $C^2$ neighborhood $U(g)$ of $g$ in $C(M,g)$ such that *every* null geodesic in $(M,g_1)$ is incomplete for *all* $g_1 \in U(g)$.

## 8.3 THE NONSPACELIKE CUT LOCUS

Recall that

DEFINITION 8.22 *The future nonspacelike cut locus* $C^+(p)$ *of* p is defined to be $C^+(p) = C_t^+(p) \cup C_N^+(p)$. The *past nonspacelike cut locus* is $C^-(p) = C_t^-(p) \cup C_N^-(p)$ and the *nonspacelike cut locus* is $C(p) = C^-(p) \cup C^+(p)$.

We mentioned in Chapter 7 that a complete noncompact Riemannian manifold admits a geodesic ray issuing from each point. Thus at each point, there is a direction such that the geodesic issuing from that point in that direction has no cut point. For strongly causal space-times, the Lorentzian analogue of this property follows similarly from the existence of past- and future-directed nonspacelike geodesic rays issuing from each point. This may be restated as the first half of the following proposition.

PROPOSITION 8.23 (a) Let $(M,g)$ be strongly causal. Then given any point $p \in M$, there is a future- and a past-directed nonspacelike tangent vector in $T_pM$ such that the geodesics issuing from p in these directions have no cut points.

(b) Let $(M,g)$ be globally hyperbolic. Given any point $p \in M$, then p has no farthest nonspacelike cut point.

*Proof.* As we have already remarked, part (a) follows immediately from Theorem 7.10. For the proof of part (b), assume $q \in M$ is a farthest cut point of p. Then q is a cut point along a maximal geodesic segment $\gamma$ from p to q. Choose a sequence of points $\{q_n\}$ such that $q \ll q_n$ for each n and $q_n \longrightarrow q$. Since (M,g) is globally hyperbolic, there exist maximal timelike geodesic segments $c_n : [0,d(p,q_n)] \longrightarrow M$ from p to $q_n$ for each n. Extend each $c_n$ to a future-inextendible geodesic. Since q is a farthest cut point of p and $d(p,q_n) \geqslant d(p,q) + d(q,q_n) > d(p,q)$, for each n the geodesic ray $c_n$ contains no cut point of p. The sequence $\{c_n\}$ has a limit curve c which is a future-directed and future-inextendible nonspacelike curve starting at p by Proposition 2.18. By passing to a subsequence if necessary, we may assume that $\{c_n\}$ converges to c in the $C^0$ topology on curves (cf. Proposition 2.21). Using the global hyperbolicity of (M,g) and $q_n \longrightarrow q$, we find that $q \in c$. If $r \in c$ and $r_n \in c_n$ with $r_n \longrightarrow r$, then $d(p,r) = \lim d(p,r_n)$. Using the upper semicontinuity of arc length for strongly causal space-times, we find that the length of c from p to r is at least as great as $\lim\sup d(p,r_n) = \lim d(p,r_n) = d(p,r)$. Thus c is a maximal geodesic ray. Since q is a cut point of p on $\gamma$, the geodesics c and $\gamma$ are distinct maximal nonspacelike geodesics containing p and q. Either Lemma 8.1 or Lemma 8.13 now yields a contradiction to the maximality of c beyond q. □

We now turn to the analogue of Klingenberg's observation that for Riemannian manifolds, if q is a closest cut point to p and q is nonconjugate to p, then there is a geodesic loop at p passing through q. For Lorentzian manifolds, however, a different result is true. If $\{q_n\} \subset C_t^+(p)$ converges to $q \in C_N^+(p)$, then $d(p,q_n) \longrightarrow 0$ in a globally hyperbolic space-time since the Lorentzian distance function is continuous. Thus it is not unreasonable to expect that for globally hyperbolic space-times, there is no closest timelike cut point q to p that is nonconjugate to p.

THEOREM 8.24 Let (M,g) be a globally hyperbolic space-time and assume that $p \in M$ has a closest future (or past) nonspacelike cut point q. Then q is either conjugate to p or else q is a *null* cut point of p.

*Proof.* Let q be a future cut point of p which is a closest cut point of p with respect to the Lorentzian distance d. Assume q is neither conjugate to p nor a null cut point of p. Then $p \ll q$ and by Theorem 8.12 there exist at least two future-directed maximal timelike geodesics $c_1$ and $c_2$ from p to q. Let $\gamma : [0,a) \longrightarrow M$ be a past-directed timelike curve starting at q. By choosing $a > 0$ sufficiently small we may assume the image of $\gamma$ lies in the chronological future of p. Then $p \ll \gamma(t) \ll q$ for $0 < t < a$ implies $d(p,q) \geqslant d(p,\gamma(t)) + d(\gamma(t),q) > d(p,\gamma(t))$ using the reverse triangle inequality. Since q is a closest cut point, the point $\gamma(t)$ comes before a cut point of p on any timelike geodesic from p to $\gamma(t)$. Thus any timelike geodesic from p to $\gamma(t)$ is maximal. Since q is not conjugate to p along $c_1$, there is a timelike geodesic from p to $\gamma(t)$ near $c_1$ for all sufficiently small t. Similarly there exists a timelike geodesic from p to $\gamma(t)$ near $c_2$ for all small t. The existence of two maximal timelike geodesics from p to $\gamma(t)$ implies $\gamma(t)$ is a cut point of p and yields a contradiction since $d(p,\gamma(t)) < d(p,q)$. □

For Riemannian manifolds, the set of unit tangent vectors in any given tangent space is compact. It is then an immediate consequence of the Riemannian analogue of Propositions 8.5 and 8.7 above that the cut locus of any point in a complete Riemannian manifold is a closed subset of M, [cf. Gromoll, Klingenberg, and Meyer (1975, pp. 170-171), Kobayashi (1967, pp. 100-101)]. For Lorentzian manifolds, the timelike cut locus is not a closed subset of M in general as the Einstein static universe (Example 4.11) shows. However, for globally hyperbolic space-times, it may be shown that the nonspacelike cut locus and the null cut locus of any point are closed subsets of M [cf. Beem and Ehrlich (1979c, Proposition 6.5)]. It may be seen by Example 8.28 given below that the assumption of global hyperbolicity is necessary for the nonspacelike cut locus to be a closed subset of M. The proof for Lorentzian manifolds is more complicated than the proof for Riemannian manifolds as a result of the lack of unit null tangent vectors.

We now turn to the proof of the closure of the nonspacelike and null cut loci for globally hyperbolic space-times. We first establish a technical lemma.

LEMMA 8.25 Let (M,g) be a strongly causal space-time. Assume $p \leqslant q_n$ and $q_n \longrightarrow q \neq p$. If $\exp_p(v_n) = q_n$, $\exp_p(v) = q$, and the directions of the nonspacelike vectors $v_n$ converge to the direction of v, then $v_n \longrightarrow v$.

*Proof*. Choose numbers $a_n > 0$ such that the vectors $w_n = a_n v_n$ converge to v. Then the sequence of points $r_n = \exp_p(w_n)$ must be defined for large n and must converge to q. Since (M,g) is causal, there is only one value $t_n$ of t such that $\exp_p(t_n w_n) = q_n$, namely $t_n = a_n^{-1}$. Using $r_n \longrightarrow q$, $q_n \longrightarrow q$, and the strong causality of (M,g) near q, we find that $t_n \longrightarrow 1$. Since $v_n = t_n^{-1} w_n$, the sequence $\{v_n\}$ converges to v. □

Fix a point $p \in M$. A tangent vector $v \in T_pM$ is a conjugate point to p in $T_pM$ if $(\exp_p)_*$ is singular at v. The conjugate points to p in $T_pM$ must form a closed subset of the domain of $\exp_p$ because $(\exp_p)_*$ is nonsingular on an open subset of $T_pM$. A point $q \in M$ is conjugate to p along a geodesic c if there is some conjugate point $v \in T_pM$ such that $\exp_p v = q$ and c is (up to reparameterization) the geodesic $\exp_p(tv)$. If $q \in M$, $p \leqslant q$, and q is conjugate to p along a nonspacelike geodesic, then we will say q is a *future nonspacelike conjugate* point of p. Past nonspacelike conjugate points are defined dually. If (M,g) is a causal space-time, then p cannot be in its own future or past nonspacelike conjugate locus because a nonspacelike geodesic passes through p at most once and $(\exp_p)_*$ is nonsingular at the origin of $T_pM$.

REMARK 8.26 It is possible even in globally hyperbolic space-times for a point to be conjugate to itself along *spacelike* geodesics, however. This happens in any Einstein static universe of dimension $\geq 3$.

LEMMA 8.27 Let $(M,g)$ be a globally hyperbolic space-time. Then the future (resp., past) nonspacelike conjugate locus in M is a closed subset of M.

*Proof.* Assume that $\{q_n\}$ is a sequence of future nonspacelike conjugate points of p with $q_n \to q$. Then $p \neq q$, $p \leq q$, and $p \leq q_n$ for each n. Let $v_n \in T_pM$ be a nonspacelike vector such that $(\exp_p)_*$ is singular at $v_n$ and $\exp_p(v_n) = q_n$. Then $c_n(t) = \exp_p(tv_n)$ is a future-directed, future-inextendible geodesic starting at p and containing $q_n$. The geodesics $c_n$ must have a limit curve $\gamma$ by Proposition 2.18. Since $(M,g)$ is globally hyperbolic, $\gamma$ must contain q. Using the strong causality of $(M,g)$ and the fact that $\gamma$ is a limit curve of nonspacelike geodesics, we find that $\gamma$ is itself a nonspacelike geodesic. Let v be the unique (nonspacelike) vector tangent to $\gamma$ at $\gamma(0) = p$ such that $\exp_p(v) = q$. Since $\gamma$ is a limit curve of $\{c_n\}$, there must be some subsequence $\{m\}$ of $\{n\}$ such that the directions of the vectors $v_m$ converge to the direction of v. Lemma 8.25 then implies that $v_m \to v$. Since the vectors $v_m$ belong to the total conjugate locus of p in $T_pM$ and this conjugate locus is a closed subset of the domain of $\exp_p$, the vector v is a conjugate point of p in $T_pM$. Because v is a future-directed nonspacelike vector, the point $q = \exp_p(v)$ is a future nonspacelike conjugate point of p in M. This establishes the closure in M of the future nonspacelike conjugate locus. The dual proof shows that the past nonspacelike conjugate locus is also closed. □

EXAMPLE 8.28 Let $M = S^1 \times \mathbb{R} = \{(\theta,t) : 0 \leq \theta \leq 2\pi,\ t \in \mathbb{R}\}$ with the flat metric $ds^2 = d\theta^2 - dt^2$. This is the two-dimensional Einstein static universe which is globally hyperbolic. There are no conjugate points in this space-time. If $p = (0,0)$, the future null cut locus $C_N^+(p)$ consists of the single point $(\pi,\pi)$. The future timelike cut locus $C_t^+(p)$ consists of $\{(\pi,t) : t > \pi\}$ and is not closed. On the other hand, $C^+(p) = C_t^+(p) \cup C_N^+(p)$ is closed.

If we delete the two points $(\pm\pi/4, 2\pi)$ from M, we obtain a strongly causal space-time $(M', g|M')$ which is not globally hyperbolic. In $(M', g|M')$ the future nonspacelike cut locus $C^+(p)$ is *not* closed.

PROPOSITION 8.29 Let (M,g) be a globally hyperbolic space-time. For any $p \in M$, the sets $C_N^+(p)$, $C_N^-(p)$, $C^+(p)$, and $C^-(p)$ are all closed subsets of M. In particular, the null cut locus and the nonspacelike cut locus of p are each closed.

*Proof*. Since the four cases are similar, we will show only $C_N^+(p)$ is closed in M. To this end, let $q_n \in C_N^+(p)$ and $q_n \longrightarrow q$. Since (M,g) is globally hyperbolic, $q \neq p$ and $q \in J^+(p)$. From the definition of $C_N^+(p)$ we see that $d(p,q_n) = 0$. Using the continuity of Lorentzian distance for globally hyperbolic space-times (Lemma 3.5), it follows that $d(p,q) = \lim d(p,q_n) = 0$. Let $\gamma$ be any nonspacelike geodesic from p to q. Then $d(p,q) = 0$ implies that $\gamma$ is a maximal null geodesic and that the cut point to p along $\gamma$ cannot come before q.

There are two cases to consider. Either infinitely many points $q_n$ are future nonspacelike conjugate to p or else no $q_n$ is future nonspacelike conjugate to p for large n. In the first case we apply Lemma 8.27 and find that if q is future nonspacelike conjugate to p along $\gamma$, then q is the cut point along $\gamma$ because the cut point along $\gamma$ comes before or at the first conjugate point along $\gamma$. If q is a future nonspacelike conjugate point to p along some other nonspacelike geodesic $\gamma'$, then $\gamma'$ must be null and Lemma 8.13 shows that q is the cut point along both $\gamma$ and $\gamma'$.

Assume now that $q_n$ is not future nonspacelike conjugate to p for all sufficiently large n. Theorem 8.15 implies that for large n there exist at least two maximal null geodesics from p to $q_n$. Thus for large n there are two future-directed null vectors $v_n$ and $w_n$ with $\exp_p(v_n) = \exp_p(w_n) = q_n$. The future-directed nonspacelike directions at p form a compact set of directions. Thus the directions determined by $\{v_n\}$ and $\{w_n\}$ have limit directions v and w respectively. If v and w determine different directions, we apply Lemma 8.14 to obtain two maximal null geodesics from p to q. In this case q is a cut point of p. On the other hand, v and w may determine the same direction. If this is so, we first note that there are constants $a > 0$ and $b > 0$ such that $av = bw$ and $\exp_p(av) = \exp_p(bw) = q$. Applying Lemma 8.25 it follows that for some subsequence $\{m\}$ of $\{n\}$ we have $v_m \longrightarrow av$ and $w_m \longrightarrow av$. However $v_m \neq w_m$ and $\exp_p(v_m) = \exp_p(w_m)$ then yield that $(\exp_p)_*$ must be singular at av. Thus q is conjugate to p along $\gamma(t) = \exp_p(tav)$ and since $d(p,q) = 0$ we find that q is the cut point of p along $\gamma(t)$. □

Surprisingly the above result holds without any assumptions about the timelike or null geodesic completeness of (M,g). In particular, the null cut locus and the nonspacelike cut locus in globally hyperbolic space-times are closed by Proposition 8.29 even though the function $s : T_{-1}M \longrightarrow \mathbb{R} \cup \{\infty\}$ may fail to be upper semicontinuous.

It is well known [cf. Kobayashi (1967, pp. 100-101)] that using the cut locus, it may be shown that a compact Riemannian manifold is the disjoint union of an open cell and a closed subset (the cut locus of a fixed $p \in M$) which is a continuous image (under $\exp_p$) of an (n - 1) sphere. Thus the cut loci inherit many of the topological properties of the compact manifold itself. For Lorentzian manifolds, cut points may be defined (using the Lorentzian distance at least) only for nonspacelike geodesics. Hence a corresponding result for space-times must describe the topology not of all of M itself, but only of $J^+(p)$ for an arbitrary $p \in M$. To obtain this decomposition, we need to assume that (M,g) is timelike geodesically complete as well as globally hyperbolic, so that the function $s : T_{-1}M \longrightarrow \mathbb{R} \cup \{\infty\}$ defined in Definition 8.3 will be continuous, not just lower semicontinuous (cf. Propositions 8.5 and 8.7).

Recall also that the future horismos $E^+(p)$ of any point $p \in M$ is given by $E^+(p) = J^+(p) - I^+(p)$ and that $C^+(p)$ denotes the future nonspacelike cut locus of p.

PROPOSITION 8.30 Let (M,g) be globally hyperbolic and timelike geodesically complete. For each $p \in M$ the set $J^+(p) - [C^+(p) \cup E^+(p)]$ is an open cell.

*Proof.* Let $B = J^+(p) - [C^+(p) \cup E^+(p)]$. Then $q \in B$ iff there is a maximal future-directed timelike geodesic which starts at p and extends beyond q. Thus $B = I^+(p) - C^+(p)$ which shows that B is open. Now $T_{-1}M|_p = \{v \in T_pM : v \text{ is future directed and } g(v,v) = -1\}$ is homeomorphic to $\mathbb{R}^{n-1}$. Let $H : T_{-1}M|_p \longrightarrow \mathbb{R}^{n-1}$ be a homeomorphism and define $\bar{s} : \mathbb{R}^{n-1} \longrightarrow \mathbb{R} \cup \{\infty\}$ by $\bar{s} = s \circ H^{-1}$. There is an induced homeomorphism of B with $\{(x,t) \in \mathbb{R}^{n-1} \times \mathbb{R} : 0 < t < \bar{s}(x)\}$ defined by $q \longrightarrow (H(v),t)$, where v is the vector in $T_{-1}M$ such that $\exp_p(sv)$ is the unique (up to reparameterization) maximal geodesic from p to q, and $\exp_p(tv) = q$ and $v \in T_{-1}M$. Let $f : [0,\infty] \longrightarrow [0,1]$ be a homeomorphism with $f(0) = 0$ and $f(\infty) = 1$. Then the map $(x,t) \longrightarrow (x, f(t)/f(\bar{s}(x)))$ shows B is homeomorphic to $\mathbb{R}^{n-1} \times (0,1)$ which establishes the proposition. □

REMARK 8.31 Since $I^+(p) \subset J^+(p)$, Proposition 8.30 yields some indirect topological information about $I^+(p)$. The Einstein static universe shows that $I^+(p)$ need not be an open cell even if (M,g) is globally hyperbolic. However, $(I^+(p), g|I^+(p))$ is globally hyperbolic whenever (M,g) is globally hyperbolic. Thus $I^+(p)$ may be expressed topologically as a product $I^+(p) = S \times \mathbb{R}$, where S is an (n - 1)-dimensional manifold (cf. Theorem 2.13).

Chapter 9

# MORSE INDEX THEORY ON LORENTZIAN MANIFOLDS

Given a nonspacelike geodesic $\gamma : [0,a) \longrightarrow M$, we have seen in Chapter 8 that if $\gamma(t_0)$ is the future cut-point to $\gamma(0)$ along $\gamma$, then for any $t < t_0$, the geodesic segment $\gamma \mid [0,t]$ is the longest nonspacelike curve from $\gamma(0)$ to $\gamma(t)$ in *all* of M. We could ask a much less stringent question: among all nonspacelike curves $\sigma$ from $\gamma(0)$ to $\gamma(t_0)$ sufficiently "close" to $\gamma$, does $L(\gamma) \geqslant L(\sigma)$? If so, $\gamma(t_0)$ comes at or before the first future conjugate point of $\gamma(0)$ along $\gamma$. The crucial difference here is between "in all of M" for cut points and "close to $\gamma$" for conjugate points. The importance of this distinction is illustrated by the fact that while no two-dimensional space-time has any null conjugate points, all null geodesics in the two-dimensional Einstein static universe have future and past null cut points.

Since only the behavior of "nearby" curves is considered in studying conjugate points, it is natural to apply similar techniques from the calculus of variations to geodesics in arbitrary Riemannian manifolds and to nonspacelike geodesics in arbitrary Lorentzian manifolds. To indicate the flavor of the Lorentzian index theory, we sketch the Morse index theory of an arbitrary (not necessarily complete) Riemannian manifold $(N,g_0)$. Let $c : [a,b] \longrightarrow N$ be a fixed geodesic segment and consider a one-parameter family of curves $\alpha_s$ starting at $c(a)$ and ending at $c(b)$. More precisely, let $\alpha : [a,b] \times (-\varepsilon,\varepsilon) \longrightarrow N$ be a continuous, piecewise smooth map with $\alpha(t,0) = c(t)$ for all $t \in [a,b]$ and $\alpha(a,s) = c(a)$, $\alpha(b,s) = c(b)$ for all $s \in (-\varepsilon,\varepsilon)$. Thus each neighboring curve

$\alpha_s : t \longrightarrow \alpha_s(t) = \alpha(t,s)$ is piecewise smooth. The *variation vector field* (or s-parameter derivative) V of this deformation is given by

$$V(t) = \frac{d}{ds}(s \longmapsto \alpha(t,s))\Big|_{s=0}$$

Since c is a smooth geodesic, one calculates that

$$\frac{d}{ds}(L_0(\alpha_s))\Big|_{s=0} = 0$$

and the second variation works out to be

$$\frac{d^2}{ds^2}(L_0(\alpha_s))\Big|_{s=0}$$

$$= \int_{t=a}^{b} \left[g_0(V',V') - g_0(R(V,c')c',V) - c'(g_0(V,c'))\right]\Big|_t \, dt + g_0(\nabla_V V,c')\Big|_a^b$$

This second variation formula naturally suggests defining an *index form*

$$I(X,Y) = \int_{t=a}^{b} [g(X',Y') - g(R(X,c')c',Y)]\Big|_t \, dt$$

on the infinite dimensional vector space $V_0^{\perp}(c)$ of piecewise smooth vector fields X along c orthogonal to c with $X(a) = X(b) = 0$. It is then shown that a necessary and sufficient condition for c to be free of conjugate points is that $I(X,X) > 0$ for all nontrivial $X \in V_0^{\perp}(c)$.

This suggests that for a geodesic $c : [a,b] \longrightarrow N$ with conjugate points, the index Ind(c) of c with respect to $I : V_0^{\perp}(c) \times V_0^{\perp}(c) \longrightarrow \mathbb{R}$ should be defined as the supremum of dimensions of all vector subspaces of $V_0^{\perp}(c)$ on which I is negative definite. Even though $V_0^{\perp}(c)$ is an infinite dimensional vector space, the Morse index theorem for arbitrary Riemannian manifolds asserts that

(1) Ind(c) is finite.

(2) Ind(c) equals the geodesic index of c, i.e., the number of conjugate points along c counted with multiplicities.

More precisely, if we let $J_t(c)$ denote the vector space of smooth vector fields Y along c satisfying the Jacobi differential equation $Y'' + R(Y,c')c' = 0$ with boundary conditions $Y(a) = Y(t) = 0$, then (2) is equivalent to the formula

$$(3)\quad \mathrm{Ind}(c) = \sum_{t\in(a,b)} \dim J_t(c)$$

Also c has only finitely many points in (a,b] conjugate to c(a).

With the Morse index theorem in hand, the homotopy type of the loop space for *complete* Riemannian manifolds may now be calculated geometrically [cf. Milnor (1963, p. 95)]. One obtains the result that if $(N,g_0)$ is a complete Riemannian manifold, and $p,q \in N$ are any pair of points which are not conjugate along any geodesic, then the loop space $\Omega_{(p,q)}$ of all continuous paths from p to q equipped with the compact-open topology has the homotopy type of a countable CW-complex which contains a cell of dimension $\lambda$ for each geodesic from p to q of index $\lambda$.

The purpose of Sections 9.1 and 9.3 is to prove the analogues of (1) through (3) for nonspacelike geodesics in arbitrary space-times. Let $c : [a,b] \longrightarrow M$ (resp., $\beta : [a,b] \longrightarrow M$) denote an arbitrary timelike (resp., null) geodesic segment in (M,g). Let $V_0^\perp(c)$ [resp., $V_0^\perp(\beta)$] denote the infinite dimensional vector space of piecewise smooth vector fields Y along c (resp., $\beta$) perpendicular to c (resp., $\beta$) with $Y(a) = Y(b) = 0$. The timelike index form $I : V_0^\perp(c) \times V_0^\perp(c) \longrightarrow \mathbb{R}$ may be defined by

$$I(X,Y) = -\int_a^b [g(X',Y') - g(R(X,c')c',Y)]\,dt$$

in analogy with the Riemannian index form. It may also be shown that $c : [a,b] \longrightarrow M$ has no conjugate points in (a,b) iff $I : V_0^\perp(c) \times V_0^\perp(c) \longrightarrow \mathbb{R}$ is negative definite.

Similarly, an index form $I : V_0^{\perp}(\beta) \times V_0^{\perp}(\beta) \longrightarrow \mathbb{R}$ may be defined by $I(X,Y) = -\int_a^b [g(X',Y') - g(R(X,\beta')\beta',Y)]\,dt$. But since $\beta$ is a null geodesic, $g(\beta',\beta') = 0$. Consequently, vector fields of the form $V(t) = f(t)\beta'(t)$ with $f : [a,b] \longrightarrow \mathbb{R}$ piecewise smooth and $f(a) = f(b) = 0$ are always in the null space of $I : V_0^{\perp}(\beta) \times V_0^{\perp}(\beta) \longrightarrow \mathbb{R}$, yet never give rise to null conjugate points.

The way around this difficulty is to consider the quotient bundle $\mathcal{X}_0(\beta)$ of $V_0^{\perp}(\beta)$ formed by identifying $Y_1$ and $Y_2$ in $V_0^{\perp}(\beta)$ if $Y_1 - Y_2 = f\beta'$ for some piecewise smooth function $f : [a,b] \longrightarrow \mathbb{R}$. The index form $I : V_0^{\perp}(\beta) \times V_0^{\perp}(\beta) \longrightarrow \mathbb{R}$ may also be projected to a quotient index form $\overline{I} : \mathcal{X}_0(\beta) \times \mathcal{X}_0(\beta) \longrightarrow \mathbb{R}$. It may then be shown that the null geodesic segment $\beta : [a,b] \longrightarrow M$ has no conjugate points in $[a,b]$ iff $\overline{I} : \mathcal{X}_0(\beta) \times \mathcal{X}_0(\beta) \longrightarrow \mathbb{R}$ is negative definite [cf. Hawking and Ellis (1973, Section 4.5), Bölts (1977, Chapters 2 and 4)].

Let $J_t(c)$ [resp., $J_t(\beta)$] denote the space of Jacobi fields along $c$ (resp., $\beta$) with $Y(a) = Y(b) = 0$. Define the *index* $\mathrm{Ind}(c)$ of $c$ [resp., $\mathrm{Ind}(\beta)$ of $\beta$] to be the supremum of dimensions of all vector subspaces of $V_0^{\perp}(c)$ [resp., $\mathcal{X}_0(\beta)$] on which $I$ (resp., $\overline{I}$) is positive definite. We establish the Morse index theorem

$$\mathrm{Ind}(c) = \sum_{t\in(a,b)} \dim J_t(c)$$

and

$$\mathrm{Ind}(\beta) = \sum_{t\in(a,b)} \dim J_t(\beta)$$

for the timelike geodesic $c : [a,b] \longrightarrow M$ and the null geodesic $\beta : [a,b] \longrightarrow M$ in Sections 9.1 and 9.3, respectively. The reason for considering the timelike and null cases separately is that the quotient index form $\overline{I} : \mathcal{X}_0(\beta) \times \mathcal{X}_0(\beta) \longrightarrow \mathbb{R}$ must be used to obtain the null, but not the timelike, Morse index theorem.

In Section 9.2, we study the theory of timelike loop spaces for globally hyperbolic space-times, summarizing some recent results of Uhlenbeck (1975). Here again, since completeness is needed to develop the Riemannian loop space theory, it is not surprising that global hyperbolicity is needed for the Lorentzian theory.

Yet a significant difference occurs between the Lorentzian and Riemannian loop spaces. Let $(N,g_0)$ be a complete Riemannian manifold with positive Ricci curvature bounded away from zero. Thus N is compact by Myers' theorem. It is shown in Milnor (1963, Theorem 19.6) that if p and q in such a Riemannian manifold are nonconjugate along any geodesic, then the loop space $\Omega_{(p,q)}$ has the homotopy type of a CW-complex having only finitely many cells in each dimension. But the loop space may still be an *infinite* CW-complex as is seen by the example of $S^n$ with the usual complete metric of constant sectional curvature. On the other hand, if (M,g) is an *arbitrary* globally hyperbolic (hence noncompact) space-time, and $p,q \in M$ are any two points with $p \ll q$ such that p and q are not conjugate along any nonspacelike geodesic, then the timelike loop space $C_{(p,q)}$ has only *finitely* many cells.

This striking difference between the Lorentzian and Riemannian loop spaces is a result of the following basic difference between Lorentzian and Riemannian manifolds. If $\gamma$ is any nonspacelike curve from p to q in an arbitrary space-time and d(p,q) is finite, then $\gamma$ has bounded Lorentzian arc length since $L(\gamma) \leqslant d(p,q)$. On the other hand, any curve $\sigma$ in the Riemannian path space satisfies $L_0(\sigma) \geqslant d_0(p,q)$, hence has arc length bounded from below but not above.

## 9.1 THE TIMELIKE MORSE INDEX THEORY

In this section, we give a detailed proof of the Morse index theorem for timelike geodesics in an arbitrary space-time. Several different approaches to Morse index theory for timelike geodesics under various causality conditions have been published by Uhlenbeck (1975), Woodhouse (1976), Everson and Talbot (1976), and Beem and Ehrlich (1979c). Here we give a treatment which parallels the proof of the Morse index theorem for an arbitrary Riemannian manifold in

Gromoll, Klingenberg, and Meyer (1975, Sections 4.5 and 4.6). A similar treatment of the results in this section through Theorem 9.22 has been given in Bölts (1977).

Let (M,g) be an *arbitrary* space-time of dimension $\geq 2$ throughout this section. We will denote by $\langle\ ,\ \rangle$ the Lorentzian metric g in this section. Also, all timelike geodesics $c : [a,b] \longrightarrow M$ will be assumed to have unit speed, i.e., $\langle c'(t),c'(t)\rangle = -1$ for all $t \in [a,b]$.

DEFINITION 9.1 A *piecewise smooth vector field Y along* (the timelike geodesic) $c : [a,b] \longrightarrow M$ is a continuous map $Y : [a,b] \longrightarrow TM$ with $Y(t) \in T_{c(t)}M$ for all $t \in [a,b]$ such that there exists some finite partition $t_0 = a < t_1 < \cdots < t_k = b$ of $[a,b]$ so that $Y \mid [t_i,t_{i+1}] : [t_i,t_{i+1}] \longrightarrow M$ is smooth for each i with $0 \leq i \leq k - 1$. Let $V^\perp(c)$ denote the $\mathbb{R}$ vector space of all piecewise smooth vector fields Y along c with $\langle Y(t),c'(t)\rangle = 0$ for all $0 \leq t \leq b$. Also let $V_0^\perp(c) = \{Y \in V^\perp(c) : Y(a) = Y(b) = 0\}$ and $N(c(t)) = \{v \in T_{c(t)}M : \langle v,c'(t)\rangle = 0\}$.

Given $Y \in V^\perp(c)$, there is a finite set of points $I_0 \subset (a,b)$ such that Y is differentiable on $[a,b] - I_0$. Define $Y' : [a,b] \longrightarrow M$ by $Y'(t) = \nabla_{\gamma'}Y(t)$ for $t \in [a,b] - I_0$ and extend to points $t_i \in I_0$ by setting

$$Y'(t_i) = \lim_{t \to t_i^-} Y'(t)$$

Thus Y' is extended to [a,b] by left continuity but is not necessarily continuous at points of $I_0$.

REMARK 9.2 Since $c'(t)$ is timelike, $N(c(t))$ is a vector space of dimension $n - 1$ consisting of spacelike tangent vectors and thus $\{v \in N(c(t)) : \langle v,v\rangle \leq 1\}$ is compact.

DEFINITION 9.3 Let $c : [a,b] \longrightarrow M$ be a timelike geodesic. Then $t_1,t_2 \in [a,b]$ with $t_1 \neq t_2$ are *conjugate with respect to* c if there is a nontrivial vector field J along c with $J(t_1) = J(t_2) = 0$ satisfying the Jacobi equation $J'' + R(J,c')c' = 0$. Here J' denotes the covariant derivative operator of vector fields along c induced by the Levi-Civita connection of < , > on M. Also $t_1 \in (a,b)$ is said to be a *conjugate point* of the geodesic segment $c : [a,b] \longrightarrow M$ if a and $t_1$ are conjugate along c. The geodesic segment c is said to have *no conjugate points* if no $t_1 \in (a,b]$ is conjugate to t = a along c. A smooth vector field J along c satisfying the differential equation $J'' + R(J,c')c' = 0$ is called a *Jacobi field along* c.

We now define the Lorentzian index form $I : V^{\perp}(c) \times V^{\perp}(c) \longrightarrow \mathbb{R}$.

DEFINITION 9.4 The *index form* $I : V^{\perp}(c) \times V^{\perp}(c) \longrightarrow \mathbb{R}$ is the symmetric bilinear form given by

$$I(X,Y) = -\int_a^b [\langle X',Y'\rangle - \langle R(X,c')c',Y\rangle]\, dt \tag{9.1}$$

for any $X,Y \in V^{\perp}(c)$.

If $X \in V^{\perp}(c)$ is smooth, we also have

$$I(X,Y) = -\langle X',Y\rangle \Big|_a^b + \int_a^b \langle X'' + R(X,c')c',Y\rangle\, dt \tag{9.2}$$

Thus if $Y \in V_0^{\perp}(c)$ and X is smooth, one obtains the formula

$$I(X,Y) = \int_a^b \langle X'' + R(X,c')c',Y\rangle\, dt \tag{9.3}$$

linking the index form to Jacobi fields.

REMARK 9.5 Given a piecewise smooth vector field $X \in V^{\perp}(c)$, we may choose a partition $a = t_0 < t_1 < t_2 < \cdots < t_k = b$ such that $X \mid [t_i,t_{i+1}]$ is smooth for each i = 0, 1, 2, ..., k - 1. Let

$$\Delta_{t_0}(X') = X'(a^+)$$

$$\Delta_{t_k}(X') = -X'(b^-)$$

and

$$\Delta_{t_i}(X') = \lim_{t \to t_i^+} X'(t) - \lim_{t \to t_i^-} X'(t)$$

for i = 1, 2, ..., k - 1. It may then be seen by applying (9.2) on each subinterval $(t_i, t_{i+1})$ that

$$I(X,Y) = \sum_{i=0}^{k} \langle \Delta_{t_i}(X'), Y \rangle + \int_a^b \langle X'' + R(X,c')c', Y \rangle \, dt \qquad (9.4)$$

This is the form of the second variation formula given in Hawking and Ellis (1973, p. 108) and Bölts (1977, pp. 86-87) [cf. Cheeger and Ebin (1975, p. 21)].

In order to give geometric applications using the index form, it is useful to make the following standard definition.

DEFINITION 9.6 Let $c : [a,b] \longrightarrow M$ be a smooth curve. A *variation* (or homotopy) of c is a smooth mapping $\alpha : [a,b] \times (-\varepsilon,\varepsilon) \longrightarrow M$, for some $\varepsilon > 0$, with $\alpha(t,0) = c(t)$ for all $t \in [a,b]$. The variation $\alpha$ is said to be a *proper variation* if $\alpha(a,s) = c(a)$ and $\alpha(b,s) = c(b)$ for all $s \in (-\varepsilon,\varepsilon)$. A continuous mapping $\alpha : [a,b] \times (-\varepsilon,\varepsilon) \longrightarrow M$ is said to be a *piecewise smooth variation* of c if there exists a finite partition $a = t_0 < t_1 < t_2 < \cdots < t_k = b$ of [a,b] such that $\alpha \mid [t_i, t_{i+1}] \times (-\varepsilon,\varepsilon)$ is a smooth variation of $c \mid [t_i, t_{i+1}]$ for each i = 0, 1, ..., k - 1.

If we set $\alpha_s(t) = \alpha(t,s)$ in Definition 9.6, then for a smooth variation $\alpha$, each curve $\alpha_s : [a,b] \longrightarrow M$ is a smooth curve and thus the mapping $s \longrightarrow \alpha_s$ is a deformation of the curve c through the "neighboring curves" $\alpha_s$. If $\alpha$ is a piecewise smooth variation, the neighboring curves $\alpha_s$ will be piecewise smooth. If $\alpha$ is a proper variation, all of the neighboring curves $\alpha_s$ begin at c(a) and end at c(b). It is customary in defining variations of timelike curves to restrict consideration to variations having the additional property that all neighboring curves $\alpha_s : [a,b] \longrightarrow M$ are timelike. However, if we use Definition 9.6, this restriction is unnecessary by the following lemma.

LEMMA 9.7 Let $\alpha : [a,b] \times (-\varepsilon,\varepsilon) \longrightarrow M$ be a piecewise smooth variation of the timelike geodesic segment $c : [a,b] \longrightarrow M$. Then there exists a constant $\delta > 0$ depending on $\alpha$ such that the neighboring curves $\alpha_s$ are timelike for all s with $|s| \leqslant \delta$.

*Proof.* We first suppose that $\alpha$ is a smooth variation. Choose any $\varepsilon_1$ with $0 < \varepsilon_1 < \varepsilon$. Then $\alpha$ is differentiable on the compact set $[a,b] \times [-\varepsilon_1,\varepsilon_1]$ by Definition 9.6. Hence by definition of differentiability, $\alpha$ extends to a smooth mapping of a larger open set containing $[a,b] \times [-\varepsilon_1,\varepsilon_1]$. Since c is a timelike geodesic, the vectors $c'(a^+)$ and $c'(b^-)$ are timelike. It follows from this and the extension of $\alpha$ to an open set containing $[a,b] \times [-\varepsilon_1,\varepsilon_1]$, that there exists a constant $\delta_1 > 0$ such that the tangent vectors $\alpha_s'(a^+)$ and $\alpha_s'(b^-)$ to the neighboring curves $\alpha_s$ are timelike for all s with $|s| < \delta_1$.

Suppose now that no $\delta > 0$ can be found such that all the curves $\alpha_s$ are timelike for $|s| < \delta$. Then we could find a sequence $s_n \longrightarrow 0$ such that the curves $\alpha_{s_n}$ failed to be timelike. Thus there would be $t_n \in [a,b]$ so that $g(\alpha_{s_n}'(t_n),\alpha_{s_n}'(t_n)) \geqslant 0$ for each n. Since $[a,b] \times [-\varepsilon_1,\varepsilon_1]$ is compact, the sequence $\{(t_n,s_n)\}$ has a point of accumulation (t,s). Since $s_n \longrightarrow 0$, this point must be of the form (t,0) and also the existence of $\delta_1$ above shows that $t \neq a$, b. But then as $g(\alpha_s'(t_n),\alpha_s'(t_n)) \geqslant 0$ for each n, it follows that $g(c'(t),c'(t)) \geqslant 0$, in contradiction to the fact that c was a timelike geodesic segment. Thus we have seen that if $\alpha : [a,b] \times (-\varepsilon,\varepsilon) \longrightarrow M$ is a smooth variation of the timelike geodesic $c : [a,b] \longrightarrow M$, then there is a constant $\delta > 0$ such that $\alpha_s'(a^+)$ and $\alpha_s'(b^-)$ are timelike vectors for all $|s| \leqslant \delta$ and the curves $\alpha_s$ are timelike for all $|s| \leqslant \delta$.

Now let $\alpha : [a,b] \times (-\varepsilon,\varepsilon) \longrightarrow M$ be a piecewise smooth variation of the timelike geodesic $c : [a,b] \longrightarrow M$. There is by Definition 9.6 a finite partition $a = t_0 < t_1 < \cdots < t_k = b$ such that $\alpha \mid [t_i,t_{i+1}] \times (-\varepsilon,\varepsilon) \longrightarrow M$ is a smooth variation of $c \mid [t_i,t_{i+1}]$. By the above paragraph, we may find a constant $\delta_i > 0$ such that for all s with $|s| \leqslant \delta_i$, the tangent vectors $\alpha_s'(t_i^+)$ and $\alpha_s'(t_{i+1}^-)$ are timelike and $\alpha_s \mid [t_i,t_{i+1}]$ is a timelike curve. Taking $\delta = \min(\delta_0,\delta_1,\ldots,\delta_{k-1})$ then yields the required $\delta$. □

REMARK 9.8 There is no result corresponding to Lemma 9.7 for variations of null geodesics (cf. Definition 9.58 ff.).

The index form may now be related to variations of timelike geodesic segments $c : [a,b] \longrightarrow M$ as follows. Given $Y \in V_0^{\perp}(c)$, define the *canonical proper variation* $\alpha : [a,b] \times (-\varepsilon,\varepsilon) \longrightarrow M$ of c by setting

$$\alpha(t,s) = \exp_{c(t)}(sY(t)) \tag{9.5}$$

It should first be noted that since $c([a,b])$ is a compact subset of M, and the differential of the exponential map $\exp_{p*}$ is nonsingular at the origin of $T_pM$ for all $p \in M$, it is possible given $c([a,b])$ to find an $\varepsilon > 0$ such that $\exp_{c(t)}(sY(t))$ is defined for all s with $|s| \leqslant \varepsilon$ for each $t \in [a,b]$. Secondly, from Definition 9.1, it follows that $\alpha(t,s)$ defined as in (9.5) is a piecewise smooth variation of c. Hence given $Y \in V_0^{\perp}(c)$, we know from Lemma 9.7 that there exists some constant $\delta > 0$ such that all the neighboring curves $\alpha_s : t \longrightarrow \alpha(t,s)$ are all timelike for all s with $-\delta < s < \delta$.

Given an arbitrary smooth variation $\alpha : [a,b] \times (-\varepsilon,\varepsilon) \longrightarrow M$ of $c : [a,b] \longrightarrow M$, the *variation vector field* V of $\alpha$ is defined to be the vector field $V(t)$ along c given by the formula

$$V(t) = \frac{d}{ds}(\alpha(t,s))\Big|_{s=0} \tag{9.6}$$

More precisely, letting $\partial/\partial s$ be the coordinate vector field on $[a,b] \times (-\varepsilon,\varepsilon)$ corresponding to the s parameter, the variation vector field is given by

$$V(t) = \alpha_* \left.\frac{\partial}{\partial s}\right|_{(t,0)} \tag{9.7}$$

For a piecewise smooth variation $\alpha : [a,b] \times (-\varepsilon,\varepsilon) \longrightarrow M$, one obtains a continuous piecewise smooth variation vector field as follows. Let $a = t_0 < t_1 < t_2 < \cdots < t_k = b$ be a partition of $[a,b]$ such that $\alpha \mid [t_i,t_{i+1}] \times (-\varepsilon,\varepsilon)$ is smooth for $i = 0, 1, \ldots, k - 1$. Given $t \in [a,b]$, choose an index i such that $t_i \leqslant t < t_{i+1}$ and set

$$V(t) = [\alpha | [t_i,t_{i+1}] \times (-\varepsilon,\varepsilon)]_* \left.\frac{\partial}{\partial s}\right|_{(t,0)}$$

The canonical variation (9.5) has the property that each curve $s \longrightarrow \alpha(t,s)$ is just the geodesic $s \longrightarrow \exp_{c(t)}(sY(t))$ which has initial direction $Y(t)$ at $s = 0$. Hence the variation vector field of the canonical variation (9.5) is just the given vector field $Y \in V_0^\perp(c)$.

If we put $L(s) := L(\alpha_s) = L(t \longrightarrow \alpha(t,s))$, then $L'(0) = d/ds\, L(s)\big|_{s=0} = 0$ since c is a smooth timelike geodesic and

$$L''(0) = \frac{d^2}{ds^2} L(s)\bigg|_{s=0} = I(Y,Y) \tag{9.8}$$

Thus if $I(Y,Y) > 0$ for some $Y \in V_0^\perp(c)$, the canonical proper variation $\alpha(t,s)$ defined by (9.5) using Y will produce timelike neighboring curves $\alpha_s$ joining $c(a)$ to $c(b)$ with $L(\alpha_s) > L(c)$ for s sufficiently small. Thus if the timelike geodesic $c : [a,b] \longrightarrow M$ is maximal [i.e., $L(c) = d(p,q)$], then $I : V_0^\perp(c) \times V_0^\perp(c) \longrightarrow \mathbb{R}$ must be negative semidefinite. Before proving the Morse index theorem for timelike geodesics, we must establish the following more precise relationship between conjugate points, Jacobi fields, and the index form. First, the null space of the index form on $V_0(c)$ consists of the *smooth* Jacobi fields in $V_0(c)$ and second, c has no conjugate points on $[a,b]$ iff the index form is negative definite on $V_0(c)$.

We first derive an elementary but important consequence of the Jacobi differential equation.

LEMMA 9.9 Let $c : [a,b] \longrightarrow M$ be a timelike geodesic segment and let $Y$ be any Jacobi field along $c$. Then $\langle Y(t), c'(t) \rangle$ is an affine function of $t$, i.e., $\langle Y(t), c'(t) \rangle = \alpha t + \beta$ for some constants $\alpha, \beta \in \mathbb{R}$.

*Proof.* First $\langle Y, c' \rangle' = \langle Y', c' \rangle + \langle Y, c'' \rangle = \langle Y', c' \rangle$ as $c'' = \nabla_{c'} c' = 0$ since $c$ is a geodesic. Differentiating again, we obtain $\langle Y, c' \rangle'' = \langle Y'', c' \rangle + \langle Y', c'' \rangle = \langle Y'', c' \rangle = -\langle R(Y, c')c', c' \rangle = 0$ by the skew symmetry of the Riemann-Christoffel tensor in the first two slots. Thus $\langle Y, c' \rangle$ is an affine function. □

COROLLARY 9.10 If $Y$ is any Jacobi field along the timelike geodesic $c : [a,b] \longrightarrow M$ and $Y(t_1) = Y(t_2) = 0$ for distinct $t_1, t_2 \in [a,b]$, then $Y \in V^{\perp}(c)$.

COROLLARY 9.11 If $Y \in V_0(c)$ is a Jacobi field, then $Y' \in V^{\perp}(c)$.

Using the canonical variation, we are now ready to derive the following geometric consequence of the existence of a conjugate point $t_0 \in (a,b)$ to $t = a$ along $c$.

PROPOSITION 9.12 Suppose that the timelike geodesic $c : [a,b] \longrightarrow M$ contains a conjugate point $t_0 \in (a,b)$ to $t = a$ along $c$. Then there exists a piecewise smooth proper variation $\alpha_s$ of $c$ such that $L(\alpha_s) > L(c)$ for all $s \neq 0$. Thus $c : [a,b] \longrightarrow M$ is not maximal.

*Proof.* In view of (9.5), (9.8), and Lemma 9.7, it is enough to construct a piecewise smooth vector field $Y \in V_0^{\perp}(c)$ with $I(Y,Y) > 0$ and let $\alpha$ be the canonical variation associated with $Y$. To this end, let $Y_1$ be a nontrivial Jacobi field along $c$ with $Y_1(a) = Y_1(t_0) = 0$. By Corollary 9.10, $Y_1 \in V^{\perp}(c)$. Hence as $Y_1(a) = Y_1(t_0) = 0$, we have $Y_1' \in V^{\perp}(c)$ by Corollary 9.11. Since $Y_1(t_0) = 0$ and $Y_1$ is a nontrivial Jacobi field, it follows that $Y_1'(t_0)$ is a (nonzero) spacelike tangent vector.

Let $I(\ ,\ )_a^s$ denote the restriction of the index form to $c \mid [a,s]$, that is,

$$I(V,W)_a^s = - \int_a^s [\langle V',W'\rangle - \langle R(V,c')c',W\rangle]\, dt$$

Then since $Y_1$ is a Jacobi field, we have from (9.2) of Definition 9.4 that

$$I(Y_1,Z)_a^s = - \langle Y_1',Z\rangle \Big|_a^s \tag{9.9}$$

for any $Z \in V^\perp(c)$.

We are now ready to construct a piecewise smooth vector field $Y \in V_0^\perp(c)$ with $I(Y,Y) > 0$. Let $\psi : [a,b] \longrightarrow \mathbb{R}$ be a smooth function with $\psi(a) = \psi(b) = 0$ and $\psi(t_0) = 1$. Also let $Z_1$ be the unique smooth parallel vector field along c with $Z_1(t_0) = -Y_1'(t_0)$. Then $Z = \psi Z_1 \in V_0^\perp(c)$. Define a 1-parameter family $Y_\varepsilon \in V_0^\perp(c)$ by

$$Y_\varepsilon(t) = \begin{cases} Y_1(t) + \varepsilon Z(t) & \text{if } a \leqslant t \leqslant t_0 \\ \\ \varepsilon Z(t) & \text{if } t_0 \leqslant t \leqslant b \end{cases}$$

Then using (9.9) we obtain

$$\begin{aligned} I(Y_\varepsilon,Y_\varepsilon) &= I(Y_\varepsilon,Y_\varepsilon)_a^{t_0} + I(Y_\varepsilon,Y_\varepsilon)_{t_0}^b \\ &= I(Y_1 + \varepsilon Z, Y_1 + \varepsilon Z)_a^{t_0} + I(\varepsilon Z,\varepsilon Z)_{t_0}^b \\ &= I(Y_1,Y_1)_a^{t_0} + 2\varepsilon I(Y_1,Z)_a^{t_0} + \varepsilon^2 I(Z,Z)_a^{t_0} + \varepsilon^2 I(Z,Z)_{t_0}^b \\ &= - \langle Y_1',Y_1\rangle \Big|_a^{t_0} - 2\varepsilon\langle Y_1',Z\rangle \Big|_a^{t_0} + \varepsilon^2 I(Z,Z) \end{aligned}$$

Since $Y_1(a) = Y_1(t_0) = 0$, this simplifies to

$$I(Y_\varepsilon, Y_\varepsilon) = -2\varepsilon\langle Y_1'(t_0), Z(t_0)\rangle + \varepsilon^2 I(Z,Z)$$

$$= 2\varepsilon\|Y_1'(t_0)\|^2 + \varepsilon^2 I(Z,Z)$$

As $Y_1'(t_0)$ is a (nonzero) spacelike tangent vector and $I(Z,Z)$ is finite, it follows that there is some $\varepsilon > 0$ such that $I(Y_\varepsilon, Y_\varepsilon) > 0$. Put the required vector field $Y = Y_\varepsilon$ for this value of $\varepsilon$. □

We now turn to an important characterization of Jacobi fields in terms of the index form. The same characterization holds for Riemannian spaces with an identical proof. It is important to note that the index form characterizes smooth Jacobi fields among all *piecewise smooth* vector fields in $V_0^\perp(c)$, not just among all smooth vector fields in $V_0^\perp(c)$.

PROPOSITION 9.13 Let $c : [a,b] \longrightarrow (M,g)$ be a timelike geodesic segment. Then for $Y \in V_0^\perp(c)$, the following are equivalent:

(a) Y is a (smooth) Jacobi field along c.

(b) $I(Y,Z) = 0$ for all $Z \in V_0^\perp(c)$.

*Proof.* First (a) implies (b) is immediate, since for smooth vector fields Y and arbitrary Z the index form may be written as

$$I(Y,Z) = -\langle Y', Z\rangle\Big|_a^b + \int_a^b \langle Y'' + R(Y,c')c', Z\rangle\, dt$$

If Y is a Jacobi field, $I(Y,Z) = -\langle Y', Z\rangle\Big|_a^b$. Thus $I(Y,Z)$ vanishes for all $Z \in V_0^\perp(c)$.

To show (b) implies (a), we first note that since c is a timelike geodesic segment and $Y \in V_0^\perp(c)$, that $\langle Y', c'\rangle = 0$ and $\langle Y'' + R(Y,c')c', c'\rangle = 0$ at all points where Y is differentiable. Taking left-hand limits, we have $\langle Y'(t_i^-), c'(t_i)\rangle = 0$ also at the finitely many points of discontinuity $a = t_0 < t_1 < t_2 < \cdots < t_k = b$ of Y. By continuity, the right-hand limit $\lim_{t\to t_i^+} \langle Y'(t), c'(t)\rangle = 0$ as well. Hence the vectors $\Delta_{t_i}(Y')$ defined in Remark 9.5 also satisfy $\langle \Delta_{t_i}(Y'), c'(t_i)\rangle = 0$. By (9.4) of Remark 9.5, the index form may be calculated as

$$I(Y,Z) = \sum_{i=0}^{k} \langle \Delta_{t_i}(Y'), Z(t_i)\rangle + \int_a^b \langle Y'' + R(Y,c')c', Z\rangle \, dt \quad (9.10)$$

Let $\phi : [a,b] \longrightarrow [0,1]$ be a smooth function with $\phi(t_0) = \phi(t_1) = \cdots = \phi(t_k) = 0$ and $\phi(t) > 0$ elsewhere. Then the vector field $Z_1 = \phi(Y'' + R(Y,c')c')$ is in $V_0^{\perp}(c)$ and $Z_1(t_i) = 0$ for all i. Since $I(Y,Z) = 0$ is assumed for all $Z \in V_0^{\perp}(c)$, we obtain from (9.10) that

$$0 = I(Y,Z_1) = \int_a^b \phi(t) \| Y'' + R(Y,c')c' \|^2 \Big|_t \, dt$$

As $Z_1$ is a spacelike vector field, smooth except at the $t_i$'s and $\phi(t) > 0$ if $t \notin \{t_i\}$, we obtain $Y''(t) + R(Y(t),c'(t))c'(t) = 0$ if $t \notin \{t_i\}$. Thus Y is a piecewise Jacobi field and formula (9.10) reduces to

$$I(Y,Z) = \sum_{i=0}^{k} \langle \Delta_{t_i}(Y'), Z(t_i)\rangle \quad (9.11)$$

Recalling from above that $\langle \Delta_{t_i}(Y'), c'(t_i)\rangle = 0$ for each i, a vector field $Z_2 \in V_0^{\perp}(c)$ may be constructed with $Z_2(t_i) = \Delta_{t_i}(Y')$ for $i = 1, \ldots, k - 1$. Then we have

$$0 = I(Y,Z_2) = \sum_{i=1}^{k-1} \| \Delta_{t_i}(Y') \|^2$$

Since all of the tangent vectors in this sum are spacelike, it follows that $\Delta_{t_i}(Y') = 0$ for $i = 1, \ldots, k - 1$. This then implies that Y' has no breaks at the $t_i$'s. Since for any $t \in [a,b]$, there is a unique Jacobi field along c with $Y(t) = v$, $Y'(t) = w$, it follows that the Jacobi fields $Y \mid [t_i, t_{i+1}]$ fit together to form a smooth Jacobi field. □

In view of Propositions 9.12 and 9.13, it should come as no surprise that the negative definiteness of the Lorentzian index form should be related to the absence of conjugate points just as the positive definiteness of the Riemannian index form is guaranteed by no conjugate points [Gromoll, Klingenberg, and Meyer (1975, p. 145)]. The negative semidefiniteness of the index form in the absence of conjugate points has been given in Hawking and Ellis (1973, Lemma 4.5.8). It has been noted in Bölts (1977, Satz 4.4.5] and Beem and Ehrlich (1979c, p. 376) that the negative definiteness of the index form in the absence of conjugate points follows "algebraically" from the semidefiniteness just as in the proof of positive definiteness for the Riemannian index form. In order to give a proof of the negative semidefiniteness of the Lorentzian index form in the absence of conjugate points, we need to obtain the Lorentzian analogues of several important results in Riemannian geometry [cf. Gromoll, Klingenberg, and Meyer (1975, pp. 132, 136-137, 140)].

For the purpose of constructing Jacobi fields, it is useful to introduce some notation for the identification of the tangent space $T_v(T_pM)$ with $T_pM$ itself by "parallel translation in $T_pM$."

DEFINITION 9.14 Given any $p \in M$ and $v \in T_pM$, the *tangent space* $T_v(T_pM)$ *to the tangent space* $T_pM$ *at* $v$ is given by

$$T_v(T_pM) = \{\phi_w : \mathbb{R} \longrightarrow T_pM\}$$

where

$$\phi_w(t) = v + tw$$

Then $T_v(T_pM)$ may intuitively be identified with $T_pM$ by identifying the image of $\phi_w$ in $T_pM$ with the vector $w$. More formally, let $\mathbb{R}$ be given the usual manifold coordinate chart $x(r) = r$ for all $r \in \mathbb{R}$, i.e., $x = \text{id}$. Then $\partial/\partial x$ is a vector field on $\mathbb{R}$. Since $\phi_w :$ $\mathbb{R} \longrightarrow T_pM$ and $\phi_w(0) = v$, we have $\phi_{w_*} : T_0\mathbb{R} \longrightarrow T_v(T_pM)$. We may then make the following definition.

DEFINITION 9.15 The *canonical isomorphism* $\tau_v : T_pM \longrightarrow T_v(T_pM)$ is given by

$$\tau_v w := \phi_{w_*} \frac{\partial}{\partial x}\bigg|_0 = \phi'(0)$$

where

$$\phi_w(t) = v + tw.$$

In particular, let $v = 0_p$, the zero vector in the tangent space $T_pM$. Then $\phi_w : \mathbb{R} \longrightarrow T_pM$ is the curve $\phi_w(t) = tw$ in $T_{0_p}(T_pM)$ and we find that $\tau_{0_p}(w) = \phi_w$. Thus $T_{0_p}(T_pM)$ is often canonically identified with $T_pM$ itself by identifying the vector $w \in T_pM$ and the map $\phi_w$. If $p \in M$ and $v \in T_pM$, then since $\exp_p : T_pM \longrightarrow M$, the definition of the differential gives

$$\exp_{p_*} : T_v(T_pM) \longrightarrow T_{\exp_p(v)}M$$

In particular, for $v = 0_p$, we have

$$\exp_{p_*} : T_{0_p}(T_pM) \longrightarrow T_pM$$

since $\exp_p(0_p) = p$. If $b = \tau_{0_p}(v) \in T_{0_p}(T_pM)$ and we define $\phi : \mathbb{R} \longrightarrow T_pM$ by $\phi(t) = tv$, then $\exp_{p_*}(b) = \exp_{p_*}(\phi_* \, \partial/\partial x|_0)$, where $\partial/\partial x$ is the usual basis vector field for $T\mathbb{R}$ defined above. Thus we obtain

$$\exp_{p_*}(b) = (\exp_{p_*} \circ \phi_*)\left(\frac{\partial}{\partial x}\bigg|_0\right) = (\exp_p \circ \phi)_*\left(\frac{\partial}{\partial x}\bigg|_0\right)$$

$$= \frac{d}{dt}(t \longrightarrow \exp_p(tv))\bigg|_{t=0} = v$$

Thus $\exp_{p_{*_{0_p}}} \circ \tau_v = \mathrm{Id}_{T_pM}$. This fact is commonly stated as "the differential of the exponential map at the origin of $T_pM$ is the identity."

The following proposition shows how the differential of the exponential map may be used to construct Jacobi fields.

PROPOSITION 9.16 Let $c : [0,b] \longrightarrow M$ be a geodesic with $c(0) = p$. Let $w \in T_pM$ be arbitrary. Then the unique Jacobi field J along c with $J(0) = 0$ and $J'(0) = w$ is given by

$$J(t) = \exp_{p*}(t\tau_{tc'(0)}w)$$

*Proof.* Set $v = c'(0)$. We may find an $\varepsilon > 0$ so that the smooth variation $\alpha : [a,b] \times (-\varepsilon,\varepsilon) \longrightarrow M$ of $c : [a,b] \longrightarrow M$ given by $\alpha(t,s) = \exp_p(t(v + sw))$ is defined. Since this is a variation of c whose s-parameter curves $t \longrightarrow \alpha(t,s)$ are geodesics, it follows that the variation vector field of this deformation is a Jacobi field. Since $\alpha_* \partial/\partial s\big|_{(t,s)} = \exp_{p*}(\tau_{t(v+sw)}(w))$, the variation vector field is just $J(t) = \exp_{p*}(\tau_{tv}tw) = \exp_{p*}(t\tau_{tv}w)$. Since $\alpha(0,s) = c(0)$ for all s, we have $J(0) = 0$ and a calculation also gives $J'(0) = w$ [cf. Gromoll, Klingenberg, and Meyer (1975, p. 132)]. □

As in the Riemannian theory, it is now possible to prove the Gauss lemma using Proposition 9.16. We first need to put a natural inner product $\langle\langle\ ,\ \rangle\rangle$ on $T_v(T_pM)$ using the given Lorentzian metric $\langle\ ,\ \rangle$ on $T_pM$ and the canonical isomorphism.

DEFINITION 9.17 The inner product $\langle\langle\ ,\ \rangle\rangle : T_v(T_pM) \times T_v(T_pM) \longrightarrow \mathbb{R}$ associated with the Lorentzian metric $\langle\ ,\ \rangle$ for M is given by $\langle\langle a,b\rangle\rangle = \langle\tau_v^{-1}(a),\tau_v^{-1}(b)\rangle$ for any $a,b \in T_v(T_pM)$.

Now we are ready to prove

THEOREM 9.18 (Gauss Lemma) Let $v \in T_pM$ be a tangent vector in the domain of definition of the exponential mapping and let $a := \tau_v(v) \in T_v(T_pM)$. Then for any $b \in T_v(T_pM)$, we have

$$\langle\langle a,b\rangle\rangle = \langle\exp_{p*}a,\exp_{p*}b\rangle \tag{9.12}$$

Thus the exponential map is a "radial isometry."

*Proof.* If $\phi(t) = tv$, then $a = \phi'(1)$. If c is the geodesic $c(t) = \exp_p(tv) = \exp_p \circ \phi(t)$, we then have $\exp_{p_*} a = c'(1)$. Also set $w := \tau_v^{-1}(b) \in T_pM$. Let Y be the unique Jacobi field along c with $Y(0) = 0$ and $Y'(0) = w$. By Proposition 9.16, we know that $Y(t) = \exp_{p_*}(t\tau_{tv}w)$. In particular, $Y(1) = \exp_{p_*}(\tau_v w) = \exp_{p_*}(b)$.

From Definition 9.17, we have $\langle\langle a,b\rangle\rangle = \langle\tau_v^{-1}(a),\tau_v^{-1}(b)\rangle = \langle v,w\rangle$. Hence the Gauss lemma is proved if we show that $\langle v,w\rangle = \langle c'(1),Y(1)\rangle$. But by Lemma 9.9, the function $f(t) = \langle c'(t),Y(t)\rangle = \alpha t + \beta$ for some constants $\alpha,\beta \in \mathbb{R}$. Since $Y(0) = 0$, we have $\beta = 0$ and $f(t) = tf'(0) = t\langle c'(0),Y'(0)\rangle = t\langle v,w\rangle$. In particular, $\langle c'(1),Y(1)\rangle = f(1) = \langle v,w\rangle$ as required. □

The Gauss lemma has the following geometric consequences. The proof of these corollaries which are given along the lines of Gromoll, Klingenberg, and Meyer (1975, pp. 137-138) in Bölts (1977, pp. 75-77) will be omitted. The use of the Gauss lemma here replaces the use of a synchronous coordinate system in Penrose (1972, p. 53).

COROLLARY 9.19 Let U be a convex normal neighborhood in M and let $c : [0,1] \longrightarrow U$ be a future-directed timelike geodesic segment from $p = c(0)$ to $q = c(1)$ in U. Then if $\beta : [0,1] \longrightarrow U$ is any future-directed timelike piecewise smooth curve from p to q, we have $L(\beta) \leq L(c)$ and $L(\beta) < L(c)$ unless $\beta$ is just a reparameterization of c.

The basic idea of the proof is that since U is convex, $\beta$ and c may be lifted to rays $\tilde{\beta} : [0,1] \longrightarrow T_pM$, $\tilde{c} : [0,1] \longrightarrow T_pM$, with $\tilde{c}(t) = tc'(0)$. Then the Gauss lemma may be applied to compare $\beta' = \exp_{p_*} \circ \tilde{\beta}$ and $c' = \exp_{p_*} \circ \tilde{c}$, and hence the lengths of $\beta$ and the geodesic c.

An alternative formulation of Corollary 9.19 is also given in Bölts (1977, pp. 75-77) as follows.

COROLLARY 9.20 Let $v \in T_pM$ be a timelike tangent vector in the domain of definition of $\exp_p$. Let $\phi : [0,1] \longrightarrow T_pM$ be the curve $\phi(t) = tv$. Let $\psi : [0,1] \longrightarrow T_pM$ be a piecewise smooth curve with $\psi(0) = \phi(0)$ and $\psi(1) = \phi(1)$ such that $\exp_p \circ \psi : [0,1] \longrightarrow M$ is a future-directed nonspacelike curve. Then $L(\exp \circ \psi) \leqslant L(\exp \circ \phi)$, and moreover,

$$L(\exp \circ \psi) < L(\exp \circ \phi)$$

provided that there is a $t_0 \in (0,1]$ such that the component b of $\psi'(t_0)$ perpendicular to $\tau_{\psi(t_0)}(\psi(t_0)/\|\psi(t_0)\|)$ satisfies $\exp_{p_*} b \neq 0$.

We are now ready to show that if the timelike geodesic segment c has no conjugate points (recall Definition 9.3), then the length of c is a local maximum.

PROPOSITION 9.21 Let $c : [a,b] \longrightarrow M$ be a future-directed timelike geodesic segment with no conjugate points and let $\alpha : [a,b] \times (-\varepsilon,\varepsilon) \longrightarrow M$ be any proper piecewise smooth variation of c. Then there exists a constant $\delta > 0$ such that the neighboring curves $\alpha_s : [a,b] \longrightarrow M$ given by $\alpha_s(t) = \alpha(t,s)$ satisfy $L(\alpha_s) \leqslant L(c)$ for all s with $|s| < \delta$. Also $L(\alpha_s) < L(c)$ if $0 < |s| < \delta$ unless the curve $\alpha_s$ is a reparameterization of c.

*Proof*. Reparameterize $\alpha$ to a variation $\alpha : [0,\beta] \times (-\varepsilon,\varepsilon) \longrightarrow M$. By Lemma 9.7, there is an $\varepsilon_1 > 0$ such that all the neighboring curves $\alpha_s$ of $\alpha \mid [0,\beta] \times (-\varepsilon_1,\varepsilon_1)$ are timelike. We may then restrict our attention to $\alpha \mid [0,\beta] \times (-\varepsilon_1,\varepsilon_1)$.

Set $p = c(0)$ and let $\phi : [0,\beta] \longrightarrow T_pM$ be the ray $\phi(t) = tc'(0)$. Since c has no conjugate points, $\exp_p$ has maximal rank at $\phi(t) \in T_pM$ for each $t \in [0,\beta]$. Thus by the inverse function theorem, there is a neighborhood of $\phi(t)$ in $T_pM$ which is mapped by $\exp_p$ diffeomorphically onto a neighborhood of $c(t)$ in M. Since $\phi([0,\beta])$ is compact in $T_pM$, we can find a finite partition $0 = t_0 < t_1 < \cdots < t_k = \beta$ and a neighborhood $U_j \supset \phi([t_j,t_{j+1}])$ in $T_pM$ for each $j = 0, 1, \ldots, k-1$ such that $h_j = \exp_p \mid U_j : U_j \longrightarrow M$ is a diffeomorphism of $U_j$

onto its image. By continuity, we may then find a constant $\delta_j > 0$ such that $\alpha([t_j,t_{j+1}] \times (-\delta_j,\delta_j)) \subset \exp_p(U_j)$ for each $j = 0, 1, \ldots, k - 1$. Set $\delta = \min\{\delta_1,\delta_2,\ldots,\delta_k\}$.

Then we may define a piecewise smooth map $\Phi : [0,\beta] \times (-\delta,\delta) \longrightarrow T_pM$ with $\exp_p \circ \Phi = \alpha$ and $\Phi(t,0) = \phi(t)$ for all $t \in [0,\beta]$ as follows. Given $(t,s) \in [0,\beta] \times [-\delta,\delta]$, choose $j$ with $t \in [t_j,t_{j+1}]$ and set $\Phi(t,s) = (h_j)^{-1}(\alpha(t,s))$.

Corollary 9.20 then implies that $L(\alpha_s) = L(\exp_p \circ \Phi) \leqslant L(\exp_p \circ \phi) = L(c)$ for each $s$ with $|s| \leqslant \delta$ and equality holds only if $\alpha_s$ is a reparameterization of $c$. □

With Proposition 9.21 in hand, we may now show that the negative definiteness of the index form on $V_0^\perp(c)$ is equivalent to the assumption that $c$ has no points conjugate to $t = a$ along $c$ in $[a,b]$.

THEOREM 9.22 For a future-directed timelike geodesic segment $c : [a,b] \longrightarrow M$ the following are equivalent:

(a) $c$ has no conjugate points in $(a,b]$.

(b) $I : V_0^\perp(c) \times V_0^\perp(c) \longrightarrow \mathbb{R}$ is negative definite.

*Proof.* (a) ⇒ (b) Suppose $Y \neq 0$ in $V_0^\perp(c)$ and $I(Y,Y) > 0$. Let $\alpha(t,s) = \exp_{c(t)}(sY(t))$ be the canonical variation associated to $Y$. Then we have $L'(0) = 0$, $L''(0) = I(Y,Y) > 0$ so that for all $s \neq 0$ sufficiently small, $L(\alpha_s) > L(c)$. But this then contradicts Proposition 9.21. Hence if $Y \neq 0$, $I(Y,Y) \leqslant 0$, so that the index form is negative semidefinite.

It remains to show that if $Y \in V_0^\perp(c)$ and $I(Y,Y) = 0$, then $Y = 0$. To this end, let $Z \in V_0^\perp(c)$ be arbitrary. By Remark 9.2, we have $Y - tZ \in V_0^\perp(c)$ for all $t \in \mathbb{R}$. Hence $I(Y - tZ, Y - tZ) \leqslant 0$ for all $t \in \mathbb{R}$ by the negative semidefiniteness of the index form just established. Since $I(Y - tZ, Y - tZ) = -2tI(Y,Z) + t^2I(Z,Z)$, it follows that $I(Y,Z) = 0$. As $Z$ was arbitrary, this then implies by Proposition 9.13 (b) that $Y$ is a Jacobi field. Since $c$ has no conjugate points, $Y = 0$.

(b) ⇒ (a) Suppose $Y$ is a Jacobi field with $Y(a) = Y(t_1) = 0$, $a < t_1 \leqslant b$. By Corollary 9.10, $Y \in V^\perp(c)$. Extend $Y$ to a nontrivial vector field $Z \in V_0^\perp(c)$ by setting

$$Z(t) = \begin{cases} Y(t) & \text{if } a \leq t \leq t_1 \\ 0 & \text{if } t_1 \leq t \leq b \end{cases}$$

Then $I(Z,Z) = I(Z,Z)_a^{t_1} + I(Z,Z)_{t_1}^b = - \langle Z',Z\rangle\Big|_a^{t_1} + 0 = 0.$ □

A consequence of Theorem 9.22 that is crucial to the proof of the timelike Morse index theorem is the following maximality property of Jacobi fields with respect to the index form for timelike geodesic segments without conjugate points. This result is dual to the minimality of Jacobi fields with respect to the index form for geodesics without conjugate points in Riemannian manifolds.

THEOREM 9.23 (Maximality of Jacobi Fields) Let $c : [a,b] \longrightarrow M$ be a timelike geodesic segment with no conjugate points and let $J \in V^\perp(c)$ be any Jacobi field. Then for any vector field $Y \in V^\perp(c)$ with $Y \neq J$ and

$$Y(a) = J(a) \qquad Y(b) = J(b) \tag{9.13}$$

we have

$$I(J,J) > I(Y,Y) \tag{9.14}$$

*Proof.* The vector field $W = J - Y \in V_0^\perp(c)$ by (9.13) and $W \neq 0$ as $Y \neq J$ by hypothesis. By Theorem 9.22, we thus have $I(W,W) < 0$. Now calculating $I(W,W)$ we obtain

$$\begin{aligned} I(W,W) &= I(Y,Y) - 2I(J,Y) + I(J,J) \\ &= I(Y,Y) + 2\langle J',Y\rangle\Big|_a^b - \langle J',J\rangle\Big|_a^b \end{aligned}$$

Since $Y(a) = J(a)$ and $Y(b) = J(b)$, we have $\langle J',Y\rangle\Big|_a^b = \langle J',J\rangle\Big|_a^b$. Thus

$$I(W,W) = I(Y,Y) + 2\langle J',J\rangle\Big|_a^b - \langle J',J\rangle\Big|_a^b$$

$$= I(Y,Y) + \langle J',J\rangle\Big|_a^b = I(Y,Y) - I(J,J)$$

As $I(W,W) < 0$, this establishes (9.14). □

Now that Theorem 9.23 is obtained, a Morse index theorem may be established. First we must define the index of any timelike geodesic $c : [a,b] \longrightarrow M$. The definition given makes sense because $V_0^\perp(c)$ is a vector space.

DEFINITION 9.24 The *extended index* $\mathrm{Ind}_0(c)$ and the *index* $\mathrm{Ind}(c)$ of the timelike geodesic $c : [a,b] \longrightarrow M$ are defined by $\mathrm{Ind}_0(c) = \mathrm{lub}\{\dim A : A$ is a vector subspace of $V_0^\perp(c)$ and $I \mid A \times A$ is positive semidefinite$\}$ and $\mathrm{Ind}(c) = \mathrm{lub}\{\dim A : A$ is a vector subspace of $V_0^\perp(c)$ and $I \mid A \times A$ is positive definite$\}$. Also let $J_t(c)$ denote the $\mathbb{R}$ vector space of smooth Jacobi fields $Y$ along $c$ with $Y(a) = Y(t) = 0$, $a < t \leq b$.

We now relate $\mathrm{Ind}(c)$ to $\mathrm{Ind}_0(c)$ and establish their finiteness in Proposition 9.25. The maximality of Jacobi fields with respect to the index form for timelike geodesics without conjugate points plays a key role in the proof of this proposition. The basic ideas involved in the proof of Proposition 9.25 and Theorem 9.27 stem from Marston Morse (1934).

PROPOSITION 9.25 Let $c : [a,b] \longrightarrow M$ be a future-directed timelike geodesic segment. Then $\mathrm{Ind}(c)$ and $\mathrm{Ind}_0(c)$ are finite and

$$\mathrm{Ind}_0(c) = \mathrm{Ind}(c) + \dim J_b(c) \tag{9.15}$$

*Proof.* The proof follows the usual method of approximating $V_0^\perp(c)$ by finite dimensional vector spaces of piecewise smooth Jacobi fields. To this end, choose a finite partition

$a = t_0 < t_1 < t_2 < \cdots < t_k = b$ so that $c \mid [t_i, t_{i+1}]$ has no conjugate points for each i with $0 \leq i \leq k - 1$. Let $J\{t_i\}$ denote the subspace of $V_0^{\perp}(c)$ consisting of all $Y \in V_0^{\perp}(c)$ such that $Y \mid [t_i, t_{i+1}]$ is a Jacobi field for each i with $0 \leq i \leq k - 1$. Since $c \mid [t_i, t_{i+1}]$ has no conjugate points for each i, it follows that $\dim J\{t_i\} = (n - 1)(k - 1)$.

We now define the approximation $\phi : V_0^{\perp}(c) \longrightarrow J\{t_i\}$ of $V_0^{\perp}(c)$ by $J\{t_i\}$ as follows. For $X \in V_0^{\perp}(c)$ let $(\phi X) \mid [t_i, t_{i+1}]$ be the unique Jacobi field along $c \mid [t_i, t_{i+1}]$ with $(\phi X)(t_i) = X(t_i)$ and $(\phi X)(t_{i+1}) = X(t_{i+1})$ for each i with $0 \leq i \leq k - 1$. Thus X is approximated by a piecewise smooth Jacobi field $\phi X$ such that $(\phi X)(t_i) = X(t_i)$ at each $t_i$, $0 \leq i \leq k$. This approximation is useful in this context as $\phi$ is index nondecreasing. More explicitly, $\phi \mid J\{t_i\}$ is just the identity map, so that $I(X,X) = I(\phi X, \phi X)$ if $X \in J\{t_i\}$. On the other hand, if $X \notin J\{t_i\}$, then

$$I(X,X) < I(\phi X, \phi X) \tag{9.16}$$

by Theorem 9.23 applied to each subinterval $[t_i, t_{i+1}]$ and summing.

We establish the following sublemma which shows the finiteness of $\mathrm{Ind}_0(c)$ and $\mathrm{Ind}(c)$ and also allows us to replace $V_0^{\perp}(c)$ by $J\{t_i\}$ in calculating these indexes.

SUBLEMMA 9.26 Let $\mathrm{Ind}_0'(c)$ and $\mathrm{Ind}'(c)$ denote the extended index and index, respectively, of $I \mid J\{t_i\} \times J\{t_i\}$. Then

$$\mathrm{Ind}_0(c) = \mathrm{Ind}_0'(c) \qquad \mathrm{Ind}(c) = \mathrm{Ind}'(c)$$

Hence $\mathrm{Ind}_0(c)$ and $\mathrm{Ind}(c)$ are finite.

*Proof.* First it is easily seen by the uniqueness of the Jacobi field J along $c \mid [t_i, t_{i+1}]$ with $J(t_i) = v$ and $J(t_{i+1}) = w$, that the map $\phi : V_0^{\perp}(c) \longrightarrow J\{t_i\}$ is $\mathbb{R}$ linear. That is, if $X_1, X_2 \in V_0^{\perp}(c)$ and $\alpha, \beta \in \mathbb{R}$, then $\phi(\alpha X_1 + \beta X_2) = \alpha\phi(X_1) + \beta\phi(X_2)$. Thus $\phi$ maps a vector subspace of $V_0^{\perp}(c)$ to a vector subspace of $J\{t_i\}$.

In order to establish the sublemma, we first show that if A is any subspace of $V_0^{\perp}(c)$ on which $I \mid A \times A$ is positive semidefinite, then $\phi \mid A \times A : A \times A \longrightarrow J\{t_i\}$ is injective. Thus suppose $X \in A$ and $\phi(X) = 0$. If $X \in J\{t_i\}$, then $\phi(X) = X$ so that $X = 0$. If $X \notin J\{t_i\}$, then $I(\phi(X),\phi(X)) > I(X,X)$ by (9.16). Thus $\phi X = 0$ implies that $I(X,X) < 0$ which contradicts the assumption that $I \mid A \times A$ is positive semidefinite. Thus if $\phi X = 0$, then $X = 0$.

Now that we know that $\phi$ is injective on subspaces of $V_0^\perp(c)$ on which $I$ is positive semidefinite, we may prove the sublemma. For if $A$ is a subspace of $V_0^\perp(c)$ on which $I \mid A \times A$ is positive semidefinite, inequality (9.16) implies that the index form of $J\{t_i\}$ is positive semidefinite on the subspace $\phi(A)$ of $J\{t_i\}$. Since $\phi$ is injective on such a subspace from above, $\dim A = \dim \phi(A)$. Hence $\mathrm{Ind}_0'(c) \geqslant \mathrm{Ind}_0(c)$. On the other hand, since $J\{t_i\}$ is a vector subspace of $V_0^\perp(c)$, we have $\mathrm{Ind}_0'(c) \leqslant \mathrm{Ind}_0(c)$. Thus $\mathrm{Ind}_0(c) = \mathrm{Ind}_0'(c)$. The same argument shows $\mathrm{Ind}(c) = \mathrm{Ind}(c')$. □

To conclude the proof of Proposition 9.25, we must establish the equality $\mathrm{Ind}_0(c) = \mathrm{Ind}(c) + \dim J_b(c)$. To this end, choose a second partition $a = s_0 < s_1 < \cdots < s_m = b$ so that $\{s_1, s_2, \ldots, s_{m-1}\} \cap \{t_1, \ldots, t_{k-1}\} = \emptyset$ and $c \mid [s_i, s_{i+1}]$ has no conjugate points for each $i$ with $0 \leqslant i \leqslant m - 1$. Let $J\{s_i\}$ denote the vector subspace of $V_0^\perp(c)$ consisting of all vector fields $Y$ such that $Y \mid [s_i, s_{i+1}]$ is a Jacobi field for each $i$ with $0 \leqslant i \leqslant m - 1$. Since the two partitions $\{s_i\}$ and $\{t_i\}$ are distinct except for $a = s_0 = t_0$ and $b = s_m = t_k$, it follows that

$$J\{t_i\} \cap J\{s_i\} = J_b(c)$$

where $J_b(c)$ denotes the vector space of all (smooth) Jacobi fields $J$ along $c$ with $J(a) = J(b) = 0$. By Corollary 9.10, we have $J_b(c) \subset V_0^\perp(c)$. Also, if $X \in J\{s_i\}$ but $X \notin J_b(c)$, we have by (9.16) that

$$I(X,X) < I(\phi X, \phi X) \tag{9.17}$$

Applying the proof of Sublemma 9.26 to the partition $\{s_i\}$ of $[a,b]$, we may choose a vector subspace $B_0'$ of $J\{s_i\}$ with $I \mid B_0' \times B_0'$ positive semidefinite and with $\mathrm{Ind}_0(c) = \dim B_0'$. Since

$\dim B_0' = \mathrm{Ind}_0(c) < \infty$, it follows that $J_b(c)$ is a vector subspace of $B_0'$. By (9.17) and the proof of Sublemma 9.26, the map

$$\phi \mid B_0' : B_0' \longrightarrow J\{t_i\}$$

is injective. Thus if we set $B_0 = \phi(B_0')$, we have $\dim B_0 = \dim B_0' = \mathrm{Ind}_0(c)$. Since $B_0$ is a finite dimensional vector space and $J_b(c)$ is a vector subspace, we may find a vector subspace B of $B_0$ such that $B_0 = B \oplus J_b(c)$, where $\oplus$ denotes the direct sum of vector spaces.

We claim now that $I \mid B \times B$ is positive definite. By construction, we know that $I \mid B_0' \times B_0'$ is positive semidefinite. Also if $0 \neq Z \in B$ and we represent $Z = \phi(X)$ with $X \in B_0'$, then $X \notin J_b(c)$. [For if $X \in J_b(c)$, we have $\phi(X) = X$ so that $Z = \phi(X) = X \in J_b(c)$ also, which is impossible since $B \cap J_b(c) = \{0\}$ by construction.] Hence $I(Z,Z) = I(\phi X, \phi X) > I(X,X)$, the last inequality by (9.17). Thus as $I \mid B_0' \times B_0'$ is positive semidefinite, we obtain $I(Z,Z) > I(X,X) \geqslant 0$ so that $I(Z,Z) > 0$. This shows that $I \mid B \times B$ is positive definite. With the notation of Sublemma 9.26, we then have $\mathrm{Ind}'(c) \geqslant \dim B$.

From the direct sum decomposition $B_0 = B \oplus J_b(c)$, we obtain the equality

$$\mathrm{Ind}_0(c) = \dim B_0 = \dim B + \dim J_b(c)$$

and we also know that $\mathrm{Ind}'(c) \geqslant \dim B$. Thus the proof of Proposition 9.25 will be complete if we show that $\mathrm{Ind}'(c) \leqslant \dim B$.

To establish this inequality, suppose $B' \subset J\{t_i\}$ is a vector subspace with $I \mid B' \times B'$ positive definite and $\dim B' = \mathrm{Ind}'(c)$. Suppose that $\dim B' > \dim B$. Then I is positive semidefinite on $B' \oplus J_b(c)$ so that $\dim(B' \oplus J_b(c)) \leqslant \mathrm{Ind}_0(c)$. On the other hand, since $\dim B' > \dim B$, we obtain

$$\dim(B' \oplus J_b(c)) > \dim(B \oplus J_b(c)) = \mathrm{Ind}_0(c)$$

This contradiction shows that $\dim B' \leqslant \dim B$, whence $\mathrm{Ind}'(c) \leqq \dim B$ as required. $\square$

We now are in a position to prove a Morse index theorem for timelike geodesic segments. The proof we give here is modeled on the proof for Riemannian spaces given in Gromoll, Klingenberg, and Meyer (1975, pp. 150-152).

THEOREM 9.27 (Timelike Morse Index Theorem) Let $c : [a,b] \longrightarrow M$ be a timelike geodesic segment and for each $t \in [a,b]$ let $J_t(c)$ denote the $\mathbb{R}$ vector space of smooth Jacobi fields $Y$ along $c$ with $Y(a) = Y(t) = 0$. Then $c$ has only finitely many conjugate points and the index $\mathrm{Ind}(c)$ and extended index $\mathrm{Ind}_0(c)$ of the index form $I : V_0^{\perp}(c) \times V_0^{\perp}(c) \longrightarrow \mathbb{R}$ are given by the formulas

$$\mathrm{Ind}(c) = \sum_{t\in(a,b)} \dim J_t(c) \tag{9.18}$$

and

$$\mathrm{Ind}_0(c) = \sum_{t\in(a,b]} \dim J_t(c) \tag{9.19}$$

respectively.

*Proof.* We first show that $\Sigma_{t\in(a,b]} \dim J_t(c)$ is a finite sum. We know that $\dim J_t(c) \geqslant 1$ iff $c(t)$ is a conjugate point of $t = a$ along $c$. We may also define imbeddings

$$i : J_t(c) \longrightarrow V_0^{\perp}(c)$$

for each $t \in [a,b]$ by

$$i(Y)(s) = \begin{cases} Y(s) & \text{for } a \leqslant s \leqslant t \\ 0 & \text{for } t \leqslant s \leqslant b \end{cases}$$

Evidently $\dim J_t(c) = \dim i(J_t(c))$ for any $t \in (a,b]$.

Recall that $\mathrm{Ind}_0(c)$ is finite by Proposition 9.25. Thus to show that $c$ has only finitely many conjugate points, it suffices to prove that if $\{t_1,\ldots,t_k\}$ are any set of pairwise distinct conjugate points to $t = a$ along $c$, then $k \leqslant \mathrm{Ind}_0(c)$. To this end, set $A_j = i(J_{t_j}(c))$ for each $j = 1, 2, \ldots, k$ and $A = A_1 \oplus \cdots \oplus A_k$. Then A is a vector subspace of $V_0^{\perp}(c)$, and decomposing $Z \in A$ as $Z = \Sigma_{j=1}^{k} \lambda_j Z_j$ with $\lambda_j \in \mathbb{R}$, $Z_j \in A_j$, we obtain $I(Z,Z) = \Sigma_{j,\ell} \lambda_j \lambda_\ell I(Z_j, Z_\ell)$. But if $t_j \leqslant t_\ell$, we obtain $I(Z_j, Z_\ell) = I(Z_j, Z_\ell)_a^{t_\ell} + I(Z_j, Z_\ell)_{t_\ell}^{b} = -\langle Z_\ell', Z_j \rangle \Big|_a^{t_\ell} + I(0, Z_\ell)_{t_\ell}^{b} = 0$ using (9.2), $Z_\ell \in A_\ell$, and $Z_j(a) = Z_j(t_\ell) = 0$. Hence $I(Z,Z) = 0$ from the symmetry of the index form. Thus $I \mid A \times A$ is positive semidefinite. Hence

$$k \leqslant \dim A = \dim A_1 + \cdots + \dim A_k \leqslant \operatorname{Ind}_0(c)$$

as required. Therefore $c : [a,b] \longrightarrow M$ has only finitely many conjugate points in $(a,b]$, which we denote by $t_1 < t_2 < \cdots < t_r$. Except for $t \in \{t_1, t_2, \ldots, t_r\}$, we have $\dim J_t(c) = 0$ and thus $\Sigma_{t \in (a,b]} \dim J_t(c)$ is a finite sum.

Since $\operatorname{Ind}_0(c) = \operatorname{Ind}(c) + \dim J_b(c)$ from Proposition 9.25, it is enough to establish the equality

$$\operatorname{Ind}_0(c) = \sum_{t \in (a,b]} \dim J_t(c) \tag{9.20}$$

to prove Theorem 9.27. Let $\mathbb{Z}$ denote the integers with the discrete topology and define $f, f_0 : (a,b] \longrightarrow \mathbb{Z}$ by $f(t) = \operatorname{Ind}(c|[0,t])$ and $f_0(t) = \operatorname{Ind}_0(c|[0,t])$. We now show that (9.20) holds if we establish the left continuity of $f$ and the right continuity of $f_0$. Here we mean that $\lim_{t_n \uparrow t} f(t_n) = f(t)$ and $\lim_{t_n \downarrow t} f_0(t_n) = f_0$. Using (9.15) it follows that $f(t) - f_0(t) = -\dim J_t(c) = 0$ if $t \notin \{t_1, \ldots, t_r\}$. Assuming we have shown $f$ is left continuous and $f_0$ is right continuous, we also have $f(t_{j+1}) = f_0(t_j)$ for each $j = 1, 2, \ldots, r-1$. Thus

$$\sum_{t \in (a,b]} \dim J_t(c) = \sum_{t \in (a,b]} [f_0(t) - f(t)]$$

$$= \sum_{j=1}^{r} [f_0(t_j) - f(t_j)] = f_0(t_r) - f(t_1)$$

By Theorem 9.22 the index form is negative definite if c has no conjugate points, so that $f(t) = 0$ for all $t < t_1$. Hence $f(t_1) = 0$ since f is left continuous. Thus $\Sigma_{t\in(a,b]} \dim J_t(c) = f_0(t_r)$. Since $f_0$ is constant on $[t_r,b]$, we have $f_0(t_r) = f_0(b)$. Hence

$$\sum_{t\in(a,b]} \dim J_t(c) = f_0(t_r) = f_0(b) = \mathrm{Ind}_0(c|[a,b])$$

which establishes (9.20).

We have thus reduced the proof of the Morse index theorem to showing that f is left continuous and $f_0$ is right continuous. First note that f and $f_0$ are nondecreasing [i.e., $f(t) \geqslant f(s)$ if $t \geqslant s$]. For suppose we fix $t_1, t_2 \in (a,b]$ with $t_1 \leqslant t_2$. Let $c_1 = c \mid [0,t_1]$ and $c_2 = c \mid [0,t_2]$. We then have an $\mathbb{R}$ linear imbedding $i : V_0^\perp(c_1) \longrightarrow V_0^\perp(c_2)$ given by

$$i(Y)(t) = \begin{cases} Y(t) & \text{for } a \leqslant t \leqslant t_1 \\ 0 & \text{for } t_1 \leqslant t \leqslant t_2 \end{cases}$$

This map has the property that $I(Y,Y) = I(i(Y),i(Y))$, where the indexes are calculated with respect to $c_1$ and $c_2$, respectively. Thus if $A \subset V_0^\perp(c_1)$ is a vector subspace on which the index form of $c_1$ is positive (semi)definite, $i(A)$ is a vector subspace of $V_0^\perp(c_2)$ on which the index form of $c_2$ is positive (semi)definite and $\dim A = \dim i(A)$. Thus $f(t_1) = \mathrm{Ind}(c|[0,t_1]) \leqslant \mathrm{Ind}(c|[0,t_2]) = f(t_2)$ and similarly $f_0(t_1) \leqslant f_0(t_2)$. Thus $f_0$ and f are nondecreasing.

To obtain the continuity properties of f and $f_0$, we fix an arbitrary $\tilde{t} \in (a,b]$ and study the continuity of f and $f_0$ at $\tilde{t}$ using the same approximation techniques as in the proof of Sublemma 9.26. Since $c([a,b])$ is a compact subset of M, there is a constant $\delta > 0$ such that for any $s_1, s_2 \in [a,b]$ with $|s_1 - s_2| < \delta$ and $s_1 \leqslant s_2$, the geodesic segment $c \mid [s_1,s_2]$ has no conjugate points. Choose a partition $a = t_0 < t_1 < t_2 < \cdots < t_k = \tilde{t}$ such that $|t_i - t_{i+1}| < \delta$ for each $i = 0, 1, 2, \ldots, k-1$. Let $J \subset [a,b]$ be an open interval containing $\tilde{t}$ with $|t - t_{k-1}| < \delta$ for all $t \in J$. For each $t \in J$, let $\tilde{J}(c_t)$ denote the finite dimensional $\mathbb{R}$ vector subspace of $V_0^\perp(c|[0,t])$ consisting of all vector fields $Y \in V_0^\perp(c|[0,t])$ such that $Y | [t_j,t_{j+1}]$ for each $j = 0, 1, \ldots, k-2$, and $Y | [t_{k-1},t]$ are Jacobi fields. By Sublemma 9.26, $f(t)$ is the index of $I$ restricted to $\tilde{J}(c_t) \times \tilde{J}(c_t)$ and $f_0(t)$ is the extended index of $I$ restricted to $\tilde{J}(c_t) \times \tilde{J}(c_t)$.

Now let $E = N(c(t_1)) \times N(c(t_2)) \times \cdots \times N(c(t_{k-1}))$. The set $E$ is closed since each $N(c(t_i)) = \{v \in T_{c(t_i)}M : \langle v,c'(t_i)\rangle = 0\}$ is a closed set of spacelike tangent vectors. We may define a Euclidean product metric $\langle\langle\ ,\ \rangle\rangle : E \times E \longrightarrow \mathbb{R}$ by $\langle\langle v,w\rangle\rangle = \Sigma_{i=1}^{k-1} \langle v_i,w_i\rangle$ for $v = (v_1,\ldots,v_{k-1})$, $w = (w_1,\ldots,w_{k-1}) \in E$. Then by Remark 9.2, $S = \{v \in E : \|v\| = 1\}$ is compact.

If $Y \in \tilde{J}(c_t)$, then $Y(0) = Y(t) = 0$ by definition. Also since $c | [t_i,t_{i+1}]$ has no conjugate points, for any $v \in N(c(t_i))$ and $w \in N(c(t_{i+1}))$ there is a unique Jacobi field $Y$ along $c$ with $Y(t_i) = v$ and $Y(t_{i+1}) = w$. Since $\langle Y,c'\rangle|_t$ is an affine function of $t$ and $\langle v,c'(t_i)\rangle = \langle w,c'(t_{i+1})\rangle = 0$, it follows that $\langle Y,c'\rangle|_t = 0$ for all $t$. Hence the map

$$\phi_t : \tilde{J}(c_t) \longrightarrow E$$

defined for $t \in J$ by

$$\phi_t(Y) = (Y(t_1),Y(t_2),\ldots,Y(t_{k-1}))$$

is an isomorphism. For each $t \in J$, we also have a quadratic form $Q_t : E \times E \longrightarrow \mathbb{R}$ given by $Q_t(u,v) = I(\phi_t^{-1}(u),\phi_t^{-1}(v))$. Sublemma 9.26 implies that $f_0(t)$ is the extended index of the quadratic form $Q_t$ on $E \times E$ and $f(t)$ is the index of $Q_t$ on $E \times E$ for each $t \in J$.

Each $Q_t$ may then be used to define a map $Q : E \times E \times J \longrightarrow \mathbb{R}$ given by $Q(u,v,t) = Q_t(u,v)$. We want to show that $Q$ is continuous in order to prove that $f$ and $f_0$ have the desired continuity properties. To this end, let $B = \{Y | [0, t_{k-1}];\ Y \in J\{c_t\}\}$. Then $B$ is isomorphic to $E$ via $\phi : B \longrightarrow E$ given by

$$\phi(Y) = (Y(t_1), Y(t_2), \ldots, Y(t_{k-1}))$$

Then

$$Q(u,v,t) = I(\phi_t^{-1}(u), \phi_t^{-1}(v))$$
$$= I(\phi^{-1}u, \phi^{-1}v) - \langle X_{u,t}, Y'_{v,t}\rangle \Big|_{t_{k-1}}^{t}$$

where $X_{u,t}$ and $Y_{v,t}$ are the Jacobi fields along c given by

$$X_{u,t} = \phi_t^{-1}(u) \mid [t_{k-1}, t]$$

and

$$Y_{v,t} = \phi_t^{-1}(v) \mid [t_{k-1}, t]$$

Since the map $(u,v) \longrightarrow I(\phi^{-1}(u), \phi^{-1}(v))$ from $E \times E \longrightarrow \mathbb{R}$ is a bilinear form, the map $(u,v,t) \longrightarrow I(\phi^{-1}(u), \phi^{-1}(v))$ is certainly continuous. By Proposition 9.16, the map $(u,v,t) \longrightarrow \langle X_{u,t}, Y'_{v,t}\rangle \Big|_{t_{k-1}}^{t}$ is continuous. This establishes the continuity of $Q : E \times E \times J \longrightarrow \mathbb{R}$.

We are finally ready to show that $f_0$ is right continuous and f is left continuous at the arbitrary $\tilde{t} \in (a,b]$. Since f is finite valued from Sublemma 9.26, we may choose a subspace A of E with $\dim A = f(t)$ and $Q(u,u,\tilde{t}) > 0$ for all $u \in A$, $u \neq 0$. Since $Q : E \times E \times J \longrightarrow \mathbb{R}$ is continuous and $S = \{u \in A : \|u\| = 1\}$ is compact, there is a neighborhood $J_0$ of $\tilde{t}$ in J such that $Q(u,u,t) > 0$ for all $t \in J_0$ and $u \in S$. Hence $Q_t \mid A \times A$ is positive definite for all $t \in J_0$. Thus $f(t) \geqslant f(\tilde{t})$ for all $t \in J_0$. Since f is nondecreasing, it follows that $f(t) = f(\tilde{t})$ for all $t \in J_0$ with $t \leqslant \tilde{t}$. Hence f is left continuous.

It remains to show the right continuity of $f_0$ at $\tilde{t}$. Let $\{s_n\}$ be a sequence in J with $s_n > t$ for all n and $s_n \longrightarrow \tilde{t}$. Since $f_0$ is nondecreasing and integer valued, we may suppose that $f_0(s_n) = k$ for all n. By the monotonicity of $f_0$, we then have $f_0(\tilde{t}) \leqslant k$. It thus remains to show that $f_0(\tilde{t}) \geqslant k$. To accomplish this, choose for each n a k-dimensional subspace $A_n$ of E such that $Q_{s_n} \mid A_n \times A_n$ is positive semidefinite. Let $\{a_1(n), a_2(n), \ldots, a_k(n)\}$ be an orthonormal basis of $A_n$ for each n. Thus $\{a_1(n), a_2(n), \ldots, a_k(n)\}$ is contained in the compact subset S of E. By the compactness of E, we may assume each $a_j(n) \longrightarrow a_j \in S$ for each j. By continuity of the inner product, it follows that the vectors $\{a_1, a_2, \ldots, a_k\}$ form an orthonormal subset of S. Let $A = \text{span}\{a_1, a_2, \ldots, a_k\}$, which is thus a k-dimensional subspace of E. Given $u \in A$, we may write $u = \Sigma_{j=1}^{k} \lambda_j a_j$. Let $u(n) = \Sigma_{j=1}^{k} \lambda_j a_j(n)$. Obviously $u(n) \longrightarrow u$ as $n \longrightarrow \infty$. Thus using the continuity of $Q : E \times E \times J \longrightarrow \mathbb{R}$, we obtain

$$Q_{\tilde{t}}(u,u) = \lim_{n\to\infty} Q(u(n), u(n), s_n) \geqslant 0$$

since $Q_{s_n} \mid A_n \times A_n$ is positive semidefinite for each n. Hence $Q_{\tilde{t}} \mid A \times A$ is positive semidefinite. Thus $f_0(\tilde{t}) \geqslant \dim A = k$ as required. □

We conclude this section with an application of the timelike Morse index theorem to the structure of the cut locus of future one-connected, globally hyperbolic space-times [cf. Beem and Ehrlich (1979c, Section 8)].

DEFINITION 9.28 A space-time (M,g) is said to be *future one-connected* if for all $p,q \in M$, any two future-directed timelike curves from p to q are homotopic through smooth future-directed timelike curves with fixed endpoints p and q.

This concept, a Lorentzian analogue for simple connectivity, has been studied in Avez (1963), Smith (1960a), and Flaherty (1975a, p. 395). The vanishing of the Lorentzian fundamental group implies that (M,g) is future one-connected. However, the simple connectivity of M as a topological space does *not* imply that (M,g) is future one-connected, as the following example of Geroch shows. Consider $\mathbb{R}^3$ with coordinates (x,y,t) and the Lorentzian metric

$ds^2 = -dt^2 + dx^2 + dy^2$. Let $T = \{(x,0,0) \in \mathbb{R}^3 : x \geqslant 0\}$ and set $M = \mathbb{R}^3 - T$ with the induced Lorentzian metric from $(\mathbb{R}^3, ds^2)$. Then M is simply connected. On the other hand, let $p = (2,0,-1)$ and $q = (2,0,1)$. Then p and q may be joined by future-directed timelike curves $\gamma_1$ and $\gamma_2$ lying on opposite sides of the positive x axis (cf. Figure 9.1). But $\gamma_1$ and $\gamma_2$ are not homotopic through future-directed timelike curves starting at p and ending at q since such a homotopy would have to go around the point (0,0,0), which would introduce spacelike curves.

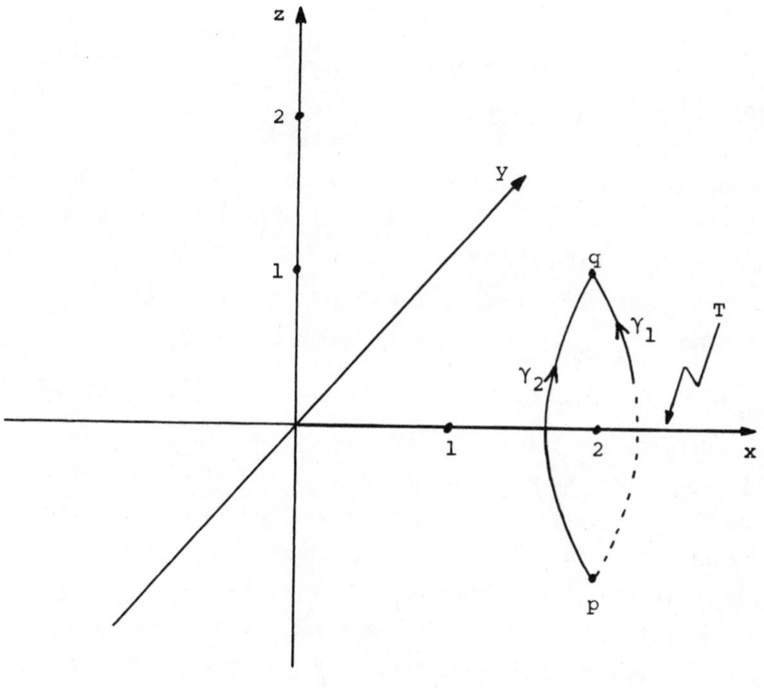


*Figure 9.1* A space-time $M = \mathbb{R}^3 - T$ which is simply connected but not future 1-connected is shown. The future-directed timelike curves $\gamma_1$ and $\gamma_2$ from p to q are not homotopic through timelike curves with endpoints p and q.

Note that if (M,g) is future one-connected, then the path space of smooth timelike curves from p to q is connected. Thus Lemma 4.11 (2) of Cheeger and Ebin (1975, p. 85) and the standard path space Morse theory [cf. Everson and Talbot (1976), Uhlenbeck (1975), Woodhouse (1976)] imply the following proposition.

PROPOSITION 9.29 Let (M,g) be future one-connected and globally hyperbolic. Fix $p \in M$ and suppose that every future-directed timelike geodesic starting at p has index 0 or index $\geq 2$. Given $q \in I^+(p)$ such that q is not conjugate to p along any future-directed timelike geodesic from p to q of index 0, namely, the unique maximal geodesic from p to q.

We are now ready to prove the Lorentzian analogue for globally hyperbolic, future one-connected space-times of a theorem of Cheeger and Ebin (1975, Theorem 5.11) on the cut locus of a complete Riemannian manifold which generalized a theorem of Crittenden (1962) for simply connected Lie groups with bi-invariant Riemannian metrics. The globally hyperbolicity is used in Theorem 9.30 to guarantee the existence of maximal geodesic segments joining chronologically related points.

THEOREM 9.30 Let (M,g) be future one-connected and globally hyperbolic. Suppose that for $p \in M$, the first future conjugate point of p along every timelike geodesic $\gamma$ with $\gamma(0) = p$ is of order $\geq 2$. Then the future timelike cut locus of p and the locus of first future conjugate points of p coincide. Thus all geodesics from p maximize up to the first future conjugate point.

*Proof.* All geodesics will be unit speed and future timelike during the course of this proof. It suffices to show that if $\gamma : [0,t] \longrightarrow M$ with $\gamma(0) = p$ has index 0 and $\gamma(t)$ is not conjugate to p along $\gamma \mid [0,t]$, then $\gamma$ is maximal. Since the set of the singular points of $\exp_p$ is closed, it follows from Theorem 9.27 that there exist $\varepsilon_1, \varepsilon_2 < 0$ such that if $\measuredangle\,(v, \gamma'(0)) < \varepsilon_1$ and $\varepsilon < \varepsilon_2$, then the future timelike geodesic $\sigma : [0, t + \varepsilon] \longrightarrow M$ with $\sigma'(0) = v$ is of index 0. Here $\measuredangle$ denotes angle measured using an auxiliary Riemannian metric.

By Sard's theorem, we may find a sequence of points $\{p_i\} \subset I^+(p)$ with $p_i \longrightarrow \gamma(t)$ such that every timelike geodesic segment from p to $p_i$ has nonconjugate endpoints. Thus, by hypothesis, every such segment has index 0 or index $\geq 2$. By Proposition 9.29, there is a unique timelike maximal geodesic segment $\gamma_i$ from p to $p_i$ of index 0.

Since $(\exp_p)_*$ is nonsingular at $t\gamma'(0)$, for i sufficiently large there are geodesic segments $\overline{\gamma}_i$ from p to $p_i$ with $\overline{\gamma}_i \longrightarrow \gamma$. Since $\measuredangle(\overline{\gamma}_i'(0), \gamma'(0)) \longrightarrow 0$, these segments have index 0 for large i. It follows that for i sufficiently large, $\overline{\gamma}_i = \gamma_i$ and hence $\overline{\gamma}_i$ is maximal. Thus $\gamma$ is maximal as a limit of maximal geodesics. □

## 9.2 THE TIMELIKE PATH SPACE OF A GLOBALLY HYPERBOLIC SPACE-TIME

In this section we discuss the Morse theory of the path space of future-directed timelike curves joining two chronologically related points in a globally hyperbolic space-time following Uhlenbeck (1975) [cf. also Woodhouse (1976)]. Both of these treatments are modeled on Milnor's exposition (1963, pp. 88-92) of the Morse theory for the path space of a complete Riemannian manifold in which the full path space is approximated by piecewise smooth geodesics. A different approach has been given to the Morse theory of nonspacelike curves in globally hyperbolic space-times by Everson and Talbot (1976, 1978). They use a result of Clarke (1970) that any four-dimensional globally hyperbolic space-time may be isometrically imbedded in a high dimensional Minkowski space-time to give a Hilbert manifold structure to a subclass of timelike curves in M.

We now turn to Uhlenbeck's treatment of the Morse theory of the path space of piecewise smooth timelike curves joining points $p \ll q$ of any globally hyperbolic space-time (M,g) of dimension $\geq 2$.

DEFINITION 9.31 Given $p,q \in (M,g)$ with $p \ll q$, let $C_{(p,q)}$ denote the space of future-directed piecewise smooth timelike curves from p to q, where two curves which differ by a parameterization are identified.

While $C_{(p,q)}$ is not a manifold modeled on a Banach space, it does possess tangent spaces consisting of piecewise smooth vector fields along the given piecewise smooth curve assumed to be parameterized by arc length. Functionals $F : C_{(p,q)} \longrightarrow \mathbb{R}$ may thus be considered from the point of view of the calculus of variations. Thus a *critical point* of F is a point at which all first variations vanish and a *critical value* is the image under F of a critical point. The functional F on $C_{(p,q)}$ is said to be a *homotopic Morse function* if for any $b > a$ not a critical value of F, the topological space $F^{-1}(-\infty,b)$ is homotopically equivalent to the space $F^{-1}(-\infty,a)$ with cells adjoined, where one cell of dimension equal to the index of the corresponding critical point is adjoined for each critical point of F with critical value in (a,b).

We will show in this section that the Lorentzian arc length functional is a homotopic Morse function for $C_{(p,q)}$ provided that p and q are nonconjugate along any nonspacelike geodesic. This result is analogous to Morse's result [cf. Milnor (1963, Theorems 16.3 and 17.3)] for complete Riemannian manifolds. Namely, if p and q are any two distinct points not conjugate along any geodesic, then the space $\Omega_{(p,q)}$ of piecewise smooth curves from p to q has the homotopy type of a countable CW-complex with a cell of dimension $\lambda$ for each geodesic from p to q of index $\lambda$.

Given that p and q must be nonconjugate along any geodesic for L to be a Morse function, it is of interest to know that such pairs of points exist. As for Riemannian spaces, conjugate points in an arbitrary Lorentzian manifold may be viewed as singularities of the differential of the exponential mapping. Hence Sard's theorem [cf. Hirsch (1976, p. 69)] may be applied. Here a subset X of a manifold is said to have *measure zero* if for every chart $(U,\phi)$, the set $\phi(U \cap X) \subset \mathbb{R}^n$ has Lebesgue measure zero in $\mathbb{R}^n$, $n = \dim M$. Also a subset of a manifold is said to be *residual* if it contains the intersection of countably many dense open sets. A residual subset of a complete metric space is dense by the Baire category theorem [cf. Kelley (1955, p. 200)]. Sard's theorem then implies the following result.

THEOREM 9.32 Let $(M,g)$ be a globally hyperbolic space-time and let $p \in M$ be arbitrary. Then the set of points of $M$ conjugate to $p$ along some geodesic has measure zero. Thus for a residual set of $q \in M$, $p$ and $q$ are nonconjugate along all geodesics between them.

Recall the following properties of the Lorentzian arc length functional $L : C_{(p,q)} \longrightarrow \mathbb{R}$. First, from the calculus of variations, the critical points of $L$ on $C_{(p,q)}$ are exactly the future-directed timelike geodesic segments from $p$ to $q$ with parameter proportional to arc length. Second, the timelike Morse index theorem 9.27 implies that the index of a critical point of $L$ is just its index as a geodesic, i.e., the number of conjugate points to $p$ along the geodesic, counting multiplicities, from $p$ up to but not including $q$.

In order to show that $L$ is a homotopic Morse function, it is necessary to approximate $C_{(p,q)}$ by a subset $M_{(p,q)}$ such that there exists a retraction of $C_{(p,q)}$ onto $M_{(p,q)}$ which increases the length functional $L$. This step corresponds to the finite dimensional approximation of the loop space in Riemannian Morse theory. The corresponding Lorentzian approximation makes crucial use of the global hyperbolicity of $(M,g)$ as will be apparent from Lemma 9.34 below. First it is useful to make the following definition.

DEFINITION 9.33 Let $p,q \in (M,g)$ with $p \ll q$. A finite collection $\{x_1,x_2,\ldots,x_j\}$ of points of $M$ is said to be a *timelike chain from* $p$ *to* $q$ if $p \ll x_1 \ll x_2 \ll \cdots \ll x_j \ll q$.

The next lemma follows from the existence of convex normal neighborhoods (cf. Section 2.2), the compactness of $J^+(p) \cap J^-(q)$ in a globally hyperbolic space-time, and Theorem 5.1.

LEMMA 9.34 Let $(M,g)$ be a globally hyperbolic space-time with a fixed globally hyperbolic time function $f : M \longrightarrow \mathbb{R}$. Let $p,q \in M$ with $p \ll q$ be given. Then there exists $\{t_1,t_2,\ldots,t_k\}$ with $f(p) < t_1 < t_2 < \cdots < t_k < f(q)$ satisfying the following properties:

(i) If $x \in f^{-1}(f(p),t_1]$ and $p \leqslant x$, there is a unique maximal future-directed nonspacelike geodesic segment from p to x.

(ii) For each i with $1 \leqslant i \leqslant k - 1$, if $x \in f^{-1}(t_i)$ and $y \in f^{-1}(t_i,t_{i+1}]$ with $p \leqslant x \leqslant y \leqslant q$, there is a unique maximal future-directed nonspacelike geodesic segment from x to y.

(iii) If $y \in f^{-1}[t_k,f(q))$ and $y \leqslant q$, there is a unique maximal future-directed nonspacelike geodesic segment from y to q.

(iv) In particular, if $\{x_1,\ldots,x_k\}$ is any timelike chain from p to q with $x_i \in f^{-1}(t_i)$ for each $i = 1, 2, \ldots, k$, then there is a unique maximal future-directed timelike geodesic segment from p to $x_1$, $x_k$ to q, and $x_i$ to $x_{i+1}$ for each i with $1 \leqslant i \leqslant k - 1$.

Since $p,q \in M$ with $p \ll q$ are given, we may now fix $\{t_1,t_2,\ldots,t_k\}$ satisfying the conditions of Lemma 9.34. Denote by $S_i$ the Cauchy surface $S_i = f^{-1}(t_i)$ for each $i = 1, 2, \ldots, k$. We may now define a space $M_{(p,q)}$ of broken timelike geodesics to approximate $C_{(p,q)}$ from the point of view of the arc length functional.

DEFINITION 9.35 Let $M_{(p,q)}$ be the space of all continuous curves $\gamma : [0,1] \longrightarrow M$ such that $\gamma(0) = p$, $\gamma(1) = q$, and $\gamma(i/(k+1)) \in S_i$ for each $i = 1, 2, \ldots, k$ and such that $\gamma \mid [i/(k+1),(i+1)/(k+1)]$ is a future-directed timelike geodesic for each $i = 0, 1, 2, \ldots, k$.

Since each $\gamma \mid [i/(k+1),(i+1)/(k+1)]$ must be the unique maximal segment joining its endpoints by Lemma 9.34, it follows that

$$M_{(p,q)} = \{(x_1,\ldots,x_k) : x_i \in S_i \text{ for } 1 \leqslant i \leqslant k, \text{ and } p \ll x_1, x_i \ll x_{i+1} \text{ for each } 1 \leqslant i \leqslant k-1, \text{ and } x_k \ll q\} \quad (9.21)$$

Since chronological future and past sets are open, $M_{(p,q)}$ viewed as in (9.21) is an open submanifold of $S_1 \times S_2 \times \cdots \times S_k$. Let $\pi_i : M_{(p,q)} \longrightarrow S_i$ be the projection map given by $\pi_i(x_1,\ldots,x_k) = x_i$ for each $i = 1, 2, \ldots, k$. Since $I^+(p) \cap I^-(q)$ has compact closure by the global hyperbolicity of M and $\pi_i(M_{(p,q)}) \subset I^+(p) \cap I^-(q)$, it follows that $\pi_i(M_{(p,q)})$ has compact closure in $S_i$ for each $i = 1, 2, \ldots, k$.

A length-nondecreasing retraction of $C_{(p,q)}$ onto $M_{(p,q)}$ may now be established along the lines of Riemannian Morse theory [cf. Milnor (1963, p. 91)].

PROPOSITION 9.36 There exists a retraction $Q_\lambda$, $0 \leqslant \lambda \leqslant 1$, of $C_{(p,q)}$ onto $M_{(p,q)}$ which is Lorentzian arc length nondecreasing.

*Proof*. Let $\gamma \in C_{(p,q)}$ be arbitrary. We may suppose that $\gamma$ is parameterized so that $f(\gamma(t)) = t$ where $f : M \longrightarrow \mathbb{R}$ is the globally hyperbolic time function fixed in Lemma 9.34. Thus $\gamma : [f(p),f(q)] \longrightarrow M$. For each $\lambda \in [0,1]$, define $Q_\lambda(\gamma) : [f(p),f(q)] \longrightarrow M$ as follows. Set $\beta = (1 - \lambda)f(p) + \lambda f(q)$ and put $t_0 = f(p)$ and $t_{k+1} = f(q)$. If $\beta$ satisfies $t_i < \beta \leqslant t_{i+1}$ for $i$ with $0 \leqslant i \leqslant k$, let $Q_\lambda(\gamma)$ be the unique broken timelike geodesic joining the points $p$ to $\gamma(t_1)$, $\gamma(t_1)$ to $\gamma(t_2)$, ..., $\gamma(t_{i-1})$ to $\gamma(t_i)$, and $\gamma(t_i)$ to $\gamma(\beta)$ successively for $t \leqslant \beta$, and let $Q_\lambda(\gamma)(t) = \gamma(t)$ for $t \geqslant \beta$ (cf. Figure 9.2). It is immediate from the definition that $Q_0(\gamma) = \gamma$ and that $Q_1(\gamma) \in M_{(p,q)}$. Since $Q_\beta(\gamma)[\gamma(t_i),\gamma(\beta)]$ is maximal from $\gamma(t_i)$ to $\gamma(t)$ by Lemma 9.34, we have

$$L(Q_\lambda(\gamma)[\gamma(t_i),\gamma(\beta)]) = d(\gamma(t_i),\gamma(\beta)) \geqslant L(\gamma|[t_i,\beta])$$

where d denotes the Lorentzian distance function of (M,g) and we have used Notational Convention 7.4. Similarly, since $Q_\lambda(\beta)$ is the unique maximal geodesic segment from $\gamma(t_j)$ to $\gamma(t_{j+1})$ for each $j$ with $0 \leqslant j \leqslant i - 1$, we have $L(Q_\lambda(\gamma)[\gamma(t_j),\gamma(t_{j+1})]) \geqslant L(\gamma|[t_j,t_{j+1}])$ for each $j$ with $0 \leqslant j \leqslant i - 1$. Summing, we obtain

$$L(Q_\lambda(\gamma)[p,\gamma(\beta)]) \geqslant L(\gamma|[f(p),\beta])$$

Since $Q_\lambda(\gamma)(t) = \gamma(t)$ for $t \geqslant \beta$, we thus have $L(Q_\lambda(\gamma)) \geqslant L(\gamma)$. Moreover, it is clear from the above argument that $L(Q_\lambda(\gamma)) = L(\gamma)$ for all $\lambda$ iff $\gamma$ is a broken timelike geodesic from p to q with breaks possible only at the $S_i$'s. In particular, $Q_\lambda \mid M_{(p,q)} = \mathrm{Id}$ for all $\lambda \in [0,1]$. Finally the continuity of the map $\lambda \longrightarrow Q_\lambda$ is

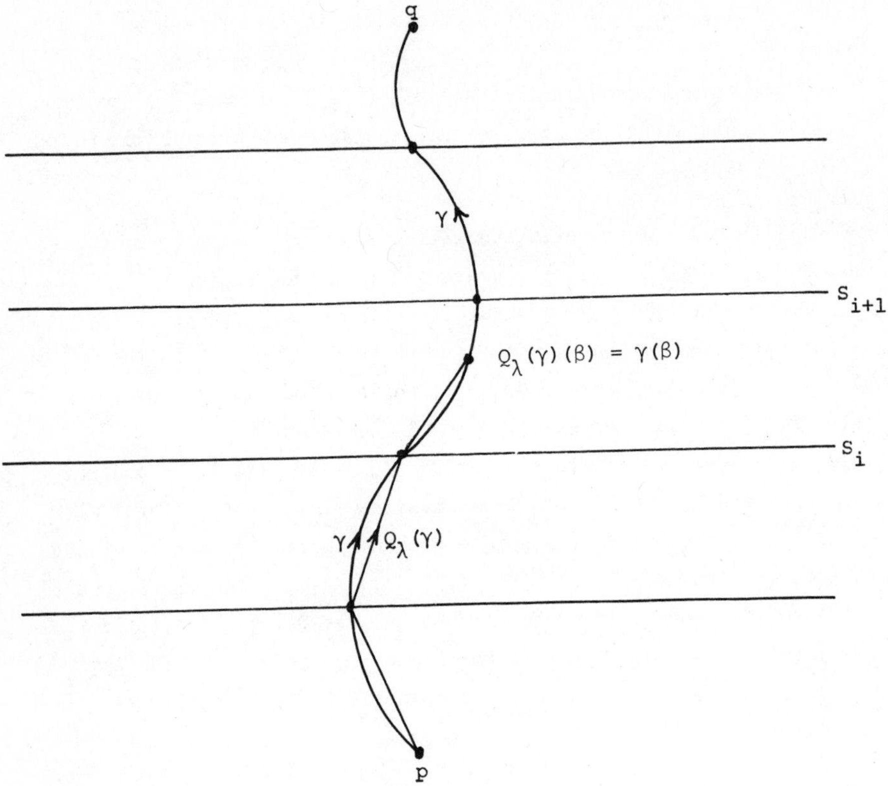


*Figure 9.2* In the proof of Proposition 9.36 the curve $Q_\lambda(\gamma)$ is used to approximate the given curve $\gamma$.

clear from the differentiable dependence of geodesics on their endpoints in convex neighborhoods. □

As in Uhlenbeck (1975, p. 79), we denote by $L_* = L \mid M_{(p,q)}$ the restriction of the Lorentzian arc length functional to the subset $M_{(p,q)}$ of $C_{(p,q)}$. Just as in the Riemannian case [cf. Milnor (1963, Theorem 16.2)], it can be seen that $L_* : M_{(p,q)} \longrightarrow \mathbb{R}$ is a faithful model of $L : C_{(p,q)} \longrightarrow \mathbb{R}$ in the following sense.

PROPOSITION 9.37 If $(M,g)$ is globally hyperbolic, then the critical points of $L_* = L \mid M_{(p,q)}$ are smooth timelike geodesic segments from p to q. These critical points are nondegenerate iff p is not conjugate to q along any timelike geodesic from p to q. Moreover, the index of each critical point is the same in $C_{(p,q)}$ and $M_{(p,q)}$, namely the index by conjugate points.

*Proof.* Identify $M_{(p,q)}$ with k-chains $\{x_1,\ldots,x_k\}$ from p to q with $x_i \in S_i$ for each i, as in (9.21). Set $x_0 = p$ and $x_{k+1} = q$. Let $\gamma_i : [0,1] \longrightarrow M$ denote the unique maximal timelike geodesic segment from $x_i$ to $x_{i+1}$ for $i = 0, 1, \ldots, k$. Then

$$L_*(x_1,\ldots,x_k) = \sum_{i=0}^{k+1} \int_0^1 \sqrt{<-\gamma_i'(t),\gamma_i'(t)>}\, dt$$

Suppose we have a smooth deformation $\{x_1(t),\ldots,x_k(t)\}$ of the given chain $\{x_1,x_2,\ldots,x_k\}$. Then since each $t \longrightarrow x_i(t)$ is a curve in $S_i$, the variation vector field V of the deformation must lie in $T_{x_i}(S_i)$ at $x_i$. Also as we are deforming the given chain, which represents a piecewise smooth timelike geodesic, through piecewise smooth timelike geodesics with discontinuities only at the $S_i$'s, the space of deformations of $\{x_1,\ldots,x_k\}$ may be identified with all vector fields Y along $\{x_1,\ldots,x_k\}$ so that $Y(x_i) \in T_{x_i}(S_i)$ for $i = 1, \ldots, k$, $Y(p) = Y(q) = 0$, and $Y \mid \gamma_i$ is a smooth Jacobi field along $\gamma_i$ for each $i = 0, \ldots, k$. By choice of the $S_i$'s as in Lemma 9.34, no $\gamma_i : [0,1] \longrightarrow M$ has any conjugate points. Thus given any $v_i \in T_{x_i}(S_i)$ and $w_i \in T_{x_{i+1}}(S_{i+1})$, there is a unique Jacobi field J along $\gamma_i : [0,1] \longrightarrow M$ with $J(0) = v_i$ and $J(1) = w_i$. Thus the space of 1-jets of deformations of the given chain $\{x_1,\ldots,x_k\}$ may simply be identified with the Cartesian product $T_{x_1}(S_1) \times T_{x_2}(S_2) \times \cdots \times T_{x_k}(S_k)$.

Let $\sigma : [a,b] \longrightarrow M$ be a smooth future-directed timelike curve parameterized so that $\sqrt{-<\sigma'(s),\sigma'(s)>} = A$ is constant for all

$s \in (a,b)$ and so that $\sigma'(a^+)$ and $\sigma'(b^-)$ are timelike tangent vectors. Let $\alpha : [a,b] \times (-\varepsilon,\varepsilon) \longrightarrow M$ be a smooth variation of $\sigma$ through nonspacelike curves with variation vector field V. If $\alpha_s : [a,b] \longrightarrow M$ denotes the curve $\alpha_s(t) = \alpha(t,s)$, the first variation formula for $L'(0) = (d/ds)L(\alpha_s)\big|_{s=0}$ is given by

$$L'(0) = \frac{-1}{A} \langle V,\sigma'\rangle \Big|_a^b + \frac{1}{A} \int_a^b \langle V,\nabla_{\sigma'}\sigma'\rangle\Big|_t \, dt \tag{9.22}$$

Thus if $\sigma$ is a timelike geodesic, $L'(0) = (-1/A)\langle V,\sigma'\rangle\Big|_a^b$.

We now apply (9.22) to calculate the first variation of a proper deformation $\alpha$ of an element $\{x_1,\ldots,x_k\} \in M_{(p,q)}$. As before, let $\gamma_i : [0,1] \longrightarrow M$ be timelike geodesics from $x_i$ to $x_{i+1}$ for $i = 0, \ldots, k$. Then $\{x_1,\ldots,x_k\}$ is represented by the piecewise smooth timelike geodesic $\gamma = \gamma_0 * \gamma_1 * \cdots * \gamma_k$. Let V be the variation vector field of $\alpha$ along $\gamma$ and set $y_i = V(x_i) \in T_{x_i}(S_i)$. As we mentioned above, the piecewise smooth Jacobi field V along $\gamma$ may then be identified with $(y_1,y_2,\ldots,y_k) \in T_{x_1}(S_1) \times \cdots \times T_{x_k}(S_k)$. Restricting V to each $\gamma_i$ and applying formula (9.22), we obtain the first variation formula

$$\delta L_*(y_1,\ldots,y_k) = \sum_{i=1}^{k} \left( \langle y_{i-1}, \frac{\gamma_i'(0)}{A_i}\rangle - \langle y_i, \frac{\gamma_i'(1)}{B_i}\rangle \right) \tag{9.23}$$

where

$$A_i = \sqrt{-\langle\gamma_i'(0),\gamma_i'(0)\rangle} \qquad B_i = \sqrt{-\langle\gamma_i'(1),\gamma_i'(1)\rangle} \qquad \text{for each } i$$

for $L_* = L \mid M_{(p,q)}$. From (9.23), it is a standard argument to see that $\delta L_*(y_1,y_2,\ldots,y_k) = 0$ for all $(y_1,y_2,\ldots,y_k) \in T_{x_1}(S_1) \times \cdots \times T_{x_k}(S_k)$ iff the tangent vectors

$$\frac{\gamma_i'(1)}{\sqrt{-\langle\gamma_i'(1),\gamma_i'(1)\rangle}} \quad \text{and} \quad \frac{\gamma_{i+1}'(0)}{\sqrt{-\langle\gamma_{i+1}'(0),\gamma_{i+1}'(0)\rangle}}$$

in $T_{x_i}(M)$ have the same projection into the subspace $T_{x_i}(S_i)$ of $T_{x_i}(M)$ for each $i = 1, \ldots, k$. This then implies that $\gamma = \gamma_0 * \gamma_1 * \cdots * \gamma_k$ can be reparameterized to be a smooth timelike geodesic from p to q.

Thus we have seen that the critical points of $L_* : M_{(p,q)} \longrightarrow \mathbb{R}$ and $L : C_{(p,q)} \longrightarrow \mathbb{R}$ coincide and are exactly the smooth timelike geodesics from p to q. By our proof of the timelike Morse index theorem (Theorem 9.27) by approximating $V_0^{\perp}(c)$ by spaces of piecewise smooth Jacobi fields (cf. Sublemma 9.26), it follows that the indices of $L_* : M_{(p,q)} \longrightarrow \mathbb{R}$ and $L : C_{(p,q)} \longrightarrow \mathbb{R}$ coincide. The timelike Morse index theorem (Theorem 9.27) then implies that a critical point c is degenerate iff p and q are conjugate along c (cf. also Proposition 9.13). □

With Proposition 9.37 in hand, the following result may then be established.

PROPOSITION 9.38 Let (M,g) be globally hyperbolic. If p and q are nonconjugate along any nonspacelike geodesic, then $L : C_{(p,q)} \longrightarrow \mathbb{R}$ and $L_* : M_{(p,q)} \longrightarrow \mathbb{R}$ have only a finite number of critical points. In particular, there are only finitely many future-directed timelike geodesics from p to q.

*Proof*. By Proposition 9.37, it is enough to show that $L_* : M_{(p,q)} \longrightarrow \mathbb{R}$ has only finitely many critical points. Suppose $L_*$ has infinitely many critical points. Then there would be an infinite sequence of smooth timelike geodesics $\{c_n\}_{n=1}^{\infty}$ from p to q. Since each $\pi_i(M_{(p,q)})$ has compact closure in $S_i$, the $\{c_n\}$'s have a subsequence which must then converge to a timelike or null geodesic c from p to q. Since c is a limit of an infinite sequence of geodesics from p to q, it follows that p is conjugate to q along c, in contradiction. □

Another interesting property of $L_* : M_{(p,q)} \longrightarrow \mathbb{R}$ is the following result.

PROPOSITION 9.39 $L_* : M_{(p,q)} \longrightarrow \mathbb{R}$ assumes its maximum on every component of $M_{(p,q)}$.

*Proof.* While $M_{(p,q)}$ is an open submanifold of $S_1 \times S_2 \times \cdots \times S_k$, its closure $\mathrm{cl}(M_{(p,q)})$ is compact in $S_1 \times \cdots \times S_k$. Also

$$L_*(\{x_1,\ldots,x_k\}) = \sum_{i=1}^{k} d(x_i,x_{i+1}) \leqslant d(p,q)$$

where $x_0 = p$ and $x_{k+1} = q$ for any timelike chain $\{x_1,x_2,\ldots,x_k\}$ from p to q. Thus as the Lorentzian distance function is finite valued for globally hyperbolic space-times, it follows that $L_*$ is bounded on each connected component U of $M_{(p,q)}$. Let $\{\gamma_n\}$ be a sequence in the connected component U of $M_{(p,q)}$ with $L_*(\gamma_n) \longrightarrow \sup\{L_*(\gamma) : \gamma \in U\}$ as $n \longrightarrow \infty$. By compactness of $\mathrm{cl}(M_{(p,q)})$ in $S_1 \times S_2 \times \cdots \times S_k$, the sequence $\{\gamma_n\}$ has a limit curve $\gamma_\infty$ in cl(U) in $S_1 \times S_2 \times \cdots \times S_k$. Since $\{\gamma_n\}$ was a maximizing sequence for $L_* \mid U$, it follows that $L_*(\gamma_\infty) \geqslant L_*(\sigma)$ for any $\sigma \in \mathrm{cl}(U)$, and $L_*(\gamma_\infty) > 0$. If $\gamma_\infty$ is represented by the element $(x_1,x_2,\ldots,x_k) \in S_1 \times \cdots \times S_k$, we have $p \leqslant x_1 \leqslant x_2 \leqslant \cdots \leqslant x_k \leqslant q$, and if $\gamma_i$ is the unique maximal geodesic from $x_i$ to $x_{i+1}$ guaranteed by Lemma 9.34 then each $\gamma_i$ is either null or timelike. If some $\gamma_i$ is null, it follows from the first variation formula (9.23) that $\gamma_\infty$ may be deformed to a curve $\gamma \in \mathrm{cl}(U)$ with $L_*(\gamma) > L_*(\gamma_\infty)$. This then contradicts the maximality of $L_* \mid \mathrm{cl}(U)$ at $\gamma_\infty$. Thus, in fact, the limit element $\gamma_\infty \in U$. □

Again, consider $M_{(p,q)}$ as a subset of $S_1 \times \cdots \times S_k$. Let $P_i : T_x(M) \longrightarrow T_x(S_i)$ denote the orthogonal projection map for any $x \in S_i$ and each i = 1, 2, ..., k. The set $M_{(p,q)}$ regarded as an open subset of $S_1 \times \cdots \times S_k$ may thus be given a Riemannian metric induced from the Lorentzian metric restricted to the spacelike Cauchy hypersurfaces $S_1, S_2, \ldots, S_k$. It then follows from formula (9.23) that at the point of $M_{(p,q)}$ represented by the broken timelike geodesic $\gamma = \gamma_0 * \gamma_1 * \cdots * \gamma_k$, with each $\gamma_i$ parameterized on [0,1] and $A_i$, $B_i$ as above, the gradient of $L_*$ is given by the formula

$$\operatorname{grad} L_* = \left(P_1\left(\frac{\gamma_1'(0)}{A_1} - \frac{\gamma_0'(1)}{B_0}\right), P_2\left(\frac{\gamma_2'(0)}{A_2} - \frac{\gamma_1'(1)}{B_1}\right), \ldots, P_k\left(\frac{\gamma_k'(0)}{A_k} - \frac{\gamma_{k-1}'(1)}{B_{k-1}}\right)\right)$$

Using this formula, Uhlenbeck (1975, pp. 80-81) then establishes the following lemma.

LEMMA 9.40 Let $\beta : (a,b) \longrightarrow M_{(p,q)}$ be a maximal integral curve for grad $L_*$. Then $b = \infty$ and $\lim_{t\to\infty} \beta(t)$ lies in the critical set of $L_*$.

With the behavior of gradient $L_*$ established, it then follows [Uhlenbeck (1975, p. 81)] that if p and q are nonconjugate along any timelike geodesic, then $L_* : M_{(p,q)} \longrightarrow \mathbb{R}$ is a Morse function for $M_{(p,q)}$. The main result now follows from this fact and Propositions 9.36 and 9.38.

THEOREM 9.41 Let (M,g) be globally hyperbolic and let p, q be any pair of points in (M,g) with $p \ll q$ and such that p and q are not conjugate along any nonspacelike geodesic from p to q. Then there are only finitely many future-directed timelike geodesics in (M,g) from p to q and the arc length functional $L : C_{(p,q)} \longrightarrow \mathbb{R}$ is a homotopic Morse function. Thus if $b > a$ are any two noncritical values of L, then $L^{-1}(-\infty,b)$ is homotopically equivalent to the space $L^{-1}(-\infty,a)$ with a cell attached for each smooth timelike geodesic $\gamma$ from p to q with $a < L(\gamma) < b$, where the dimension of the attached cell is the (geodesic) index of $\gamma$. Thus $C_{(p,q)}$ has the homotopy type of a finite CW-complex with a cell of dimension $\lambda$ for each smooth future-directed timelike geodesic $\gamma$ from p to q of index $\lambda$.

Note that in Theorem 9.41 the topology of $C_{(p,q)}$ is not related to the given manifold topology. But Uhlenbeck (1975, Theorem 3) has shown for a class of globally hyperbolic space-times satisfying a metric growth condition [cf. Uhlenbeck (1975, p. 72)] that the homotopy of the loop space of M itself may be calculated geometrically as follows. Let $(M,g)$ be a globally hyperbolic space-time satisfying Uhlenbeck's metric growth condition. Then there is a class of smooth timelike curves $\gamma : [0,\infty) \longrightarrow M$ with the following property. For any such timelike curve $\gamma$, there is a residual set of points $p \in M$ such that the loop space of M is homotopic to a cell complex with a cell for each null geodesic from p to $\gamma$ where the dimension of the cell corresponds to the conjugate point index of the geodesic.

It will be clear from the proof of Proposition 9.42 below that the finiteness of the homotopy type of the timelike loop space $C_{(p,q)}$ in Theorem 9.41 follows from the assumption that p is not conjugate to q along any null geodesic together with the fact that $L(\gamma) \leqslant d(p,q) < \infty$ for all $\gamma \in \Omega_{(p,q)}$. For complete Riemannian manifolds $(N,g_0)$ on the other hand, it is known from work of Serre (1951) that if N is not acyclic [i.e., $H_i(N;\mathbb{Z}) \neq 0$ for some $i > 0$], then the loop spaces $\Omega_{(p,q)}$ are infinite CW-complexes for all $p,q \in N$. Thus if p and q are nonconjugate along any geodesic, there are infinitely many geodesics $\{c_n : [0,1] \longrightarrow N\}$ from p to q. Since p and q are nonconjugate along any geodesic and $L_0(\gamma) \geqslant d_0(p,q) > 0$ for all $\gamma \in \Omega_{(p,q)}$, it may be seen that $L_0(c_n) \longrightarrow \infty$ as $n \longrightarrow \infty$.

For completeness, we now give a different proof from Proposition 9.38 of the existence of only finitely many critical points for $L : C_{(p,q)} \longrightarrow \mathbb{R}$. Instead of using the finite dimensional approximation of $C_{(p,q)}$ by $M_{(p,q)}$, we work directly with $C_{(p,q)}$ using the existence of nonspacelike limit curves as in Section 2.3.

PROPOSITION 9.42 Let $(M,g)$ be globally hyperbolic and suppose that $p,q \in M$ with $p \ll q$ are chosen such that p and q are not conjugate along any future-directed nonspacelike geodesic. Then there are only finitely many timelike geodesics from p to q.

*Proof*. Suppose that there are infinitely many future-directed timelike geodesic segments $c_n : [0,1] \longrightarrow M$ in $C_{(p,q)}$. Using Corollary 2.19 and the arguments of Section 2.3 we obtain a nonspacelike geodesic $c : [0,1] \longrightarrow M$ with $c(0) = p$ and $c(1) = q$ which is a limit curve of the sequence $\{c_n\}$ and such that a subsequence of $\{c_n\}$ converges to c in the $C^0$ topology on curves. Since a subsequence of the pairwise distinct tangent vectors $\{c_n'(0)\}$ converges to $c'(0)$, it follows that q is conjugate to p along c, in contradiction. □

Suppose that it is only assumed that p and q are not conjugate along any timelike geodesic in the hypotheses of Proposition 9.42. If there are infinitely many timelike geodesics $\{c_n\}$ in $C_{(p,q)}$, we then have $L(c_n) \longrightarrow 0$ as $n \longrightarrow \infty$ and the limit curve c in the proof of Proposition 9.42 is a null geodesic such that q is conjugate to p along c. In particular, c contains a future null cut point to p (cf. Section 8.2). Thus, Theorem 9.41 or the proof of Proposition 9.42 also yields the following result which applies, in particular, to the Friedman cosmological models with $p = \Lambda = 0$.

COROLLARY 9.43 Suppose that (M,g) is globally hyperbolic and that $p,q \in M$ are chosen such that $p \ll q$, p is not conjugate to q along any timelike geodesic, and the future null cut locus of p in M is empty. Then there are only finitely many timelike geodesics from p to q.

*Proof.* If there were infinitely many timelike geodesics from p to q, there would be a null geodesic c from p to q such that q is conjugate to p along c. But then p contains a future null cut point along c, in contradiction. □

If dim M = 2, then (M,g) contains no null conjugate points (cf. Lemma 9.45). Thus we also have the following result.

COROLLARY 9.44 Suppose that (M,g) is any two-dimensional globally hyperbolic space-time and that $p,q \in M$ are chosen such that $p \ll q$ and q is not conjugate to p along any timelike geodesic. Then there are only finitely many timelike geodesics from p to q.

## 9.3 THE NULL MORSE INDEX THEORY

This section is devoted to the proof of a Morse index theorem for null geodesic segments $\beta : [0,1] \longrightarrow (M,g)$ in an arbitrary space-time. The appropriate index form to use, however, is not the standard index form defined on piecewise smooth vector fields orthogonal to $\beta'$, but rather its projection to the quotient bundle formed by identifying vector fields differing by a multiple of $\beta'$. The idea to use the quotient bundle as the domain of definition for the null index form is implicitly contained in the discussion of variation of arc length for null geodesics in Hawking and Ellis (1973, Section 4.5) and is further developed in Bölts (1977). The first part of this section develops the basic theory of the index form along the lines of Chapters 2 and 4 of Bölts (1977). In the second part of this section, we give a detailed proof of the Morse index theorem for null geodesics sketched in Beem and Ehrlich (1979d).

Instead of working with the energy functional, Uhlenbeck (1975, Theorem 4.5) has constructed a Morse theory for nonspacelike curves in globally hyperbolic space-times as follows. Choosing a globally hyperbolic splitting $M = S \times (a,b)$ as in Theorem 2.13, Uhlenbeck projected nonspacelike curves $\gamma(t) = (c_1(t), c_2(t))$ onto the second factor and showed that the functional $J(\gamma) = \int [c_2'(t)]^2\, dt$ yielded an index theory.

It should be mentioned at the outset that the null index theory, unlike the timelike index theory, is interesting only if $\dim M \geq 3$ for the following reason.

LEMMA 9.45 No null geodesic $\beta$ in any two-dimensional Lorentzian manifold has any null conjugate points.

*Proof*. Let $\beta : (a,b) \longrightarrow (M,g)$ be an arbitrary null geodesic. Suppose that $J$ is a Jacobi field along $\beta$ with $J(t_1) = J(t_2) = 0$ for some $t_1 \neq t_2$ in $(a,b)$. Just as in the proof of Lemma 9.9, we have $\langle J,\beta'\rangle'' = -\langle R(J,\beta')\beta',\beta'\rangle = 0$ so that $\langle J(t),\beta'(t)\rangle = 0$ for all $t \in (a,b)$. Since spacelike and null tangent vectors are never orthogonal when dim M = 2, the space of vector fields Y along $\beta$ perpendicular to $\beta'$ is spanned by $\beta'$ itself. Hence $J(t) = f(t)\beta'(t)$ for some smooth function $f : (a,b) \longrightarrow \mathbb{R}$. The Jacobi equation then becomes $0 = J'' + R(J,\beta')\beta' = f''\beta' + fR(\beta',\beta')\beta' = f''\beta'$ by the skew symmetry of the curvature tensor in the first two slots. Hence $f''(t) = 0$ for all $t \in (a,b)$. Since $J(t_1) = J(t_2) = 0$, it follows that $f = 0$. Thus $J = 0$ as required. □

In view of Lemma 9.45, we will let (M,g) be an arbitrary space-time of dimension $\geqslant 3$ throughout this section and let $\beta : [a,b] \longrightarrow M$ be a fixed null geodesic segment in M. If we let $V^{\perp}(\beta)$ denote the $\mathbb{R}$ vector space of all piecewise smooth vector fields Y along $\beta$ with $\langle Y(t),\beta'(t)\rangle = 0$ for all $t \in [a,b]$, then $\beta'(t)$ and $t\beta'(t)$ are both Jacobi fields in $V^{\perp}(\beta)$. Further, suppose we consider an index form $I : V_0^{\perp}(\beta) \times V_0^{\perp}(\beta) \longrightarrow \mathbb{R}$ defined by

$$I(X,Y) = -\int [\langle X',Y'\rangle - \langle R(X,\beta')\beta',Y\rangle]\, dt$$

in analogy to the index form (9.1) for timelike geodesics in Section 9.1. Then $A = \{f(t)\beta'(t) : f : [a,b] \longrightarrow \mathbb{R}$ is a smooth function with $f(a) = f(b) = 0\}$ is an infinite dimensional vector space such that $I(Y,Y) = 0$ for all $Y \in A$. Thus the extended index of $\beta$ defined using the index form $I : V_0^{\perp}(\beta) \times V_0^{\perp}(\beta) \longrightarrow \mathbb{R}$ is always infinite. Also while $I(f\beta',Y) = 0$ for any $Y \in V_0^{\perp}(\beta)$ and $f\beta' \in A$, the vector field $f\beta'$ is not a Jacobi field unless $f'' = 0$. Thus the relationships we derived for the index form of a timelike geodesic in Section 9.1 linking Jacobi fields, conjugate points, and the definiteness of the index form fail to hold for $I : V_0^{\perp}(\beta) \times V_0^{\perp}(\beta) \longrightarrow \mathbb{R}$.

The crux of the difficulty is that $\beta'$ and $t\beta'$ are both Jacobi fields in $V^{\perp}(\beta)$. Thus by ignoring vector fields in $A$, it is possible to define an index form for null geodesics nicely related not only to the second variation formula for the energy functional but also to conjugate points and Jacobi fields. This may be accomplished by working with the quotient bundle $V^{\perp}(\beta)/[\beta']$ rather than $V^{\perp}(\beta)$ itself. Using this quotient space, an index form $\overline{I}$ may be defined so that $\beta$ has no conjugate points iff $\overline{I}$ is negative definite [cf. Hawking and Ellis (1973, Proposition 4.5.11), Bölts (1977, Satz 4.5.5)]. Also, a Morse index theorem for null geodesic segments in arbitrary space-times may be obtained for the index form $\overline{I}$ [cf. Beem and Ehrlich (1979d)].

Since we are interested in studying conjugate points along null geodesics, it is important to note that Lemma 9.9 and Corollaries 9.10 and 9.11 carry over to the null geodesic case with exactly the same proofs.

LEMMA 9.46 Let $\beta : [a,b] \longrightarrow (M,g)$ be a null geodesic segment and let Y be any Jacobi field along $\beta$. Then $\langle Y(t),\beta'(t)\rangle$ is an affine function of t. Thus if $Y(t_1) = Y(t_2) = 0$ for distinct $t_1, t_2 \in [a,b]$, then $\langle Y(t),\beta'(t)\rangle = 0$ for all $t \in [a,b]$.

Accordingly, we may restrict our attention to the following spaces of vector fields.

DEFINITION 9.47 Let $V^{\perp}(\beta)$ denote the $\mathbb{R}$ vector space of all piecewise smooth vector fields Y along $\beta$ with $\langle Y(t),\beta'(t)\rangle = 0$ for all $t \in [a,b]$. Let $V_0^{\perp}(\beta) = \{Y \in V^{\perp}(\beta) : Y(a) = Y(b) = 0\}$. Also set $N(\beta(t)) = \{v \in T_{\beta(t)}M : \langle v,\beta'(t)\rangle = 0\}$ and let

$$N(\beta) = \bigcup_{a \leq t \leq b} N(\beta(t))$$

For any $Y \in V^{\perp}(\beta)$, the vector field $Y' \in V^{\perp}(\beta)$ may be defined using left-hand limits just as in Section 9.1. Since $\beta$ is a smooth null geodesic, $\beta'(t) \in N(\beta(t))$ for all $t \in [a,b]$. Accordingly, following Bölts (1977, pp. 39-44), we make the following algebraic construction. Since $N(\beta(t))$ is a vector space and

$$[\beta'(t)] = \{\lambda\beta'(t) : \lambda \in \mathbb{R}\} \tag{9.24}$$

is a vector subspace of $N(\beta(t))$ for each $t \in [a,b]$, we may define the quotient vector space

$$G(\beta(t)) = N(\beta(t))/[\beta'(t)] \tag{9.25}$$

and the quotient bundle

$$G(\beta) = N(\beta)/[\beta'] = \bigcup_{a\leq t\leq b} G(\beta(t)) \tag{9.26}$$

Elements of $G(\beta(t))$ are cosets of the form $v + [\beta'(t)]$, where $v \in N(\beta(t))$ and the vector subspace $[\beta'(t)]$ is the zero element of $G(\beta(t))$ for each $t \in [a,b]$. Also $v + [\beta'(t)] = w + [\beta'(t)]$ in $G(\beta(t))$ iff $v = w + \lambda\beta'(t)$ for some $\lambda \in \mathbb{R}$. We may define a natural projection map $\pi : N(\beta(t)) \longrightarrow G(\beta(t))$ by

$$\pi(v) = v + [\beta'(t)] \tag{9.27}$$

The projection map $\pi$ on each fiber induces a projection map $\pi : N(\beta) \longrightarrow G(\beta)$ given by $\pi(Y) = Y + [\beta']$, i.e., $\pi(Y)(t) = Y(t) + [\beta'(t)] \in G(\beta(t))$ for each $t \in [a,b]$.

This quotient bundle construction may be given a (non-unique) geometric realization as follows. Let $n \in T_{\beta(0)}M$ be a null tangent vector with $\langle n,\beta'(0)\rangle = -1$. Parallel translate $n$ along $\beta$ to obtain a null parallel field $\eta$ along $\beta$ with $\langle\eta(t),\beta'(t)\rangle = -1$ for all $t \in [a,b]$. Choose spacelike tangent vectors $e_1,e_2,\ldots,e_{n-2} \in T_{\beta(0)}M$ such that $\langle n,e_j\rangle = \langle\beta'(0),e_j\rangle = 0$ for all $j = 1, 2, \ldots, n - 2$ and $\langle e_i,e_j\rangle = \delta_{ij}$ for all $j = 1, 2, \ldots, n - 2$. Extend by parallel translation to spacelike parallel vector fields $E_1,E_2,\ldots,E_{n-2} \in V^{\perp}(\beta)$ and set

$$V(\beta(t)) = \{\sum_{j=1}^{n-2} \lambda_j E_j(t) : \lambda_j \in \mathbb{R} \text{ for } 1 \leq j \leq n - 2\} \tag{9.28}$$

Then $V(\beta(t))$ is a vector subspace of $N(\beta(t))$ consisting of spacelike tangent vectors and we have a direct sum decomposition

$$N(\beta'(t)) = [\beta'(t)] \oplus V(\beta(t)) \tag{9.29}$$

for each $t \in [a,b]$. Set $V(\beta) = \cup_{a\leq t\leq b} V(\beta(t))$ and let $V_0(\beta) = \{Y \in V(\beta) : Y(a) = Y(b) = 0\}$. If $\beta'(t)$ and $\eta(t)$ are given, $V(\beta)$ is independent of the particular choice of $\{e_1,e_2,\ldots,e_{n-2}\}$ in $T_{\beta(0)}M$. However, if $n \in T_{\beta(0)}M$ and hence $\eta$ are changed, a different direct sum decomposition (9.29) may arise since the given Lorentzian metric $g$ restricted to $N(\beta(t))$ is degenerate. Nonetheless, we may regard $V(\beta(t))$ as being a geometric realization of the quotient bundle $G(\beta)$ via the map $Z \longrightarrow Z + [\beta']$ from $V(\beta)$ into $G(\beta)$. It is easily checked that this map is an isomorphism since (9.29) is a direct sum decomposition.

We may define the inverse isomorphism

$$\theta : G(\beta(t)) \longrightarrow V(\beta(t)) \tag{9.30}$$

for each $t \in [a,b]$ as follows. Given $v \in G(\beta(t))$, choose any $x \in N(\beta(t))$ with $\pi(x) = v$. Decomposing $x$ uniquely as $x = \lambda\beta'(t) + v_0$ with $v_0 \in V(\beta(t))$ by (9.29), set $\theta(v) = v_0$. If any other $x_1 \in N(\beta(t))$ with $\pi(x_1) = v$ had been chosen, we would still have $x_1 = \mu\beta'(t) + v_0$ with the same $v_0 \in V(\beta(t))$ for some $\mu \in \mathbb{R}$. Thus $\theta$ is well defined.

As a first step towards defining the index form $\overline{I}$ on the quotient bundle $G(\beta)$, we show how the Lorentzian metric, covariant derivative, and curvature tensor of $(M,g)$ may be projected to $G(\beta)$. First given any $v, w \in G(\beta(t))$, choose $x,y \in N(\beta(t))$ with $\pi(x) = v$ and $\pi(y) = w$. Define $\overline{g}(v,w)$ by

$$\overline{g}(v,w) = g(x,y) \tag{9.31}$$

Suppose we had chosen $x_1, y_1 \in N(\beta(t))$ with $\pi(x_1) = v$ and $\pi(y_1) = w$. Then $x = x_1 + \lambda\beta'(t)$, $y = y_1 + \mu\beta'(t)$ for some $\lambda,\mu \in \mathbb{R}$ and thus $g(x,y) = g(x_1,y_1) + \lambda g(y_1,\beta'(t)) + \mu g(x_1,\beta'(t)) + \mu\lambda g(\beta'(t),\beta'(t)) = g(x_1,y_1)$. Hence $\overline{g}$ is well defined. It is also easily checked that $\overline{g}(v,w) = g(\theta(v),\theta(w))$ for all $v,w \in G(\beta(t))$. Hence the metric $\overline{g}$ on $G(\beta(t))$ may be identified with the given Lorentzian metric $g$ on $V(\beta(t))$. Since $g \mid V(\beta(t)) \times V(\beta(t))$ is positive definite for each $t \in [a,b]$, the induced metric $\overline{g} : G(\beta(t)) \times G(\beta(t)) \longrightarrow \mathbb{R}$ thus has the following important property.

REMARK 9.48 For each $t \in [a,b]$, the metric $\overline{g} : G(\beta(t)) \times G(\beta(t)) \longrightarrow \mathbb{R}$ is positive definite.

We now extend the covariant derivative operator acting on vector fields along $\beta$ to a covariant derivative operator for sections of $G(\beta)$. We first introduce the following notation.

DEFINITION 9.49 Let $\mathfrak{X}(\beta)$ denote the piecewise smooth sections of the quotient bundle $G(\beta)$ and let $\mathfrak{X}_0(\beta) = \{W \in \mathfrak{X}(\beta) : W(a) = [\beta'(a)]$ and $W(b) = [\beta'(b)]\}$.

Given $V \in \mathfrak{X}(\beta)$, choose $X \in V^{\perp}(\beta)$ with $\pi(X) = V$. Set

$$V'(t) = \nabla_{\beta'}V(t) = \pi(\nabla_{\beta'}X(t)) \tag{9.32}$$

If $X_1 \in V^{\perp}(\beta)$ also satisfies $\pi(X_1) = V$, then $X_1 = X + f\beta'$ for some $f : [a,b] \longrightarrow \mathbb{R}$ and we obtain $X_1' = X' + f'\beta'$ since $\beta$ is a geodesic. Thus $\pi(X_1'(t)) = \pi(X'(t))$ for all $t \in [a,b]$. Hence the covariant derivative for $\mathfrak{X}(\beta)$ given by (9.32) is well defined. It may also be checked that this covariant differentiation is compatible with the metric $\bar{g}$ for $G(\beta)$ and satisfies the usual properties of a covariant derivative.

Given $V \in \mathfrak{X}(\beta)$, choose $X \in V^{\perp}(\beta)$ with $\pi(X) = V$. We then have $X(t) = f(t)\beta'(t) + \theta(V)(t)$ with $\theta$ as in (9.30). According to (9.32), we may thus calculate $V'(t)$ using $\theta(V)$ as $V'(t) = \pi(\nabla_{\beta'}(\theta(V))(t))$. Now $\theta(V)$ satisfies $\langle\beta'(t),\theta(V)(t)\rangle = \langle\eta(t),\theta(V)(t)\rangle = 0$ for all $t \in [a,b]$. Since $\beta$ is a geodesic and $\eta$ is parallel along $\beta$, we obtain on differentiating that $\langle\beta'(t),\nabla_{\beta'}\theta(V)(t)\rangle = \langle\eta(t),\nabla_{\beta'}\theta(V)(t)\rangle = 0$ for all $t \in [a,b]$. Thus $\nabla_{\beta'}\theta(V) \in V(\beta)$. Hence we obtain

$$\theta(V'(t)) = (\theta(V))'(t) \qquad \text{for all } t \in [a,b] \tag{9.33}$$

where the differentiation on the left-hand side is in $G(\beta)$ and on the right-hand side is in $V(\beta)$. Thus if we identify $G(\beta)$ and $V(\beta)$ using $\theta$, covariant differentiation in $G(\beta)$ and in $V(\beta)$ is the same.

To project Jacobi fields and the Jacobi differential equation to $G(\beta)$, it is necessary to define a curvature endomorphism

$$\overline{R}(\cdot,\beta'(t))\beta'(t) : G(\beta(t)) \longrightarrow G(\beta(t))$$

for each $t \in [a,b]$. This may be done as follows. Given $v \in G(\beta(t))$, choose any $x \in N(\beta(t))$ with $\pi(x) = v$ and set

$$\overline{R}(v,\beta'(t))\beta'(t) = \pi(R(x,\beta'(t))\beta'(t)) \tag{9.34}$$

This definition is easily seen to be independent of the choice of $x \in N(\beta(t))$ with $\pi(x) = v$ since $R(\beta',\beta')\beta' = 0$. If $v = \pi(x)$ and $w = \pi(y)$ with $x,y \in N(\beta(t))$, it follows from (9.31) and (9.34) that

$$\overline{g}(\overline{R}(v,\beta'(t))\beta'(t),w) = g(R(x,\beta'(t))\beta'(t),y) \tag{9.35}$$

Finally from the symmetry properties of $g(R(x,y)z,w)$, we obtain

$$\overline{g}(\overline{R}(v,\beta'(t))\beta'(t),w) = \overline{g}(\overline{R}(w,\beta'(t))\beta'(t),v) \tag{9.36}$$

for all $v,w \in G(\beta(t))$.

We are now ready to define Jacobi classes in $G(\beta)$, [cf. Bölts (1977, pp. 43-44)].

DEFINITION 9.50 A smooth section $V \in \mathfrak{X}(\beta)$ is said to be a *Jacobi class* in $G(\beta)$ if $V$ satisfies the Jacobi differential equation

$$V'' + \overline{R}(V,\beta')\beta' = [\beta'] \tag{9.37}$$

where the covariant differentiation is given by (9.32) and the curvature endomorphism $\overline{R}$ by (9.34).

Given a Jacobi class $W \in \mathfrak{X}(\beta)$ with $W(a) \neq [\beta'(a)]$ and $W(b) \neq [\beta'(b)]$, we will see in the next series of lemmas that there is a two-parameter family $J_{\lambda,\mu}$ of Jacobi fields of the form $J_{\lambda,\mu} = J + \lambda\beta' + \mu t\beta'$ in $V^{\perp}(\beta)$, $\lambda,\mu \in \mathbb{R}$, with $\pi(J_{\lambda,\mu}) = W$. Nonetheless, it should be emphasized that given a Jacobi class $W \in \mathfrak{X}(\beta)$, there may be *no* Jacobi field $J$ in any geometric realization $V(\beta)$ for $G(\beta)$ with $\pi(J) = W$. On the other hand, there will always exist a Jacobi field $J \in V^{\perp}(\beta)$ with $\pi(J) = W$. But it may be necessary for $J$ to have a component in $[\beta']$. The reason for this is made precise in Lemma 9.52. In the next series of lemmas, we will use the geometric realizations $V^{\perp}(\beta) = [\beta'] \oplus V(\beta)$ defined in (9.29).

LEMMA 9.51 Let W be a Jacobi class of vector fields in G(β). Then there is a smooth Jacobi field $Y \in V^{\perp}(\beta)$ with $\pi(Y) = W$. Conversely, if Y is a Jacobi field in $V^{\perp}(\beta)$, then $\pi(Y)$ is a Jacobi class in G(β).

*Proof.* The second part of the lemma is clear, for if $Y'' + R(Y,\beta')\beta' = 0$, then $0 = \pi(Y'' + R(Y,\beta')\beta') = (\pi(Y))'' + \overline{R}(\pi(Y),\beta')\beta'$.

It remains to establish the first part of the lemma. Given the Jacobi class W, let $Y_1$ be a smooth vector field in the geometric realization V(β) with $\pi(Y_1) = W$. Since $W'' + \overline{R}(W,\beta')\beta' = [\beta']$ in $\mathfrak{X}(\beta)$, there is a smooth function $f : [a,b] \longrightarrow \mathbb{R}$ such that $Y_1'' + R(Y_1,\beta')\beta' = f\beta'$. Let $h : [a,b] \longrightarrow \mathbb{R}$ be a smooth function with $h'' = f$ and set $Y = Y_1 - h\beta'$. Then $\pi(Y) = W$ and we obtain

$$\begin{aligned} f\beta' &= Y_1'' + R(Y_1,\beta')\beta' \\ &= (Y + h\beta')'' + R(Y + h\beta',\beta')\beta' \\ &= Y'' + h''\beta' + R(Y,\beta')\beta' \\ &= Y'' + f\beta' + R(Y,\beta')\beta' \end{aligned}$$

Therefore $0 = Y'' + R(Y,\beta')\beta'$ as required. □

A more precise relationship between Jacobi fields in the geometric realization V(β) of G(β) and Jacobi classes in $\mathfrak{X}(\beta)$ is given by the following lemma.

LEMMA 9.52 Let W be a Jacobi class in $\mathfrak{X}(\beta)$. Then there is a Jacobi field $J \in V(\beta)$ with $\pi(J) = W$ iff the geometric realization $\theta(W)$ of W in V(β) satisfies the condition $R(\theta(W),\beta')\beta'\big|_t \in V(\beta(t))$ for all $t \in [a,b]$.

*Proof.* If $J \in V(\beta)$ and $\pi(J) = W$, we have $\theta(W) = J$. But as J is a Jacobi field, we then obtain

$$R(\theta(W),\beta')\beta' = R(J,\beta')\beta' = -J'' \in V(\beta)$$

since $J \in V(\beta)$.

Now suppose that $R(\theta(W),\beta')\beta' \in V(\beta)$ and $J = \theta(W)$. Then $R(\theta(W),\beta')\beta' = R(J,\beta')\beta'$. Since $W'' + \overline{R}(W,\beta')\beta' = [\beta']$ in G(β), we know that J must satisfy a differential equation of the form

$J'' + R(J,\beta')\beta' = f\beta'$ in $V^\perp(\beta)$. But if $R(\theta(V),\beta')\beta' = R(J,\beta')\beta' \in V(\beta)$, the vector field $J'' + R(J,\beta')\beta'$ is in $V(\beta)$. Hence by the decomposition (9.29), $J'' + R(J,\beta')\beta' = 0$. □

For the purpose of studying conjugate points, it is necessary to prove the following refinement of Lemma 9.51 [cf. Bölts (1977, pp. 43-44)].

LEMMA 9.53 Let $W \in \mathcal{J}(\beta)$ be a Jacobi class with $W(a) = [\beta'(a)]$ and $W(b) = [\beta'(b)]$. Then there is a unique Jacobi field $Z \in V^\perp(\beta)$ with $\pi(Z) = W$ and $Z(a) = Z(b) = 0$.

*Proof.* From Lemma 9.51, we know that there exists a Jacobi field $Y \in V^\perp(\beta)$ with $\pi(Y) = W$. However, we do not know that $Y(a) = Y(b) = 0$. But for any constants $\lambda,\mu \in \mathbb{R}$, the vector field $Y + \lambda\beta' + \mu t\beta'$ is also a Jacobi field in $V^\perp(\beta)$ with $\pi(Y + \lambda\beta' + \mu t\beta') = W$. Since $\pi(Y) = W$ and $W(a) = [\beta'(a)]$, $W(b) = [\beta'(b)]$, we know that $Y(a) = c_1\beta'(a)$ and $Y(b) = c_2\beta'(b)$ for some constants $c_1, c_2 \in \mathbb{R}$. Choosing $\lambda = (c_2a - c_1b)(b - a)^{-1}$ and $\mu = b^{-1}[(c_1b - c_2a)(b - a)^{-1} - c_2]$, it follows easily that $Z = Y + \lambda\beta' + \mu t\beta'$ satisfies $Z(a) = Z(b) = 0$.

For the uniqueness, suppose that $Z_1$ is a second Jacobi field in $V^\perp(\beta)$ with $\pi(Z_1) = W$ and $Z_1(a) = Z_1(b) = 0$. Then $X = Z_1 - Z$ is a Jacobi field of the form $X = h\beta'$ with $X(a) = X(b) = 0$. Since $0 = X'' + R(X,\beta')\beta' = h''\beta' + hR(\beta',\beta')\beta' = h''\beta'$, it follows that $h$ is an affine function. As $h(a) = h(b) = 0$, we must have $h = 0$. Therefore $Z_1 = Z$ as required. □

With Lemma 9.53 in hand, it is now possible to make the following definition. Recall also that if $J$ is any Jacobi field along $\beta$ with $J(t_1) = J(t_2) = 0$ for $t_1 \neq t_2$, then $J \in V^\perp(\beta)$ (cf. Lemma 9.46).

DEFINITION 9.54 Let $\beta : [a,b] \longrightarrow (M,g)$ be a null geodesic. For $t_1 \neq t_2$ in $[a,b]$, $t_1$ and $t_2$ are said to be *conjugate along* $\beta$ if there exists a Jacobi class $W \neq [\beta']$ in $\mathcal{J}(\beta)$ with $W(t_1) = [\beta'(t_1)]$ and $W(t_2) = [\beta'(t_2)]$. Also $t_1 \in (a,b]$ is said to be a *conjugate point of* $\beta$ if $t = a$ and $t_1$ are conjugate along $\beta$. Let

$$J_t(\beta) = \{\text{Jacobi fields } Y \text{ along } \beta : Y(a) = Y(t) = 0\}$$

and

$$\overline{J}_t(\beta) = \{\text{Jacobi classes } W \in \mathfrak{X}(\beta) : W(a) = [\beta'(a)] \text{ and } W(t) = [\beta'(t)]\}$$

Then $J_t(\beta) \subset V^\perp(\beta)$ and $t_1$ and $t_2$ are conjugate along $\beta$ in the sense of Definition 9.54 iff there exists a nontrivial Jacobi field J along $\beta$ with $J(t_1) = J(t_2) = 0$. Also we obtain the following important result from the uniqueness in Lemma 9.53.

COROLLARY 9.55 The natural projection map $\pi : J_t(\beta) \longrightarrow \overline{J}_t(\beta)$ is an isomorphism for each $t \in (a,b]$. Thus $\overline{J}_t(\beta)$ is finite dimensional and also $\dim J_t(\beta) = \dim \overline{J}_t(\beta)$ for all $t \in (a,b]$.

We are now ready to study the index form of the null geodesic $\beta : [a,b] \longrightarrow (M,g)$. For the geometric interpretation of the index form, it will be useful to introduce a functional analogous to the arc length functional for timelike geodesics, namely the energy functional. The reason for using energy rather than arc length is simply that the derivative of the function $f(x) = \sqrt{x}$ does not exist at $x = 0$, but $\sqrt{-g(\beta'(t),\beta'(t))} = 0$ for all $t \in [a,b]$ if $\beta$ is a null geodesic.

DEFINITION 9.56 Let $\gamma : [c,d] \longrightarrow (M,g)$ be a smooth nonspacelike curve. The smooth mapping $E_\gamma : [c,d] \longrightarrow \mathbb{R}$ given by

$$E_\gamma(t) = \frac{1}{2}\int_{s=c}^{t} \|\gamma'(s)\|^2 \, ds = -\frac{1}{2}\int_{s=c}^{t} g(\gamma'(s),\gamma'(s)) \, ds$$

is called the *energy function* of $\gamma$ and the number $E(\gamma) = E_\gamma(d)$ is called the *energy* of $\gamma$.

The energy of a piecewise smooth nonspacelike curve $\gamma$ is calculated by summing the energies over the intervals on which $\gamma$ is smooth just as in formula (3.1) for the arc length functional. If we let $L_\gamma(t) = L(\gamma|[c,t])$, then the Cauchy-Schwarz inequality yields

$$L_\gamma^2(d) \leqslant 2(d - c)E_\gamma(d) \tag{9.38}$$

Also, equality holds in (9.38) iff $\|\gamma'(t)\|$ is constant. Thus equality holds for null or timelike geodesics.

Recall that $V^\perp(\beta)$ denotes the space of piecewise smooth vector fields Y along β that are orthogonal to β' and $V_0^\perp(\beta) = \{Y \in V^\perp(\beta) : Y(a) = Y(b) = 0\}$.

DEFINITION 9.57 The index form $I : V^\perp(\beta) \times V^\perp(\beta) \longrightarrow \mathbb{R}$ of the null geodesic $\beta : [a,b] \longrightarrow M$ is given by

$$I(X,Y) = -\int_a^b (\langle X',Y'\rangle - \langle R(X,\beta')\beta',Y\rangle)\, dt \tag{9.39}$$

for any $X,Y \in V^\perp(\beta)$.

Formula (9.39) may be integrated by parts just as in the timelike case (cf. Remark 9.5) to give the alternate but equivalent definition of the index form used in Hawking and Ellis (1973, p. 114) and Bölts (1977, p. 110). Explicitly,

$$I(X,Y) = \int_a^b \langle X'' + R(X,\beta')\beta',Y\rangle\, dt + \sum_{i=0}^{k} \langle \Delta_{t_i}(X'),Y\rangle \tag{9.40}$$

where a partition $a = t_0 < t_1 < t_2 < \cdots < t_k = b$ has been chosen such that $X \mid [t_i,t_{i+1}]$ is smooth for $i = 0, 1, 2, \ldots, k - 1$ and

$$\Delta_{t_0}(X') = X'(a^+)$$

$$\Delta_{t_k}(X') = - X'(b^-)$$

and

$$\Delta_{t_i}(X') = \lim_{t\to t_i^+} X'(t) - \lim_{t\to t_i^-} X'(t)$$

for $i = 1, 2, \ldots, k - 1$.

If $\alpha$ is a variation of the null geodesic $\beta$ such that the neighboring curves are all null, then all derivatives of the energy functional vanish at $s = 0$ since $E(\alpha_s) = 0$ for all s. Thus it is customary to restrict attention to the following class of variations.

DEFINITION 9.58 A piecewise smooth variation $\alpha : [a,b] \times (-\varepsilon,\varepsilon) \longrightarrow M$ of a piecewise smooth nonspacelike curve $\beta : [a,b] \longrightarrow M$ is said to be *admissible* if all the neighboring curves $\alpha_s : [a,b] \longrightarrow M$ given by $\alpha_s(t) = \alpha(t,s)$ are timelike for each $s \neq 0$ in $(-\varepsilon,\varepsilon)$.

Now suppose that for $W \in V^{\perp}(\beta)$, there exists an admissible variation $\alpha : [a,b] \times (-\varepsilon,\varepsilon) \longrightarrow M$ with variation vector field $\alpha_* \partial/\partial s|_{(t,0)} = W(t)$ for each $t \in [a,b]$. Let $E(\alpha_s)$ be the energy of the neighboring curve $\alpha_s : [a,b] \longrightarrow M$ for each $s \in (-\varepsilon,\varepsilon)$. Then $d/ds\, E(\alpha_s)|_{s=0} = 0$ since $\beta$ is a geodesic and

$$\frac{d^2}{ds^2} E(\alpha_s)\Big|_{s=0} = I(W,W) \tag{9.41}$$

As $E(\beta) = E(\alpha_0) = 0$, we must have $d^2/ds^2\, E(\alpha_s)|_{s=0} \geqslant 0$. Thus a necessary condition for $W \in V^{\perp}(\beta)$ to be the variation vector field of some admissible variation $\alpha$ of $\beta$ is that $I(W,W) \geqslant 0$. Note also that if the future null cut point to $\beta(a)$ along $\beta$ comes after $\beta(b)$, then there are no admissible proper variations of $\beta$ (cf. Corollary 3.14 and Section 8.2).

It is immediate from (9.39) that if $X \in V_0^{\perp}(\beta)$ is a vector field of the form $X(t) = f(t)\beta'(t)$ for any piecewise smooth function $f : [a,b] \longrightarrow \mathbb{R}$ with $f(a) = f(b) = 0$ and $Y \in V_0^{\perp}(\beta)$ is arbitrary, then $I(X,Y) = 0$. Thus the index form $I : V_0^{\perp}(\beta) \times V_0^{\perp}(\beta) \longrightarrow \mathbb{R}$ is never negative definite and, even worse, always has an infinite dimensional null space. This suggests that the index form should be projected to $G(\beta)$ [cf. Hawking and Ellis (1973, p. 114), Bölts (1977, p. 111)]. Recall that the notation $\mathfrak{X}(\beta)$ was introduced for piecewise smooth sections of $G(\beta)$ and $\mathfrak{X}_0(\beta) = \{W \in \mathfrak{X}(\beta) : W(a) = [\beta'(a)], W(b) = [\beta'(b)]\}$.

DEFINITION 9.59 The index form $\overline{I} : \mathcal{X}(\beta) \times \mathcal{X}(\beta) \longrightarrow \mathbb{R}$ is given by

$$\overline{I}(V,W) = - \int_{t=a}^{b} [\overline{g}(V',W') - \overline{g}(\overline{R}(V,\beta')\beta',W)]\, dt \tag{9.42}$$

where $V,W \in \mathcal{X}(\beta)$ and $\overline{g}$, $\overline{R}$ and the covariant differentiation of sections of $G(\beta)$ are given by (9.31), (9.34), and (9.32), respectively.

Just as for the index form $I : V^{\perp}(\beta) \times V^{\perp}(\beta) \longrightarrow \mathbb{R}$, formula (9.42) may be integrated by parts to give the formula

$$\overline{I}(V,W) = \int_{a}^{b} \overline{g}(V'' + \overline{R}(V,\beta')\beta', W)\, dt + \sum_{j=0}^{k} \overline{g}(W(t_j), \Delta_{t_j}(V')) \tag{9.43}$$

where a partition $t_0 = a < t_1 < t_2 < \cdots < t_k = b$ has been chosen such that $V \mid [t_i,t_{i+1}]$ is smooth for $i = 0, 1, \ldots, k - 1$. In particular, if $V$ is a smooth section of $G(\beta)$ we obtain

$$\overline{I}(V,W) = - \overline{g}(V',W)\Big|_a^b + \int_a^b \overline{g}(V'' + \overline{R}(V,\beta')\beta', W)\, dt \tag{9.44}$$

and, if $V$ is a Jacobi class in $\mathcal{X}(\beta)$, we have

$$\overline{I}(V,W) = - \overline{g}(V',W)\Big|_a^b \tag{9.45}$$

Also, since vector fields of the form $f(t)\beta'(t)$ are in the null space of $I : V^{\perp}(\beta) \times V^{\perp}(\beta) \longrightarrow \mathbb{R}$, for any $X,Y \in V^{\perp}(\beta)$ with $\pi(X) = V$ and $\pi(Y) = W$, it follows that

$$\overline{I}(V,W) = I(X,Y) \tag{9.46}$$

where the index on the left-hand side is calculated in $\mathcal{X}(\beta)$ and on the right-hand side in $V^{\perp}(\beta)$.

We saw in Section 9.1 that for timelike geodesic segments, the null space of the index form consists precisely of the smooth Jacobi fields vanishing at both endpoints. We now establish the analogous result for the index form $\overline{I}$ on the quotient space $\mathcal{X}_0(\beta)$ for an arbitrary null geodesic segment $\beta : [a,b] \longrightarrow M$.

THEOREM 9.60 For piecewise smooth vector classes $W \in \mathfrak{X}_0(\beta)$, the following are equivalent:

(a) W is a smooth Jacobi class in $\mathfrak{X}_0(\beta)$.

(b) $\overline{I}(W,Z) = 0$ for all $Z \in \mathfrak{X}_0(\beta)$.

*Proof.* First (a) implies (b) is clear from formula (9.45). For the purpose of showing (b) implies (a), fix the unique piecewise smooth vector field Y in the geometric realization $V(\beta)$ for $G(\beta)$ given by (9.28) with $\pi(Y) = W$ and $Y(a) = Y(b) = 0$. Let $a = t_0 < t_1 < t_2 < \cdots < t_{k-1} < t_k = b$ be a finite partition of $[a,b]$ such that $Y \mid [t_j, t_{j+1}]$ is smooth for $j = 0, 1, \ldots, k-1$. Let $\rho : [a,b] \longrightarrow \mathbb{R}$ be a smooth function with $\rho(t_j) = 0$ for each j with $0 \leqslant j \leqslant k$ and $\rho(t) > 0$ otherwise. Set $Z = \pi(\rho(Y'' + R(Y,\beta')\beta')) \in \mathfrak{X}_0(\beta)$. Then from (9.43), we obtain

$$0 = \overline{I}(W,Z) = \int_{t=a}^{b} \rho(t)\overline{g}(W'' + \overline{R}(W,\beta')\beta', W'' + \overline{R}(W,\beta')\beta')\Big|_t \, dt$$

Remembering that $\overline{g}$ is positive definite (cf. Remark 9.48), we obtain $W''(t) + \overline{R}(W(t),\beta'(t))\beta'(t) = [\beta'(t)]$ except possibly at the $t_j$'s. Thus $W \mid [t_j,t_{j+1}]$ is a smooth Jacobi class for each j. To show that W fits together at the $t_j$'s to form a smooth Jacobi class on all of $[a,b]$, it is enough to show that the vector field Y in the geometric realization $V(\beta)$ representing W is a $C^1$ vector field. First observe that since $W''(t) + \overline{R}(W(t),\beta'(t))\beta'(t) = [\beta'(t)]$ except possibly at the $t_j$'s, we have using (9.43) that

$$0 = \overline{I}(W,Z) = \sum_{j=1}^{k-1} \overline{g}(Z(t_j),\Delta_{t_j}(W'))$$

for any $Z \in \mathfrak{X}_0(\beta)$. Since $Y \in V(\beta)$, we have $\langle Y(t),\beta'(t)\rangle = \langle Y(t),\eta(t)\rangle = 0$ for all $t \in [a,b]$. Thus $\langle Y'(t),\beta'(t)\rangle = \langle Y'(t),\eta(t)\rangle = 0$ for $t \notin \{t_0,t_1,\ldots,t_k\}$ and it follows by continuity that $\Delta_{t_j}(Y') \in V(\beta(t_j))$ for each $j = 1, 2, \ldots, k-1$. Let $X \in V(\beta)$ be a smooth vector field with $X(a) = X(b) = 0$ and $X(t_j) = \Delta_{t_j}(Y')$ for each $j = 1, \ldots, k-1$. Set $Z = \pi(X) \in \mathfrak{X}_0(\beta)$. Then we obtain

$$0 = \overline{I}(W,Z) = \sum_{j=1}^{k-1} \langle \Delta_{t_j}(Y'), \Delta_{t_j}(Y') \rangle$$

which yields $\Delta_{t_j}(Y') = 0$ for $j = 1, 2, \ldots, k - 1$. Hence $Y'$ is continuous at the $t_j$'s as required. □

Notice that in the first part of the proof of (b) ⇒ (a) of Theorem 9.60, we do *not* know that $Y'' + R(Y,\beta')\beta' \in V(\beta)$ even though $Y \in V(\beta)$. Thus we cannot conclude that $Y$ is a Jacobi field except at the $t_j$'s. This is precisely the point in passing to the quotient bundle $G(\beta)$ in which $W'' + \overline{R}(W,\beta')\beta'$ lies in $G(\beta)$ by construction and the induced metric $\overline{g}$ is positive definite on $G(\beta) \times G(\beta)$.

The aim of the next portion of this section is to prove the important result that the index form $\overline{I} : \mathfrak{X}_0(\beta) \times \mathfrak{X}_0(\beta) \longrightarrow \mathbb{R}$ is negative definite iff there are no conjugate points to $t = a$ along $\beta$ in $[a,b]$. A proof of this result, which is implicitly given in Propositions 4.5.11 and 4.5.12 of Hawking and Ellis (1973), is given in complete detail in Bölts (1977, pp. 117-123). Because the proof of negative definiteness differs considerably from the corresponding proof (cf. Theorem 9.22) for timelike geodesics, we briefly outline the proof in the timelike case in order to clarify the differences. Recall that we first showed that arbitrary proper piecewise smooth variations $\alpha$ of a timelike geodesic segment have timelike neighboring curves $\alpha_s$ for sufficiently small $s$. It was then a consequence of the Gauss lemma and the inverse function theorem that if $c : [a,b] \longrightarrow M$ is a future-directed timelike geodesic segment without conjugate points, for any proper piecewise smooth variation $\alpha$ of $c$, the neighboring curves $\alpha_s$ satisfy $L(\alpha_s) \leqslant L(c)$ for all $s$ sufficiently small. This implied that the index form is negative semidefinite in the absence of conjugate points. Algebraically, we were then able to prove the negative definiteness. Conversely, if $c$ had a conjugate point in $[a,b]$, we produced a piecewise smooth vector field $Z \in V_0^{\perp}(c)$ of zero index using a nontrivial Jacobi field guaranteed by the existence of a conjugate point to $t = a$.

For null geodesic segments, on the other hand, it is necessary to work with admissible variations to get a sensible theory. But from the theory of null cut points, we know that if $\beta(b)$ comes before the future null cut point of $\beta(a)$ along $\beta$, there are no future-directed timelike curves in M from $\beta(a)$ to $\beta(b)$. Thus the geometric argument of lifting in the absence of conjugate points and using the Gauss lemma cannot be used to obtain the negative semidefiniteness of the index form $\overline{I}$ in the absence of conjugate points. Instead, it is necessary to work directly with the Jacobi fields themselves. This is most conveniently done by using Jacobi tensors.

On the other hand, the proof that if $\overline{I} < 0$, there are no conjugate points may be done just as in the Riemannian and timelike geodesic cases. Considerably more complicated proofs are presented in Hawking and Ellis (1973, Proposition 4.5.12) and Bölts (1977, Satz 4.5.3) because these authors wish to obtain the result that if $\beta : [a,b] \longrightarrow M$ has a conjugate point $\beta(t)$ to $\beta(a)$ with $t \in (a,b)$, then there is a timelike curve from $\beta(a)$ to $\beta(b)$ "close" to $\beta$. The variation vector field W of the proper admissible variation of $\beta$ used to prove this result projects to a vector field $\overline{W} = \pi(W) \in \mathfrak{X}_0(\beta)$, $\overline{W} \neq [\beta']$ which satisfies $\overline{I}(\overline{W},\overline{W}) \geq 0$ by (9.46) and hence demonstrates the remaining half of Theorem 9.69. For completeness, we will also give the proof of this Proposition 4.5.12 of Hawking and Ellis. Notice that this result proves that the null cut point of $\beta$ comes at or before the first null conjugate point of $\beta$.

In order to prove that $\overline{I}$ is negative definite if $\beta$ has no conjugate points, we now give a brief description of the theory of Jacobi and Lagrange tensors [cf. Bölts (1977, pp. 45-49)]. A description of these tensors from the point of view of the $\nabla$ notation of this monograph was first given in Eschenburg (1975) for Riemannian manifolds and also published in Eschenburg and O'Sullivan (1976) where these tensors were used to study the divergence of geodesics in complete Riemannian manifolds.

Recall that given the null geodesic $\beta : [a,b] \longrightarrow M$, we had chosen a null parallel vector field $\eta$ along $\beta$ with $g(\eta(t),\beta'(t)) = -1$ for all $t \in [a,b]$ and then chosen spacelike parallel fields $\{E_1,\ldots,E_{n-2}\}$ along $\beta$ satisfying $g(E_i(t),E_j(t)) = \delta_{ij}$ and $g(E_i(t),\beta'(t)) = g(E_i(t),\eta(t)) = 0$ for all i, j and all $t \in [a,b]$. Let $E_0 = \beta'$ below.

A (1,1) tensor field $A(t)$ of $V^\perp(\beta)$ is a linear map

$$A = A(t) : N(\beta(t)) \longrightarrow N(\beta(t))$$

for each $t \in [a,b]$. Thus if $v \in N(\beta(t))$, then $A(t)(v) \in N(\beta(t))$. A composite endomorphism $RA(t) : N(\beta(t)) \longrightarrow N(\beta(t))$ may be defined by setting

$$RA(t)(v) = R(A(t)(v),\beta'(t))\beta'(t) \in N(\beta(t)) \tag{9.47}$$

for any $v \in N(\beta(t))$. The tensor field $A(t)$ may be defined to be smooth (resp., piecewise smooth) if the maps

$$t \longrightarrow A(E_j)(t) \qquad [a,b] \longrightarrow V^\perp(\beta)$$

for each $j = 0, 1, 2, \ldots, n - 2$ are smooth (resp., piecewise smooth). Writing

$$A(E_j) = \sum_{i=1}^{n-2} f_j^i E_i + f_j \beta' \tag{9.48}$$

where $f_j^i, f_j : [a,b] \longrightarrow \mathbb{R}$, we may define the (1,1) tensor field $A'(t) : N(\beta(t)) \longrightarrow N(\beta(t))$ by

$$A'(E_j) = \sum_{i=1}^{n-2} (f_j^i)' E_i + f_j' \beta'$$

for $j = 0, 1, 2, \ldots, n - 2$. With these rules, it follows that composition of (1,1) tensors may be identified with matrix multiplication of the component functions $f_j^i$ and $f_j$. Thus if $A = A(t)$ and $B = B(t)$ are (1,1) tensor fields along $\beta$, we have

$$(AB)' = A'B + AB' \tag{9.49}$$

Applying (9.49) to the formula $AA^{-1} = \mathrm{Id}$ in the case that $A(t) : N(\beta(t)) \longrightarrow N(\beta(t))$ is nonsingular for all $t \in [a,b]$, we obtain for nonsingular (1,1) tensor fields the differentiation rule

$$(A^{-1})' = - A^{-1}A'A^{-1} \tag{9.50}$$

For the purpose of the null index theory, however, we need to consider (1,1) tensor fields not on $V^{\perp}(\beta)$ but rather on the quotient bundle $G(\beta)$. A (1,1) tensor field $\overline{A} = \overline{A}(t) : G(\beta(t)) \longrightarrow G(\beta(t))$ is a linear map for each $t \in [a,b]$ which maps vector classes to vector classes. Using the projection of the covariant derivative to $\mathfrak{X}(\beta)$ [cf. Eq. (9.32)], we may differentiate piecewise smooth (1,1) tensor fields in $G(\beta(t))$ and obtain the formulas

$$(\overline{A}\overline{B})' = \overline{A}'\overline{B} + \overline{A}\overline{B}' \tag{9.51}$$

and

$$(\overline{A}^{-1})' = - (\overline{A})^{-1}\overline{A}'(\overline{A})^{-1} \tag{9.52}$$

provided $\overline{A}$ is nonsingular. Since the projected metric $\overline{g} : G(\beta(t)) \times G(\beta(t)) \longrightarrow \mathbb{R}$ is positive definite, we may also define the *adjoint* $\overline{A}^* = \overline{A}^*(t)$ to the (1,1) tensor field $\overline{A}(t)$ for $G(\beta(t))$ by the formula

$$\overline{g}(\overline{A}(W),Z) = \overline{g}(\overline{A}^*(Z),W) \tag{9.53}$$

for all vector classes $Z,W \in G(\beta(t))$, all $t \in [a,b]$. We may also define the composed endomorphism $\overline{R}\overline{A} : G(\beta(t)) \longrightarrow G(\beta(t))$ by

$$\overline{R}\overline{A}(W) = \overline{R}(\overline{A}(W),\beta')\beta' \tag{9.54}$$

where $\overline{R}$ is the projected curvature operator given by (9.34). Also the *kernel* $\ker(\overline{A}(t))$ of the (1,1) tensor field $\overline{A}(t) : G(\beta(t)) \longrightarrow G(\beta(t))$ is the vector space

$$\ker(\overline{A}(t)) = \{w \in G(\beta(t)) : \overline{A}(t)(w) = [\beta'(t)]\} \tag{9.55}$$

A (1,1) tensor field $\overline{A}(t) : G(\beta(t)) \longrightarrow G(\beta(t))$ is the zero tensor field, written $\overline{A} = 0$, if $\overline{A}(t)(w) = [\beta'(t)]$ for all $w \in G(\beta(t))$, all $t \in [a,b]$.

With these preliminaries out of the way, we are ready to define Jacobi and Lagrange tensor fields.

DEFINITION 9.61 A smooth (1,1) tensor field $\overline{A} : G(\beta) \longrightarrow G(\beta)$ is said to be a *Jacobi tensor* if $\overline{A}$ satisfies the conditions

$$\overline{A}'' + \overline{R}\overline{A} = 0 \tag{9.56}$$

and

$$\ker(\overline{A}(t)) \cap \ker(\overline{A}'(t)) = \{[\beta'(t)]\} \tag{9.57}$$

for all $t \in [a,b]$.

Condition (9.56) has the consequence that if $Y \in \mathfrak{X}(\beta)$ is any parallel vector class along $\beta$, then $J = \overline{A}(Y)$ is a Jacobi class along $\beta$. This follows since

$$\begin{aligned} J'' + \overline{R}(J,\beta')\beta' &= \overline{A}''(Y) + \overline{R}\overline{A}(Y) \\ &= (\overline{A}'' + \overline{R}\overline{A})(Y) = 0 \end{aligned}$$

using $Y' = 0$. Condition (9.57) has the following implication: if $Y_1, \ldots, Y_{n-2}$ are linearly independent parallel sections of $\mathfrak{X}(\beta)$, then $\overline{A}(Y_1), \ldots, \overline{A}(Y_{n-2})$ are linearly independent Jacobi sections in the following sense. If $\lambda_1, \ldots, \lambda_{n-2}$ are real constants such that

$$\sum_{j=1}^{n-2} \lambda_j \overline{A}(Y_j)(t) = [\beta'(t)]$$

for all $t \in [a,b]$, then $\lambda_1 = \lambda_2 = \cdots = \lambda_{n-2} = 0$.

The converse of the above construction may be used to show that Jacobi tensors exist. For let $E_1, \ldots, E_{n-2}$ be the spacelike parallel fields spanning $V(\beta)$ chosen in (9.28). Let $\overline{E}_j = \pi(E_j)$ be the corresponding parallel vector class in $\mathfrak{X}(\beta)$. Also let $\overline{J}_1, \overline{J}_2, \ldots, \overline{J}_{n-2}$ be the Jacobi classes along $\beta$ with initial conditions $\overline{J}_i(a) = [\beta'(a)]$ and $\overline{J}_i'(a) = \overline{E}_i(a)$ for $i = 1, \ldots, n - 2$. A Jacobi tensor $\overline{A}$ satisfying the initial conditions $\overline{A}(a) = 0$, $\overline{A}'(a) = \mathrm{Id}$ may then be defined by requiring that

$$\overline{J}_i = \overline{A}(\overline{E}_i) \tag{9.58}$$

for each i = 1, ..., n - 2 and extending to all G(β) by linearity. Since the $\overline{J}_i$'s are Jacobi classes and $\overline{E}_i$'s are parallel classes in $\mathfrak{X}(\beta)$, it follows that $\overline{A}$ satisfies (9.56). To check that $\overline{A}$ satisfies (9.57), suppose that $\overline{A}(t)(\overline{v}) = \overline{A}'(t)(\overline{v}) = [\beta'(t)]$ for some $v \in G(\beta(t))$. Choose the unique $v \in V(\beta(t))$ with $\pi(v) = \overline{v}$, and write $v = \Sigma_{i=1}^{n-2} \lambda_i E_i(t)$. Then $\overline{v} = \Sigma_{i=1}^{n-2} \lambda_i \overline{E}_i(t)$ and we obtain $[\beta'(t)] = \overline{A}(\overline{v}) = \Sigma_{i=1}^{n-2} \lambda_i \overline{J}_i(t)$ and

$$[\beta'(t)] = \overline{A}'(\overline{v}) = \sum_{i=1}^{n-2} \lambda_i \overline{A}'(\overline{E}_i)\Big|_t = \sum \lambda_i \overline{J}_i'(t)$$

Thus $\overline{J} = \Sigma_{i=1}^{n-2} \lambda_i \overline{J}_i$ is a Jacobi class in G(β) with $\overline{J}(t) = \overline{J}'(t) = [\beta'(t)]$. Hence $\overline{J} = [\beta']$. Thus $\Sigma_{i=1}^{n-2} \lambda_i \overline{E}_i(s) = [\beta'(s)]$ for all $s \in [a,b]$, contradicting the linear independence of the parallel classes $\overline{E}_1, \ldots, \overline{E}_{n-2}$. Therefore $\ker(\overline{A}(s)) \cap \ker(\overline{A}'(s)) = [\beta'(s)]$ for all $s \in [a,b]$ as required.

It is also useful to note the following result.

LEMMA 9.62 $\ker(\overline{A}(s_0)) \cap \ker(\overline{A}'(s_0)) = [\beta'(s_0)]$ for some $s_0 \in [a,b]$ iff $\overline{A}$ satisfies condition (9.57).

*Proof.* Suppose $v \in \ker(\overline{A}(s_0)) \cap \ker(\overline{A}'(s_0))$, $v \neq [\beta'(s_0)]$. Let Y be the parallel class in $\mathfrak{X}(\beta)$ with $Y(s_0) = v$. Then $J = \overline{A}(Y)$ is a Jacobi class satisfying $J(s_0) = \overline{A}(v) = [\beta'(s_0)]$ and $J'(s_0) = \overline{A}'(v) = [\beta'(s_0)]$. Hence $J = [\beta']$ from which it follows that $Y(t) \in \ker(\overline{A}(t)) \cap \ker(\overline{A}'(t))$ for all $t \in [a,b]$. □

The following two lemmas are not difficult to obtain from the relationship between parallel classes, Jacobi classes, and Jacobi tensors indicated above [cf. Bölts (1977, p. 28 and p. 49)].

LEMMA 9.63 $\beta(t_0)$ and $\beta(t_1)$ are conjugate along the null geodesic $\beta : [a,b] \longrightarrow M$ iff there exists a Jacobi tensor $\overline{A} : G(\beta) \longrightarrow G(\beta)$ with $\overline{A}(t_0) = 0$, $\overline{A}'(t_0) = \mathrm{Id}$, and $\ker(\overline{A}(t_1)) \neq \{[\beta'(t_1)]\}$.

LEMMA 9.64 Let $\beta : [a,b] \longrightarrow M$ be a null geodesic without conjugate points. Then there exists a unique smooth (1,1) tensor field $\overline{A} : G(\beta) \longrightarrow G(\beta)$ satisfying the differential equation $\overline{A}'' + \overline{RA} = 0$ with given boundary conditions $\overline{A}(a) : G(\beta(a)) \longrightarrow G(\beta(a))$ and $\overline{A}(b) : G(\beta(b)) \longrightarrow G(\beta(b))$.

*Proof.* Let $\mathcal{J}$ denote the space of (1,1) tensor fields in $G(\beta)$ satisfying the differential equation $\overline{A}'' + \overline{RA} = 0$. A linear map $\phi : \mathcal{J} \longrightarrow L(G(\beta(a))) \times L(G(\beta(b)))$ may then be defined by $\phi(\overline{A}) = (\overline{A}(a), \overline{A}(b))$. Since $\beta : [a,b] \longrightarrow M$ has no conjugate points, $\phi$ is injective. For if $\phi(\overline{A}) = (0,0)$ and $Y$ is any parallel vector class in $\mathfrak{X}(\beta)$, then $J = \overline{A}(Y)$ is a Jacobi class with $J(a) = [\beta'(a)]$ and $J(b) = [\beta'(b)]$ so that $J = [\beta']$. Since $\phi$ is an injective linear map, it follows that $\phi$ is surjective since $\dim \mathcal{J} = \dim(L(G(\beta(a))) \times L(G(\beta(b))))$. □

Even though $\overline{R} = \overline{R}^*$ and $(\overline{A}')^* = (\overline{A}^*)'$, the adjoint $\overline{A}^*$ of a Jacobi tensor A is not necessarily a Jacobi tensor since $(\overline{RA})^* \neq \overline{RA}^*$ in general. Rather $(\overline{RA})^* = \overline{A}^*\overline{R}$. Nonetheless, Jacobi tensors and their adjoints have the following useful property which is conveniently stated using the Wronskian tensor field [cf. Eschenburg and O'Sullivan (1976, p. 226), Bölts (1977, p. 48)].

DEFINITION 9.65 Let $\overline{A}$ and $\overline{B}$ be two Jacobi tensors along $G(\beta)$. Then their *Wronskian* $W(\overline{A},\overline{B})$ is the (1,1) tensor field along $G(\beta)$ given by

$$W(\overline{A},\overline{B}) = (\overline{A}')^*\overline{B} - \overline{A}^*\overline{B}' \tag{9.59}$$

It follows from the fact that $\overline{R}^* = \overline{R}$ and equations (9.51) and (9.56) that if $\overline{A}$ and $\overline{B}$ are any two Jacobi tensor fields along $G(\beta)$, then $[W(\overline{A},\overline{B})]' = 0$. Thus $W(\overline{A},\overline{B})$ is a constant tensor field. It is then natural to consider the following subclass of Jacobi tensors.

DEFINITION 9.66 A Jacobi tensor field $\overline{A}$ along $G(\beta)$ is said to be a *Lagrange tensor field* if $W(\overline{A},\overline{A}) = 0$.

For the proof of Proposition 9.68, we need the following consequence of Definition 9.66.

LEMMA 9.67 Let $\overline{A}$ be a Jacobi tensor along $G(\beta)$. If $\overline{A}(s_0) = 0$ for some $s_0 \in [a,b]$, then $\overline{A}$ is a Lagrange tensor and, in particular,

$$(\overline{A}')^* \overline{A} = \overline{A}^* \overline{A}' \tag{9.60}$$

*Proof.* We know that $W(\overline{A},\overline{A})$ is a constant tensor already. But if $\overline{A}(s_0) = 0$, $\overline{A}^*(s_0) = 0$ also and hence $W(\overline{A},\overline{A})(s_0) = 0$. Thus $W(\overline{A},\overline{A}) = 0$ as required. □

We are now ready to prove the following important proposition.

PROPOSITION 9.68 Let $\beta : [a,b] \longrightarrow M$ be a null geodesic without conjugate points to $t = a$ in $(a,b]$. Then $\overline{I}(W,W) < 0$ for all $W \in \mathfrak{X}_0(\beta)$, $W \neq [\beta']$.

*Proof.* Let $\overline{A}$ be the Jacobi tensor along $G(\beta)$ with initial conditions $\overline{A}(a) = 0$ and $\overline{A}'(a) = \mathrm{Id}$. Since $\beta$ has no conjugate points, Lemma 9.63 implies that $\ker(\overline{A}(t)) = \{[\beta'(t)]\}$ for all $t \in (a,b]$.

Now let $W \in \mathfrak{X}_0(\beta)$ be arbitrary. Since $\overline{A}$ is nonsingular in $(a,b]$ and $W(a) = [\beta'(a)]$, we may find $Z \in \mathfrak{X}(\beta)$ with $W = \overline{A}(Z)$. By the rules for covariant differentiation of tensor fields, we obtain

$$\overline{A}'(Z) = [\overline{A}(Z)]' - \overline{A}(Z') \tag{9.61}$$

and

$$\overline{A}''(Z) = (\overline{A}'(Z))' - \overline{A}'(Z') \tag{9.62}$$

Now we are ready to calculate $\overline{I}(W,W)$. First

$$\begin{aligned} I(W,W) &= -\int_a^b [\overline{g}(W',W') - \overline{g}(\overline{R}(W,\beta')\beta',W)]\,dt \\ &= -\int_a^b [\overline{g}([\overline{A}(Z)]',[\overline{A}(Z)]') - \overline{g}(\overline{RA}(Z),\overline{A}(Z))]\,dt \\ &= -\int_a^b [\overline{g}(\overline{A}'(Z),\overline{A}'(Z)) + 2\overline{g}(\overline{A}'(Z),\overline{A}(Z')) + \overline{g}(\overline{A}(Z'),\overline{A}(Z')) \\ &\quad + \overline{g}(\overline{A}''(Z),A(Z))]\,dt \end{aligned}$$

using (9.56) and (9.61). Now

$$\begin{aligned}\bar g(\bar A''(Z),\bar A(Z)) &= \bar g([\bar A'(Z)]',\bar A(Z)) - \bar g(\bar A'(Z'),\bar A(Z))\\ &= [\bar g(\bar A'(Z),\bar A(Z))]' - \bar g(\bar A'(Z),[\bar A(Z)]') - \bar g(\bar A'(Z'),\bar A(Z))\\ &= [\bar g(\bar A'(Z),\bar A(Z))]' - \bar g(\bar A'(Z),\bar A(Z')) - \bar g(\bar A'(Z),\bar A'(Z))\\ &\quad - \bar g(\bar A'(Z'),\bar A(Z))\end{aligned}$$

Substituting into the above formula for $\bar I(W,W)$ then yields

$$\begin{aligned}\bar I(W,W) &= -\bar g(\bar A'(Z),\bar A(Z))\Big|_a^b\\ &\quad - \int_a^b [\bar g(\bar A(Z'),\bar A(Z')) + \bar g(\bar A'(Z),\bar A(Z')) - \bar g(\bar A'(Z'),\bar A(Z))]\,dt\\ &= -\bar g(\bar A'(Z),W)\Big|_a^b - \int_a^b [\bar g(\bar A(Z'),\bar A(Z')) + \bar g(\bar A^*\bar A'(Z),Z')\\ &\quad - \bar g(Z',(\bar A')^*\bar A(Z))]\,dt\end{aligned}$$

Now the first term vanishes since $W(a) = [\beta'(a)]$ and $W(b) = [\beta'(b)]$. Thus we obtain

$$\begin{aligned}\bar I(W,W) &= -\int_a^b \{\bar g(\bar A(Z'),\bar A(Z')) + \bar g([\bar A^*\bar A' - (\bar A')^*\bar A](Z),Z')\}\,dt\\ &= -\int_a^b \bar g(\bar A(Z'),\bar A(Z'))\,dt\end{aligned}$$

using (9.60). If $Z'(t) = 0$ for all $t$, then $Z$ is parallel along $\beta$ and hence $W = \bar A(Z)$ would be a Jacobi vector class along $\beta$ with $W(a) = [\beta'(a)]$ and $W(b) = [\beta'(b)]$. Since $\beta$ has no conjugate points in $(a,b]$, this is impossible. Hence from the positive definiteness of $\bar g$ and the nonsingularity of $\bar A(t)$ for $t \in (a,b]$, we obtain $\bar I(W,W) < 0$ as required. □

This proposition has the well known geometric consequence in general relativity that if $\beta : [a,b] \longrightarrow (M,g)$ is free of conjugate points, then no "small" variation of $\beta$ gives a timelike curve from p to q [cf. Hawking and Ellis (1973, p. 115)].

We are now ready to prove the following theorem.

THEOREM 9.69 Let $\beta : [a,b] \longrightarrow M$ be a null geodesic segment. Then the following are equivalent.

(a) $\beta$ has no conjugate points in $(a,b]$.

(b) $\overline{I}(W,W) < 0$ for all $W \in \mathfrak{X}_0(\beta)$, $W \neq [\beta']$.

*Proof.* We have already shown (a) implies (b) in Proposition 9.68. To show (b) implies (a), we suppose $\beta(a)$ is conjugate to $\beta(t_0)$ with $0 < t_0 \leqslant b$ and produce a nontrivial vector class $W \in \mathfrak{X}_0(\beta)$ with $\overline{I}(W,W) \geqslant 0$. If $\beta(a)$ is conjugate to $\beta(t_0)$, we know that there is a nontrivial Jacobi class $Z \in \mathfrak{X}(\beta)$ with

$$Z(a) = [\beta'(a)] \qquad Z(t_0) = [\beta'(t_0)]$$

Set

$$W(t) = \begin{cases} Z(t) & a \leqslant t \leqslant t_0 \\ [\beta'(t)] & t_0 \leqslant t \leqslant b \end{cases}$$

Then

$$\begin{aligned} \overline{I}(W,W) &= \overline{I}(W,W)_a^{t_0} + \overline{I}(W,W)_{t_0}^{b} \\ &= -\,\overline{g}(Z',Z)\Big|_a^{t_0} + 0 = 0 \end{aligned}$$

using (9.45) and $Z(a) = [\beta'(a)]$, $Z(t_0) = [\beta'(t_0)]$. Thus (b) $\Rightarrow$ (a) holds. □

As in the Riemannian and timelike cases (cf. Theorem 9.23), Theorem 9.69 has the following consequence which is essential to the proof of the Morse index theorem for null geodesics.

THEOREM 9.70 (Maximality of Jacobi Classes) Let $\beta : [a,b] \longrightarrow M$ be a null geodesic segment with no conjugate points and let $J \in \mathfrak{X}(\beta)$

be any Jacobi class. Then for any vector class $Y \in \mathfrak{X}(\beta)$ with $Y \neq J$ and

$$Y(a) = J(a) \qquad Y(b) = J(b) \tag{9.63}$$

we have

$$\overline{I}(J,J) > \overline{I}(Y,Y) \tag{9.64}$$

*Proof.* This may be established just as in Theorem 9.23 using (9.45) and Theorem 9.69. □

Since the canonical variation (9.5) of Section 9.1 applied to a vector field $W \in V^{\perp}(\beta)$ is *not* necessarily an admissible variation of β in the sense of Definition 9.58, Theorem 9.69 does *not* imply the existence of a timelike curve from β(a) to β(b) provided that $t = a$ is conjugate to some $t_0 \in (a,b)$ along β. We thus give a separate proof of this result in Theorem 9.72 [cf. Hawking and Ellis (1973, pp. 115-116), Bölts (1977, pp. 117-121)]. It is first helpful to derive conditions for a proper variation of a null geodesic to be admissible. For convenience, we will assume that $\beta : [0,1] \longrightarrow M$ so that the formulas in the proof of Theorem 9.72 will be simpler.

Now let $\alpha : [0,1] \times (-\varepsilon,\varepsilon) \longrightarrow M$ be a piecewise smooth proper variation of β such that each neighboring curve $\alpha_s(t) := \alpha(t,s)$ is future timelike for $s \neq 0$. Let $\partial/\partial t$ and $\partial/\partial s$ denote the coordinate vector fields on $[0,1] \times (-\varepsilon,\varepsilon)$ and put $V = \alpha_* \, \partial/\partial s$ and $T = \alpha_* \, \partial/\partial t$. Then $T|_{(t,0)} = \beta'(t)$ and $V|_{(t,0)}$ is called the *variation vector field* of the variation α of β. Since α is a proper variation, $V|_{(0,0)} = V|_{(1,0)} = 0$. Also,

$$\frac{d}{ds}\,(g(T,T))\Big|_{(t,0)} = 0 \tag{9.65}$$

since $g(T,T)|_{(t,s)} < 0$ for $s \neq 0$ and $g(T,T)|_{(t,0)} = g(\beta'(t),\beta'(t)) = 0$. Calculating,

$$\frac{d}{ds}\,(g(T,T))\Big|_{(t,0)} = 2g(\nabla_{\partial/\partial s}T,T)\Big|_{(t,0)} = 2g(\nabla_{\partial/\partial t}V,T)\Big|_{(t,0)}$$

$$= 2 \frac{d}{dt} (g(V,T))\Big|_{(t,0)} - 2g(V,\nabla_{\partial/\partial t}T)\Big|_{(t,0)}$$

$$= 2 \frac{d}{dt} (g(V,T))\Big|_{(t,0)} - 2g(V\big|_{(t,0)},\nabla_{\partial/\partial t}\, \beta'\big|_t)$$

Thus since $\beta$ is a geodesic, we obtain

$$\frac{d}{ds} (g(T,T))\Big|_{(t,0)} = 2 \frac{d}{dt} (g(V,T))\Big|_{(t,0)}$$

Now $f(t) = g(V,T)\big|_{(t,0)}$ is a piecewise smooth function with $f(0) = f(1) = 0$. Hence if $f(t) \neq 0$ for some $t \in [0,1]$, there exists a $t_0 \in (0,1)$ with $f'(t_0) > 0$. Then

$$\frac{d}{ds} (g(T,T))\Big|_{(t_0,0)} = 2f'(t_0) > 0$$

in contradiction to (9.65). Hence we obtain as a first necessary condition for $\alpha$ to be an admissible deformation of $\beta$ that

$$g(V,T)\Big|_{(t,0)} = 0 \tag{9.66}$$

for all $t \in [0,1]$. Thus $V \in V^\perp(\beta)$. Consequently

$$\frac{d}{ds} (g(T,T))\Big|_{(t,0)} = 2 \frac{d}{dt} (g(V,T))\Big|_{(t,0)} = 0 \tag{9.67}$$

for all $t \in [0,b]$. It then follows from (9.67) that the neighboring curves $\alpha_s$ of the variation will be timelike provided that the variation field satisfies (9.66), (9.67), and the condition that $(d^2/ds^2)(g(T,T))\big|_{(t,0)} < -c < 0$ for all $t \in (0,1)$. As above,

$$\frac{d}{ds} (g(T,T)) = 2 \frac{d}{dt} (g(V,T)) - 2g(V,\nabla_{\partial/\partial t}T)$$

Hence

$$\frac{d^2}{ds^2} (g(T,T))\Big|_{(t,0)} = 2 \frac{d}{ds}\left[\frac{d}{dt} (g(V,T))\right]\Big|_{(t,0)} - 2 \frac{d}{ds} (g(V,\nabla_{\partial/\partial t}T))\Big|_{(t,0)}$$

$$= 2 \frac{d}{dt}\left[\frac{d}{ds} g(V,T)\right]\Big|_{(t,0)} - 2g(\nabla_{\partial/\partial s}V,\nabla_{\partial/\partial t}T)\Big|_{(t,0)}$$

$$- 2g(V,\nabla_{\partial/\partial s}\nabla_{\partial/\partial t}T)\Big|_{(t,0)}$$

Using the identity $R(V,T)T = \nabla_{\partial/\partial s}\nabla_{\partial/\partial t}T - \nabla_{\partial/\partial t}\nabla_{\partial/\partial s}T = \nabla_{\partial/\partial s}\nabla_{\partial/\partial t}T - \nabla_{\partial/\partial t}\nabla_{\partial/\partial t}V$ and $\nabla_{\partial/\partial t}T\big|_{(t,0)} = \nabla_{\partial/\partial t}\beta'\big|_t = 0$, we obtain

$$\begin{aligned}\frac{d^2}{ds^2}(g(T,T))\Big|_{(t,0)} &= 2\frac{d}{dt}[g(\nabla_{\partial/\partial s}V,T) + g(V,\nabla_{\partial/\partial s}T)]\Big|_{(t,0)} \\ &\quad - 2g(V,\nabla_{\partial/\partial t}\nabla_{\partial/\partial t}V + R(V,T)T)\Big|_{(t,0)} \\ &= 2\frac{d}{dt}[g(\nabla_{\partial/\partial s}V,T) + g(V,\nabla_{\partial/\partial t}V)]\Big|_{(t,0)} \\ &\quad - 2g(V,\nabla_{\partial/\partial s}\nabla_{\partial/\partial t}T)\Big|_{(t,0)}\end{aligned}$$

This calculation yields the following lemma.

LEMMA 9.71 A sufficient condition for the piecewise smooth proper variation $\alpha : [0,1] \times (-\varepsilon,\varepsilon) \longrightarrow M$ of the null geodesic $\beta : [0,1] \longrightarrow M$ to be an admissible variation [i.e., the curves $\alpha_s(t) = \alpha(t,s)$ are timelike for $s \neq 0$] for all $s \neq 0$ sufficiently small is that the variation vector field $V(t) = \alpha_* \, \partial/\partial s\big|_{(t,0)}$ satisfies the conditions

$$g(V(t),\beta'(t)) = 0 \qquad \text{for all } t \in [0,1] \tag{9.68}$$

$$\frac{d}{ds}(g(T,T))\Big|_{(t,0)} = 0 \qquad \text{for all } t \in [0,1] \tag{9.69}$$

$$\frac{d}{dt}[g(\nabla_{\partial/\partial s}V,\beta') + g(V,V')]\Big|_{(t,0)} - g(V,V'' + R(V,\beta')\beta')\Big|_t < -c < 0 \tag{9.70}$$

for all $t \in (0,1)$ at which $V$ is smooth.

We are now ready to prove the desired result [cf. Hawking and Ellis (1973, p. 115)].

THEOREM 9.72 Let $\beta : [0,1] \longrightarrow M$ be a null geodesic. If $\beta(t_0)$ is conjugate to $\beta(0)$ along $\beta$ for some $t_0 \in (0,1)$, then there is a timelike curve from $\beta(0)$ to $\beta(1)$.

*Proof.* We will suppose that $t_0 > 0$ is the first conjugate point of $\beta(0)$ along $\beta$. It is enough to show that for some $t_2$ with $t_0 < t_2 \leq 1$, there is a future-directed timelike curve from $\beta(0)$ to $\beta(t_2)$. For then we have $\beta(0) \ll \beta(t_2) \leq \beta(1)$ whence $\beta(0) \ll \beta(1)$. Thus there exists a timelike curve from $\beta(0)$ to $\beta(1)$.

Since $\beta(t_0)$ is conjugate to $\beta(0)$ along $\beta$, there exists a smooth nontrivial Jacobi class $W \in \mathfrak{X}(\beta)$ with $W(0) = [\beta'(0)]$ and $W(t_0) = [\beta'(t_0)]$. We may write

$$W(t) = f(t)\hat{W}(t) \tag{9.71}$$

where $\hat{W}$ is a smooth vector class along $\beta$ with $\overline{g}(\hat{W},\hat{W}) = 1$ and $f : [0,1] \longrightarrow \mathbb{R}$ a smooth function. Since $t_0$ is the first conjugate point along $\beta$, $f(0) = f(t_0) = 0$ and changing $\hat{W}$ to $-\hat{W}$ if necessary, we may assume that $f(t) > 0$ for all $t \in (0,t_0)$. Since $W$ is a nontrivial Jacobi class and $W(t_0) = [\beta'(t_0)]$, we have $W'(t_0) \neq [\beta'(t_0)]$. Thus from the formula $W'(t_0) = f'(t_0)\hat{W}(t_0) + f(t_0)\hat{W}'(t_0) = f'(t_0)\hat{W}(t_0)$, we obtain $f'(t_0) \neq 0$. Thus we may choose $t_1 \in (t_0,1]$ such that $W(t) \neq [\beta'(t)]$ and $f(t) < 0$ for all $t \in (t_0,t_1]$.

With $t_1$ as above, the idea of the proof is now to show that there exists a $t_2 \in (t_0,t_1]$ such that there is an admissible proper variation $\alpha : [0,t_2] \times (-\varepsilon,\varepsilon) \longrightarrow M$ of $\beta \mid [0,t_2]$. Then the neighboring curves $\alpha_s$ of the variation $\alpha$ will be timelike curves from $\beta(0)$ to $\beta(t_2)$ for $s \neq 0$. This will then imply that $\beta(0) \ll \beta(1)$ as required. To this end, we want to deform $W \mid [0,t_1]$ to a vector class $\overline{Z} \mid [0,t_2]$ with $\overline{g}(\overline{Z},\overline{Z}'' + \overline{R}(\overline{Z},\beta')\beta') > 0$ so that if $Z \in V^{\perp}(\beta)$ is an appropriate lift of $\overline{Z} \in \mathfrak{X}(\beta|[0,t_2])$ and $\alpha$ is a variation of $\beta \mid [0,t_2]$ with variation vector field $Z$, then conditions (9.69) and (9.70) of Lemma 9.71 will be satisfied.

Consider a vector class of the form

$$\overline{Z}(t) = [b(e^{at} - 1) + f(t)]\hat{W}(t)$$

with $b = -f(t_1)(e^{at_1} - 1)^{-1} \in \mathbb{R}$ and $a > 0$ in $\mathbb{R}$ chosen such that $(a^2 + \mathrm{glb}\{h(t) : t \in [0,t_1]\}) > 0$, where

$$h(t) = \bar{g}(\hat{W}'' + \bar{R}(\hat{W},\beta')\beta',\hat{W})\Big|_t$$

Because W is a Jacobi class, we obtain

$$\begin{aligned} 0 &= \bar{g}(W'' + \bar{R}(W,\beta')\beta',\hat{W}) \\ &= f'' + 2f'\bar{g}(\hat{W}',\hat{W}) + f\bar{g}(\hat{W}'',\hat{W}) + f\bar{g}(\bar{R}(\hat{W},\beta')\beta',\hat{W}) \\ &= f'' + 2f'\bar{g}(\hat{W}',\hat{W}) + fh \end{aligned}$$

But since $\bar{g}(\hat{W}',\hat{W}) = 1/2(g(\hat{W},\hat{W}))' = 1/2(1)' = 0$, we obtain the formula $f'' = -fh$. Returning to consideration of the vector class $\bar{Z}$, first note that by choice of the constants a and b, we have $\bar{Z}(0) = [\beta'(0)]$ and $\bar{Z}(t_1) = [\beta'(t_1)]$. In view of formula (9.70), we wish to have $\bar{g}(\bar{Z},\bar{Z}'' + \bar{R}(\bar{Z},\beta')\beta') > 0$ also. Setting $r(t) = b(e^{at} - 1) + f(t)$, remembering that $\bar{g}(\hat{W}',\hat{W}) = 0$, and differentiating yields $\bar{g}(\bar{Z},\bar{Z}'' + \bar{R}(\bar{Z},\beta')\beta') = r(r'' + rh) = r[be^{at}(a^2 + h) - bh + f'' + fh]$. Since $f'' = -fh$, we obtain

$$\bar{g}(\bar{Z},\bar{Z}'' + \bar{R}(\bar{Z},\beta')\beta')\Big|_t = r(t)b\{e^{at}[a^2 + h(t)] - h(t)\}$$

Now $b = -f(t_1)(e^{at_1} - 1) > 0$ as $f(t_1) < 0$; so the expression $b\{e^{at}[a^2 + h(t)] - h(t)\} > 0$ for all $t \in [0,t_1]$. Thus $\bar{g}(\bar{Z},\bar{Z}'' + \bar{R}(\bar{Z},\beta')\beta')\big|_t > 0$ provided $r(t) > 0$. Since $f(t) > 0$ for $t \in (0,t_0)$, we have $r(t) > 0$ for $t \in (0,t_0]$. By continuity, there is some $t_2 > t_0$ with $r(t) > 0$ for $t \in [t_0,t_2)$ and $r(t_2) = 0$. If $t_2 \geqslant t_1$, then in fact $t_2 = t_1$ since $r(t_1) = 0$ by construction and we let $t_2 = t_1$ below. If $t_2 < t_1$, then the vector class $\bar{Z} \mid [0,t_2]$ will satisfy $\bar{Z}(t_2) = [\beta'(t_2)]$ since $r(t_2) = 0$ and also $\bar{g}(\bar{Z},\bar{Z}'' + \bar{R}(\bar{Z},\beta')\beta')\big|_t > 0$ for all $t \in (0,t_2)$. We now work with $\beta \mid [0,t_2]$.

Let $\tilde{Z} \in V^{\perp}(\beta|[0,t_2])$ satisfy $\pi(\tilde{Z}) = \bar{Z}$. Since $\bar{Z}(0) = [\beta'(0)]$ $\bar{Z}(t_2) = [\beta'(t_2)]$, we have $\tilde{Z}(0) = \mu\beta'(0)$ and $\tilde{Z}(t_2) = \lambda\beta'(t_2)$ for some constants $\mu,\lambda \in \mathbb{R}$. Set $Z = \tilde{Z} - \mu\beta' + [(\mu - \lambda)/t_2]t\beta'$. Then $Z(0) = Z(t_2) = 0$ and $\pi(Z) = \bar{Z}$. Consequently

$$g(Z'' + R(Z,\beta')\beta',Z)\Big|_t = \overline{g}(\overline{Z}'' + \overline{R}(\overline{Z},\beta')\beta',\overline{Z})\Big|_t > 0 \tag{9.72}$$

for all $t \in (0,t_2)$. Choose a constant $\varepsilon > 0$ so that

$$\varepsilon < \mathrm{glb}\left\{g(Z''(t) + R(Z(t),\beta'(t))\beta'(t),Z(t)) \;:\; t \in \left[\frac{t_2}{4}, \frac{3t_2}{4}\right]\right\} \tag{9.73}$$

which is possible in view of (9.72). Now define a function $\rho$ : $[0,t_2] \longrightarrow \mathbb{R}$ by

$$\rho(t) = \begin{cases} -\varepsilon t & 0 \leq t \leq \frac{t_2}{4} \\ \varepsilon(t - \frac{t_2}{2}) & \frac{t_2}{4} \leq t \leq \frac{3t_2}{4} \\ \varepsilon(t_2 - t) & \frac{3t_2}{4} \leq t \leq t_2 \end{cases}$$

We now have a given vector field $Z \in V_0^{\perp}(\beta|[0,t_2])$ and a given function $\rho : [0,t_2] \longrightarrow \mathbb{R}$. Recall that we had fixed a pseudo-orthonormal frame field $E_1, E_2, \ldots, E_{n-2}, \eta, \beta'$ for $\beta$ in (9.28). We now need to find a proper variation $\alpha : [0,t_2] \times (-\varepsilon,\varepsilon) \longrightarrow M$ of $\beta \mid [0,t_2]$ satisfying the initial conditions

$$\alpha_* \frac{\partial}{\partial s}\Big|_{(t,0)} = Z(t) \tag{9.74}$$

and

$$\nabla_{\partial/\partial s}\alpha_* \frac{\partial}{\partial s}\Big|_{(t,0)} = [g(Z,Z')\big|_t - \rho(t)]\eta(t) \tag{9.75}$$

for all $t \in [0,t_2]$. Thus we wish to specify the first and second derivatives of the curves $s \longrightarrow \alpha(t,s)$ for each $t \in [0,t_2]$. The existence of such a deformation is guaranteed by the theory of differential equations applied to (9.74) and (9.75) written out in terms of a Fermi coordinate system for the geodesic $\beta$ defined by the pseudo-orthonormal frame $E_1, E_2, \ldots, E_{n-2}, \eta, \beta'$.

Given the proper variation $\alpha$ of $\beta \mid [0,t_2]$ satisfying (9.74) and (9.75) and setting $T = \alpha_*\partial/\partial t$ and $V = \alpha_*\partial/\partial s$ as above, it follows that

$$g(\nabla_{\partial/\partial s} V, \beta')\Big|_{(t,0)} + g(V,V')\Big|_{(t,0)} = \rho(t)$$

Hence

$$\frac{d}{dt}(g(\nabla_{\partial/\partial s} V, \beta') + g(V,V'))\Big|_{(t,0)} = \rho'(t) = \begin{cases} -\varepsilon & 0 \leqslant t < \frac{t_2}{4} \\ +\varepsilon & \frac{t_2}{4} \leqslant t \leqslant \frac{3t_2}{4} \\ -\varepsilon & \frac{3t_2}{4} < t \leqslant t_2 \end{cases}$$

Thus in view of (9.73), the variation $\alpha$ of $\beta \mid [0,t_2]$ satisfies condition (9.70) of Lemma 9.71. Applying 9.71, we find that this variation produces timelike curves $\alpha_s$ from $\beta(0)$ to $\beta(t_2)$ for small $s \neq 0$ as required. □

COROLLARY 9.73 The null cut point of $\beta : [0,a) \longrightarrow M$ comes at or before the first null conjugate point.

We are at last ready to turn to the proof of the Morse index theorem for null geodesics. The proof parallels that of the timelike Morse index theorem, Theorem 9.27. In view of Theorem 9.69, the index $\mathrm{Ind}(\beta)$ and the extended index $\mathrm{Ind}_0(\beta)$ of $\beta$ with respect to the index form $\overline{I} : \mathfrak{X}_0(\beta) \times \mathfrak{X}_0(\beta) \longrightarrow \mathbb{R}$ should be defined as follows.

DEFINITION 9.74 The *index* $\mathrm{Ind}(\beta)$ and *extended index* $\mathrm{Ind}_0(\beta)$ of $\beta$ with respect to the index form $\overline{I} : \mathfrak{X}_0(\beta) \times \mathfrak{X}_0(\beta) \longrightarrow \mathbb{R}$ are given by

$$\mathrm{Ind}(\beta) = \mathrm{lub}\{\dim A : A \text{ is a subspace of } \mathfrak{X}_0(\beta) \text{ and } \overline{I} \mid A \times A \text{ is positive definite}\}$$

and

$$\mathrm{Ind}_0(\beta) = \mathrm{lub}\{\dim A : A \text{ is a subspace of } \mathfrak{X}_0(\beta) \text{ and } I \mid A \times A \text{ is positive semidefinite}\}$$

respectively.

As a preliminary step toward establishing the null index theorem, we need the following lemma.

LEMMA 9.75 If $\beta \mid [s,t]$ is free of conjugate points, then given any $\overline{v} \in G(\beta(s))$ and $\overline{w} \in G(\beta(t))$, there is a unique Jacobi class $Z \in \mathfrak{X}(\beta)$ with $Z(s) = \overline{v}$ and $Z(t) = \overline{w}$.

*Proof.* Let $v \in V(\beta(s))$ and $w \in V(\beta(t))$ satisfy $\pi(v) = \overline{v}$ and $\pi(w) = \overline{w}$. By the nonconjugacy hypothesis, there is a unique Jacobi field $J \in V^{\perp}(\beta)$ with $J(s) = v$ and $J(t) = w$. Then $Z = \pi(J)$ is a Jacobi class in $\mathfrak{X}(\beta)$ with $Z(s) = \overline{v}$ and $Z(t) = \overline{w}$.

Suppose now that $Z_1$ is a second Jacobi class in $\mathfrak{X}(\beta)$ with $Z_1(s) = \overline{v}$ and $Z_1(t) = \overline{w}$. By Lemma 9.51, there is a Jacobi field $J_1 \in V^{\perp}(\beta)$ with $\pi(J_1) = Z_1$. Since $Z_1(s) = \overline{v}$ and $Z_1(t) = \overline{w}$, it follows that $J_1(s) = v + c_1\beta'(s)$ and $J_1(t) = w + c_2\beta'(t)$ for some constants $c_1, c_2 \in \mathbb{R}$. It is then elementary to check that constants $\lambda, \mu \in \mathbb{R}$ may be found so that the Jacobi field $J_2(\tau) = J_1(\tau) + \lambda\beta'(\tau) + \mu\tau\beta'(\tau) \in V^{\perp}(\beta)$ satisfies $J_2(s) = v$ and $J_2(t) = w$. By the nonconjugacy assumption, we must then have $J_2 = J$. But this implies that $Z = \pi(J) = \pi(J_2) = \pi(J_1) = Z_1$ as required. □

We are now ready to establish the finiteness of the index and extended index.

PROPOSITION 9.76 $\mathrm{Ind}_0(\beta)$ and $\mathrm{Ind}(\beta)$ are both finite and are related by the equality $\mathrm{Ind}_0(\beta) = \mathrm{Ind}(\beta) + \dim \overline{J}_b(\beta)$.

*Proof.* Given the fixed null geodesic $\beta : [a,b] \longrightarrow M$, choose a finite partition $a = t_0 < t_1 < t_2 < \cdots < t_k = b$ so that $\beta \mid [t_j, t_{j+1}]$ is free of conjugate points for each $j$. Let $J\{t_j\}$ denote the vector space of continuous vector classes $W$ along $\beta$ such that $W \mid [t_j, t_{j+1}]$ is a smooth Jacobi class for each $j$. It is immediate from the uniqueness in Lemma 9.75 that $\overline{J}\{t_j\}$ is finite dimensional. We may then define a finite dimensional approximation $\phi : \mathfrak{X}_0(\beta) \longrightarrow \overline{J}\{t_j\}$ in analogy with the Riemannian and timelike Lorentzian index theory. Namely, given $W \in \mathfrak{X}_0(\beta)$, let $\phi(W)$ be the piecewise smooth Jacobi class in $\overline{J}\{t_j\}$ such that for each $j$, the vector class

$\phi(W) \mid [t_j,t_{j+1}]$ is the unique Jacobi class in $\beta \mid [t_j,t_{j+1}]$ with $\phi(W)(t_j) = W(t_j)$ and $\phi(W)(t_{j+1}) = W(t_{j+1})$. Then $\phi \mid \overline{J}\{t_j\} = \mathrm{Id}$ and (9.64) applied to each subinterval $[t_j,t_{j+1}]$ yields the inequality

$$\overline{I}(\phi(W),\phi(W)) > \overline{I}(W,W) \tag{9.76}$$

if $W \in \mathfrak{X}_0(\beta)$, $W \notin \overline{J}\{t_j\}$. Using (9.76), it may be shown just as in the proof of Sublemma 9.26 in section 9.1 that

If A is a subspace of $\mathfrak{X}_0(\beta)$ on which $\overline{I} \mid A \times A$ is positive semidefinite, then $\phi : A \longrightarrow \overline{J}\{t_j\}$ is injective (9.77)

Now let $\mathrm{Ind}'(\beta)$ and $\mathrm{Ind}_0'(\beta)$ denote the index and extended index respectively of the index form $\overline{I} : \overline{J}\{t_j\} \times \overline{J}\{t_j\} \longrightarrow \mathbb{R}$ restricted to $\overline{J}\{t_j\}$. Suppose that A is a subspace of $\mathfrak{X}_0(\beta)$ on which $\overline{I} \mid A \times A$ is positive semidefinite. Using (9.76), it is easily seen that $\overline{I} : \overline{J}\{t_j\} \times \overline{J}\{t_j\} \longrightarrow \mathbb{R}$ is positive semidefinite on the subspace $\phi(A)$ of $\overline{J}\{t_j\}$. Also by (9.77), $\dim A = \dim \phi(A)$. Hence $\mathrm{Ind}_0'(\beta) \geqslant \mathrm{Ind}_0(\beta)$. On the other hand, as $\overline{J}\{t_j\} \subset \mathfrak{X}_0(\beta)$, we have $\mathrm{Ind}_0'(\beta) \leqslant \mathrm{Ind}_0(\beta)$. Thus $\mathrm{Ind}_0'(\beta) = \mathrm{Ind}_0(\beta)$. A similar proof establishes that $\mathrm{Ind}'(\beta) = \mathrm{Ind}(\beta)$. Since $\overline{J}\{t_j\}$ is finite dimensional, these equalities imply that $\mathrm{Ind}(\beta)$ and $\mathrm{Ind}_0(\beta)$ are finite dimensional.

It remains to show that $\mathrm{Ind}_0(\beta) = \mathrm{Ind}(\beta) + \dim \overline{J}_b(\beta)$. To this end, choose a second finite partition $s_0 = a < s_1 < s_2 < \cdots < s_{\ell-1}$ $s_\ell = b$ such that $\{s_1,s_2,\ldots,s_{\ell-1}\} \cap \{t_1,t_2,\ldots,t_{k-1}\} = \emptyset$ and such that $\beta \mid [s_i,s_{i+1}]$ has no conjugate points for each i. Then $\overline{J}\{s_i\} \cap \overline{J}\{t_j\} = \overline{J}_b(\beta)$. Thus if $X \in \overline{J}\{s_i\}$, $X \notin J_b(\beta)$, it follows from (9.76) that

$$\overline{I}(\phi(X),\phi(X)) > \overline{I}(X,X) \tag{9.78}$$

From the preceding part of the proof of this proposition applied to the partition $\{s_i\}$ and associated finite dimensional vector space $\overline{J}\{s_i\}$, we may choose a vector subspace $B_0'$ of $\overline{J}\{s_i\}$ so that $\overline{I} \mid B_0' \times B_0'$ is positive semidefinite and $\mathrm{Ind}_0(\beta) = \dim B_0' < \infty$. Note that $\overline{J}_b(\beta)$ must be a subspace of $B_0'$. For otherwise there is a nontrivial vector subspace V of $\overline{J}_b(\beta)$ with $V \cap B_0' = \{[\beta']\}$. Then $B_0'' = B_0' \oplus V$ would be a subspace of $\overline{J}\{s_i\}$ on which $\overline{I} \mid B_0'' \times B_0''$ is positive semidefinite, but dim $B_0''$ > dim $B_0'$, in contradiction to our choice of $B_0'$.

Regarding $B_0'$ as a subspace of $\mathcal{X}_0(\beta)$, we then know from (9.77) that $\phi \mid B_0' : B_0' \longrightarrow \overline{J}\{t_j\}$ is injective. Thus if we set $B_0 = \phi(B_0')$, we have dim $B_0$ = dim $B_0'$ = $\text{Ind}_0(\beta)$. Since $\phi \mid \overline{J}_b(\beta) = \text{Id}$, we also have that $\overline{J}_b(\beta)$ is a subspace of $B_0$. Choose a vector subspace B of $B_0$ such that $B_0 = B \oplus \overline{J}_b(\beta)$. Using (9.78), it may be established just as in the proof of Proposition 9.25 of Section 9.1 that $\overline{I} \mid B \times B$ is positive definite. Hence $\text{Ind}'(\beta) \geq \dim B$.

But $\text{Ind}_0(\beta) = \text{Ind}_0'(\beta) = \dim B_0 = \dim B + \dim \overline{J}_b(\beta)$. Thus the proof will be complete if we show that $\text{Ind}'(\beta) \leq \dim B$. Hence suppose that B' is a subspace of $\overline{J}\{t_i\}$ with $\overline{I} \mid B' \times B'$ positive definite and $\text{Ind}'(\beta) = \dim(B') > \dim B$. As $\overline{I} \mid B' \times B'$ is positive definite, $B' \cap \overline{J}_b(\beta) = \{[\beta']\}$. Hence $\overline{I}$ is positive semidefinite on the direct sum $B' \oplus \overline{J}_b(\beta)$ so that $\dim B' + \dim \overline{J}_b(\beta) \leq \text{Ind}_0'(\beta)$. On the other hand, $\dim B' + \dim \overline{J}_b(\beta) > \dim B + \dim \overline{J}_b(\beta) = \text{Ind}_0'(\beta)$ in contradiction. Hence $\text{Ind}'(\beta) \leq \dim B$ as required, completing the proof of the proposition. □

Now that we have obtained Proposition 9.76, it is straightforward to see that the proof of Theorem 9.27 of Section 9.1 may be applied to the index form $\overline{I} : G(\beta) \times G(\beta) \longrightarrow \mathbb{R}$ and the projected positive definite metric $\overline{g}$ to yield the equalities

$$\text{Ind}(\beta) = \sum_{t \in (a,b)} \dim \overline{J}_t(\beta)$$

and

$$\text{Ind}_0(\beta) = \sum_{t \in (a,b]} \dim \overline{J}_t(\beta)$$

Since $\dim \overline{J}_t(\beta) = \dim J_t(\beta)$ by Corollary 9.55, we have thus established the following Morse index theorem for null geodesics.

THEOREM 9.77 Let $\beta : [a,b] \longrightarrow M$ be a null geodesic in an arbitrary space-time. Let $\overline{I} : \mathfrak{X}_0(\beta) \times \mathfrak{X}_0(\beta) \longrightarrow \mathbb{R}$ be the index form on piecewise smooth sections of the quotient bundle $G(\beta)$ defined in (9.42). Then $\beta$ has only finitely many conjugate points and the index $\mathrm{Ind}(\beta)$ and extended index $\mathrm{Ind}_0(\beta)$ of $\overline{I} : \mathfrak{X}_0(\beta) \times \mathfrak{X}_0(\beta) \longrightarrow \mathbb{R}$ are related to the geodesic index of the null geodesic $\beta$ by the formulas

$$\mathrm{Ind}(\beta) = \sum_{t\in(a,b)} \dim J_t(\beta)$$

and

$$\mathrm{Ind}_0(\beta) = \sum_{t\in(a,b]} \dim J_t(\beta)$$

where $J_t(\beta)$ denotes the vector space of Jacobi fields $Y$ along $\beta$ with $Y(a) = Y(t) = 0$.

Chapter 10

# SOME RESULTS IN GLOBAL LORENTZIAN GEOMETRY

In Chapter 10, we apply the techniques of the preceding chapters to obtain Lorentzian analogues of two remarkable results in global Riemannian geometry. The first, the Bonnet-Myers diameter theorem, asserts that if a complete Riemannian manifold N has everywhere positive Ricci curvature bounded away from zero, then N is compact, has finite diameter, and has finite fundamental group. The second result, the Hadamard-Cartan theorem, is that if a complete Riemannian manifold has everywhere nonpositive sectional curvature, its universal covering manifold is diffeomorphic to $\mathbb{R}^n$ and thus the higher homotopy groups $\pi_i(N,*) = 0$ for $i \geqslant 2$. In addition, the universal covering space with the pullback Riemannian metric has the property that any two points may be joined by exactly one geodesic up to reparameterization.

In Section 10.1, we consider the Lorentzian analogue of the Bonnet-Myers theorem and in so doing, study the timelike diameter of space-times. The *timelike diameter* diam (M,g) of a space-time (M,g) is given by

$$\text{diam}\,(M,g) = \sup\{d(p,q) : p,q \in M\}$$

Classes of space-times with finite timelike diameter, including the "Wheeler universes," have been studied in general relativity [cf. Tipler (1977c, p. 500)].

If a complete Riemannian manifold has finite diameter, it is compact by the Hopf-Rinow theorem. Even so, all geodesics have *infinite* length as a result of the metric completeness. But for space-times $(M,g)$, since $L(\gamma) \leq d(p,q)$ for all future-directed nonspacelike curves $\gamma$ from p to q, *every* timelike geodesic must satisfy $L(\gamma) \leq \text{diam}\,(M,g)$. Thus if a space-time $(M,g)$ has finite timelike diameter, all timelike geodesics have finite length and hence are incomplete. In particular, a space-time $(M,g)$ with finite timelike diameter is timelike geodesically incomplete.

Since we have used the convention $(-, +, \ldots, +)$ for the Lorentzian metric instead of $(+, -, \ldots, -)$, curvature conditions of positive (resp., negative) sectional curvature for Riemannian manifolds translate as curvature conditions of negative (resp., positive) timelike sectional curvature for Lorentzian manifolds.

Using the timelike index theory developed in Section 9.1, we obtain the following Lorentzian analogue of the Bonnet-Myers theorem for complete Riemannian manifolds. Let $(M,g)$ be a globally hyperbolic space-time with either (a) all nonspacelike Ricci curvatures positive and bounded away from zero, or (b) all timelike sectional curvatures negative and bounded away from zero. Then $(M,g)$ has finite timelike diameter.

In Section 10.2, we give Lorentzian versions of two well-known comparison theorems in Riemannian geometry, the index comparison theorem and the Rauch comparison theorem. Using the latter of these two results, we are able to give an easy proof (Corollary 10.12) of the basic fact that in a space-time with everywhere nonnegative timelike sectional curvatures, the differential $\exp_{p*}$ of the exponential map

$$\exp_{p*} : T_v(T_pM) \longrightarrow T_{\exp_p(v)}M$$

is norm nondecreasing on nonspacelike tangent vectors.

Finally, in Section 10.3 we give an analogue of the Hadamard-Cartan theorem for future one-connected space-times. Here $(M,g)$ is said to be *future one-connected* if for any $p,q \in M$ with $p \ll q$, any two smooth future-directed timelike curves from p to q are homotopic through smooth future-directed timelike curves with endpoints p and q. Using the Morse theory of the timelike path space $C_{(p,q)}$ from

Section 9.2, it may be shown that if (M,g) is a future one-connected globally hyperbolic space-time with no nonspacelike conjugate points, then given any $p,q \in M$ with $p \ll q$, there is exactly one future-directed timelike geodesic segment (up to parameterization) from p to q.

## 10.1 THE TIMELIKE DIAMETER

Motivated by the concept of the diameter of a complete Riemannian manifold, the following analogue has been considered for arbitrary space-times [cf. Beem and Ehrlich (1979c, Section 9)].

DEFINITION 10.1 The *timelike diameter* diam (M,g) of the space-time (M,g) is defined to be

$$\text{diam } (M,g) = \sup\{d(p,q) : p,q \in M\}$$

A similar concept has been used by Tipler (1977a, p. 17) in studying singularity theory in general relativity. Physically, the timelike diameter represents the supremum of possible proper times any particle could possibly experience in the given space-time. A space-time of finite timelike diameter is singular (recall Definition 5.3) in a striking way.

REMARK 10.2 If diam $(M,g) < \infty$, then *all* timelike geodesics have length $\leq$ diam (M,g) and are thus incomplete.

*Proof*. Suppose $c : (a,b) \longrightarrow M$ is a timelike geodesic with $L(c) >$ diam (M,g). We may then find $s,t \in (a,b)$, $s < t$, such that $L(c|[s,t]) >$ diam (M,g). But then

$$d(c(s),c(t)) \geq L(c|[s,t]) > \text{diam } (M,g)$$

which is impossible. □

From a physical point of view the most interesting space-times of finite timelike diameter are the Wheeler universes [cf. Tipler (1977c, p. 500)]. In particular, the "closed" Friedmann cosmological models are examples of Wheeler universes.

For a complete Riemannian manifold $(N,g_0)$ the diameter is finite if the manifold is compact. In this case, we may always find two points of N whose distance realizes the diameter. On the other hand, for space-times with finite timelike diameter, the diameter is never achieved.

PROPOSITION 10.3 Let $(M,g)$ be an arbitrary space-time and suppose that there exist points $p,q \in M$ such that $d(p,q) = \operatorname{diam}(M,g)$. Then $d(p,q) = \infty$.

*Proof.* Suppose $d(p,q) = \operatorname{diam}(M,g) < \infty$. Let $q' \in I^+(q)$ be arbitrary. Then

$$d(p,q') \geqslant d(p,q) + d(q,q') > d(p,q) = \operatorname{diam}(M,g)$$

in contradiction. □

Recalling that globally hyperbolic space-times satisfy the finite distance condition, it follows from Proposition 10.3 that

COROLLARY 10.4 (1) The timelike diameter is never realized by any pair of points in a space-time of finite timelike diameter.
(2) The timelike diameter is never achieved in a globally hyperbolic space-time.

We now prove the Lorentzian analogue (Theorem 10.9) of Bonnet's theorem and Myers' theorem for complete Riemannian manifolds [cf. Cheeger and Ebin (1975, pp. 27-28)]. Similar results have been given by Avez (seminar lecture), Flaherty (unpublished), Uhlenbeck (1975, Theorem 5.4 and Corollary 5.5), and Beem and Ehrlich (1979c, Theorem 9.5). Also, Theorem 10.9 is contained implicitly in stronger results using the Raychaudhuri equation needed for singularity theory in general relativity [cf. Section 11.2 or Hawking and Ellis (1973, Section 4.4)].

DEFINITION 10.5 A *timelike 2-plane* $\sigma$ is a two-dimensional subspace of $T_pM$ for some $p \in M$ which is spanned by a spacelike and timelike tangent vector.

Recall that the sectional curvature $K(\sigma)$ of the timelike 2 plane $\sigma$ may be calculated by choosing a basis $\{v,w\}$ for $\sigma$ consisting of a timelike and a spacelike tangent vector and setting

$$K(\sigma) = \frac{g(R(v,w)w,v)}{g(v,v)g(w,w) - [g(v,w)]^2}$$

REMARK 10.6 Rather than considering all sectional curvatures, it is essential to restrict attention to timelike sectional curvatures for the following reason. If $(M,g)$ is a space-time of dimension $\geq 3$ and the sectional curvature function of $(M,g)$ is either bounded from above for all nonsingular 2 planes or bounded from below for all nonsingular 2 planes, then $(M,g)$ has constant sectional curvature [Kulkarni (1979)]. Here a 2 plane $\sigma$ is said to be *nonsingular* if $g(v,v)g(w,w) - [g(v,w)]^2 \neq 0$ for some basis $\{v,w\}$ for $\sigma$. On the other hand, space-times with all *timelike* sectional curvatures $\leq -k^2$ or all timelike sectional curvatures $\geq k^2$ exist. But Harris (1979a) has shown that if all timelike sectional curvatures are bounded both from above and below, then $(M,g)$ has constant sectional curvature. Thus no obvious analogue for the Lorentzian sectional curvature exists for pinched Riemannian manifolds [cf. Cheeger and Ebin (1975, p. 118) for Riemannian pinching].

With the signature convention $(-, +, \ldots, +)$ used here for Lorentzian metrics, curvature conditions in Riemannian geometry for complete Riemannian manifolds of positive (resp., negative) sectional curvature tend to correspond to theorems for globally hyperbolic space-times of negative (resp., positive) sectional curvature. On the other hand, if we change this signature convention to $(+, -, \ldots, -)$ by setting $(M,\hat{g}) = (M,-g)$, then $K(\hat{g}) = -K(g)$. Thus Riemannian theorems for positive (resp., negative) sectional curvature correspond to Lorentzian theorems for positive (resp., negative) sectional curvature for $(M,\hat{g})$, [cf. for instance Flaherty (1975a, pp. 395-396) where the convention $(+, -, \ldots, -)$ is used]. But whether $g$ or $\hat{g}$ is chosen as the Lorentzian metric for $M$, $\mathrm{Ric}(g) = \mathrm{Ric}(\hat{g})$.

It is convenient to isolate part of the proof of Theorem 10.9 in the following proposition. Recall the notation $V_0^{\perp}(c)$ from Definition 9.1.

PROPOSITION 10.7 Let (M,g) be an arbitrary space-time of dimension $n \geqslant 2$. Suppose that (M,g) satisfies either the curvature condition

(a) All timelike planes have sectional curvature $\leqslant -k < 0$,

or the curvature condition

(b) $\mathrm{Ric}(g)(v,v) \geqslant (n-1)k > 0$ for all unit timelike tangent vectors $v \in TM$.

Then if $c : [0,b] \longrightarrow M$ is any timelike geodesic with $L(c) \geqslant \pi/\sqrt{k}$, the geodesic segment c has a pair of conjugate points.

*Proof.* Since curvature condition (a) implies curvature condition (b) by taking the trace, we will prove that condition (b) implies the desired conclusion. For convenience, we will suppose that $c : [0,L] \longrightarrow M$ is parameterized as a unit speed timelike geodesic with length L. Set $E_n(t) = c'(t)$ for all $t \in [0,L]$ and let $\{E_1,\ldots,E_{n-1}\}$ be $n-1$ spacelike parallel vector fields along c such that $\{E_1(t),E_2(t),\ldots,E_n(t)\}$ forms a Lorentzian orthonormal basis of $T_{c(t)}M$ for each $t \in [0,L]$. Set $W_i(t) = \sin(\pi t/L)E_i(t)$, so that $W_i \in V_0^{\perp}(c)$. Using (9.3) of Definition 9.4, we obtain

$$I(W_i,W_i) = \int_{t=0}^{L} \sin^2 \frac{\pi t}{L}\left[g(R(E_i,c')c',E_i)\Big|_t - \frac{\pi^2}{L^2}\right]dt$$

Hence

$$\sum_{i=1}^{n-1} I(W_i,W_i) = \int_{t=0}^{L} \sin^2 \frac{\pi t}{L}\left[\mathrm{Ric}(c'(t),c'(t)) - \frac{(n-1)\pi^2}{L^2}\right]dt$$

If $\mathrm{Ric}(c'(t),c'(t)) \geqslant (n-1)k$ for all $t \in [0,L]$ and $L \geqslant \pi/\sqrt{k}$, we find that $\Sigma_{i=1}^{n-1} I(W_i,W_i) \geqslant 0$. Hence $I(W_i,W_i) \geqslant 0$ for some $i \in \{1,2,\ldots,n-1\}$. On the other hand, if $c \mid [0,L]$ is free of conjugate points, then $I(W_i,W_i) < 0$ for each i by Theorem 9.22. Hence c has a pair of conjugate points if $L \geqslant \pi/\sqrt{k}$ as required. □

A slight variant of Proposition 10.7 may also be proved similarly using the timelike Morse index theorem (Theorem 9.27).

PROPOSITION 10.8 Let $(M,g)$ be an arbitrary space-time of dimension n satisfying either (or both) of the curvature conditions of Proposition 10.7. If $c : [a,b] \longrightarrow M$ is any timelike geodesic with $L(c) > \pi/\sqrt{k}$, then $t = a$ is conjugate along c to some $t_0 \in (a,b)$ and hence c is not maximal.

*Proof.* Suppose that diam $(M,g) > \pi/\sqrt{k}$. We may then find
If $L(c) > \pi/\sqrt{k}$, we obtain this time that

$$\sum_{i=1}^{n-1} I(W_i,W_i) > 0$$

Hence $I(W_i,W_i) > 0$ for some i. Thus $\mathrm{Ind}(c) > 0$. By the timelike Morse index theorem (Theorem 9.27), $\dim J_t(c) \neq 0$ for some $t \in (a,b)$. This completes the proof. □

Now we are ready to give the Lorentzian analogue of the Bonnet-Myers diameter theorem for complete Riemannian manifolds.

THEOREM 10.9 Let $(M,g)$ be a globally hyperbolic space-time of dimension n satisfying either of the following curvature conditions:

(a) All timelike sectional curvatures are $\leqslant -k < 0$.

(b) $\mathrm{Ric}(v,v) \geqslant (n-1)k > 0$ for all unit timelike vectors $v \in TM$.

Then $\mathrm{diam}(M,g) \leqslant \pi/\sqrt{k}$.

*Proof.* Suppose that diam $(M,g) > \pi/\sqrt{k}$. We may then find $p,q \in M$ with $d(p,q) > \pi/\sqrt{k}$ by definition of diam $(M,g)$. Since $(M,g)$ is globally hyperbolic, there exists a maximal timelike geodesic segment $c : [0,1] \longrightarrow M$ with $c(0) = p$, $c(1) = q$. But as $L(c) = d(p,q) > \pi/\sqrt{k}$, the geodesic segment c is not maximal by Proposition 10.8, in contradiction. □

It is clear that an analogue of Proposition 10.8 could be obtained for null geodesics using the null index theory developed in

Section 9.3. On the other hand, a stronger result may be obtained using the Raychaudhuri effect [cf. Hawking and Ellis (1973, p. 101)]. Thus we refer the reader to Section 11.4 for a discussion of these results rather than pursuing this analogy any further here [cf. also Harris (1979a)].

## 10.2 LORENTZIAN COMPARISON THEOREMS

For use in Section 10.3 as well as for their own sake, we now present the timelike analogues of two important tools in global Riemannian geometry, the index comparison theorem [Gromoll, Klingenberg, and Meyer (1975, p. 174)] and the Rauch comparison theorem [Gromoll, Klingenberg, and Meyer (1975, p. 178) or Cheeger and Ebin (1975, p. 29)]. The results in this section except for Corollary 10.12 have been published in Beem and Ehrlich (1979c, Section 9). Actually, there are two versions of the Rauch comparison theorem, often called *Rauch Theorem I* and *Rauch Theorem II*, that are useful in global Riemannian geometry [cf. Cheeger and Ebin (1975, Theorems 1.28 and 1.29, respectively)]. The result (Theorem 10.11) given in this section is the Lorentzian analogue of Rauch theorem I. Harris (1979a, 1979b) has given proofs of Lorentzian analogues for both Rauch theorems I and II and using Rauch Theorem II has given a Lorentzian version of Toponogov's comparison theorem [cf. Cheeger and Ebin (1975, p. 42)] for timelike geodesic triangles in certain classes of space-times. Using this result, Harris (1979a, 1979b) has also obtained a Lorentzian analogue of Toponogov's diameter theorem [cf. Cheeger and Ebin (1975, p. 110)] (cf. Appendix D).

In the rest of this section, let $(M_1, g_1)$ and $(M_2, g_2)$ be arbitrary space-times with dimension $M_1 \leq$ dimension $M_2$. Also let $c_i : [0,L] \longrightarrow M_i$ be *unit speed* future-directed timelike geodesic segments Throughout this section, we will denote both the index form on $V^\perp(c_1)$ and $V^\perp(c_2)$ by I. Also, during the proofs, we will denote both Lorentzian metrics $g_1$ and $g_2$ by $\langle\ ,\ \rangle$. The index form I for a timelike geodesic c and the index Ind(c) and extended index $\mathrm{Ind}_0(c)$

of c are defined in Section 9.1, formula (9.1) and Definition 9.24, respectively.

We first need to define an isomorphism

$$\phi : V^{\perp}(c_1) \longrightarrow V^{\perp}(c_2)$$

so that $g_2(\phi X(t), \phi X(t)) = g_1(X(t), X(t))$ for all $t \in [0,L]$. This may be done following the usual parallel translation construction in Riemannian geometry. We first define an isometry

$$\phi_t : T_{c_1(t)}M_1 \longrightarrow T_{c_2(t)}M_2$$

as follows. Let

$$P_t : T_{c_1(t)}M_1 \longrightarrow T_{c_1(0)}M_1$$

denote the Lorentzian inner product-preserving isomorphism of parallel translation along $c_1$. Explicitly, given $v \in T_{c_1(t)}M_1$, let Y be the unique parallel field along $c_1$ with $Y(t) = v$ and set $P_t(v) = Y(0)$. Similarly, let

$$Q_t : T_{c_2(t)}M_2 \longrightarrow T_{c_2(0)}M_2$$

denote parallel translation along $c_2$. Choose an injective Lorentzian inner product-preserving linear map

$$i : (T_{c_1(0)}M_1, g_1|_{c_1(0)}) \longrightarrow (T_{c_2(0)}M_2, g_2|_{c_2(0)})$$

where $i(c_1'(0)) = c_2'(0)$. Then the map

$$\phi_t : (T_{c_1(t)}M_1, g_1|_{c_1(t)}) \longrightarrow (T_{c_2(t)}M_2, g_2|_{c_2(t)})$$

given by $\phi_t = Q_t^{-1} \circ i \circ P_t$ is an isometry since parallel translation preserves the Lorentzian structures. We may then define the map

$$\phi : V^{\perp}(c_1) \longrightarrow V^{\perp}(c_2)$$

as follows. Given $X \in V^{\perp}(c_1)$, define $\phi X \in V^{\perp}(c_2)$ by $(\phi X)(t) = \phi_t(X(t))$. It follows as in the Riemannian proof that $(\phi X)' = \phi(X')$, where the first covariant differentiation is in $M_2$ and the second is in $M_1$.

Let $G_{2,t}(c_i)$ denote the set of all timelike planes $\sigma$ containing $c_i'(t)$ for $i = 1, 2$. There is then an induced map

$$\phi_t : G_{2,t}(c_1) \longrightarrow G_{2,t}(c_2)$$

defined as follows. If $\sigma \in G_{2,t}(c_1)$, we may write $\sigma = \text{span}\{v, c_1'(t)\}$ where v is spacelike. Put $\phi_t(\sigma) = \text{span}\{\phi_t(v), c_2'(t)\} \in G_{2,t}(c_2)$.

We are now ready to state the timelike version of the index comparison theorem.

THEOREM 10.10 (Timelike Index Comparison Theorem) Let $(M_1, g_1)$ and $(M_2, g_2)$ be space-times with $\dim M_1 \leqslant \dim M_2$ and let $c_1 : [0,\beta] \longrightarrow M_1$ and $c_2 : [0,\beta] \longrightarrow M_2$ be unit speed future-directed timelike geodesics. Suppose for all t with $0 \leqslant t \leqslant \beta$ and all timelike planes $\sigma \in G_{2,t}(c_1)$ that the sectional curvature $K_{M_1}(\sigma) \geqslant K_{M_2}(\phi_t \sigma)$. Then for any $X \in V^{\perp}(c_1)$ we have

(1) $I(X,X) \leqslant I(\phi X, \phi X)$.

(2) $\text{Ind}(c_1) \leqslant \text{Ind}(c_2)$.

(3) $\text{Ind}_0(c_1) \leqslant \text{Ind}_0(c_2)$.

*Proof.* Recalling that $\langle c_1', c_1' \rangle = -1$ and $\langle X, c_1' \rangle = 0$, we obtain

$$\langle R(X,c_1')c_1', X\rangle = - \langle X,X\rangle K(X,c_1')$$

Similar formulas hold for $\phi X$ and $c_2'$. Thus

$$\begin{aligned} I(\phi X, \phi X) &= \int_0^\beta [-\langle(\phi X)', (\phi X)'\rangle + \langle R(\phi X, c_2')c_2', \phi X\rangle]\, dt \\ &= \int_0^\beta [-\langle\phi(X'), \phi(X')\rangle + \langle R(\phi X, c_2')c_2', \phi X\rangle]\, dt \\ &\geqslant \int_0^\beta [-\langle X', X'\rangle + \langle R(X, c_1')c_1', X\rangle]\, dt = I(X,X) \quad \square \end{aligned}$$

We may now obtain the more powerful timelike Rauch comparison theorem using the timelike index comparison theorem. Recall first that since $c_i : [0,L] \longrightarrow M_i$ are timelike geodesic segments, the vector fields in $V^{\perp}(c_i)$ are all spacelike vector fields.

THEOREM 10.11 (Timelike Rauch Comparison Theorem) Let $(M_1,g_1)$ and $(M_2,g_2)$ be space-times with dimension $M_1 \leqslant$ dimension $M_2$. Let $c_1 : [0,L] \longrightarrow M_1$ and $c_2 : [0,L] \longrightarrow M_2$ be future-directed timelike unit speed geodesic segments. Suppose for all $t \in [0,L]$ and any $\sigma \in G_{2,t}(c_1)$ that

$$K_{M_1}(\sigma) \geqslant K_{M_2}(\phi_t \sigma)$$

Let $Y_1 \in V^{\perp}(c_1)$ and $Y_2 \in V^{\perp}(c_2)$ be Jacobi fields on $M_1$ and $M_2$ respectively satisfying the initial conditions

$$Y_1(0) = Y_2(0) = 0 \qquad (10.1)$$

$$g_1(Y_1'(0),Y_1'(0)) = g_2(Y_2'(0),Y_2'(0)) \qquad (10.2)$$

If $c_2$ has no conjugate points in $(0,L)$, then

$$g_1(Y_1(t),Y_1(t)) \geqslant g_2(Y_2(t),Y_2(t)) \qquad (10.3)$$

for all $t \in (0,L]$. In particular, $c_1$ has no conjugate points in $(0,L)$.

*Proof*. This may be given along the lines of Gromoll, Klingenberg, and Meyer (1975, pp. 180-181), except for the proof of inequality (7), p. 181, which must be modified as follows. Let $Z \in V^{\perp}(c_2)$ be the unique Jacobi field along $c_2$ with $Z(0) = 0$ and $Z(t_0) = \phi Y_1(t_0)$, $\phi$ as above. We must show that

$$\langle Y_1(t_0),Y_1'(t_0)\rangle \geqslant \langle Z(t_0),Z'(t_0)\rangle$$

To this end, let $c_3 = c_1 \mid [0,t_0]$ and $c_4 = c_2 \mid [0,t_0]$. Then

$$\begin{aligned}\langle Y_1(t_0),Y_1'(t_0)\rangle &= - I(Y_1|c_3,Y_1|c_3)\\ &\geqslant - I((\phi Y_1)|c_4,(\phi Y_1)|c_4)\end{aligned}$$

the above inequality by the timelike index comparison theorem [Theorem 10.10 (1)]

$$\geqslant - I(Z|c_4, Z|c_4)$$

the above inequality by the maximality of Jacobi fields with respect to the index form in the absence of conjugate points (Theorem 9.23)

$$= \langle Z(t_0), Z'(t_0) \rangle \quad \square$$

It may also be assumed in Theorem 10.11 that $c_2$ has no conjugate points in $[0,L]$. Then $c_1$ would also have no conjugate points in $[0,L]$ by Theorem 10.10 (3).

For Riemannian manifolds of nonpositive sectional curvature, equipping the tangent space with the "flat metric" (cf. Definition 9.17), it may be shown that the exponential map does not decrease the length of tangent vectors [cf. Bishop and Crittenden (1964, p. 178, Theorem 2 (i)) for a precise statement]. A simple proof of this fact may be given using the Rauch comparison theorem and comparing Jacobi fields on the given Riemannian manifold to those in $\mathbb{R}^n$. We will now use the timelike Rauch comparison theorem to prove the analogous result for space-times of nonnegative timelike sectional curvature [cf. Flaherty (1975a, p. 397)]. Intuitively, Corollary 10.12 below expresses the fact that if all timelike sectional curvatures of $(M,g)$ are positive, then future-directed timelike geodesics emanating from a given point of M spread apart faster than "corresponding" geodesics in Minkowski space-time. Recall that the canonical isomorphism $\tau_v$ has been defined in Section 9.1, Definition 9.15.

COROLLARY 10.12 Let $(M,g)$ be a space-time with everywhere nonnegative timelike sectional curvature and let $v \in T_pM$ be a given future-directed timelike tangent vector with $g(v,v) = -1$. Then for any future-directed nonspacelike tangent vector $w \in T_pM$, the vector $b = \tau_v(w) \in T_v(T_pM)$ satisfies the inequality

$$g(\exp_{p*} b, \exp_{p*} b) \geqslant g(w,w) = \langle\langle b,b \rangle\rangle$$

*Proof.* We first prove the inequality for $b = \tau_v(w) \in T_v(T_pM)$ with $g(v,w) = 0$ by applying the timelike Rauch comparison theorem with $(M_1,g_1) = (M,g)$ and $(M_2,g_2)$ the Minkowski space-time $(\mathbb{R}^n_1, g_0)$ with $n = \dim M$. Setting $c_1(t) = \exp_p tv$, let $Y_1 \in V^\perp(c)$ be the unique Jacobi field with $Y_1(0) = 0$, $Y_1'(0) = w$. By Proposition 9.16, we have $Y_1(1) = \exp_{p*} b$.

Now let $c_2 : [0,1] \longrightarrow \mathbb{R}^n_1$ be an arbitrary unit speed timelike geodesic and choose $\overline{w} \in N(c_2(0))$ with $g_0(\overline{w},\overline{w}) = g_1(w,w)$. Let $Y_2 \in V^\perp(c_2)$ be the unique Jacobi field with $Y_2(0) = 0$ and $Y_2'(0) = \overline{w}$. Then $Y_2(t) = tP_t(\overline{w})$, where $P_t$ denotes the Lorentzian parallel translation along $c_2$ from $c_2(0)$ to $c_2(t)$. Applying Theorem 10.11, we obtain

$$\begin{aligned} g_1(\exp_{p*} b, \exp_{p*} b) &= g_1(Y_1(1), Y_1(1)) \\ &\geqslant g_0(Y_2(1), Y_2(1)) \\ &= g_0(P_1(\overline{w}), P_1(\overline{w})) \\ &= g_0(\overline{w},\overline{w}) = g_1(w,w) = \langle\langle b,b\rangle\rangle \end{aligned}$$

as required.

Returning to the general case, we may decompose $w = w_1 + w_2$ where $w_1 = \lambda v$ for some $\lambda > 0$ and $g(v,w_2) = 0$. Set $b_i = \tau_v(w_i)$ for $i = 1, 2$ so that $b = b_1 + b_2$. We now calculate

$$\begin{aligned} g(\exp_{p*} b, \exp_{p*} b) &= g(\exp_{p*} b_1, \exp_{p*} b_1) + 2g(\exp_{p*} b_1, \exp_{p*} b_2) \\ &\quad + g(\exp_{p*} b_2, \exp_{p*} b_2) \end{aligned}$$

Applying the Gauss lemma (Theorem 9.18) to the first two terms, we obtain

$$\begin{aligned} g(\exp_{p*} b, \exp_{p*} b) &= \langle\langle b_1,b_1\rangle\rangle + 2\langle\langle b_1,b_2\rangle\rangle \\ &\quad + g(\exp_{p*} b_2, \exp_{p*} b_2) \\ &= \langle\langle b_1,b_1\rangle\rangle + g(\exp_{p*} b_2, \exp_{p*} b_2) \end{aligned}$$

as $\langle\langle b_1,b_2\rangle\rangle = g(w_1,w_2) = 0$. Now applying the first part of the proof to the last term, we have

$$g(\exp_{p*} b, \exp_{p*} b) \geqslant \langle\langle b_1,b_1\rangle\rangle + \langle\langle b_2,b_2\rangle\rangle$$
$$= \langle\langle b_1 + b_2, b_1 + b_2\rangle\rangle = \langle\langle b,b\rangle\rangle$$

as required. □

## 10.3 LORENTZIAN HADAMARD-CARTAN THEOREMS

We first prove a basic result linking conjugate points and timelike sectional curvature [cf. Flaherty (1975a, Proposition 2.1)--here Flaherty uses the convention (+, -, ..., -) for the Lorentzian metric so that his sectional curvature condition has the opposite sign from ours].

PROPOSITION 10.13 Let $(M,g)$ be a space-time with everywhere nonnegative timelike sectional curvatures. Then no nonspacelike geodesic has any conjugate points.

*Proof.* First let $c : [0,a) \longrightarrow M$ be an arbitrary future-directed unit speed timelike geodesic. Recall from Corollary 9.10 that if $X$ is a Jacobi field along $c$ with $X(0) = X(t_0) = 0$ for some $t_0 \in (0,a)$, then $X \in V^{\perp}(c)$. Thus we may restrict our attention to Jacobi fields $J \in V^{\perp}(c)$ with $J(0) = 0$. Since $J \in V^{\perp}(c)$ and $c$ is a timelike geodesic, $J' \in V^{\perp}(c)$ also. Consider the smooth function $f(t) = g(J(t),J'(t))$. Differentiating yields

$$\begin{aligned} f'(t) &= g(J'(t),J'(t)) + g(J(t),J''(t)) \\ &= g(J'(t),J'(t)) - g(J(t),R(J(t),c'(t))c'(t)) \\ &= g(J'(t),J'(t)) + g(J(t),J(t))K(J(t),c'(t)) \geqslant 0 \end{aligned}$$

for all $t \in [0,a)$. Now if $J(t_0) = 0$ for some $t_0 \in [0,a)$, then $f(0) = f(t_0) = 0$. Thus as $f$ is nondecreasing, $f(t) = 0$ for all $t \in [0,t_0]$. Hence $0 = f'(0) = g(J'(0),J'(0))$ as $J(0) = 0$ from which we conclude that $J'(0) = 0$. Therefore $J = 0$ and no $t_0 \in [0,a)$ is conjugate to $t = 0$ along $c$.

We now treat the case that $\beta : [0,a) \longrightarrow M$ is a null geodesic using the null index form. Let $t_0 \in [0,a)$ be arbitrary. We will show that $\overline{I} : \mathfrak{X}_0(\beta|[0,t_0]) \times \mathfrak{X}_0(\beta|[0,t_0]) \longrightarrow \mathbb{R}$ is negative definite. Hence no $s \in (0,t_0]$ is conjugate to $t = 0$ along $\beta$ by Theorem 9.69.

If $W \in \mathfrak{X}_0(\beta)$ is a smooth parallel vector class, then $W = [\beta']$ since $W(0) = [\beta'(0)]$. Thus if $W \in \mathfrak{X}_0(\beta)$ is not a smooth parallel vector class, we have $\overline{g}(W'(s),W'(s)) > 0$ for some $s \in (0,t_0)$. Also $\overline{g}(W'(t),W'(t)) \geqslant 0$ for all $t \in [0,t_0]$ since $\overline{g}$ is positive definite. Now consider $\overline{g}(\overline{R}(W(t),\beta'(t))\beta'(t),W(t))$ for any fixed $t \in [0,t_0]$. If $W(t) = [\beta'(t)]$, then $\overline{g}(\overline{R}(W(t),\beta'(t))\beta'(t),W(t)) = 0$. Otherwise, we may find a spacelike tangent vector $w$ perpendicular to $\beta'(t)$ with $\pi(w) = W(t)$. Then $\overline{g}(\overline{R}(W(t),\beta'(t))\beta'(t),W(t)) = g(R(w,\beta'(t))\beta'(t),w)$ using formula (9.35) of Section 9.3. Since $w$ is spacelike, we may find a sequence of timelike 2 planes $\sigma_n = \{w_n,v_n\}$ with $w_n$ spacelike, $v_n$ timelike, $g(w_n,v_n) = 0$, and $w_n \longrightarrow w$, $v_n \longrightarrow \beta'(t)$, so that $\sigma_n \longrightarrow \{w,\beta'(t)\}$. We then obtain by continuity

$$\begin{aligned}\overline{g}(\overline{R}(W(t),\beta'(t))\beta'(t),W(t)) &= g(R(w,\beta'(t))\beta'(t),w)\\ &= \lim_{n\to\infty} g(R(w_n,v_n)v_n,w_n)\\ &= \lim_{n\to\infty} K(w_n,v_n)g(w_n,w_n)g(v_n,v_n) \leqslant 0\end{aligned}$$

since $g(w_n,w_n) > 0$, $g(v_n,v_n) < 0$, and $K(v_n,w_n) \geqslant 0$ for each $n$. Thus in either case, $\overline{g}(\overline{R}(W(t),\beta'(t))\beta'(t),W(t)) \leqslant 0$. Hence, provided $W \neq [\beta']$ we have

$$\overline{I}(W,W) = \int_{t=0}^{t_0} (-\overline{g}(W',W') + \overline{g}(\overline{R}(W,\beta')\beta',W))\Big|_t \, dt < 0$$

as required. □

Motivated by a standard definition in global Riemannian geometry [cf. O'Sullivan (1974), Gulliver (1975)], we make the following definition.

DEFINITION 10.14 The space-time $(M,g)$ is said to have *no future timelike conjugate points* if for any future-directed timelike geodesic $c : [0,a) \longrightarrow (M,g)$, no nontrivial Jacobi field in $V^{\perp}(c)$ vanishes more than once.

In view of Lemma 9.46, similar definitions may be formulated for space-times with no future null conjugate points or no future nonspacelike conjugate points. Proposition 10.13 guarantees that if $(M,g)$ is a space-time with everywhere nonnegative timelike sectional curvature, then $(M,g)$ has no future nonspacelike conjugate points.

Lorentzian manifolds with nonnegative timelike sectional curvature or with no future timelike conjugate points may be characterized in terms of the behavior of their Jacobi fields. A similar characterization applies to Riemannian manifolds [cf. O'Sullivan (1974, Proposition 4)].

PROPOSITION 10.15

(a) $(M,g)$ has everywhere nonnegative timelike sectional curvature iff

$$\frac{d^2}{dt^2}(g(Y(t),Y(t)) \geq 0$$

for for every Jacobi field $Y \in V^{\perp}(c)$ along any future-directed timelike geodesic $c$.

(b) $(M,g)$ has no future timelike conjugate points iff $g(Y(t),Y(t)) > 0$ for all $t > 0$ where $Y \in V^{\perp}(c)$ is any nontrivial Jacobi field with $Y(0) = 0$ along any future-directed timelike geodesic $c$.

*Proof.* (a) Assume that $(M,g)$ has everywhere nonnegative timelike sectional curvature. Let $Y \in V^{\perp}(c)$ be a Jacobi field along the unit speed timelike geodesic $c$. Then $Y' \in V^{\perp}(c)$ also and we obtain

$$\frac{d^2}{dt^2}(g(Y,Y)) = 2g(Y',Y') - 2g(R(Y,c')c',Y)$$

$$= 2g(Y',Y') + 2g(Y,Y)K(Y,c') \geq 0$$

Conversely, let $\{v,w\}$ be a future timelike and spacelike tangent vector respectively with $g(v,v) = -1$, $g(w,w) = 1$, and $g(v,w) = 0$ spanning an arbitrary timelike 2 plane. Let $c(t) = \exp(tv)$ and let $Y \in V^{\perp}(c)$ be the Jacobi field with initial conditions $Y(0) = w$ and $Y'(0) = 0$. Then we have by hypothesis,

$$0 \leqslant \frac{d^2}{dt^2}(g(Y,Y))\Big|_{t=0} = -\ 2g(R(Y(0),c'(0))c'(0),Y(0))$$

$$= -\ 2g(R(w,v)v,w) = 2K(v,w)$$

since the first term in the differentiation vanishes as $Y'(0) = 0$. Thus $K(v,w) \geqslant 0$ as required.

(b) This is clear from Definition 10.14. □

Using the timelike index theory of Sections 9.1 and 9.2, we now give the following version of a Lorentzian Hadamard-Cartan theorem for globally hyperbolic space-times. This proof is similar to the Morse theory proof of the Hadamard-Cartan theorem for complete Riemannian manifolds [cf. Milnor (1963, p. 102)]. Recall that a space-time is said to be *future 1-connected* (Definition 9.28) if any two smooth future-directed timelike curves from p to q are homotopic through (smooth) future-directed timelike curves with fixed endpoints p and q.

THEOREM 10.16 Let $(M,g)$ be a future 1-connected globally hyperbolic space-time with no future nonspacelike conjugate points. Then given any $p,q \in M$ with $p \ll q$, there is exactly one future-directed timelike geodesic (up to reparameterization) from p to q.

*Proof.* Since $(M,g)$ is globally hyperbolic, there exists a maximal future-directed timelike geodesic from p to q. Since there are no future nonspacelike conjugate points, any future-directed geodesic from p to q has index 0 by Theorem 9.27. Thus the timelike path space $C_{(p,q)}$ has the homotopy type of a CW-complex with a cell of dimension 0, i.e., a point, for each future-directed timelike geodesic from p to q. On the other hand, since M is future 1-connected,

$C_{(p,q)}$ is connected, hence consists of a single point. Thus there is at most one future-directed timelike geodesic from p to q. □

A similar result was obtained by Uhlenbeck (1975, Theorem 5.3) for globally hyperbolic space-times satisfying a metric growth condition [Uhlenbeck (1975, p. 72)] and the curvature condition $g(R(v,w)w,v) \leq 0$ for all future-directed null vectors v and vectors w with $g(v,w) = 0$ at every point of M. Namely, M can be covered by a space which is topologically Minkowski (i.e., Euclidean) space.

Flaherty (1975a, p. 398) has also shown that if (M,g) is future 1-connected, future nonspacelike complete, and has everywhere nonnegative timelike sectional curvatures, then the exponential map $\exp_p$ regularly embeds the future cone in $T_pM$ at each point p into M. To obtain this result, Flaherty used a lifting argument to show that under these hypotheses, if $v,w \in T_pM$ are any two future-directed timelike tangent vectors with $\exp_p v = \exp_p w$, then $v = w$. Thus future 1-connected, future nonspacelike complete space-times with nonnegative timelike sectional curvatures satisfy the conclusion of Theorem 10.16. On the other hand, Flaherty showed (1975b, p. 200) that any future 1-connected, future nonspacelike complete space-time with everywhere nonnegative timelike sectional curvatures is also globally hyperbolic.

Chapter 11

# SINGULARITIES

A common assumption made in studying Riemannian manifolds is that the spaces under consideration are Cauchy complete or, equivalently, geodesically complete. This assumption seems reasonable since a large number of important Riemannian manifolds are complete.

The situation for Lorentzian manifolds is quite different. A large number of the more important Lorentzian manifolds used as models in general relativity fail to be geodesically complete. Also, the problem of completeness is further complicated by the fact, observed in earlier chapters, that there are a number of inequivalent forms of completeness for Lorentzian manifolds.

In this chapter we will be concerned with establishing theorems which guarantee the nonspacelike geodesic incompleteness of a large class of space-times. These space-times contain at least one non-spaceline geodesic which is both inextendible and incomplete. Such a geodesic has an endpoint $\overline{p}$ in the causal boundary $\partial_c M$ which may be thought of as being outside the space-time, but not at infinity. For example, if $\gamma$ is a future-inextendible and future-incomplete timelike geodesic which has $\overline{p} \in \partial_c M$ as a future endpoint, then $\gamma$ corresponds to the path of a "freely falling" test particle which falls to the edge of the universe (at $\overline{p}$) in finite time.

It has been known for a long time in general relativity that a number of important space-times are nonspacelike incomplete. None-theless, this incompleteness was thought to be caused by the symmetry of these models. Thus it was felt that nonspacelike completeness was a reasonable assumption for physically realistic space-times. The argument for this assumption was based on physical intuition which was evidently unjustified, with hindsight [cf. Tipler, Clarke, and Ellis (1980, Chapter 4)].

If $(M,g)$ is an inextendible space-time, which has an inextendible nonspacelike geodesic which is incomplete, then $(M,g)$ is said to have a singularity. The purpose of this chapter is to establish several singularity (i.e., incompleteness) theorems.

Before beginning our study of singularity theory, we pause to explain why this theory works for all space-times of dimension $\geq 3$ but *not* for space-times of dimension 2. It is simply the fact noted at the beginning of Section 9.3 that *no* null geodesic in a two-dimensional space-time contains any conjugate points. Yet a key argument in proving singularity theorems is showing that certain curvature conditions force every complete nonspacelike geodesic to contain a pair of conjugate points.

Familiarity with the notations and some of the basic properties of Jacobi fields treated in Sections 9.1 and 9.3 will be assumed in this chapter.

## 11.1 JACOBI TENSORS

As we saw in Section 9.3 (Definition 9.61 ff.), Jacobi tensors are a convenient way of studying conjugate points. Given a timelike geodesic segment $c : [a,b] \longrightarrow M$, let $N(c(t))$ denote the $(n - 1)$-dimensional subspace of $T_{c(t)}M$ consisting of tangent vectors orthogonal to $c'(t)$ as in Definition 9.1. A (1,1)-tensor field $A(t)$ on $V^{\perp}(c)$ is a linear map

$$A = A(t) : N(c(t)) \longrightarrow N(c(t))$$

for each $t \in [a,b]$. Furthermore, a composite endomorphism $RA(t) : N(c(t)) \longrightarrow N(c(t))$ may be defined by

$$RA(t)(v) = R(A(t)(v), c'(t))c'(t)$$

The *adjoint* $A^*(t)$ of $A(t)$ is defined by requiring that

$$g(A(t)(w),v) = g(A^*(t)(v),w)$$

for all $v,w \in N(c(t))$.

A smooth (1,1)-tensor field $A(t)$ on $V^\perp(c)$ is said to be a *Jacobi tensor field* if

$$A'' + RA = 0$$

and

$$\ker (A(t)) \cap \ker (A'(t)) = \{0\}$$

for all $t \in [a,b]$. If $Y$ is a parallel vector field along $c$ and $A$ is a Jacobi tensor on $V^\perp(c)$, then the vector field $J = A(Y)$ satisfies the differential equation $J'' + R(J,c')c' = 0$ and hence is a Jacobi field. The condition $\ker (A(t)) \cap \ker (A'(t)) = \{0\}$ for all $t \in [a,b]$ guarantess that if $Y$ is a nonzero parallel field along $c$, then $J = A(Y)$ is a nontrivial Jacobi field. Suppose that $A$ is a Jacobi tensor on $V^\perp(c)$ with $A(a) = 0$. If $A(t_0)(v) = 0$ for some $t_0 \in (a,b]$ and $0 \neq v \in N(c(t_0))$, then letting $Y$ be the unique parallel field along $c$ with $Y(t_0) = v$, we find that $J = A(Y)$ is a Jacobi field with $J(a) = J(t_0) = 0$.

A Jacobi tensor $A$ is said to be a *Lagrange tensor field* if

$$(A')^*A - A^*A' = 0$$

for all $t \in [a,b]$. As in the proof of Lemma 9.67, it may be shown that a Jacobi tensor field $A$ is a Lagrange tensor field if $A(t_0) = 0$ for some $t_0 \in [a,b]$.

REMARK 11.1 Let $c : [a,b] \longrightarrow M$ be a unit speed timelike geodesic and suppose $E_1, \ldots, E_n$ is a parallelly propagated orthonormal basis along $c$ with $E_n = c'$. Then $N(c(t))$ is the span of $E_1, \ldots, E_{n-1}$ and each Jacobi vector field $J$ along $c$ which is everywhere orthogonal to $c'$ may be expressed in terms of $E_1, \ldots, E_{n-1}$. Thus $J$ may be represented as a column vector with $(n - 1)$ components. Using this representation, let $J_i = J_i(t)$ be the column vector corresponding to the Jacobi field J along c which satisfies $J(t_0) = 0$ and $J'(t_0) = E_i(t_0)$. Let

$$A(t) = [J_1(t), \ldots, J_{n-1}(t)]$$

be the $(n - 1) \times (n - 1)$ matrix with $J_i(t)$ for the ith column. This matrix A(t) is a representation of a Lagrange tensor field along c. Using this same basis $E_1, \ldots, E_{n-1}$, the adjoint $A^*(t)$ is represented by the transpose of A(t). The space of Jacobi fields which vanish at $t_0$ and which have derivatives orthogonal to c' at $t_0$ may be identified with the span of the columns of A. Thus conjugate points of $c(t_0)$ along c are exactly the points where det A(t) = 0. Hence det A(t) has isolated zeroes on [a,b]. Also the multiplicity of a conjugate point $t = t_1$ to $t_0$ along c is just the nullity of $A(t_1) : N(c(t_1)) \longrightarrow N(c(t_1))$.

The fact that $A(t_0) = 0$ and $A'(t_0) = E$ is essential to the above discussion. Lagrange tensors along a timelike geodesic $c : J \longrightarrow (M,g)$ may be constructed which are singular at distinct $t_0, t_1 \in J$, yet c has no conjugate points. For example, let (M,g) be $\mathbb{R}^3$ with the Lorentzian metric $ds^2 = -dx^2 + dy^2 + dz^2$ and let $c(t) = (t,0,0)$. Let $E_1 = \partial/\partial y$ and $E_2 = \partial/\partial z$. Then if A is the Jacobi tensor along c with the matrix representation

$$A(t) = \begin{bmatrix} t & 0 \\ 0 & t - 1 \end{bmatrix}$$

with respect to $E_1 \circ c$ and $E_2 \circ c$, we have $A' = E$ and $A^* = A$ so that $(A')^*A - A^*A' = 0$ and A is a Lagrange tensor. Evidently, $A(0)(E_1(c(0))) = 0$ and $A(1)(E_2(c(1))) = 0$. But c has no conjugate points since $(\mathbb{R}^3, ds^2)$ is Minkowski 3-space.

We now define the expansion, vorticity and shear of a Jacobi tensor A along the timelike geodesic $c : [a,b] \longrightarrow M$. As before, $E = E(t)$ will represent the (1,1) tensor field on $V^\perp(c)$ such that $E(t) = \mathrm{id} : N(c(t)) \longrightarrow N(c(t))$ for each t.

DEFINITION 11.2 Let A be a Jacobi tensor field along a timelike geodesic and set $B = A'A^{-1}$ at points where $A^{-1}$ is defined.

(a) The *expansion* $\theta$ is defined by

$$\theta = \mathrm{tr}(B)$$

(b) The *vorticity tensor* $\omega$ is defined by

$$\omega = \frac{1}{2}(B - B^*)$$

(c) The *shear tensor* $\sigma$ is defined by

$$\sigma = \frac{1}{2}(B + B^*) - \frac{\theta}{n - 1} E$$

Using an orthonormal basis of parallel fields for $V^{\perp}(c)$ and matrix algebra, it may be shown that

$$\theta = \mathrm{tr}(A'A^{-1}) = (\det A)^{-1}(\det A)'$$

Thus if A is a Jacobi tensor field with $A(t_0) = 0$, $A'(t_0) = E$ and $|\theta(t)| \longrightarrow \infty$ as $t \longrightarrow t_1$, then $\det A(t_1) = 0$ and $t = t_1$ is conjugate to $t_0$ along c.

We now calculate the derivative of $B = A'A^{-1}$ using $(A^{-1})' = -A^{-1}A'A^{-1}$. First we have

$$B' = (A'A^{-1})' = A''A^{-1} - A'A^{-1}A'A^{-1} = -R - BB \tag{11.1}$$

Using $\theta = \mathrm{tr}(B)$ and $B = \omega + \sigma + [\theta/(n - 1)]E$, we obtain

$$\begin{aligned}
\theta' &= \mathrm{tr}(B') \\
&= -\mathrm{tr}(R) - \mathrm{tr}(BB) \\
&= -\mathrm{tr}(R) - \mathrm{tr}((\omega + \sigma + \frac{\theta}{n - 1} E)^2) \\
&= -\mathrm{tr}(R) - \mathrm{tr}(\omega^2 + \sigma^2 + \frac{\theta^2}{(n - 1)^2} E) \\
&= -\mathrm{tr}(R) - \mathrm{tr}(\omega^2) - \mathrm{tr}(\sigma^2) - \frac{\theta^2}{n - 1}
\end{aligned}$$

where we have used $\mathrm{tr}(\omega) = \mathrm{tr}(\sigma) = \mathrm{tr}(\omega\sigma) = 0$. Using the orthonormal basis $E_1, \ldots, E_n$ along c with $E_n = c'$, we find that

$$\operatorname{tr}(R) = \sum_{i=1}^{n-1} g(R(E_i,c')c',E_i) = \sum_{i=1}^{n} g(E_i,E_i)g(R(E_i,c')c',E_i)$$
$$= \operatorname{Ric}(c',c')$$

This yields the *Raychaudhuri equation* for Jacobi tensors along timelike geodesics:

$$\theta' = -\operatorname{Ric}(c',c') - \operatorname{tr}(\omega^2) - \operatorname{tr}(\sigma^2) - \frac{\theta^2}{n-1}$$

Definition 11.2 implies that the shear tensor $\sigma$ is self-adjoint for arbitrary Jacobi tensor fields A. Thus if $E_1, \ldots, E_n$ is an orthonormal basis at $c(t)$ with $E_n = c'(t)$, we may represent $\sigma$ as a symmetric matrix $[\sigma_{ij}]$ with respect to $E_1, \ldots, E_{n-1}$. Consequently,

$$\operatorname{tr}(\sigma^2) = \operatorname{tr}([\sum_k \sigma_{ik}\sigma_{kj}])$$
$$= \sum_{i,k} \sigma_{ik}\sigma_{ki}$$
$$= \sum_i \sum_k \sigma_{ik}^2 \geqslant 0$$

Thus $\operatorname{tr}(\sigma^2) = 0$ iff $\sigma = 0$.

If A is a Lagrange tensor field as well as a Jacobi tensor field, then the tensor $B = A'A^{-1}$ has the following property.

LEMMA 11.3 If A is a Lagrange tensor field, then $B = A'A^{-1}$ is self-adjoint.

*Proof.* The equation $A^*A' = A'^*A$ implies that

$$B = A'A^{-1} = A^{*-1}A^{*\prime} = B^* \qquad \square$$

Lemma 11.3 then has the following consequence.

COROLLARY 11.4 If A is a Lagrange tensor field, then the vorticity $\omega = (1/2)(B - B^*)$ vanishes along c.

We thus obtain the *vorticity-free Raychaudhuri equation* for Lagrange tensor fields along timelike geodesics:

$$\theta' = -\mathrm{Ric}(c',c') - \mathrm{tr}(\sigma^2) - \frac{\theta^2}{n-1} \tag{11.2}$$

We now consider the Raychaudhuri equation for a null geodesic $\beta : [a,b] \longrightarrow M$. As discussed in Section 9.3 we use the quotient bundle $G(\beta) = N(\beta)/[\beta']$ along $\beta$ rather than $N(\beta)$. Recall that a smooth (1,1) tensor field $\overline{A} : G(\beta) \longrightarrow G(\beta)$ is said to be a *Jacobi tensor* along the null geodesic $\beta$ if

$$\overline{A}'' + \overline{RA} = 0$$

and

$$\ker(\overline{A}(t)) \cap \ker(\overline{A}'(t)) = \{[\beta'(t)]\}$$

for all $t \in [a,b]$ (cf. Definition 9.61). We proceed in the null case in much the same manner as in the timelike case remembering that we work modulo $\beta'$ in $G(\beta)$ and that $\dim G(\beta(t)) = n - 2$. Also the adjoint $\overline{A}^*$ of $\overline{A}$ is defined by

$$\overline{g}(\overline{A}w,v) = \overline{g}(\overline{A}^*(v),w)$$

where $\overline{g}$ is the positive definite metric on $G(\beta)$ given by formula (9.31) of Section 9.3.

DEFINITION 11.5 Let $\overline{A}$ be a Jacobi tensor along a null geodesic $\beta$ and set $\overline{B} = \overline{A}'\overline{A}^{-1}$ at points where $\overline{A}^{-1}$ is defined.

(a) The *expansion* $\overline{\theta}$ is defined by

$$\overline{\theta} = \mathrm{tr}(\overline{B}) = (\det \overline{A})^{-1}(\det \overline{A})'$$

(b) The *vorticity tensor* $\overline{\omega}$ is defined by

$$\overline{\omega} = \frac{1}{2}(\overline{B} - \overline{B}^*)$$

(c) The *shear tensor* $\overline{\sigma}$ is defined by

$$\overline{\sigma} = \frac{1}{2}(\overline{B} + \overline{B}^*) - \frac{\overline{\theta}}{n-2}\,\overline{E}$$

Using the same reasoning as in the timelike case we may obtain

$$\overline{B}' = -\overline{R} - \overline{B}\,\overline{B} \tag{11.3}$$

and

$$\overline{\theta}' = -\mathrm{tr}(\overline{R}) - \mathrm{tr}(\overline{\omega}^2) - \mathrm{tr}(\overline{\sigma}^2) - \frac{\overline{\theta}^2}{n-2}$$

We may calculate $\mathrm{tr}(\overline{R})$ as follows. Let $V(\beta)$ denote the geometric realization for $G(\beta)$ constructed as in (9.28) of Section 9.3 and let $\{Y_1, \cdots, Y_{n-2}\}$ be an orthonormal basis for $V(\beta)$ at every point of $\beta$. Extend $\{Y_1, \cdots, Y_{n-2}\}$ to an orthonormal basis $\{Y_1, \ldots, Y_n\}$ along $\beta$, where $Y_n$ is timelike and $\beta' = (Y_{n-1} + Y_n)/\sqrt{2}$. Then we have

$$\begin{aligned} &g(R(Y_{n-1},\beta')\beta',Y_{n-1}) - g(R(Y_n,\beta')\beta',Y_n) \\ &= 2^{-1}g(R(Y_{n-1},Y_n)Y_n,Y_{n-1}) - 2^{-1}g(R(Y_n,Y_{n-1})Y_{n-1},Y_n) = 0 \end{aligned}$$

using the basic properties of the curvature tensor. Consequently, we obtain

$$\begin{aligned} \mathrm{tr}(\overline{R}) &= \sum_{i=1}^{n-2} \overline{g}(\overline{R}(\pi(Y_i),\beta')\beta',\pi(Y_i)) \\ &= \sum_{i=1}^{n-2} g(R(Y_i,\beta')\beta',Y_i) \\ &= \sum_{i=1}^{n} g(Y_i,Y_i)g(R(Y_i,\beta')\beta',Y_i) = \mathrm{Ric}(\beta',\beta') \end{aligned}$$

This yields the *Raychaudhuri equation* for Jacobi tensors along null geodesics:

$$\overline{\theta}' = -\mathrm{Ric}(\beta',\beta') - \mathrm{tr}(\overline{\omega}^2) - \mathrm{tr}(\overline{\sigma}^2) - \frac{\overline{\theta}^2}{n-2} \tag{11.4}$$

The same reasoning as in the proof of Lemma 11.3 shows we may simplify the above equation when $\overline{A}$ is a Lagrange tensor field (i.e., when $\overline{A}^*\overline{A}' = \overline{A}'^*\overline{A}$).

LEMMA 11.6 If $\overline{A}$ is a Lagrange tensor field, then the vorticity tensor $\overline{\omega}$ vanishes along $\beta'$.

We thus obtain the *vorticity-free Raychaudhuri equation* for Lagrange tensor fields along null geodesics:

$$\bar{\theta}' = -\mathrm{Ric}(\beta',\beta') - \mathrm{tr}(\bar{\sigma}^2) - \frac{\bar{\theta}^2}{n-2} \tag{11.5}$$

## 11.2 THE GENERIC AND STRONG ENERGY CONDITIONS

In this section, we show that if (M,g) is a space-time of dimension at least 3 which satisfies the generic and strong energy conditions, then every complete nonspacelike geodesic contains a pair of conjugate points [cf. Hawking and Penrose (1970, p. 539)]. The timelike and null cases are handled separately. Similar treatments of the material in this section may be found in Bölts (1977), Hawking and Ellis (1973, pp. 96-101) and Eschenburg and O'Sullivan (1976).

We first state the definitions of the generic condition and the strong energy condition.

DEFINITION 11.7 A timelike geodesic $c : (a,b) \longrightarrow (M,g)$ is said to satisfy the *generic condition* if there exists some $t_0 \in (a,b)$ such that the curvature endomorphism

$$R(-,c'(t_0))c'(t_0) : V^{\perp}(c(t_0)) \longrightarrow V^{\perp}(c(t_0))$$

is not identically zero. A null geodesic $\beta : (a,b) \longrightarrow (M,g)$ is said to satisfy the *generic condition* if there exists some $t_0 \in (a,b)$ such that the curvature endomorphism

$$\bar{R}(-,\beta'(t_0))\beta'(t_0) : G(\beta(t_0)) \longrightarrow G(\beta(t_0))$$

of the quotient space $G(\beta(t_0))$ is not identically zero. The space-time (M,g) is said to satisfy the *generic condition* if each inextendible nonspacelike geodesic satisfies this condition.

In Appendix B, we show that this formulation of the generic condition is equivalent to the usual definition given in general relativity; namely, a nonspacelike geodesic c with tangent vector W satisfies the generic condition if there is some point of c at which

$$W^c W^d W_{[a} R_{b]cd[e} W_{f]} \neq 0$$

DEFINITION 11.8 A space-time satisfies the *strong energy condition* if $\text{Ric}(v,v) \geqslant 0$ for all nonspacelike tangent vectors $v \in TM$.

By continuity, the curvature condition of Definition 11.8 is equivalent to the *timelike convergence condition* of Hawking and Ellis (1973, p. 95) that $\text{Ric}(v,v) \geqslant 0$ for all timelike $v \in TM$. Hawking and Ellis (1973, p. 95) also call the curvature condition $\text{Ric}(w,w) \geqslant 0$ for all null $w \in TM$ the *null convergence condition*. In Hawking and Ellis (1973, p. 89), a four-dimensional space-time (M,g) with energy-momentum tensor T (cf. Appendix C) is said to satisfy the *weak energy condition* if $T(v,v) \geqslant 0$ for all timelike $v \in TM$. If the Einstein equations hold for the four-dimensional space-time (M,g) and T with cosmological constant $\Lambda$, then the condition $\text{Ric}(v,v) \geqslant 0$ for all timelike $v \in TM$ implies that

$$T(v,v) \geqslant \left(\frac{\text{tr } T}{2} - \frac{\Lambda}{8\pi}\right) g(v,v)$$

for all timelike $v \in TM$. Hence in Hawking and Ellis (1973, p. 95), the four-dimensional space-time (M,g) and energy-momentum tensor T are said to satisfy the *strong energy condition* if $T(v,v) \geqslant (\text{tr } T/2)g(v,v)$ for all timelike $v \in TM$. When $\dim M = 4$ and $\Lambda = 0$, this may be seen to be equivalent to the condition $\text{Ric}(v,v) \geqslant 0$ for all timelike $v \in TM$ (cf. Appendix C). In Hawking and Penrose (1970, p. 539), the condition $\text{Ric}(v,v) \geqslant 0$ for all unit timelike vectors $v \in TM$ is called the *energy condition*. In Frankel (1979) and Lee (1975), the same definition of strong energy condition is used as in our Definition 11.8. A discussion of the physical interpretation of these curvature conditions in general relativity may be found in Hawking and Ellis (1973, Section 4.3).

As we have just noted above, if (M,g) satisfies the timelike convergence condition or the strong energy condition, then (M,g) satisfies the null convergence condition. In view of several rigidity theorems associated with curvature conditions in Riemannian geometry [cf. Cheeger and Ebin (1975, pp. v and vi)], it is natural to consider the implications of the curvature condition $\mathrm{Ric}(w,w) = 0$ for all null vectors $w \in TM$. Applying linear algebraic arguments to each tangent space, Dajczer and Nomizu (1979) have obtained the rigidity result that if $\dim M \geqslant 3$ and $\mathrm{Ric}(w,w) = 0$ for all null vectors $w \in TM$, then $(M,g)$ is Einstein, i.e., $\mathrm{Ric} = \lambda g$ for some constant $\lambda \in \mathbb{R}$. Thus if $(M,g)$ is not Einstein, there are some nonzero null Ricci curvatures. Suppose further that $(M,g)$ is globally hyperbolic with a smooth globally hyperbolic time function $h : M \longrightarrow \mathbb{R}$ such that for some Cauchy hypersurface $S = h^{-1}(t_0)$, all null Ricci curvatures $\mathrm{Ric}(g)(w,w) > 0$ if $\pi(w) \in S$. If $(M,g)$ also satisfies the null convergence condition, then $M$ admits a metric $g_1$ globally conformal to $g$ such that the globally hyperbolic space-time $(M,g_1)$ satisfies the curvature condition $\mathrm{Ric}(g_1)(w,w) > 0$ for all null vectors $w \in TM$ [cf. Beem and Ehrlich (1978, p. 174, Theorem 7.1)].

An essential step in proving that any complete timelike geodesic in a space-time satisfying the generic and strong energy conditions contains a pair of conjugate points is the following proposition.

PROPOSITION 11.9 Let $c : J \longrightarrow (M,g)$ be an inextendible timelike geodesic satisfying $\mathrm{Ric}(c'(t),c'(t)) \geqslant 0$ for all $t \in J$. Let $A$ be a Lagrange tensor field along $c$. Suppose that the expansion $\theta(t) = \mathrm{tr}(A'(t)A^{-1}(t))$ has a negative (resp., positive) value $\theta_1 = \theta(t_1)$ at $t_1 \in J$. Then $\det A(t) = 0$ for some $t$ in the interval from $t_1$ to $t_1 - (n-1)/\theta_1$ [resp., some $t$ in the interval from $t_1 - (n-1)/\theta_1$ to $t_1$] provided that $t \in J$.

*Proof.* Since $\theta = (\det A)'(\det A)^{-1}$, we have $\det A(t_0) = 0$ provided that $|\theta| \longrightarrow \infty$ as $t \not\longrightarrow t_0$. Thus we need only show that $|\theta| \longrightarrow \infty$ on the above intervals. Put

$$s_1 = \frac{n-1}{\theta_1}$$

The vorticity-free Raychaudhuri equation (11.2) for timelike geodesics and the condition $\mathrm{Ric}(c',c') \geqslant 0$ yield the inequality

$$\frac{d\theta}{dt} \leqslant -\frac{\theta^2}{n-1}$$

In the case that $\theta_1 < 0$, integrating this inequality from $t_1$ to $t > t_1$, we obtain

$$\theta(t) \leqslant \frac{n-1}{t+s_1-t_1}$$

for $t \in [t_1, t_1 - s_1)$. Hence $|\theta(t)|$ becomes infinite for some $t \in (t_1, t_1 - s_1)$ provided that $c(t)$ is defined. In the case that $\theta_1 > 0$, we obtain for $t \in (t_1 - s_1, t_1]$ that

$$\theta(t) \geqslant \frac{n-1}{t+s_1-t_1}$$

Hence $|\theta(t)|$ again becomes infinite for some $t \in (t_1 - s_1, t_1)$ provided that $c(t)$ is defined. □

We now show that a timelike geodesic in a space-time that satisfies the strong energy condition and the generic condition must either be incomplete or else have a pair of conjugate points.

PROPOSITION 11.10 Let $(M,g)$ be an arbitrary space-time of dimension $\geqslant 2$. Suppose that $c : \mathbb{R} \longrightarrow (M,g)$ is a complete timelike geodesic which satisfies $\mathrm{Ric}(c'(t),c'(t)) \geqslant 0$ for all $t \in \mathbb{R}$. If $R(-,c'(t_1))c'(t_1) : N(c(t_1)) \longrightarrow N(c(t_1))$ is not zero for some $t_1 \in \mathbb{R}$, then $c$ has a pair of conjugate points.

We first prove four lemmas needed for the proof of Proposition 11.10 [cf. Bölts (1977, pp. 30-37)].

LEMMA 11.11 Let $c : [a,b] \longrightarrow (M,g)$ be a timelike geodesic without conjugate points. Then there is a unique (1,1) tensor field $A$ on $V^{\perp}(c)$ which satisfies the differential equation $A'' + RA = 0$ with given boundary conditions $A(a)$ and $A(b)$.

*Proof.* Let S be the vector space of (1,1) tensor fields A on $V^{\perp}(c)$ with $A'' + RA = 0$ and let $L(N(c(t)))$ denote the set of linear endomorphisms of $N(c(t))$. Define a linear transformation $\phi : S \longrightarrow L(N(c(a))) \times L(N(c(b)))$ by

$$\phi(A) = (A(a), A(b))$$

Since $\dim S = \dim (L(N(c(a)))) + \dim (L(N(c(b)))) = 2(n - 1)^2$, it is only necessary to show that $\phi$ is injective in order to prove that $\phi$ is an isomorphism and establish the existence of a unique solution A. Assume that $\phi(A) = (A(a), A(b)) = (0,0)$. If $Y(t)$ is any parallel vector field along c, then $J(t) = A(t)Y(t)$ is a Jacobi field with $J(a) = J(b) = 0$. Thus $J = 0$. On the other hand, since $Y(t)$ was an arbitrary parallel field, this implies that $A(t) = 0$ which shows $\phi$ is injective and establishes the lemma. □

Now let $c : [t_1, \infty) \longrightarrow (M,g)$ be a timelike geodesic without conjugate points and fix $s \in (t_1, \infty)$. Then by Lemma 11.11, there exists a unique (1,1) tensor field on $V^{\perp}(c)$ which we will denote by $D_s$, satisfying the differential equation $D_s'' + RD_s = 0$ with initial conditions $D_s(t_1) = E$ and $D_s(s) = 0$. As $D_s(t_1) = E$, we have $\ker (D_s(t_1)) \cap \ker (D_s'(t_1)) = \{0\}$. Thus $D_s$ is a Jacobi tensor field (cf. Lemma 9.62). Also since $D_s(s) = 0$, it follows that $D_s$ is a Lagrange tensor field. It will also be shown during the course of the proof of Lemma 11.12, that if A is the Lagrange tensor field on $V^{\perp}(c)$ with $A(t_1) = 0$ and $A'(t_1) = E$, then $D_s'(s) = -(A^*)^{-1}(s)$.

LEMMA 11.12 Let $c : [t_1, \infty) \longrightarrow M$ be a timelike geodesic without conjugate points. Let A be the unique Lagrange tensor on $V^{\perp}(c)$ with $A(t_1) = 0$ and $A'(t_1) = E$. Then for each $s \in (t_1, \infty)$ the Lagrange tensor $D_s$ on $V^{\perp}(c)$ with $D_s(t_1) = E$ and $D_s(s) = 0$ satisfies the equation

$$D_s(t) = A(t) \int_t^s (A^* A)^{-1}(\tau)\, d\tau$$

for all $t \in (t_1, s]$. Thus $D_s(t)$ is nonsingular for $t \in (t_1, s)$.

*Proof*. Set $X(t) = A(t) \int_t^s (A^*A)^{-1}(\tau)\, d\tau$. It suffices to show that $X'' + RX = 0$, $X(s) = D_s(s) = 0$, and $X'(s) = D_s'(s)$.
0 and $X'(s) = D_s'(s)$.

We first check that $X'' + RX = 0$. Differentiating, we obtain

$$X'(t) = A'(t) \int_t^s (A^*A)^{-1}(\tau)\, d\tau - A(t)(A^*A)^{-1}(t)$$

$$= A'(t) \int_t^s (A^*A)^{-1}(\tau)\, d\tau - (A^*)^{-1}(t)$$

Hence

$$X''(t) = A''(t) \int_t^s (A^*A)^{-1}(\tau)\, d\tau - A'(t)(A^*A)^{-1}(t)$$

$$- ((A^*)^{-1})'(t)$$

$$= A''(t) \int_t^s (A^*A)^{-1}(\tau)\, d\tau - A'(t)A^{-1}(t)(A^*)^{-1}(t)$$

$$+ (A^*)^{-1}(A^*)'(A^*)^{-1}(t)$$

But since A is a nonsingular Lagrange tensor, $(A^*)' = A^*A'A^{-1}$ so that $(A^*)^{-1}(A^*)'(A^*)^{-1} = A'A^{-1}(A^*)^{-1}$ and we obtain

$$X''(t) = A''(t) \int_t^s (A^*A)^{-1}(\tau)\, d\tau$$

We then have

$$X''(t) + R(t)X(t) = [A''(t) + R(t)A(t)] \int_t^s (A^*A)^{-1}(\tau)\, d\tau = 0$$

since $A''(t) + R(t)A(t) = 0$. Thus X satisfies the Jacobi differential equation.

Setting $t = s$, we obtain

$$X(s) = A(s) \int_s^s (A^*A)^{-1}(\tau)\, d\tau = 0$$

and

$$X'(s) = A'(s) \int_s^s (A^*A)^{-1}(\tau)\, d\tau - A(s)(A^*A)^{-1}(s) = -(A^*)^{-1}(s)$$

Thus it remains to check that $D_s'(s) = -(A^*)^{-1}(s)$. But using $R^* = R$, we obtain

$$\begin{aligned}[(A^*)'D_s - A^*D_s']' &= (A^*)''D_s + (A^*)'D_s' - (A^*)'D_s' - A^*D_s'' \\ &= (A^*)''D_s - A^*D_s'' \\ &= -A^*R^*D_s + A^*RD_s = 0\end{aligned}$$

Thus $(A^*)'D_s - A^*D_s'$ is parallel along c. At $t = t_1$, the initial conditions $A(t_1) = 0$ and $A'(t_1) = E$ for A imply that $A^*(t_1) = 0$ and $(A^*)'(t_1) = (A')^*(t_1) = E$. Hence as $D_s(t_1) = E$, we obtain

$$((A^*)'D_s - A^*D_s')(t_1) = E$$

Hence $((A^*)'D_s - A^*D_s')(t) = E$ for all t. Setting $t = s$ we have

$$E = ((A^*)'D_s - A^*D_s')(s) = -(A^*D_s')(s)$$

which implies that $D_s'(s) = -(A^*)^{-1}(s) = X'(s)$. Therefore since $D_s$ and X both satisfy $A'' + RA = 0$ and have the same values and first derivatives at $t = s$, the tensors must agree for all t.

Finally, the nonsingularity of $D_s(t)$ for $t \in (t_1,s)$ follows from the formula

$$D_s(t) = A(t) \int_t^s (A^*A)^{-1}(\tau)\, d\tau$$

since $(A^*A)^{-1}(t)$ is a positive definite, self-adjoint tensor field for all $t > t_1$. □

Note that while the integral representation of the Lagrange tensor $D_s$ along c satisfying $D_s(t_1) = E$ and $D_s(s) = 0$ given in Lemma 11.12 was proven only for $t \in (t_1,s]$, if c is defined for all

$t \in \mathbb{R}$ and has no conjugate points, then $D_s(t)$ is defined for all $t \in \mathbb{R}$.

We now show that if $c : [a,\infty) \longrightarrow (M,g)$ is a timelike geodesic without conjugate points, then the above tensor fields $D_s$ converge to a Lagrange tensor field D as $s \longrightarrow \infty$. This construction parallels the construction of stable Jacobi fields in certain classes of complete Riemannian manifolds without conjugate points [cf. Eschenburg and O'Sullivan (1976, pp. 227 ff.), Green (1958), E. Hopf (1948, p. 48)].

LEMMA 11.13 Let $c : [a,\infty) \longrightarrow (M,g)$ be a timelike geodesic without conjugate points. For $t_1 > a$ and $s \in [a,\infty) - \{t_1\}$, let $D_s$ be the Lagrange tensor field along c determined by $D_s(t_1) = E$ and $D_s(s) = 0$. Then $D(t) = \lim_{s\to\infty} D_s(t)$ is a Lagrange tensor field. Furthermore, D(t) is nonsingular for all t with $t_1 < t < \infty$.

*Proof*. We first show that $D_s'(t_1)$ has a self-adjoint limit as $s \longrightarrow \infty$. Since $D_s$ is a Lagrange tensor we have $(D_s'^* D_s)(t_1) = (D_s^* D_s')(t_1)$. Using $D_s(t_1) = E$, we obtain $D_s'^*(t_1) = D_s'(t_1)$. Thus the limit of $D_s'(t_1)$ must be a self-adjoint linear map which we will denote by $D'(t_1) : N(c(t_1)) \longrightarrow N(c(t_1))$ if it exists. Consequently, we need only show that for each $y \in N(c(t_1))$ the value of $g(D_s'(t_1)y,y)$ converges to some value $g(D'(t_1)y,y)$.

We will show that the function $s \longrightarrow g(D_s'(t_1)y,y)$ is monotone increasing for all s with $t_1 < s < \infty$ and is bounded from above by $g(D_a'(t_1)y,y)$ to establish the existence of this limit. To this end, assume that $t_1 < r < s$. Then by Lemma 11.12 we have

$$D_s'(t) = A'(t) \int_t^s (A^*A)^{-1}(\tau)\, d\tau - (A^*)^{-1}(t)$$

Thus for $t \in (t_1,s)$ we obtain

$$g(D_s'(t)Y(t),Y(t)) = g((A'(t) \int_t^s (A^*A)^{-1}(\tau)\, d\tau)(Y(t)),Y(t)) - g((A^*)^{-1}(t)Y(t),Y(t))$$

where A is the Lagrange tensor field along c satisfying $A(t_1) = 0$, $A'(t_1) = E$ and $Y(t)$ is the parallel vector field along c with $Y(t_1) = y$. Thus for t with $t_1 < t < r$, it follows that

$$g(D_s'(t)Y(t),Y(t)) - g(D_r'(t)Y(t),Y(t))$$

is given by

$$g((A'(t)\int_r^s (A^*A)^{-1}(\tau)\, d\tau)(Y(t)),Y(t))$$

Letting $t \longrightarrow t_1^+$ and using $Y(t_1) = y$ and $A'(t_1) = E$, we then have

$$g(D_s'(t_1)y,y) - g(D_r'(t_1)y,y) = g((\int_r^s (A^*A)^{-1}(\tau)\, d\tau)(Y(t_1)),Y(t_1))$$

Since Y is parallel along c, it may be checked by choosing an orthonormal basis of parallel fields for $V^{\perp}(c)$ that

$$g((\int_r^s (A^*A)^{-1}(\tau)\, d\tau)(Y(t_1)),Y(t_1)) = \int_r^s g((A^*A)^{-1}(\tau)(Y(\tau)),Y(\tau))\, d\tau$$

Since $(A^*A)^{-1} = A^{-1}A^{*-1}$, this may be written as

$$\int_r^s g((A^*)^{-1}(\tau)Y(\tau),(A^*)^{-1}(\tau)Y(\tau))\, d\tau$$

which must be positive because $(A^*)^{-1}(\tau)Y(\tau)$ is a spacelike vector in $N(c(\tau))$ for each $\tau \in [r,s]$. Thus

$$g(D_s'(t_1)y,y) - g(D_r'(t_1)y,y) > 0$$

and the map $s \longrightarrow g(D_s'(t_1)y,y)$ is monotone for all $s > t_1$ as required.

We now show that $g(D_s'(t_1)y,y) < g(D_a'(t_1)y,y)$ for all $s > t_1$ and any $y \in N(c(t_1))$. Again let Y be the unique parallel field along c with $Y(t_1) = y$. Let J be the piecewise smooth Jacobi field along $c \mid [a,s]$ given by

$$J(t) = \begin{cases} D_a(t)Y(t) & \text{for } a \leq t < t_1 \\ \\ D_s(t)Y(t) & \text{for } t_1 \leq t \leq s \end{cases}$$

Also let $J_a = J \mid [a,t_1]$ and $J_s = J \mid [t_1,s]$. Then $J(a) = J(s) = 0$ and J is well defined at $t = t_1$ since $D_a(t_1) = D_s(t_1) = E$. Using the index form I for $c \mid [a,s]$ given in Section 9.1, Definition 9.4, we obtain

$$\begin{aligned} I(J,J) &= I(J,J)_a^{t_1} + I(J,J)_{t_1}^s \\ &= -g(J_a'(t_1),J_a(t_1)) + g(J_s'(t_1),J_s(t_1)) \\ &= -g(D_a'(t_1)Y(t_1),Y(t_1)) + g(D_s'(t_1)Y(t_1),Y(t_1)) \\ &= -g(D_a'(t_1)y,y) + g(D_s'(t_1)y,y) \end{aligned}$$

where we have used formula (9.2) of Definition 9.4 and $D_a(t_1) = D_s(t_1) = E$. Since $J(a) = J(s) = 0$ and c has no conjugate points in $[a,\infty)$, we have $I(J,J) < 0$ by Theorem 9.22. Thus

$$g(D_s'(t_1)y,y) < g(D_a'(t_1)y,y)$$

for all $s > t_1$ and we conclude that the self-adjoint tensor $D'(t_1) = \lim_{s\to+\infty} D_s'(t_1)$ exists.

Now we define D(t) by setting D(t) equal to the unique Jacobi tensor along c which satisfies $D(t_1) = E$ and $D'(t_1) = \lim_{s\to+\infty} D_s'(t_1)$. Since D(t) and $D_s(t)$ both satisfy the differential equation $A'' + RA = 0$ and the initial conditions of $D_s$ approach the initial conditions of D as $s \longrightarrow \infty$, it follows that $D(t) = \lim_{s\to+\infty} D_s(t)$ and $D'(t) = \lim_{s\to+\infty} D_s'(t)$ for all $t \in [a,\infty)$. This implies that the limit D(t) of the Lagrange tensors $D_s(t)$ must also be a Lagrange tensor.

The last statement of the lemma now follows using the representation

$$D(t) = A(t) \int_t^{\infty} (A^*A)^{-1}(\tau)\, d\tau$$

and the fact that $(A^*A)^{-1}(t)$ is a positive definite self-adjoint tensor field for all $t > t_1$. □

We now divide the Lagrange tensors with $A(t_1) = E$ along a complete timelike geodesic $c : (-\infty,\infty) \longrightarrow (M,g)$ with $\mathrm{Ric}\,(c',c') \geqslant 0$ and $R(-,c'(t_1))c'(t_1) \neq 0$ for some $t_1 \in \mathbb{R}$ into two classes $L_+$ and $L_-$ as follows [cf. Bölts (1977, p. 36), Hawking and Ellis (1973, p. 98)]. Put

$$L_+ = \{A : A \text{ is a Lagrange tensor with } A(t_1) = E \text{ and } \mathrm{tr}(A'(t_1)) \geqslant 0\}$$

and

$$L_- = \{A : A \text{ is a Lagrange tensor with } A(t_1) = E \text{ and } \mathrm{tr}(A'(t_1)) \leqslant 0\}$$

LEMMA 11.14 Let $c : \mathbb{R} \longrightarrow (M,g)$ be a complete timelike geodesic such that $\mathrm{Ric}(c',c') \geqslant 0$ and $R(-,c'(t_1))c'(t_1) \neq 0$ for some $t_1 \in \mathbb{R}$. Then each $A \in L_-$ satisfies $\det A(t) = 0$ for some $t > t_1$ and each $A \in L_+$ satisfies $\det A(t) = 0$ for some $t < t_1$.

*Proof.* If $A \in L_-$, then $\theta(t_1) = \mathrm{tr}(A'(t_1)A^{-1}(t_1)) = \mathrm{tr}(A'(t_1)) \leqslant 0$. Using the vorticity-free Raychaudhuri equation (11.2) for timelike geodesics with $\mathrm{Ric}(c',c') \geqslant 0$ and $\mathrm{tr}(\sigma^2) \geqslant 0$ we find $\theta'(t) \leqslant 0$ for all $t$. Thus $\theta(t) \leqslant 0$ for all $t \geqslant t_1$. If $\theta(t_0) < 0$ for some $t_0 > t_1$, then the result for A follows from Proposition 11.9. Assume therefore that $\theta(t) = 0$ for $t \geqslant t_1$. This implies $\theta'(t) = 0$ for $t \geqslant t_1$ which yields $\mathrm{tr}(\sigma^2) = 0$. Hence $\sigma = 0$ for $t \geqslant t_1$ since $\sigma$ is self-adjoint. Using $\theta = 0$ and the self-adjointness of B, we thus have $B = \sigma = 0$ which by Eq. (11.1) implies that $R = -B^2 - B' = 0$ for $t \geqslant t_1$, in contradiction to $R(t_1) \neq 0$.

If $A \in L_+$, the proof is similar. □

We now come to the

*Proof of Proposition 11.10.* Let $c : \mathbb{R} \longrightarrow (M,g)$ be a complete timelike geodesic with $\mathrm{Ric}(c'(t),c'(t)) \geq 0$ for all $t \in \mathbb{R}$ and with $R(-,c'(t_1))c'(t_1) \neq 0$ for some $t_1 \in \mathbb{R}$. Suppose that c has no conjugate points. Then let $D = \lim_{s\to\infty} D_s$ be the Lagrange tensor field on $V^\perp(c)$ with $D(t_1) = E$ constructed in Lemma 11.13. Since $c \mid [t_1,\infty)$ has no conjugate points, $D(t)$ is nonsingular for all $t \geq t_1$. Thus $D \notin L_-$ by Lemma 11.14. Hence $D \in L_+$ and, moreover, $\mathrm{tr}\, D'(t_1) > 0$ as $D \notin L_-$. Since $D'(t_1) = \lim_{s\to\infty} D_s'(t_1)$, there exists an $s > t_1$ such that $\mathrm{tr}(D_s'(t_1)) > 0$. Hence by Lemma 11.14, there exists a $t_2 < t_1$ and a nonzero tangent vector $v \in N(c(t_2))$ such that $D_s(t_2)(v) = 0$. Recall also from the proof of Lemma 11.12 that $D_s(s) = 0$ but $D_s'(s) = (A^*)^{-1}(s)$ is nonsingular. Therefore, if we let $Y \in V^\perp(c)$ be the unique parallel field along c with $Y(t_2) = v$, then $J = D_s(Y)$ is a nontrivial Jacobi field along c with $Y(t_2) = Y(s) = 0$, in contradiction. □

COROLLARY 11.15 Let (M,g) be a space-time of dimension $\geq 2$ which satisfies the strong energy condition and the generic condition. Then each timelike geodesic of (M,g) is either incomplete or else has a pair of conjugate points.

We now consider the existence of conjugate points on null geodesics. The methods and results for null geodesics are much the same as for timelike geodesics except that $\dim M \geq 3$ is necessary since $\dim G(\beta) = \dim M - 2$ and null geodesics in two-dimensional space-times are free of conjugate points. First, using the vorticity-free Raychaudhuri equation (11.5) for null geodesics, the following analogue of Proposition 11.9 may be established with the same type of reasoning as in the timelike case.

PROPOSITION 11.16 Let (M,g) be an arbitrary space-time of dimension $\geq 3$. Suppose that $\beta : J \longrightarrow (M,g)$ is an inextendible null geodesic satisfying $\mathrm{Ric}(\beta'(t),\beta'(t)) \geq 0$ for all $t \in J$. Let $\bar{A}$ be a Lagrange tensor field along $\beta$ such that the expansion $\bar{\theta}(t) = \mathrm{tr}(\bar{A}'(t)\bar{A}^{-1}(t)) = [\det \bar{A}(t)]^{-1}[\det \bar{A}(t)]'$ has the negative (resp., positive) value

$\overline{\theta}_1 = \overline{\theta}(t_1)$ at $t_1 \in J$. Then $\det \overline{A}(t) = 0$ for some t in the interval from $t_1$ to $t_1 - (n-2)/\overline{\theta}_1$ [resp., some t in the interval from $t_1 - (n-2)/\overline{\theta}_1$ to $t_1$] provided that $t \in J$.

The null analogue of Proposition 11.10 may also be established for all space-times of dimension $\geqslant 3$ using $\overline{A}$, $\overline{B}$, $\overline{\theta}$, $\overline{\sigma}$, etc., in place of the corresponding A, B, θ, σ, etc., used in the proof for timelike geodesics.

PROPOSITION 11.17 Let $\beta : \mathbb{R} \longrightarrow (M,g)$ be a complete null geodesic with $\mathrm{Ric}(\beta'(t),\beta'(t)) \geqslant 0$ for all $t \in \mathbb{R}$. If $\dim M \geqslant 3$ and if $R(-,\beta'(t))\beta'(t) : G(\beta(t)) \longrightarrow G((\beta(t))$ is nonzero for some $t_1 \in \mathbb{R}$, then β has a pair of conjugate points.

Combining this result with Corollary 11.15, we obtain the following theorem.

THEOREM 11.18 Let (M,g) be a space-time of dimension $\geqslant 3$ which satisfies the strong energy condition and the generic condition. Then each nonspacelike geodesic in (M,g) is either incomplete or else has a pair of conjugate points. Thus every nonspacelike geodesic in (M,g) without conjugate points is incomplete.

The material presented in this section may also be treated within the framework of conjugate points and oscillation theory in ordinary differential equations [cf. Tipler (1977d), Chicone and Ehrlich (1980)]. In this approach, the Raychaudhuri equation is transformed by a change of variables to the differential equation

$$x''(t) + F(t)x(t) = 0$$

with

$$F(t) = \frac{1}{m}[\mathrm{Ric}(\gamma'(t),\gamma'(t)) + 2\sigma^2(t)]$$

where $m = n - 1$ if γ is timelike and $m = n - 2$ if γ is null.

## 11.3 FOCAL POINTS

The concept of a conjugate point along a geodesic can be generalized to the notion of a focal point of a submanifold. Let H be a nondegenerate submanifold of the space-time (M,g). At each $p \in H$ the tangent space $T_pH$ may be naturally identified with the vectors of $T_pM$ which are tangent to H at p. The normal space $T_p^\perp H$ consists of all vectors orthogonal to H at p. Since H is nondegenerate, $T_p^\perp H \cap T_pH = \{0_p\}$ for all $p \in H$. We will denote the exponential map restricted to the normal bundle $T^\perp H$ by $\exp^\perp$. Then the vector $X \in T_p^\perp H$ is said to be a *focal point* of H if $(\exp^\perp)_*$ is singular at X. The corresponding point $\exp^\perp(X)$ of M is said to be a focal point of H along the geodesic segment $\exp^\perp(tX)$. When H is a single point, then $T_p^\perp H = T_pM$ and a focal point is just an ordinary conjugate point.

Focal points may also be defined using Jacobi fields and the second fundamental form [cf. Bishop and Crittenden (1964, p. 225)]. This approach will be used in this section following the treatment given in Bölts (1977). Jacobi fields are used to measure the separation (or deviation) of nearby geodesics. For example, when a point q is conjugate to p along a geodesic c, geodesics which start at p with initial tangent close to c' at p will tend to focus at q up to second order. They need not actually pass through q, but must pass close to q. In studying submanifolds one may take a congruence of geodesics orthogonal to the submanifold and use Jacobi fields to measure the separation of geodesics in this congruence. If p is a focal point along a geodesic c which is orthogonal to the submanifold H, then some geodesics close to c and orthogonal to H tend to focus at p. This is illustrated for the Euclidean plane with the usual positive definite metric in Figure 11.1 and for Lorentzian manifolds in Figure 11.2.

In Section 2.5 we defined the second fundamental form $S_n$ : $T_pH \times T_pH \longrightarrow \mathbb{R}$ in the direction n, the second fundamental form $S : T_p^\perp H \times T_pH \times T_pH \longrightarrow \mathbb{R}$ and the second fundamental form operator $L_n : T_pH \longrightarrow T_pH$ (cf. Definition 2.35). Recall that

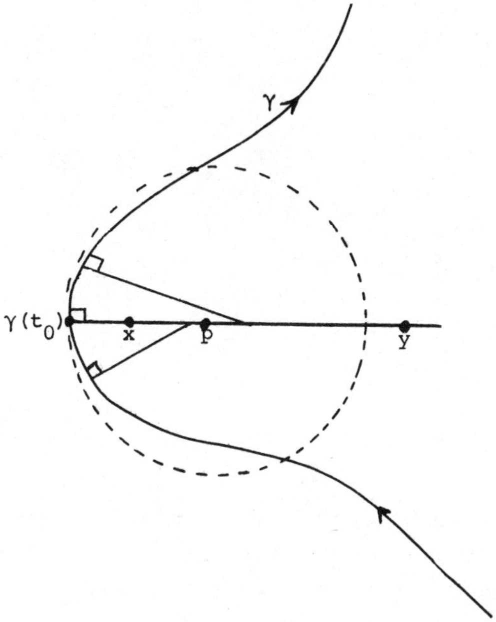


*Figure 11.1* A curve in the Euclidean plane has its centers of curvature as its focal points. The osculating circle to the curve $\gamma$ at $t = t_0$ is shown. The segment from $\gamma(t_0)$ to the center p of the osculating circle contains x in its interior. The point y lies beyond p on the ray from $\gamma(t_0)$ through p. For some interval $t_0 - \varepsilon_1 < t < t_0 + \varepsilon_1$ the closest point of $\gamma$ to x is $\gamma(t_0)$. On the other hand, for some interval $t_0 - \varepsilon_2 < t < t_0 + \varepsilon_2$ the farthest point of $\gamma(t)$ from y is $\gamma(t_0)$. Furthermore, the straight lines which are orthogonal to $\gamma$ near $\gamma(t_0)$ tend to focus at p.

$S(n,x,y) = S_n(x,y) = S_n(y,x)$ and $g(L_n(x),y) = S_n(x,y) = g(\nabla_X Y,n)$ for $n \in T_p H^\perp$ and $x,y \in T_p H$, where X and Y are local vector field extensions of x, y.

In this section we will primarily be concerned with the operator $L_n : T_p H \longrightarrow T_p H$. Note that a vector field $\eta$ which is orthogonal to H at all points of H defines a (1,1) tensor field $L_\eta$ on H. We will first consider spacelike hypersurfaces. If the timelike normal vector field $\eta$ on H satisfies $g(\eta,\eta) = -1$, then we may calculate $L_\eta$ as follows.

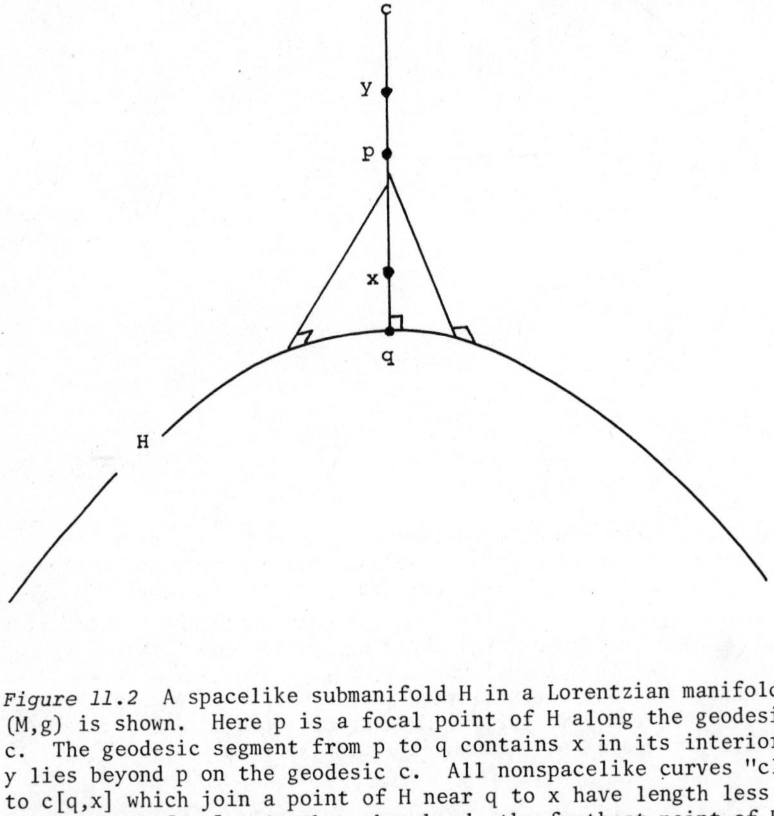


*Figure 11.2* A spacelike submanifold H in a Lorentzian manifold (M,g) is shown. Here p is a focal point of H along the geodesic c. The geodesic segment from p to q contains x in its interior and y lies beyond p on the geodesic c. All nonspacelike curves "close" to c[q,x] which join a point of H near q to x have length less than or equal to c[q,x]. On the other hand, the farthest point of H near q to y is not q. There are points close to q on H which may be joined to y by timelike curves longer than c[q,y]. Furthermore, there is at least one curve $\gamma$ on H through q such that a family of geodesics orthogonal to H with respect to the given Lorentzian metric and starting on $\gamma$ near q tend to focus at p up to second order.

LEMMA 11.19 Let H be a spacelike hypersurface with unit timelike normal field $\eta$. If x is a vector tangent to H, then $L_\eta(x) = -\nabla_x \eta$.

*Proof.* Since $g(\eta,\eta) = -1$ we have $0 = x(g(\eta,\eta)) = 2g(\nabla_x \eta,\eta)$ which shows that $\nabla_x \eta$ is tangent to H. Now if Y is any vector field tangent to H, then $g(\eta,Y) = 0$. Thus $0 = x(g(\eta,Y)) = g(\nabla_x \eta,Y) + g(\eta,\nabla_x Y)$ so that $g(\nabla_x Y,\eta) = g(-\nabla_x \eta,Y)$. Consequently, $g(L_\eta(x),Y) = g(\nabla_x Y,\eta) = g(-\nabla_x \eta,Y)$. Since Y was arbitrary, the result follows. □

Given a unit normal field $\eta$ for the spacelike hypersurface H, the collection of unit speed timelike geodesics orthogonal to H with initial direction $\eta(q)$ at $q \in H$ determines a congruence of timelike geodesics. Let c be a timelike geodesic of this congruence intersecting H at q and let J denote the variation vector field along c of a 1-parameter subfamily of the congruence. Then J is a Jacobi field which measures the rate of separation of geodesics in the 1-parameter subfamily from c. Since the geodesics of the congruence are all orthogonal to H, it may be shown using Lemma 11.19 that J satisfies the initial condition

$$J'(q) = -L_{\eta(q)} J$$

This suggests the following definition of a focal point to a spacelike hypersurface in terms of Jacobi fields.

DEFINITION 11.20 Let c be a timelike geodesic which is orthogonal to the spacelike hypersurface H at q. A point p on c is said to be a *focal point of H along c* if there is a nontrivial Jacobi field J along c such that J is orthogonal to c', vanishes at p and satisfies $J' = -L_\eta J$ at q.

Suppose that A is a Jacobi tensor along the timelike geodesic c which satisfies A = E and $A' = -L_\eta A = -L_\eta$ at the point q, where c intersects the spacelike hypersurface H. Then every Jacobi field J orthogonal to c which satisfies $J' = -L_\eta J$ at q may be expressed as J = AY where Y = Y(t) is a parallel vector field along c which is orthogonal to c. Since there are n - 1 linearly independent parallel vector fields orthogonal to c, there is an (n - 1)-dimensional vector space of Jacobi fields along c which satisfy $J' = -L_\eta J$ at q. We now show that such a Jacobi tensor A satisfying $A = E$ and $A' = -L_\eta$ at q is in fact a Lagrange tensor field.

LEMMA 11.21 Suppose that A is a Jacobi tensor field along the timelike geodesic c. Let c be orthogonal to H at $t_1$ and let $L_\eta$ be the second fundamental form operator on H. If $A(t_1) = E$ and $A'(t_1) = -L_\eta A(t_1)$, then A is a Lagrange tensor field.

*Proof.* The second fundamental form $S_\eta$ at $\eta \in T^\perp H$ is symmetric which implies that $L_\eta$ is self-adjoint at q since $g(L_\eta(x),y) = S_\eta(x,y) = S_\eta(y,x) = g(L_\eta(y),x)$. Thus at $t_1$ we have that $A'(t_1) = -L_\eta A(t_1) = -L_\eta$ is self-adjoint. Hence $A^*{}'(t_1) = A'(t_1)$. Using $A(t_1) = A^*(t_1) = E$, it follows that $A^*{}'(t_1)A(t_1) = A^*(t_1)A'(t_1)$. Thus A is a Lagrange tensor field as required. □

The tensor $B = A'A^{-1}$, expansion $\theta$, and shear $\sigma$ of Lagrange tensors A satisfying the conditions of Lemma 11.21 may be defined as in Section 11.1, Definition 11.2. As before, the expansion $\theta$ of the Lagrange tensor A along c satisfies the vorticity free Raychaudhuri equation (11.2) for timelike geodesics. We now establish the following analogue for spacelike hypersurfaces of Proposition 11.9.

PROPOSITION 11.22 Let (M,g) be an arbitrary space-time of dimension $\geq 2$. Suppose that $c : J \to (M,g)$ is an inextendible timelike geodesic which satisfies $\mathrm{Ric}(c'(t),c'(t)) \geq 0$ for all $t \in J$ and is orthogonal to the spacelike hypersurface H at $q = c(t_1)$. If $-\mathrm{tr}(L_\eta)$ has the negative (resp., positive) value $\theta_1$ at q, then there is a focal point t to H on the interval from $t_1$ to $t_1 - (n - 1)/\theta_1$, provided that $t \in J$.

*Proof.* The second fundamental form $S_\eta$ at $\eta \in T^\perp H$ is sym-
E and $A'(t_1) = -L_{\eta(q)}$. Then $\theta_1 = \theta(t_1) = -\mathrm{tr}(L_\eta)$. Hence by Proposition 11.9 the tensor A is singular on the interval from $t_1$ to $t_1 - (n - 1)/\theta_1$. Thus the result follows from the remarks following Definition 11.20. □

For the purpose of studying focal points to submanifolds, it will be helpful to have the second variation formula for the arc length functional at hand. We thus give a derivation of the first and second variation formulas for completeness. Consider a piecewise smooth variation $\alpha : [a,b] \times (-\varepsilon,\varepsilon) \longrightarrow (M,g)$ of a piecewise smooth timelike curve $c : [a,b] \longrightarrow (M,g)$. Hence $\alpha(t,0) = c(t)$ for all $t \in [a,b]$ and there is a finite partition $a = t_0 < t_1 < \cdots < t_k = b$ such that $\alpha \mid [t_{i-1},t_i] \times (-\varepsilon,\varepsilon)$ is a smooth variation of $c \mid [t_{i-1},t_i]$ for each $i = 1, 2, \ldots, k$ (cf. Definition 9.6). We will also assume that the neighboring curves $\alpha_s = \alpha(-,s) : [a,b] \longrightarrow (M,g)$ are timelike for all $s$ with $-\varepsilon < s < \varepsilon$, (cf. Lemma 9.7). As in Section 9.1, for $t$ with $t_{i-1} \leqslant t < t_i$, we define the variation vector field $V$ of $\alpha$ along $c$ by

$$V(t) = (\alpha|[t_{i-1},t_i] \times (-\varepsilon,\varepsilon))_* \left.\frac{\partial}{\partial s}\right|_{(t,0)} = \frac{d}{ds}\left(\alpha(t,s)\right)\Big|_{s=0}$$

Also, set

$$\Delta_{t_i}(Y') = Y'(t_i^+) - Y'(t_i^-) \qquad \text{for } i = 1, \ldots, k - 1$$

and

$$\Delta_{t_k}(Y') = -Y'(t_k^-) \qquad \Delta_{t_0}(Y') = Y'(t_0)$$

for any piecewise smooth vector field $Y(t)$ along $c$ which is smooth on each subinterval $(t_{i-1},t_i)$.

As in Section 9.1, we will use $L(s) = L(\alpha_s)$ to denote the length of the curve $t \longmapsto \alpha(t,s)$. The first variation formula for the arc length functional may then be derived as usual.

PROPOSITION 11.23 Let $c : [a,b] \longrightarrow (M,g)$ be a unit speed timelike curve which is piecewise smooth. If $\alpha : [a,b] \times (-\varepsilon,\varepsilon) \longrightarrow (M,g)$ is a variation of $c$ through timelike curves, then

$$L'(0) = \int_a^b g(V,c'')\Big|_{(t,0)}\, dt + \sum_{i=0}^{k} g(V(t_i),\Delta_{t_i}(c'))$$

*Proof.* If $L_i : (-\varepsilon,\varepsilon) \longrightarrow \mathbb{R}$ denotes the arc length function of $\alpha \mid [t_{i-1},t_i] \times (-\varepsilon,\varepsilon)$, then $L(s) = \Sigma_{i=1}^{k} L_i(s)$ and

$$L_i(s) = \int_{t_{i-1}}^{t_i} \sqrt{-g\left(\alpha_* \frac{\partial}{\partial t}, \alpha_* \frac{\partial}{\partial t}\right)}\ dt$$

Thus

$$\frac{dL_i}{ds} = \int_{t_{i-1}}^{t_i} \frac{1}{2}\left[-g\left(\alpha_* \frac{\partial}{\partial t}, \alpha_* \frac{\partial}{\partial t}\right)\right]^{-1/2}\left[-2g\left(\nabla_{\partial/\partial t}\left(\alpha_* \frac{\partial}{\partial s}\right), \alpha_* \frac{\partial}{\partial t}\right)\right] dt \tag{11.6}$$

On the other hand, since

$$\frac{d}{dt}\, g\left(\alpha_* \frac{\partial}{\partial s}, \alpha_* \frac{\partial}{\partial t}\right) = g\left(\nabla_{\partial/\partial t}\, \alpha_* \frac{\partial}{\partial s}, \alpha_* \frac{\partial}{\partial t}\right) + g\left(\alpha_* \frac{\partial}{\partial s}, \nabla_{\partial/\partial t}\, \alpha_* \frac{\partial}{\partial t}\right)$$

and $g\left(\alpha_* \frac{\partial}{\partial t}, \alpha_* \frac{\partial}{\partial t}\right)\Big|_{(t,0)} = -1$, it follows that

$$\left.\frac{dL_i}{ds}\right|_{s=0} = \int_{t_{i-1}}^{t_i} g\left(\alpha_* \frac{\partial}{\partial s}, \nabla_{\partial/\partial t}\, \alpha_* \frac{\partial}{\partial t}\right) dt - \int_{t_{i-1}}^{t_i} \frac{d}{dt}\, g\left(\alpha_* \frac{\partial}{\partial s}, \alpha_* \frac{\partial}{\partial t}\right) dt$$

Thus we have

$$L_i'(0) = \int_{t_{i-1}}^{t_i} g(V,c'')\ dt - [g(V,c')]_{t_{i-1}^+}^{t_i^-}$$

which implies that

$$L'(0) = \int_a^b g(V,c'')\ dt + \sum_{i=0}^{k} g(V(t_i),\Delta_{t_i}(c'))$$

as required. □

If $c : [a,b] \longrightarrow M$ is a timelike geodesic, then $c'' = \nabla_{c'}c' = 0$ and $\Delta_{t_i}(c') = 0$ for all $i = 1, \ldots, k - 1$. Thus for a timelike geodesic, the first variation formula simplifies as follows.

COROLLARY 11.24 If $c : [a,b] \longrightarrow (M,g)$ is a unit speed timelike geodesic segment and $\alpha : [a,b] \times (-\varepsilon,\varepsilon) \longrightarrow (M,g)$ is a variation of c, then

$$L'(0) = -g(V,c')\Big|_a^b$$

Now let H be a spacelike hypersurface and assume that $c : [a,b] \longrightarrow M$ is a unit speed timelike curve with $c(a) \in H$. In studying focal points of the submanifold H, attention may be restricted to variations $\alpha : [a,b] \times (-\varepsilon,\varepsilon) \longrightarrow (M,g)$ of c which start on H [i.e., $\alpha(a,s) \in H$ for all $s \in (-\varepsilon,\varepsilon)$] and which end at c(b) [i.e., $\alpha(b,s) = c(b)$ for all $s \in (-\varepsilon,\varepsilon)$]. For these variations $V(b) = 0$ and V(a) is tangent to H at c(a). Proposition 11.23 then yields the following first variation formula:

$$L'(0) = \int_a^b g(V,c'')\Big|_{(t,0)} dt + \sum_{i=1}^{k-1} g(V(t_i),\Delta_{t_i}(c')) + g(V,c')\Big|_{(a,0)} \tag{11.7}$$

Given a spacelike hypersurface H without boundary in (M,g), fix $q \in M - H$. Consider the collection (possibly empty) of all timelike curves which join some point of H to q. If this collection contains a longest curve $c : [a,b] \longrightarrow (M,g)$, then c must be a smooth timelike geodesic by the usual arguments that timelike geodesics locally maximize arc length and using the first variational formula to see that c has no corners. Thus assume that the unit speed timelike geodesic $c : [a,b] \longrightarrow (M,g)$ is the longest curve from H to q. If $\alpha : [a,b] \times (-\varepsilon,\varepsilon) \longrightarrow (M,g)$ is a variation of c through timelike curves with $\alpha(a,s) \in H$ and $\alpha(b,s) = q$ for all s with $-\varepsilon \leqslant s \leqslant \varepsilon$, then using $V(b) = 0$ and Corollary 11.24 we have

$$L'(0) = g(V(a),c'(a))$$

On the other hand, $L'(0) = 0$ since c is of maximal length from H to q. Thus V(a) is orthogonal to c'(a). Since variations $\alpha$ as above may be constructed with $V(a) \in T_{c(a)}H$ arbitrary, it follows that c must be orthogonal to H at c(a). Thus we have obtained the following standard result. Note also that it is necessary for H to have no boundary in order for the extremal c to be perpendicular to H at c(a).

PROPOSITION 11.25 Let H be a spacelike hypersurface without boundary in (M,g) and assume that $c : [a,b] \longrightarrow (M,g)$ is a timelike curve from H to the point $q = c(b) \notin H$ which is of maximal length among all timelike curves from H to q. Then c is a timelike geodesic segment and c is orthogonal to H at c(a).

The variation vector field $V(t) = \alpha_*(\partial/\partial s)\big|_{(t,0)}$ of a variation $\alpha$ along c may have discontinuities in its derivative at the t-parameter values $\{t_1, t_2, \ldots, t_{k-1}\}$ at which $\alpha$ may fail to be smooth. Thus the normal component $N = \alpha_*\partial/\partial s + g(\alpha_*\partial/\partial s, \alpha_*\partial/\partial t)\alpha_*\partial/\partial t$ of V along c may also fail to be smooth at these parameter values. We now derive the second variation formula for L''(0) in terms of N and V [cf. Bölts (1977, pp. 86-86), Hawking and Ellis (1973, p. 108)].

PROPOSITION 11.26 Let $c : [a,b] \longrightarrow (M,g)$ be a unit speed timelike geodesic segment and let $\alpha : [a,b] \times (-\varepsilon,\varepsilon) \longrightarrow (M,g)$ be a piecewise smooth variation of c which is smooth on each set $(t_{i-1},t_i) \times (-\varepsilon,\varepsilon)$ for the partition $a = t_0 < t_1 < \cdots < t_k = b$ of [a,b]. Let $V(t) = \alpha_*\partial/\partial s\big|_{(t,0)}$ denote the variation vector field of $\alpha$ along c and set

$$\begin{aligned} N(t) &= V(t) + g(V(t),c'(t))c'(t) \\ &= \alpha_* \frac{\partial}{\partial s}\bigg|_{(t,0)} + g\left(\alpha_* \frac{\partial}{\partial s}, \alpha_* \frac{\partial}{\partial t}\right)\alpha_* \frac{\partial}{\partial t}\bigg|_{(t,0)} \end{aligned}$$

Then

$$\begin{aligned} L''(0) = &\int_a^b g(N'' + R(V,c')c', N)\big|_{(t,0)}\, dt \\ &+ \sum_{i=0}^{k} g(N(t_i),\Delta_{t_i}(N')) - g(\nabla_{\partial/\partial s}V,c')\Big|_b^a \end{aligned}$$

*Proof*. Setting $L_i = L \mid [t_{i-1},t_i]$ and recalling that

$$\frac{dL_i}{ds} = \int_{t_{i-1}}^{t_i} \left[-g\left(\alpha_* \frac{\partial}{\partial t}, \alpha_* \frac{\partial}{\partial t}\right)\right]^{-1/2} \left[-g\left(\nabla_{\partial/\partial t}\left(\alpha_* \frac{\partial}{\partial s}\right), \alpha_* \frac{\partial}{\partial t}\right)\right] dt$$

we obtain

$$\frac{d^2L_i}{ds^2} = \int_{t_{i-1}}^{t_i} \frac{d}{ds} \left\{\left[-g\left(\alpha_* \frac{\partial}{\partial t}, \alpha_* \frac{\partial}{\partial t}\right)\right]^{-1/2} \left[-g\left(\nabla_{\partial/\partial t}\alpha_* \frac{\partial}{\partial s}, \alpha_* \frac{\partial}{\partial t}\right)\right]\right\} dt$$

Differentiating the expression under the integral sign and using the identity $\nabla_{\partial/\partial s}\alpha_* \frac{\partial}{\partial t} - \nabla_{\partial/\partial t}\alpha_* \frac{\partial}{\partial s} = \alpha_*\left[\frac{\partial}{\partial s}, \frac{\partial}{\partial t}\right] = 0$ yields the integrand

$$\frac{-g(\nabla_{\partial/\partial s}\nabla_{\partial/\partial t}\alpha_*\partial/\partial s,\alpha_*\partial/\partial t) - g(\nabla_{\partial/\partial t}\alpha_*\partial/\partial s,\nabla_{\partial/\partial s}\alpha_*\partial/\partial t)}{[-g(\alpha_*\partial/\partial t,\alpha_*\partial/\partial t)]^{1/2}}$$

$$+ \frac{g(\nabla_{\partial/\partial t}\alpha_*\partial/\partial s,\alpha_*\partial/\partial t)g(\nabla_{\partial/\partial t}\alpha_*\partial/\partial s,\alpha_*\partial/\partial t)}{[-g(\alpha_*\partial/\partial t,\alpha_*\partial/\partial t)]^{1/2}g(\alpha_*\partial/\partial t,\alpha_*\partial/\partial t)} \tag{11.8}$$

for $(t,s) \in (t_{i-1},t_i) \times (-\varepsilon,\varepsilon)$. Furthermore, we have

$$\begin{aligned}\nabla_{\partial/\partial t}\, g\left(\alpha_* \frac{\partial}{\partial s}, \alpha_* \frac{\partial}{\partial t}\right)\alpha_* \frac{\partial}{\partial t}\Bigg|_{(t,0)} &= \frac{d}{dt}\, g\left(\alpha_* \frac{\partial}{\partial s}, \alpha_* \frac{\partial}{\partial t}\right)\alpha_* \frac{\partial}{\partial t}\Bigg|_{(t,0)} \\ &\quad + g\left(\alpha_* \frac{\partial}{\partial s}, \alpha_* \frac{\partial}{\partial t}\right)\nabla_{\partial/\partial t}\alpha_* \frac{\partial}{\partial t}\Bigg|_{(t,0)} \\ &= \frac{d}{dt}\, g\left(\alpha_* \frac{\partial}{\partial s}, \alpha_* \frac{\partial}{\partial t}\right)\alpha_* \frac{\partial}{\partial t}\Bigg|_{(t,0)}\end{aligned}$$

since $\nabla_{\partial/\partial t}\alpha_*\partial/\partial t\big|_{(t,0)} = \nabla_{\partial/\partial t}c'(t) = 0$ as c is a smooth geodesic. Also

$$\begin{aligned}g\left(\nabla_{\partial/\partial t}N, \alpha_* \frac{\partial}{\partial t}\right)\Bigg|_{(t,0)} &= \frac{d}{dt}\left(g\left(N, \alpha_* \frac{\partial}{\partial t}\right)\right)\Bigg|_{(t,0)} \\ &\quad - g\left(N, \nabla_{\partial/\partial t}\, \alpha_* \frac{\partial}{\partial t}\right)\Bigg|_{(t,0)} = 0\end{aligned}$$

since c is a geodesic and $g(N, \alpha_*\, \partial/\partial t)\big|_{(t,0)} = 0$ for all $t \in (t_{i-1},t_i)$. Using these last two calculations, we then obtain

$$g\left(\nabla_{\partial/\partial t}N,\ \nabla_{\partial/\partial t}\left[g\left(\alpha_*\frac{\partial}{\partial s},\ \alpha_*\frac{\partial}{\partial t}\right)\alpha_*\frac{\partial}{\partial t}\right]\right)\Bigg|_{(t,0)}$$

$$= \frac{d}{dt}\left(g\left(\alpha_*\frac{\partial}{\partial s},\ \alpha_*\frac{\partial}{\partial t}\right)\right)\Bigg|_{(t,0)} g\left(\nabla_{\partial/\partial t}N,\ \alpha_*\frac{\partial}{\partial t}\right)\Bigg|_{(t,0)} = 0$$

Thus since

$$\alpha_*\frac{\partial}{\partial s}\Bigg|_{(t,0)} = N(t) - g\left(\alpha_*\frac{\partial}{\partial s},\ \alpha_*\frac{\partial}{\partial t}\right)\alpha_*\frac{\partial}{\partial t}\Bigg|_{(t,0)}$$

it follows from these calculations that

$$g\left(\nabla_{\partial/\partial t}\alpha_*\frac{\partial}{\partial s}, \nabla_{\partial/\partial s}\alpha_*\frac{\partial}{\partial t}\right)\Bigg|_{(t,0)}$$

$$= g\left(\nabla_{\partial/\partial t}\ \alpha_*\frac{\partial}{\partial s},\ \nabla_{\partial/\partial t}\ \alpha_*\frac{\partial}{\partial s}\right)\Bigg|_{(t,0)}$$

$$= g\left(\nabla_{\partial/\partial t}N,\ \nabla_{\partial/\partial t}N\right)\Bigg|_{(t,0)} - 2g\left(\nabla_{\partial/\partial t}N,\ \nabla_{\partial/\partial t}\left[g\left(\alpha_*\frac{\partial}{\partial s},\ \alpha_*\frac{\partial}{\partial t}\right)\alpha_*\frac{\partial}{\partial t}\right]\right)\Bigg|_{(t,0)}$$

$$+\ g\left(\nabla_{\partial/\partial t}\left[g\left(\alpha_*\frac{\partial}{\partial s},\ \alpha_*\frac{\partial}{\partial t}\right)\alpha_*\frac{\partial}{\partial t}\right],\ \nabla_{\partial/\partial t}\left[g\left(\alpha_*\frac{\partial}{\partial s},\ \alpha_*\frac{\partial}{\partial t}\right)\alpha_*\frac{\partial}{\partial t}\right]\right)\Bigg|_{(t,0)}$$

$$= g\left(\nabla_{\partial/\partial t}N,\ \nabla_{\partial/\partial t}N\right)\Bigg|_{(t,0)} + \left[\frac{d}{dt}\ g\left(\alpha_*\frac{\partial}{\partial s},\ \alpha_*\frac{\partial}{\partial t}\right)\right]^2 g\left(\alpha_*\frac{\partial}{\partial t},\ \alpha_*\frac{\partial}{\partial t}\right)\Bigg]^2\Bigg|_{(t,0)}$$

$$= g\left(\nabla_{\partial/\partial t}N,\ \nabla_{\partial/\partial t}N\right)\Bigg|_{(t,0)} - \left[g\left(\nabla_{\partial/\partial t}\left(\alpha_*\frac{\partial}{\partial s}\right),\ \alpha_*\frac{\partial}{\partial t}\right)\right]^2\Bigg|_{(t,0)}$$

Substituting this expression for $g(\nabla_{\partial/\partial t}\alpha_*\partial/\partial s, \nabla_{\partial/\partial s}\alpha_*\partial/\partial t)\big|_{(t,0)}$ in (11.8) and using $g(\alpha_*\partial/\partial t, \alpha_*\partial/\partial t)\big|_{(t,0)} = -1$, we obtain

$$\frac{d}{ds}\left\{\left[-g\left(\alpha_*\frac{\partial}{\partial t},\ \alpha_*\frac{\partial}{\partial t}\right)\right]^{-1/2}\left[-g\left(\nabla_{\partial/\partial t}\alpha_*\frac{\partial}{\partial s},\ \alpha_*\frac{\partial}{\partial t}\right)\right]\right\}\Bigg|_{(t,0)}$$

$$= -g\left(\nabla_{\partial/\partial s}\nabla_{\partial/\partial t}\ \alpha_*\frac{\partial}{\partial s},\ \alpha_*\frac{\partial}{\partial t}\right)\Bigg|_{(t,0)} - g\left(\nabla_{\partial/\partial t}N,\ \nabla_{\partial/\partial t}N\right)\Bigg|_{(t,0)}$$

This yields

$$L_i''(0) = \int_{t_{i-1}}^{t_i}\left[-g\left(\nabla_{\partial/\partial s}\nabla_{\partial/\partial t}\ \alpha_*\frac{\partial}{\partial s},\ \alpha_*\frac{\partial}{\partial t}\right)\Bigg|_{(t,0)} - g\left(\nabla_{\partial/\partial t}N,\ \nabla_{\partial/\partial t}N\right)\Bigg|_{(t,0)}\right] dt$$

Also

$$R\left(\alpha_* \frac{\partial}{\partial t}, \alpha_* \frac{\partial}{\partial s}\right)\alpha_* \frac{\partial}{\partial s} = \nabla_{\partial/\partial t}\nabla_{\partial/\partial s}\left(\alpha_* \frac{\partial}{\partial s}\right) - \nabla_{\partial/\partial s}\nabla_{\partial/\partial t}\left(\alpha_* \frac{\partial}{\partial s}\right)$$

since

$$\left[\alpha_* \frac{\partial}{\partial t}, \alpha_* \frac{\partial}{\partial s}\right] = \alpha_* \left[\frac{\partial}{\partial t}, \frac{\partial}{\partial s}\right] = 0$$

Since $c(t) = \alpha(t,0)$ is a geodesic,

$$g\left(\nabla_{\partial/\partial t}\nabla_{\partial/\partial s} \alpha_* \frac{\partial}{\partial s}, \alpha_* \frac{\partial}{\partial t}\right)\bigg|_{(t,0)} = \frac{d}{dt} g\left(\nabla_{\partial/\partial s} \alpha_* \frac{\partial}{\partial s}, \alpha_* \frac{\partial}{\partial t}\right)\bigg|_{(t,0)}$$

Therefore,

$$\begin{aligned} L_i''(0) = \int_{t_{i-1}}^{t_i} & \left[ g\left(R\left(\alpha_* \frac{\partial}{\partial t}, \alpha_* \frac{\partial}{\partial s}\right)\alpha_* \frac{\partial}{\partial s}, \alpha_* \frac{\partial}{\partial t}\right)\bigg|_{(t,0)} \right. \\ & \left. - g\left(\nabla_{\partial/\partial t}N, \nabla_{\partial/\partial t}N\right)\bigg|_{(t,0)}\right] dt \\ & - g\left(\nabla_{\partial/\partial s} \alpha_* \frac{\partial}{\partial s}, \alpha_* \frac{\partial}{\partial t}\right)\bigg|_{t_{i-1}}^{t_i} \end{aligned}$$

The equation

$$\begin{aligned} -g(\nabla_{\partial/\partial t}N, \nabla_{\partial/\partial t}N)\Big|_{(t,0)} &= -\frac{d}{dt} g(N, \nabla_{\partial/\partial t}N)\Big|_{(t,0)} \\ &\quad + g(N, \nabla_{\partial/\partial t}\nabla_{\partial/\partial t}N)\Big|_{(t,0)} \end{aligned}$$

and $V(t) = \alpha_*\partial/\partial s\big|_{(t,0)}$ together with $L = \Sigma_{i=1}^{k} L_i$ yield

$$\begin{aligned} L''(0) &= \int_a^b [g(R(c',V)V,c')\big|_{(t,0)} + g(N,N'')\big|_{(t,0)}]\, dt \\ &\quad - \sum_{i=1}^{k} g(N,N')\bigg|_{t_{i-1}^+}^{t_i^-} - g(\nabla_{\partial/\partial s}V, c')\bigg|_a^b \\ &= \int_a^b [g(R(V,c')c',N - g(V,c')c')\big|_t + g(N,N'')\big|_t]\, dt \\ &\quad + \sum_{i=0}^{k} g(N,\Delta_{t_i}(N')) - g(\nabla_{\partial/\partial s}V, c')\bigg|_a^b \end{aligned}$$

since $V = N - g(V,c')c'$. Thus

$$L''(0) = \int_a^b g(N'' + R(V,c')c', N)\Big|_t \, dt + \sum_{i=0}^{k} g(N(t_i),\Delta_{t_i}(N')) - g(\nabla_{\partial/\partial s}V,c')\Big|_a^b$$

where

$$\nabla_{\partial/\partial s}V\Big|_{(t,0)} = \nabla_{\partial/\partial s}\,\alpha_*\partial/\partial s\Big|_{(t,0)}$$ as required. □

Proposition 11.26 has the following consequence.

COROLLARY 11.27 Let H be a spacelike hypersurface and assume that $c : [a,b] \longrightarrow (M,g)$ is a unit speed timelike geodesic which is orthogonal to H at $p = c(a) \in H$. Suppose that $\alpha : [a,b] \times (-\varepsilon,\varepsilon) \longrightarrow (M,g)$ is a variation of c such that $\alpha(a,s) \in H$ and $\alpha(b,s) = q = c(b)$ for all s with $-\varepsilon \leqslant s \leqslant \varepsilon$. If $V = \alpha_*\partial/\partial s\big|_{(t,0)}$ and $N = V + g(V,c')c'$ then

$$L''(0) = \int_a^b g(N'' + R(V,c')c', N)\Big|_t \, dt + \sum_{i=1}^{k-1} g(N(t_i),\Delta_{t_i}(N')) + g(N,N')\Big|_p + g(L_{c'}(N),N)\Big|_p$$

where $L_{c'}$ is the second fundamental form operator of H at p.

*Proof.* In view of Proposition 11.26 and the equation

$$\sum_{i=0}^{k} g(N(t_i),\Delta_{t_i}(N')) = \sum_{i=1}^{k-1} g(N(t_i),\Delta_{t_i}(N')) + g(N,N')\Big|_p$$

it is only necessary to show that $-g(\nabla_{\partial/\partial s}V,c')\big|_a^b$ is equal to $g(L_{c'}(N),N)$. To this end, we first note that $\alpha(b,s) = q$ implies that $\alpha_*\partial/\partial s\big|_{(b,s)} = 0$ for all s which yields $g(\nabla_{\partial/\partial s}V,c')\big|_b = 0$. Also $\alpha(a,s) \in H$ for all s with $-\varepsilon < s < \varepsilon$ implies that $\alpha_*\partial/\partial s\big|_{(a,s)}$ is tangential to H for all s and hence $N(a) = \alpha_*\partial/\partial s\big|_{(a,0)}$. Let $\gamma(s) = \alpha(a,s)$ for all s with $-\varepsilon < s < \varepsilon$. Extend the vector

$N(a) \in T_pH$ to a local vector field $X$ along $H$ with $X \circ \gamma(s) = \alpha_* \partial/\partial s\big|_{(a,s)}$ for all $s$ with $-\varepsilon < s < \varepsilon$. Then

$$g(L_{c'(a)}(N),N) = g(\nabla_{X(a)}X, c'(a))$$

by Definition 2.35. Also let $\eta$ be a unit normal field to $H$ near $p$ with $\eta(p) = c'(a)$. Then we have

$$\begin{aligned} g(\nabla_{\partial/\partial s}V, c'(a)) &= g\left(\nabla_{\partial/\partial s}\alpha_* \frac{\partial}{\partial s}, \eta \circ \alpha\right)\bigg|_{(a,0)} \\ &= \frac{d}{ds}\left(g\left(\alpha_* \frac{\partial}{\partial s}, \eta \circ \alpha\right)\right)\bigg|_{(a,0)} - g\left(\alpha_* \frac{\partial}{\partial s}, \nabla_{\partial/\partial s}\eta \circ \alpha\right)\bigg|_{(a,0)} \end{aligned}$$

But $g(\alpha_*\partial/\partial s, \eta \circ \alpha)\big|_{(a,s)} = 0$ for all $s$ since $\alpha_*\partial/\partial s\big|_{(a,s)}$ is tangential to $H$. Thus we obtain

$$\begin{aligned} g(\nabla_{\partial/\partial s}V, c'(a)) &= -g\left(\alpha_* \frac{\partial}{\partial s}, \nabla_{\partial/\partial s}\eta \circ \alpha\right)\bigg|_{(a,0)} \\ &= -g(N(a), \nabla_{N(a)}\eta) \\ &= -g(X, \nabla_X \eta)\big|_{c(a)} \\ &= -X\big|_p(g(X,\eta)) + g(\nabla_X X\big|_p, \eta(a)) \\ &= 0 + g(\nabla_X X\big|_p, c'(a)) \\ &= g(L_{c'(a)}(N),N) \end{aligned}$$

as required. Here we have used the fact that $X\big|_p(g(X,\eta))$ may be calculated as

$$X\big|_p(g(X,\eta)) = \frac{d}{ds}(g(X \circ \gamma(s), \eta(s)))\Big|_{s=0} = \frac{d}{ds}(0) = 0$$

since $X\big|_p = \alpha_*\partial/\partial s\big|_{(a,0)}$. □

In view of Corollary 11.27, the index $I_H(V,V)$ of a vector field $V$ along a timelike geodesic orthogonal to a spacelike hypersurface should be defined as follows [cf. Bölts (1977, p. 94)].

DEFINITION 11.28 Let $c : [a,b] \longrightarrow (M,g)$ be a unit speed timelike geodesic which is orthogonal to a spacelike hypersurface $H$ at $c(a)$. Assume that $Z$ is a piecewise smooth vector field along $c$ which is orthogonal to c. If $Z(a) \neq 0$ and $Z(b) = 0$, then the index of Z with respect to H is given by

$$I_H(Z,Z) = I(Z,Z) + g(L_{c'}(Z),Z)\Big|_a$$

where

$$I(Z,Z) = \int_a^b g(Z'' + R(Z,c')c',Z)\Big|_t \, dt + \sum_{i=0}^{k-1} g(Z(t_i),\Delta_{t_i}(Z'))$$

Let $\alpha : [a,b] \times (-\varepsilon,\varepsilon) \longrightarrow (M,g)$ be a variation of the timelike geodesic c and assume the variation vector field $V = \alpha_* \partial/\partial s\big|_{(t,0)}$ satisfies the conditions of Definition 11.28. Then equation (9.4) of Section 9.1 together with Corollary 11.27 yield $L''(0) = I_H(V,V)$. This may also be written

$$L_{**}(V) = I_H(V,V)$$

Using the index form $I_H$, it may now be shown that a timelike geodesic orthogonal to a spacelike hypersurface H fails to maximize arc length to H after the first focal point. Recall that $V^\perp(c)$ consists of piecewise smooth vector fields along c which are orthogonal to c.

PROPOSITION 11.29 Let $c : [a,b] \longrightarrow M$ be a unit speed timelike geodesic orthogonal to a spacelike hypersurface H (without boundary) at a point $p = c(a) \in H$. If for some $t_k \in (a,b)$ the point $c(t_k)$ is a focal point to H along c, then there is a variation vector field $Z \in V^\perp(c)$ with Z(a) tangential to H and $Z(b) = 0$, such that $I_H(Z,Z) > 0$. Consequently, there are timelike curves from H to c(b) which are longer than c.

*Proof.* By hypothesis there exists a nontrivial Jacobi field $J_1$ along c with $J_1$ orthogonal to c, $J_1(t_k) = 0$ and $J_1'(a) = -L_{c'(a)}J_1(a)$. Define a piecewise smooth Jacobi field J along c by

$$J(t) = \begin{cases} J_1(t) & \text{if } a \leq t \leq t_k \\ 0 & \text{if } t_k \leq t \leq b \end{cases}$$

Since $\Delta_{t_k}(J') \neq 0$ we may construct a smooth vector field V orthogonal to c such that $V'(a) = V(a) = V(b) = 0$ and $g(V(t_k), \Delta_{t_k}(J')) = -1$. Define a vector field Z in $V^{\perp}(c)$ by

$$Z = \frac{1}{r} J - rV \qquad \text{for } r \in \mathbb{R} - \{0\}$$

Then $Z'(a) = -L_{c'(a)} Z(a)$ and the index $I_H(Z,Z)$ is given by

$$\begin{aligned}
I_H(Z,Z) &= I(Z,Z) + g(L_{c'}(Z),Z)\big|_a \\
&= I(Z,Z) + g(L_{c'}(r^{-1}J - rV), r^{-1}J - rV)\big|_a \\
&= I(Z,Z) + r^{-2} g(L_{c'}J,J)\big|_a \\
&= I(Z,Z) + r^{-2} g(-J',J)\big|_a \\
&= r^{-2} I(J,J) + r^2 I(V,V) - 2I(J,V) + r^{-2} g(-J',J)\big|_a \\
&= r^2 I(V,V) - 2I(J,V)
\end{aligned}$$

Here we have used

$$I(J,J) = \sum_{i=0}^{k-1} g(\Delta_{t_i}(J'), J(t_i)) = g(J',J)\big|_a$$

Since J is a piecewise smooth Jacobi field, equation (9.4) of Section 9.1 implies that

$$I(J,V) = g(V(t_k), \Delta_{t_k}(J')) = -1$$

Consequently,

$$I_H(Z,Z) = r^2 I(V,V) + 2$$

which shows that for sufficiently small $r \neq 0$ the index satisfies

$$I_H(Z,Z) > 0$$

This last inequality and the condition $Z'(a) = -L_{c'(a)}(Z(a))$ imply that there exist small variations of c with variation vector field Z which join H to c(b) and have length greater than c. □

We now turn our attention to focal points along null geodesics which are orthogonal to (n - 2)-dimensional spacelike submanifolds. If H is a spacelike (n - 2)-dimensional submanifold, then the induced metric on $T_pH$ is positive definite and the induced metric on $T_p^{\perp}H$ is a two-dimensional Minkowski metric for each $p \in H$. Thus there are exactly two null lines through the origin in $T_p^{\perp}H$. Since the time orientation of (M,g) induces a time orientation on $T_p^{\perp}H$, there are thus two well-defined future null directions in $T_p^{\perp}H$ for each $p \in H$. Locally we may then choose a smooth pseudo-orthonormal basis of vector fields $E_1, \cdots, E_{n-1}, E_n$ on H such that $E_{n-1}$ and $E_n$ are future-directed null vectors in $T_p^{\perp}H$ for $p \in H$. That is,

$$g(E_i,E_j) = \delta_{ij} \qquad \text{if } 1 \leqslant i, j \leqslant n - 2$$

$$g(E_i,E_{n-1}) = g(E_i,E_n) = 0 \qquad \text{if } 1 \leqslant i \leqslant n - 2$$

$$g(E_n,E_n) = g(E_{n-1},E_{n-1}) = 0$$

and

$$g(E_n,E_{n-1}) = -1$$

The null vector fields $E_{n-1}$ and $E_n$ defined locally on H give rise to second fundamental forms $L_{E_{n-1}}$ and $L_{E_n}$, respectively, which are locally defined (1,1) tensor fields on H.

If $\beta : [a,b] \longrightarrow (M,g)$ is a future-directed null geodesic with $\beta'(a) = E_n(p)$ [or $\beta'(a) = E_{n-1}(p)$] at $p \in H$, the tangent vectors $E_1(p), \ldots, E_n(p)$ at p may then be parallel translated along β to give a pseudo-orthonormal basis along β which will also be denoted by $E_1, \ldots, E_n$. The set of vectors normal to $\beta'(t)$ is thus the space $N(\beta(t))$ spanned by $E_1, \ldots, E_{n-2}, \beta'$, and we may form the quotient space $G(\beta(t)) = N(\beta(t))/[\beta'(t)]$ with corresponding quotient bundle $G(\beta)$ as in Section 9.3. If $\pi : N(\beta) \longrightarrow G(\beta)$ denotes the projection map, then $\pi \mid T_pH : T_pH \longrightarrow G(\beta(a))$ is a vector space isomorphism. Hence we may project the second fundamental form operator $L_{E_i} : T_pH \longrightarrow T_pH$ to an operator $\overline{L}_{E_i} : G(\beta(a)) \longrightarrow G(\beta(a))$ by setting

$$\overline{L}_{E_i} = \pi \circ L_{E_i} \circ (\pi|T_pH)^{-1} \qquad \text{for } i = n-1, n$$

Let $\alpha : [a,b] \times (-\varepsilon,\varepsilon) \longrightarrow (M,g)$ be a variation of the null geodesic $\beta$ such that $\alpha(a,s) \in H$ for all s with $-\varepsilon < s < \varepsilon$. If $V = \alpha_*\partial/\partial s$ and $W = \alpha_*\partial/\partial t$, we will also require that $W(a,s) = E_n(\alpha(a,s))$ for all s with $-\varepsilon < s < \varepsilon$. Thus the neighboring curve $\alpha(-,s)$ starts at $\alpha(a,s)$ on H perpendicular to H and has as initial direction the null vector $E_n(\alpha(a,s))$ for all s.

We now calculate $V'(a)$.

LEMMA 11.30 Let $V = \alpha_*\partial/\partial s$ be tangential to H for t = a and s arbitrary as above. Then $V'(a) = -L_{\beta'(a)}(V(a)) + \lambda\beta'(a)$ for some constant $\lambda \in \mathbb{R}$.

*Proof.* Using $[V,W] = [\alpha_*\partial/\partial s, \alpha_*\partial/\partial t] = \alpha_*[\partial/\partial s, \partial/\partial t] = 0$, we have $\nabla_V W = \nabla_W V = \nabla_{\beta'} V$ at $t = a$, $s = 0$.

On the other hand, $W(a,s) = E_n(\alpha(a,s))$ and $g(E_n,E_n) = 0$ yield $0 = V(g(W,W)) = 2g(\nabla_V W,W) = 2g(\nabla_W V,W) = 2g(\nabla_{\beta'}V,\beta')$ at $t = a$, $s = 0$. Thus

$$\nabla_{\beta'(a)}V \in N(\beta(a)) = T_pH \oplus [\beta'(a)]$$

Since $L_{\beta'(a)}(V(a)) \in T_{\beta(a)}H$ and $\nabla_{\beta'(a)}V \in N(\beta(a))$, it thus suffices to show that

$$g(L_{\beta'(a)}(V(a)),y) = -g(\nabla_{\beta'(a)}V,y)$$

for any $y \in T_{\beta(a)}H$ to establish the result. To calculate $g(L_{\beta'(a)}(V(a)),y)$, extend y to a local vector field Y along the curve $s \longrightarrow \alpha(a,s)$ that is tangent to H. Then we have

$$\begin{aligned} g(L_{\beta'(a)}(V(a)),y) &= g(\nabla_V Y|_{(a,0)},\beta'(a)) \\ &= g(\nabla_V Y,W)\Big|_{(a,0)} \\ &= \alpha_*\frac{\partial}{\partial s}\Big|_{(a,0)}(g(Y,W)) - g(Y,\nabla_V W)\Big|_{(a,0)} \\ &= 0 - g(Y,\nabla_W V)\Big|_{(a,0)} \\ &= -g(y,\nabla_{\beta'(a)}V) \end{aligned}$$

where we have used $\alpha_* \partial/\partial s\big|_{(a,0)}(g(Y,W)) = 0$ since $g(Y,W) = g(Y,E_n) = 0$ along the curve $s \longrightarrow \alpha(a,s)$. □

If V is to be a Jacobi field measuring the separation of a congruence of null geodesics perpendicular to H, then the last result implies that the vector class $\overline{V} = \pi(V)$ along β should satisfy the initial condition

$$\overline{V}'(a) = -\overline{L}_{\beta'(a)}\overline{V}(a)$$

at $p = \beta(a) \in H$. Here $\overline{L}_{\beta'} = \pi \circ L_{\beta'} \circ (\pi|T_pH)^{-1}$ as above. This motivates the following definition of focal point along a null geodesic perpendicular to H [cf. Hawking and Ellis (1973, p. 102), Bölts (1977, p. 56)].

DEFINITION 11.31 Let H be a spacelike submanifold of dimension (n - 2) and suppose that $\beta : [a,b] \longrightarrow (M,g)$ is a null geodesic orthogonal to H at $p = \beta(a)$. Then $t_0 \in (a,b]$ is said to be a *focal point* of H along β if there is a smooth Jacobi class $\overline{J}$ in $G(\beta)$ such that $\overline{J}'(a) = -\overline{L}_{\beta'(a)}\overline{J}(a)$ and $\overline{J}(t_0) = 0$.

We noted above that a timelike geodesic orthogonal to a spacelike hypersurface fails to be the longest nonspacelike curve to the hypersurface after the first focal point (cf. Proposition 11.29). A similar result holds for a null geodesic which is orthogonal to an (n - 2)-dimensional spacelike submanifold [cf. Hawking and Ellis (1973, p. 116), Bölts (1977, p. 123)]. We state this result as the following proposition.

PROPOSITION 11.32 Let H be a spacelike submanifold of (M,g) of dimension n - 2 and let $\beta : [a,b] \longrightarrow (M,g)$ be a null geodesic orthogonal at β(a) to H. If $t_0 \in (a,b)$ is a focal point of H along β, then there is a timelike curve from H to β(b). Thus β does not maximize the distance to H after the first focal point.

A simple example of a focal point is shown in Figure 11.3.

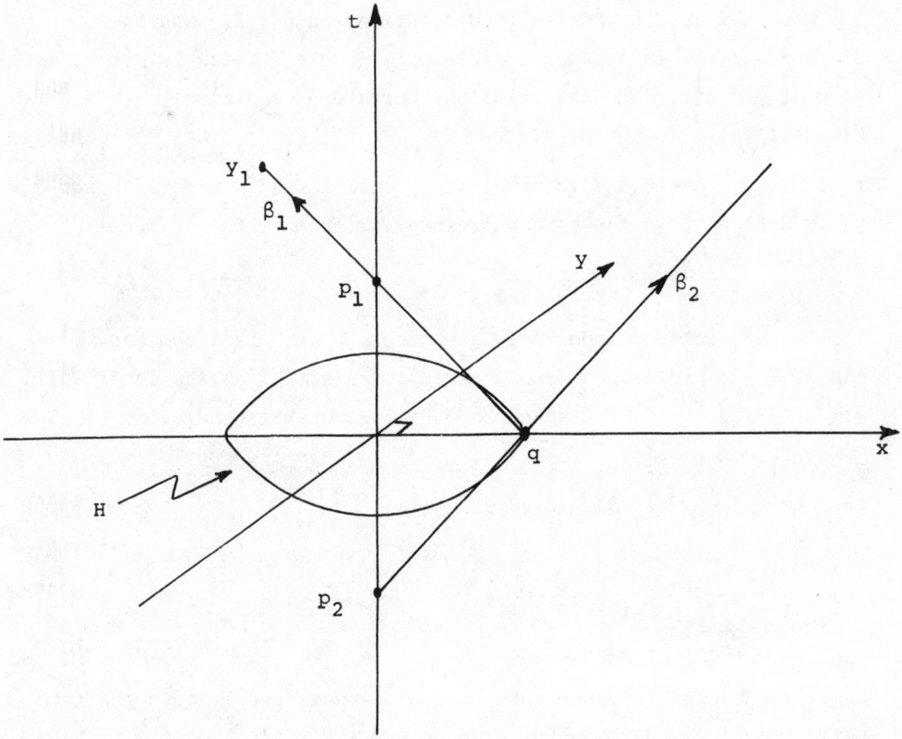


*Figure 11.3* Let (M,g) be three-dimensional Minkowski space-time with the usual metric $ds^2 = -dt^2 + dx^2 + dy^2$. Let H be a circle of radius a in the xy plane. Then H is a spacelike submanifold of codimension two. For each $q \in H$, there are exactly two future directed null geodesics $\beta_1$ and $\beta_2$ through q which are orthogonal to H. The focal point of H along $\beta_1$ is $p_1 = (0,0,a)$ and the focal point along $\beta_2$ is $p_2 = (0,0,-a)$. The point $y_1$ lies on $\beta_1$ beyond $p_1$. All of H - {q} is contained in the chronological past of $y_1$. Thus there are timelike curves from points on H arbitrarily close to q to $y_1$ and $\beta_1[q,y_1]$ does not realize the distance from H to $y_1$.

Using the same type of reasoning as in Proposition 11.22, the following result may also be established.

PROPOSITION 11.33 Let $(M,g)$ be a space-time of dimension $n \geq 3$ and let $H$ be a spacelike submanifold of dimension $n - 2$. Suppose that $\beta : J \longrightarrow (M,g)$ is an inextendible null geodesic which is orthogonal to $H$ at $p = \beta(t_1)$ and satisfies the curvature condition $\mathrm{Ric}(\beta'(t),\beta'(t)) \geq 0$ for all $t \in J$. If $-\mathrm{tr}(L_{\beta'(t_1)})$ has the negative (resp., positive) value $\theta_1$ at $p$, then there is a focal point $t_0$ to $H$ along $\beta$ in the parameter interval from $t_1$ to $t_1 - (n - 2)/\theta_1$ provided that $t_0 \in J$.

A particularly important case occurs when $H$ is a compact spacelike $(n - 2)$-dimensional submanifold which satisfies the condition $(\mathrm{tr}\, L_{E_n})\cdot(\mathrm{tr}\, L_{E_{n-1}}) > 0$ at each point [cf. Hawking and Ellis (1973, p. 262)]. Here if $\{e_1,\ldots,e_{n-2}\}$ is an orthonormal basis for $T_pH$, then $\mathrm{tr}\, L_{E_n}$ may be calculated as

$$\mathrm{tr}\, L_{E_n} = \sum_{i=1}^{n-2} g(L_{E_n}(e_i),e_i)$$

DEFINITION 11.34 Suppose that $H$ is a compact spacelike submanifold without boundary of $(M,g)$ of dimension $n - 2$. Let $E_n$ and $E_{n-1}$ be linearly indepedent future-directed null vector fields on $H$ as above. Assume that $L_1$ and $L_2$ are the second fundamental forms on $H$ corresponding to $E_n$ and $E_{n-1}$, respectively. Then $H$ is said to be a *closed trapped surface* if $\mathrm{tr}\, L_1$ and $\mathrm{tr}\, L_2$ are both either always positive or always negative on $H$.

A related concept is that of a trapped set [cf. Hawking and Penrose (1970, pp. 534-537)]. Recall that the future horismos of a set $A$ is defined by $E^+(A) = J^+(A) - I^+(A)$.

DEFINITION 11.35 A nonempty achronal set A is said to be *future* (resp., *past*) *trapped* if $E^+(A)$ [resp., $E^-(A)$] is compact. A *trapped set* is a set which is either future trapped or past trapped.

In general, a closed trapped surface need not be a trapped set and vice versa. However, in Proposition 11.45 we will show that under certain conditions, the existence of a closed trapped surface implies either null incompleteness or the existence of a trapped set. In establishing Proposition 11.45, we will need the following corollary to Proposition 11.33.

COROLLARY 11.36 Let (M,g) be a space-time of dimension $\geqslant 3$ which satisfies the condition $\text{Ric}(v,v) \geqslant 0$ for all null vectors $v \in TM$. If (M,g) contains a closed trapped surface H, then either (1) or (2) or both is true:

(1) At least one of the sets $E^+(H)$ or $E^-(H)$ is compact.

(2) (M,g) is null incomplete.

*Proof.* Assume that (M,g) is null complete and that $\text{tr } L_1 > 0$ and $\text{tr } L_2 > 0$ for all $q \in H$. Consider all future-directed null geodesics which start at some point of H and have initial direction either $E_{n-1}$ or $E_n$ at this point. Each such geodesic contains a geodesic segment which goes from a point $q \in H$ to a first focal point p to H. Using Proposition 11.33 and the compactness of H, it follows that the union of all such null geodesic segments from H to a focal point is contained in a compact set K consisting of null geodesic segments starting on H. Now if $r \in E^+(H)$, then r can be joined to H by a past-directed null geodesic but not by a past-directed timelike curve. Thus $r \in K$. Hence $E^+(H) \subset K$. To show that $E^+(H)$ is closed, let $\{x_n\}$ be a sequence of points of $E^+(H)$. This sequence has a limit point $x \in K$. From the definition of K we have $x \in J^+(H)$. If $x \in I^+(H)$, then the open set $I^+(H)$ must contain some elements of the sequence $\{x_n\}$, contradicting $x_n \in E^+(H)$ for all n. Thus $x \notin I^+(H)$ which yields $x \in E^+(H)$. This shows that $E^+(H)$ is a closed subset of the compact set K and hence is compact.

If we assume that (M,g) is null complete and that tr $L_1 < 0$, tr $L_2 < 0$ for all $q \in H$, then similar arguments show $E^-(H)$ is compact. This establishes the corollary. □

We now define the Cauchy developments of a closed subset S of (M,g) [cf. Hawking and Ellis (1973, pp. 201-205)].

DEFINITION 11.37 Let S be a closed subset of (M,g). The *future* (resp., *past*) *Cauchy development* $D^+(S)$ [resp., $D^-(S)$] consists of all points $p \in M$ such that every past- (resp., future-) inextendible nonspacelike curve through p intersects S. The *Cauchy development* of S is $D(S) = D^+(S) \cup D^-(S)$ (Figure 11.4).

Closed achronal sets have played an important role in causality theory and singularity theory in general relativity. They have the following property [cf. Hawking and Ellis (1973, pp. 209, 268), Hawking and Penrose (1970, p. 537)].

PROPOSITION 11.38 If S is a closed achronal set in the space-time (M,g), then $\mathrm{Int}(D(S)) = D(S) - \partial D(S)$ is globally hyperbolic.

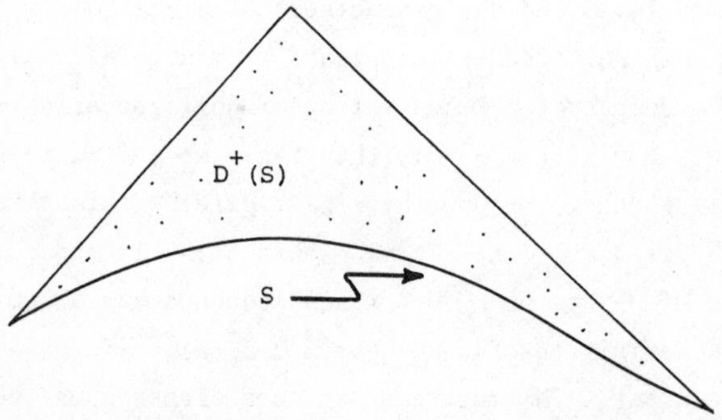


*Figure 11.4* The future Cauchy development $D^+(S)$ of S consists of all points p such that every past inextendible nonspacelike curve through p intersects S.

## 11.4 THE EXISTENCE OF SINGULARITIES

In this section, we give the proofs of several singularity theorems in general relativity. In particular, we prove the main theorem of Hawking and Penrose (1970, p. 538). Our approach is somewhat different from Hawking and Penrose (1970) in that we show $(M,g)$ is causally disconnected if it contains a trapped set and then apply Theorem 6.3 of Beem and Ehrlich (1979a, p. 172).

The basic technique in proving nonspacelike incompleteness is to first use physical or geometric assumptions on $(M,g)$ to construct an inextendible nonspacelike geodesic which is maximal and hence contains no conjugate points. If $(M,g)$ has dimension $\geq 3$ and satisfies the generic condition and strong energy condition, this geodesic must then be incomplete by Theorem 11.18.

We first show that a chronological space-time with a sufficient number of conjugate points is strongly causal [cf. Hawking and Penrose (1970, p. 536), Lerner (1972, p. 41)].

PROPOSITION 11.39 If $(M,g)$ is a chronological space-time such that each inextendible null geodesic has a pair of conjugate points, then $(M,g)$ is strongly causal.

*Proof.* Assume that strong causality fails at $p \in M$. Let $U$ be a convex normal neighborhood of $p$ and let $V_k \subset U$ be a sequence of neighborhoods which converge to $p$. Since strong causality fails at $p$, for each $k$ there is a future-directed nonspacelike curve $\gamma_k$ which starts in $V_k$ leaves $U$ and returns to $V_k$. Using Proposition 2.18, one may obtain an inextendible nonspacelike limit curve $\gamma$ through $p$ of the sequence $\{\gamma_k\}$. No two points of $\gamma$ are chronologically related since otherwise one could obtain a closed timelike curve and $(M,g)$ is chronological. Thus $\gamma$ is a null geodesic. This yields a contradiction since by hypothesis each null geodesic has conjugate points and thus contains points which may be joined by timelike curves. $\square$

Proposition 11.39 and Theorem 11.18 then imply the following result.

PROPOSITION 11.40 Let (M,g) be a chronological space-time of dimension $\geq 3$ which satisfies the generic condition and the strong energy condition. Then (M,g) is either strongly causal or null incomplete.

We may now prove the following singularity theorem. The concept of a causally disconnected space-time has been given in Section 7.3, Definition 7.11.

THEOREM 11.41 Let (M,g) be a chronological space-time of dimension $\geq 3$ which is causally disconnected. If (M,g) satisfies the generic condition and the strong energy condition, then (M,g) is nonspacelike incomplete.

*Proof*. Assume all nonspacelike geodesics of (M,g) are complete. By Proposition 11.39, the space-time (M,g) is strongly causal and by Theorem 11.18 every nonspacelike geodesic has conjugate points. On the other hand, Theorem 7.13 yields the existence of an inextendible maximal nonspacelike geodesic. But then this geodesic has no conjugate points, in contradiction. □

Recall that the *future* (resp., *past*) *horismos* of a subset S of (M,g) is given by $E^+(S) = J^+(S) - I^+(S)$ [resp., $E^-(S) = J^-(S) - I^-(S)$]. The achronal set S is said to be *future trapped* (resp., *past trapped*) if $E^+(S)$ [resp., $E^-(S)$] is compact. We now give conditions under which the existence of a trapped set implies causal disconnection.

PROPOSITION 11.42 Let (M,g) be a chronological space-time of dimension $\geq 3$ such that each inextendible null geodesic has a pair of conjugate points. If (M,g) contains a future- (resp., past-) trapped set S, then (M,g) is causally disconnected by $E^+(S)$ [resp., $E^-(S)$].

*Proof*. First Proposition 11.39 shows that (M,g) is strongly causal. If S is assumed to be future trapped, then Corollary 7.16 yields a future-inextendible timelike curve $\gamma$ in the future Cauchy development $D^+(E^+(S))$. Extend $\gamma$ to a future- and past-inextendible timelike curve in (M,g), still denoted by $\gamma$. Then $\gamma$ intersects the achronal set $E^+(S)$ in exactly one point r. As in the proof of

Proposition 7.18, we choose two sequences $\{p_n\}$ and $\{q_n\}$ on $\gamma$ which diverge to infinity and satisfy $p_n \ll r \ll q_n$ for each n. To show that $\{p_n\}$, $\{q_n\}$, and $E^+(S)$ causally disconnect (M,g), we must show that for each n, every nonspacelike curve $\lambda : [0,1] \longrightarrow M$ with $\lambda(0) = p_n$ and $\lambda(1) = q_n$ meets $E^+(S)$. Given $\lambda$, extend $\lambda$ to a past inextendible curve $\tilde{\lambda}$ by traversing $\gamma$ up to $p_n$ and then traversing $\lambda$ from $p_n$ to $q_n$. As $q_n \in D^+(E^+(S))$, the curve $\tilde{\lambda}$ must intersect $E^+(S)$. Since $\gamma$ meets $E^+(S)$ only at r, it follows that $\lambda$ intersects $E^+(S)$. This establishes the proposition if S is future trapped. A similar argument may be used if S is past trapped. □

Theorem 11.41 and Proposition 11.42 now imply the main theorem of Hawking and Penrose (1970, p. 538).

THEOREM 11.43 No space-time (M,g) of dimension $\geqslant 3$ can satisfy all of the following three requirements together:

(a) (M,g) contains no closed timelike curves.

(b) Every inextendible nonspacelike geodesic in (M,g) contains a pair of conjugate points.

(c) There exists a future- or past-trapped set S in (M,g).

This result of Hawking and Penrose implies the following result which is more similar to Theorem 11.41.

THEOREM 11.44 Let (M,g) be a chronological space-time of dimension $\geqslant 3$ which satisfies the generic condition and the strong energy condition. If (M,g) contains a trapped set, then (M,g) is nonspacelike incomplete.

Recall that a closed trapped surface is a compact spacelike submanifold of dimension n - 2 for which the trace of both null second fundamental forms $L_1$ and $L_2$ is either always positive or always negative (cf. Definition 11.34).

PROPOSITION 11.45 Let (M,g) be a strongly causal space-time of dimension $\geqslant 3$ which satisfies the condition $\mathrm{Ric}(v,v) \geqslant 0$ for all null vectors $v \in TM$. If $(M,g)$ contains a closed trapped surface $H$, then at least one of (1) or (2) is true:

(1) $(M,g)$ contains a trapped set.

(2) $(M,g)$ is null incomplete.

*Proof.* Assume that $(M,g)$ is null complete. Corollary 11.36 then implies that one of $E^+(H)$ or $E^-(H)$ is compact. We consider the case that $E^+(H)$ is compact and define $S$ by $S = E^+(H) \cap H$. We will show that $S$ is a future-trapped set. Notice that $S$ is achronal since $E^+(H)$ is achronal and $S$ is compact as the intersection of two compact sets.

Since $E^+(H) = J^+(H) - I^+(H)$, the set $S$ will be nonempty iff $H$ contains some points which are not in $I^+(H)$. But if $H$ were contained in $I^+(H)$, there would be a finite cover of the compact set $H$ by open sets $I^+(p_1), \ldots, I^+(p_n)$ with all $p_i \in H$. However, this would imply the existence of a closed timelike curve in $(M,g)$ (cf. the proof of Proposition 2.6) which would contradict the strong causality of $(M,g)$. Hence $S \neq \emptyset$.

In order to show that $S$ is a future-trapped set, it is sufficient to prove that $E^+(S) = E^+(H)$. We will demonstrate this by showing that $I^+(S) = I^+(H)$ and $J^+(S) = J^+(H)$. To this end, cover the compact set $H$ with a finite number of open sets $U_1, \ldots, U_k$ of $(M,g)$ such that each $U_i$ is a convex normal neighborhood and no nonspacelike curve which leaves $U_i$ ever returns.

Since $H$ is a spacelike submanifold, we may also assume that each $U_i \cap H$ is achronal by choosing the $U_i$ sufficiently small. Clearly, $I^+(S) \subset I^+(H)$ since $S \subset H$. To show $I^+(H) \subset I^+(S)$, assume that $q \in I^+(H) - I^+(S)$. Then there exists $p_1 \in H$ with $p_1 \ll q$. Now $p_1 \in U_{i(1)} \cap H$ for some index $i(1)$. Since $q \notin I^+(S)$, we have $p_1 \notin S$ and hence $p_1 \notin E^+(H)$. Thus there exists $p_2 \in H$ with $p_2 \ll p_1$. Since $U_{i(1)} \cap H$ is achronal, $p_2 \notin U_{i(1)}$. Now $p_2 \in U_{i(2)} \cap H$ for some $i(2) \neq i(1)$. Again $q \notin I^+(S)$ yields $p_2 \notin E^+(H)$. Thus there exists $p_3 \in H$ with $p_3 \ll p_2$. Furthermore, by construction of the sets $U_i$ we have $p_3 \notin U_{i(1)} \cap U_{i(2)}$. Thus $p_3 \in U_{i(3)} \cap H$ for some $i(3)$ different from $i(1)$ and $i(2)$. Continuing in this fashion, we obtain an infinite sequence $p_1, p_2, \ldots$ in H with corresponding sets $U_{i(1)}, U_{i(2)}, U_{i(3)}, \ldots$ such that $i(j_1) \neq i(j_2)$ if $j_1 \neq j_2$. This contradicts the finiteness of the number of sets $U_i$ of the given cover. Hence $I^+(S) = I^+(H)$. It remains to show that $J^+(S) = J^+(H)$. First $J^+(S) \subset J^+(H)$ as $S \subset H$. Thus assume that $q \in J^+(H) - J^+(S)$. Then $q \notin I^+(S) = I^+(H)$ and hence there is a future-directed null curve from some point $p \in H$ to the point q. Since $p \in H$ and $p \notin I^+(H)$ as $p \leqslant q$ and $q \notin I^+(H)$, we have $p \in E^+(H)$. Thus $p \in E^+(H) \cap H = S$ which yields $q \in J^+(S)$, in contradiction. This shows that S is a future-trapped set.

A similar argument shows that if (M,g) is null complete and if $E^-(H)$ is compact, then $S = E^-(H) \cap H$ is a past-trapped set. Thus the proposition is established. □

It is possible for a trapped set to consist of just a single point. One way this may arise is as follows [cf. Hawking and Penrose (1970, p. 543), Hawking and Ellis (1973, pp. 266-267)]. Let (M,g) be a space-time of dimension $\geqslant 3$ which satisfies the curvature condition $\mathrm{Ric}(v,v) \geqslant 0$ for all null vectors $v \in TM$. Suppose that there exists a point p such that on every future-directed null geodesic $\beta : [0,a) \longrightarrow (M,g)$ with $\beta(0) = p$, the expansion $\overline{\theta}$ of the Lagrange tensor field $\overline{A}$ on $G(\beta)$ with $\overline{A}(0) = 0$, $\overline{A}'(0) = E$ becomes negative for some $t_1 > 0$. Intuitively, each future-directed null geodesic emanating from p has a point $\beta(t_1)$ in the future of p at which all future-directed null geodesics are reconverging. Thus provided that the given null geodesic β can be extended to the parameter value $t_1 - (n-2)/\overline{\theta}(t_1)$, β has a future null conjugate point to $t = t_1$ along β. Hence $\beta(t) \in I^+(p)$ for all $t \geqslant t_1 - (n-2)/\overline{\theta}(t_1)$. Since the set of null directions at p is a compact set, it follows that $E^+(p) = J^+(p) - I^+(p)$ is compact provided that (M,g) is null geodesically complete. Thus we have obtained the following result.

PROPOSITION 11.46 Let (M,g) be a space-time of dimension $\geqslant 3$ with $\mathrm{Ric}(v,v) \geqslant 0$ for all null vectors $v \in TM$. Suppose that there exists a point $p \in M$ such that on every future-directed null geodesic β from $p = \beta(0)$, the expansion $\overline{\theta}$ of the Lagrange tensor field $\overline{A}$ on $G(\beta)$ with $\overline{A}(0) = 0$, $\overline{A}'(0) = E$ becomes negative for some $t_1 > 0$. Then at least one of (1) or (2) holds:

(1) $\{p\}$ is a trapped set, i.e., $E^+(p) = J^+(p) - I^+(p)$ is compact.

(2) $(M,g)$ is null incomplete.

We now consider the case of a compact connected spacelike hypersurface S in a space-time $(M,g)$. If S is achronal, then $E^+(S) = S$ and S is a trapped set. On the other hand, it may happen that S is not achronal. In fact, an example may easily be constructed of a compact spacelike hypersurface S with $S \subset I^+(S)$ and hence $E^+(S) = \emptyset$ (cf. Figure 11.5).

A compact spacelike hypersurface S which is not achronal may be used to construct an achronal compact spacelike hypersurface $\tilde{S}$ in a covering manifold $\tilde{M}$ of M [cf. Geroch (1970), Hawking (1967, p. 194), O'Neill (1981)]. Thus if a space-time has a compact spacelike hypersurface, then there is a covering manifold of the given space-time which has a trapped set. But in proving the nonspacelike incompleteness of $(M,g)$, we may work with covering manifolds just as well as $(M,g)$ since $(M,g)$ is nonspacelike incomplete iff each covering manifold of $(M,g)$ equipped with the pullback metric is nonspacelike incomplete.

These observations on covering spaces together with Theorem 11.44, Proposition 11.45, and Proposition 11.46 yield the following theorem [cf. Hawking and Penrose (1970, p. 544)].

THEOREM 11.47 Let $(M,g)$ be a chronological space-time of dimension $\geqslant 3$ which satisfies the generic condition and the strong energy condition. Then the space-time $(M,g)$ is nonspacelike incomplete if any of the following three conditions are satisfied:

(1) $(M,g)$ has a closed trapped surface.

(2) $(M,g)$ has a point p such that each null geodesic starting at p is reconverging somewhere in the future (or past) of p.

(3) $(M,g)$ has a compact spacelike hypersurface.

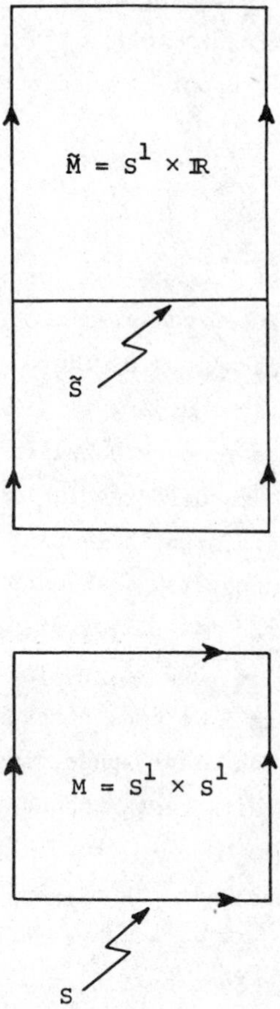


*Figure 11.5* Let $M = S^1 \times S^1$ be given the Lorentzian metric $ds^2 = d\theta_1^2 - d\theta_2^2$. The set $S = \{(\theta_1, 0) : \theta_1 \in S^1\}$ is a compact spacelike submanifold of codimension 1 such that $E^+(S) = \emptyset$. Here S is not a trapped set because it is not achronal. However, (M,g) has a covering manifold $(\tilde{M}, \tilde{g})$ which contains a trapped set $\tilde{S}$ which is diffeomorphic to S.

REMARK 11.48 Conditions (1) and (2) of Theorem 11.47 are reasonable cosmological assumptions. Robertson-Walker space-times with physically reasonable energy momentum tensors, positive energy density, and $\Lambda = 0$ have closed trapped surfaces [cf. Hawking and Ellis (1973, p. 353)] and thus satisfy (1). There is some astronomical evidence for (2) [cf. Hawking and Ellis (1973, p. 355)].

## 11.5 SMOOTH BOUNDARIES

In this section we consider the relationship between causal disconnection, nonspacelike geodesic incompleteness and points of the causal boundary $\partial_c M$ (cf. Section 5.4) at which the boundary is differentiable. Many of the more important space-times studied in general relativity have causal boundaries which are differentiable at a large number of points. For example, the differentiable part of the causal boundary of Minkowski space-time consists of $\mathscr{I}^+$ and $\mathscr{I}^-$ (cf. Figure 4.4). Since these sets correspond to null hypersurfaces, it is thus natural to call the points of $\mathscr{I}^+$ and $\mathscr{I}^-$ null boundary points. Penrose (1968) has used conformal methods to study smooth boundary points of Minkowski space-time and other space-times.

In general, we consider a space-time $(M,g)$ with causal boundary $\partial_c M$ and let $M^*$ denote the causal completion $M \cup \partial_c M$ of $(M,g)$. This completion may be given a Hausdorff topology such that the original topology on M agrees with the topology induced on M as a subset of $M^*$ [cf. Hawking and Ellis (1973, pp. 220-221) or Section 5.4].

Assume that $\bar{p} \in \partial_c M$ and let $U^* = U^*(\bar{p})$ be a neighborhood of $\bar{p}$ in $M^*$. Denote by $(U,g)$ the metric g restricted to the set

$$U = U^* \cap M$$

A *conformal representation* of $U^*(\bar{p})$ will be a space-time $(M',g')$ and a homeomorphic embedding $f : U^* \longrightarrow M'$ such that

(1) $f \mid U$ is a smooth map.

(2) There is a smooth function $\Omega : U \longrightarrow \mathbb{R}$ such that $\Omega > 0$ and $\Omega g = f^* f'$ on U (Figure 11.6).

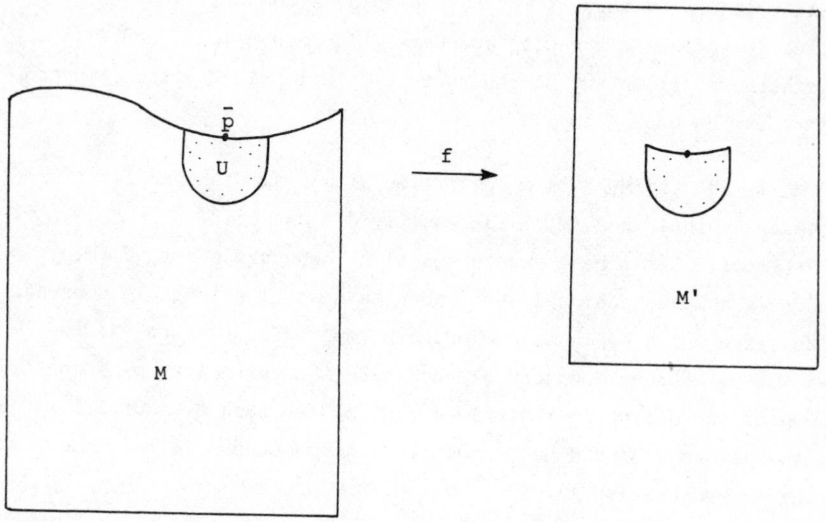


*Figure 11.6* The space-time (M,g) has a causal boundary point $\overline{p}$ and $U^*(\overline{p})$ is a neighborhood of $\overline{p}$ in the causal completion $M^* = M \cup \partial_c M$ of (M,g). The homeomorphic embedding $f : U^* \longrightarrow M'$ is a smooth conformal map on $U = U^*(\overline{p}) \cap M$.

If the conformal representation $f : U^* \longrightarrow M'$ maps $U^*$ to a smooth manifold with boundary, then we will say that $\overline{p}$ is a *smooth boundary point*.

DEFINITION 11.49 Let $U^*(\overline{p})$ have a smooth conformal representation $f : U^* \longrightarrow M'$ such that f(U) is a smooth manifold with a smooth boundary $\partial(f(U))$ in M'. Then the point $\overline{p}$ is said to be a *smooth spacelike* (resp., *null*, *timelike*) *boundary point* if the corresponding boundary $\partial(f(U))$ is a spacelike (resp., null, timelike) hypersurface in (M',g').

If $\gamma : [a,b) \longrightarrow U$ is a curve in M such that $\gamma(t) \longrightarrow \overline{p} \in M^* - U$ as $t \longrightarrow b$, then the curve $\gamma$ is said to have the boundary point $\overline{p} \in M^*$ as an endpoint. Also if $\overline{p}$ is a smooth spacelike boundary point, then it is not hard to find a compact set K in M such that all inextendible nonspacelike curves with one endpoint at $\overline{p}$ must intersect K. In fact, K may be chosen as a compact, achronal spacelike hypersurface with boundary (cf. Figures 11.7 and 11.8). Furthermore, given any neighborhood $U^*(\overline{p})$ of $\overline{p}$ in $M^*$, the compact set K may be chosen to lie in $U = U^*(\overline{p}) \cap M$.

LEMMA 11.50 If (M,g) is a space-time with a smooth spacelike boundary point, then (M,g) is causally disconnected.

*Proof.* Let $\overline{p}$ be a smooth spacelike boundary point of (M,g). Choose a compact achronal set K such that any inextendible nonspacelike curve which has $\overline{p}$ as one endpoint must meet K. Let $\gamma : (-\infty,\infty) \longrightarrow M$ be an inextendible nonspacelike curve with $\overline{p}$ as one endpoint and define $p_n = \gamma(-n)$, $q_n = \gamma(n)$ for each n. For all large n, the points $p_n$ and $q_n$ are causally disconnected by K; cf. the proof of Proposition 11.42. □

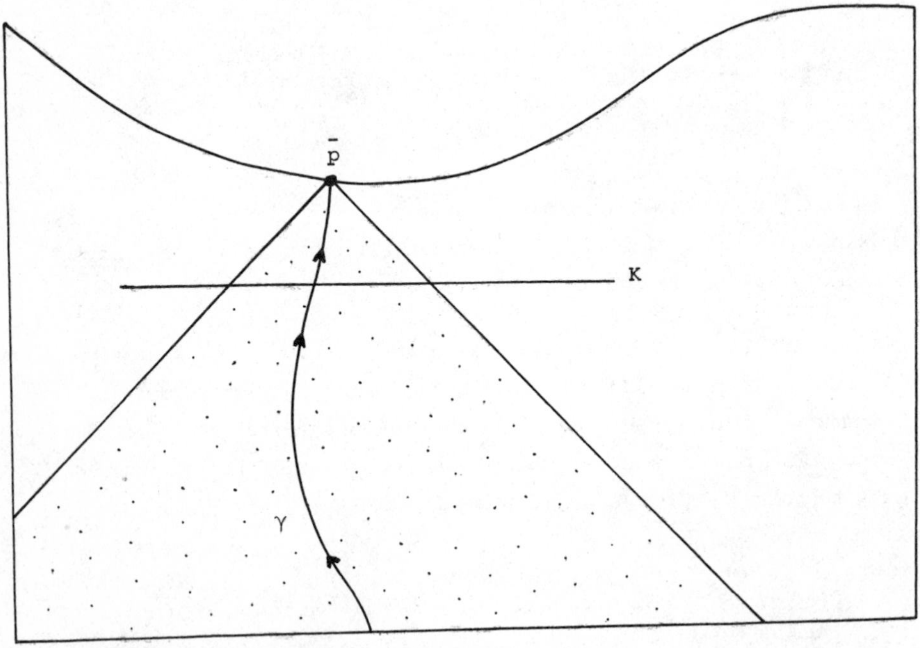


*Figure 11.7* A space-time with a smooth spacelike boundary point $\overline{p}$ is shown. The compact set K is chosen such that any nonspacelike curve $\gamma$ with one endpoint at $\overline{p}$ must intersect K.

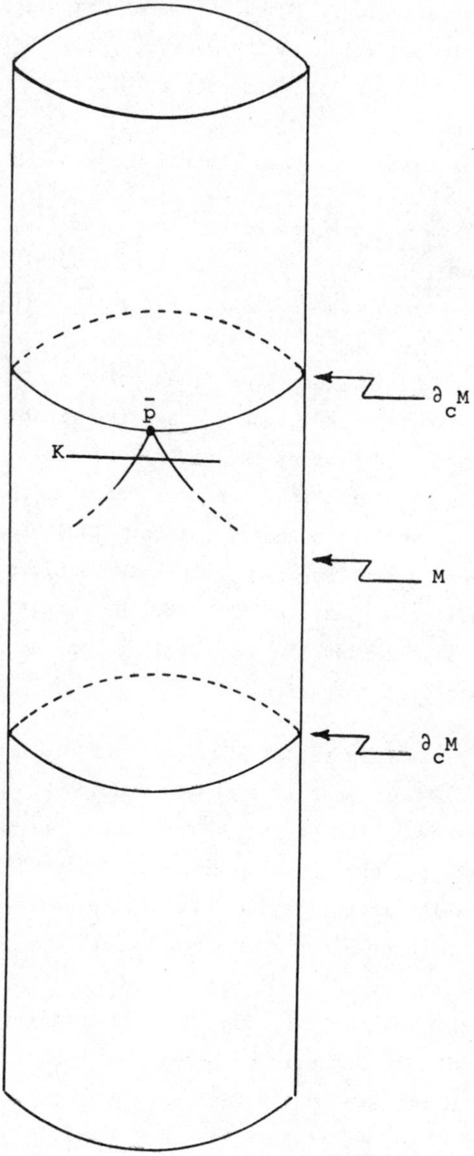


*Figure 11.8* A closed Robertson-Walker cosmological model conformally represented as a subset of the Einstein static universe is shown. The set K is a causally disconnecting set.

THEOREM 11.51 Let (M,g) be a chronological space-time of dimension $\geq 3$ which satisfies the generic condition and the strong energy condition. If (M,g) has a smooth spacelike boundary point, then (M,g) is nonspacelike incomplete.

*Proof.* This follows from Lemma 11.50 and Theorem 11.41. □

Notice that if (M,g) has one smooth spacelike (resp., null, timelike) boundary point, then (M,g) has uncountably many smooth spacelike (resp., null, timelike) boundary points. For if $\overline{p} \in \partial_c M$ corresponds to the point $x \in \partial(f(U))$ under the given conformal representation $f : U^* \longrightarrow M'$ of Definition 11.49, then points $y \in \partial(f(U))$ close to x will also represent smooth spacelike (resp., null, timelike) boundary points in $\partial_c M$ under the given conformal representation and $\partial(f(U))$ has dimension n - 1 in M'. Hence M contains uncountably many smooth spacelike boundary points. Thus using the fact that a causally disconnecting set K may be chosen arbitrarily close to any smooth spacelike boundary point $\overline{p}$ and the result that a limit of maximal curves is maximal in a strongly causal space-time, we may also establish the following result.

THEOREM 11.52 Let (M,g) be a strongly causal space-time of dimension $\geq 3$. Assume that (M,g) has a smooth spacelike boundary point and satisfies the generic condition and the strong energy condition. Then for *each* smooth spacelike boundary point $\overline{p}$ there is an incomplete nonspacelike geodesic which is inextendible and has $\overline{p}$ as an endpoint. Thus (M,g) has an uncountable number of incomplete, inextendible nonspacelike geodesics.

*Proof.* First, it follows from the preceding remarks that (M,g) contains an incomplete nonspacelike geodesic for each snooth non-spacelike boundary point. But we have just noted that if (M,g) contains one spacelike boundary point, it contains uncountably many spacelike boundary points. □

Appendix A

# CONNECTIONS AND CURVATURE

Let $(M,g)$ be a (pseudo-) Riemannian manifold of arbitrary signature $(-, \ldots, -, +, \ldots, +)$. Then there exists a unique affine connection $\nabla$ on $M$ which is both compatible with the metric $g$ and torsion free. This connection, which is called the Levi-Civita connection of $(M,g)$, satisfies the same formal relations whether $(M,g)$ is Riemannian or pseudo-Riemannian.

Thus geodesics, curvature, Ricci curvature, scalar curvature, and sectional curvature may all be defined for pseudo-Riemannian manifolds exactly as for Riemannian manifolds. The only difficulty which arises is that the sectional curvature is not defined on degenerate sections of the tangent space when $(M,g)$ is not of constant curvature. In fact, the sectional curvature is only bounded near degenerate sections in the case of constant curvature (cf. Section A.2).

In this appendix we give a brief discussion of affine connections, pseudo-Riemannian manifolds and curvature tensors. We include local coordinate representations using the natural basis $\partial/\partial x^1, \ldots, \partial/\partial x^n$ associated to an arbitrary chart $(U,x)$ for $M$. We also show that if $(M,g)$ is a two-dimensional Lorentzian manifold, then $\mathrm{Ric}(v,v) = 0$ for all null vectors $v \in TM$.

## A.1 AFFINE CONNECTIONS

Let $\mathfrak{X}(p)$ denote the set of smooth vector fields each of whose domains includes the point $p \in M$ and let $\mathfrak{X}(M)$ denote the smooth vector fields defined on all of M. An *affine connection* at p is a function $v \longrightarrow \nabla_v$ which associates to each $v \in T_pM$ a map

$$\nabla_v : \mathfrak{X}(p) \longrightarrow T_pM$$

with the properties that

$$\nabla_v(X + Y) = \nabla_v X + \nabla_v Y \tag{A.1}$$

$$\nabla_{fv+gw}(X) = f\nabla_v X + g\nabla_w X \tag{A.2}$$

and

$$\nabla_v(fX) = f\nabla_v X + v(f)X \tag{A.3}$$

Here f and g are smooth functions whose domains include p, and $v,w \in T_pM$, $X,Y \in \mathfrak{X}(p)$.

An affine connection on M is a function which associates to each $p \in M$ an affine connection $\nabla$ at p such that for all $X,Y \in \mathfrak{X}(M)$ the map

$$\nabla_X Y : p \longrightarrow \nabla_{X(p)} Y$$

is a smooth vector field on M.

The vector $\nabla_{X(p)} Y = \nabla_X Y\big|_p$ at the point $p \in M$ depends only on X(p) and on the values of Y along any smooth curve which passes through p and has tangent X(p) at p [cf. Hicks (1965, p. 57)]. To see this, let $E_1, \ldots, E_n$ be smooth vector fields defined near p which form a basis of the tangent space at each point in a neighborhood of p. Then $X(p) = \Sigma X^i(p)E_i(p)$ and $Y = \Sigma Y^i E_i$. Hence

$$\begin{aligned} \nabla_X Y\big|_p &= \nabla_{X(p)}(\Sigma Y^i E_i) \\ &= \sum_{i=1}^{n} Y^i(p)\nabla_{X(p)} E_i + \sum_{i=1}^{n} X(p)(Y_i)E_i(p) \end{aligned}$$

$$= \sum_{i,j=1}^{n} X^j(p)Y^i(p)\nabla_{E_j(p)}E_i + \sum_{i=1}^{n} X(p)(Y^i)E_i(p)$$

Thus $X^j(p)$, $Y^i(p)$, and $X(p)(Y^i)$ determine $\nabla_X Y|_p$ completely if the $\nabla_{E_j(p)}E_i$'s are known.

Given an affine connection $\nabla$ on M and a curve $\gamma : [a,b] \longrightarrow M$, we may define parallel translation of vector fields along $\gamma$. Here a *vector field* Y *along* $\gamma$ is a smooth mapping $Y : [a,b] \longrightarrow TM$ such that $Y(t) \in T_{\gamma(t)}M$ for each $t \in [a,b]$. For $t_0 \in [a,b]$ we may locally extend Y to a smooth vector field defined on a neighborhood of $\gamma(t_0)$. Then we may consider the vector field $\nabla_{\gamma'(t)}Y$ along $\gamma$ defined in this neighborhood. The preceding arguments show that this vector field along $\gamma$ is independent of the local extension and consequently $\nabla_{\gamma'}Y$ is a well-defined vector field along $\gamma$. A vector field Y along $\gamma$ which satisfies $\nabla_{\gamma'}Y(t) = 0$ for all $t \in [a,b]$ is said to move by *parallel translation* along $\gamma$. A *geodesic* $c : (a,b) \longrightarrow M$ is a smooth curve of M such that the tangent vector c' moves by parallel translation along c. In other words, c is a geodesic if

$$\nabla_{c'}c' = 0 \tag{A.4}$$

A *pregeodesic* is a smooth curve c which may be reparameterized to be a geodesic. Any parameter for which c is a geodesic is called an *affine parameter*. If s and t are two affine parameters for the same pregeodesic, then $s = at + b$ for some constants $a,b \in \mathbb{R}$. A pregeodesic is said to be *complete* if for some affine parameterization (hence for all affine parameterizations) the domain of the parameterization is all of $\mathbb{R}$.

Equation (A.4) may also be expressed as a system of linear differential equations. To this end we let $(U,(x^1,\ldots,x^n))$ be local coordinates on M and let $\partial/\partial x^1, \ldots, \partial/\partial x^n$ denote the *natural basis* with respect to these coordinates. The coefficients $\Gamma_{ij}^k$ of $\nabla$ with respect to $(x^1,\ldots,x^n)$ are defined by

$$\nabla_{\partial/\partial x^i} \frac{\partial}{\partial x^j} = \sum_{k=1}^{n} \Gamma^k_{ij} \frac{\partial}{\partial x^k} \tag{A.5}$$

Using these coefficients we may write equation (A.4) as the system

$$\frac{d^2x^k}{dt^2} + \sum_{i,j=1}^{n} \Gamma^k_{ij} \frac{dx^i}{dt}\frac{dx^j}{dt} = 0 \tag{A.6}$$

The connection coefficients may also be used to give a local representation of the action of $\nabla$. If the vector fields A and B have local representations

$$A = \sum_{i=1}^{n} a^i(x) \frac{\partial}{\partial x^i}$$

and

$$B = \sum_{i=1}^{n} b^i(x) \frac{\partial}{\partial x^i}$$

then $\nabla_A B$ has a local representation

$$\nabla_A B = \sum_{k=1}^{n} \left( \sum_{i=1}^{n} a^i \frac{\partial b^k}{\partial x^i} + \sum_{i,j=1}^{n} \Gamma^k_{ij} a^i b^j \right) \frac{\partial}{\partial x^k} \tag{A.7}$$

The *torsion tensor* T of $\nabla$ is a function which assigns to each point $p \in M$ the f-bilinear mapping $\mathfrak{X}(p) \times \mathfrak{X}(p) \longrightarrow \mathfrak{X}(p)$ given by

$$T(X,Y) = \nabla_X Y - \nabla_Y X - [X,Y] \tag{A.8}$$

Here $[X,Y]$ denotes the Lie bracket of X and Y which is given by $[X,Y](f) = X(Y(f)) - Y(X(f))$ for any smooth function f. The value $T(X,Y)|_p$ only depends on the connection $\nabla$ and the values $X(p)$, $Y(p)$. Consequently, T determines a bilinear map $T_pM \times T_pM \longrightarrow T_pM$ at each point $p \in M$.

Using the skew-symmetry of the Lie bracket, it is easily seen that $T(X,Y) = -T(Y,X)$ and hence T is skew. Since $[\partial/\partial x^i, \partial/\partial x^j] = 0$ for all $1 \leq i,j \leq n$, it follows that

$$T\left(\frac{\partial}{\partial x^i}, \frac{\partial}{\partial x^j}\right) = \sum_{k=1}^{n} (\Gamma^k_{ij} - \Gamma^k_{ji}) \frac{\partial}{\partial x^k} \tag{A.9}$$

Thus the torsion tensor provides a measure of the nonsymmetry of the connection coefficients. Clearly, $T = 0$ iff these coefficients are symmetric in their subscripts. An affine connection $\nabla$ with $T = 0$ is said to be *torsion free* or *symmetric*.

The *curvature* R of $\nabla$ is a function which assigns to each $p \in M$ and pair $X,Y \in \mathfrak{X}(p)$ an f-linear map $R(X,Y) : \mathfrak{X}(p) \longrightarrow \mathfrak{X}(p)$ given by

$$R(X,Y)Z = \nabla_X\nabla_Y Z - \nabla_Y\nabla_X Z - \nabla_{[X,Y]}Z \tag{A.10}$$

Curvature thus provides a measure of the noncommutativity of $\nabla_X$ and $\nabla_Y$. It should be noted that some authors use the sign convention

$$R(X,Y)Z = -\nabla_X\nabla_Y Z + \nabla_Y\nabla_X Z + \nabla_{[X,Y]}Z$$

for the curvature and consequently modify the definitions of certain of the curvature quantities given below accordingly.

Since the map $(X,Y,Z) \longrightarrow R(X,Y)Z$ from $\mathfrak{X}(p) \times \mathfrak{X}(p) \times \mathfrak{X}(p)$ to $\mathfrak{X}(p)$ is f-trilinear, it follows that R represents a tensor field and that $R(X,Y)Z\big|_p$ depends only on $\nabla$, $X(p)$, $Y(p)$, and $Z(p)$. Hence if $\mathfrak{X}$ $x,y,z \in T_pM$, one may locally extend these vectors to corresponding vector fields X, Y, Z and define $R(x,y)z = R(X,Y)Z\big|_p$.

The *curvature tensor* is a (1,3) tensor field which is also denoted by R. If $\omega \in T_p^*M$ is a cotangent vector at p and $x,y,z \in T_pM$, then one defines

$$R(\omega,x,y,z) = (\omega, R(x,y)z) = \omega(R(x,y)z) \tag{A.11}$$

In local coordinates, we have

$$R = \sum_{i,j,k,m} R^i_{jkm} \frac{\partial}{\partial x^i} \otimes dx^j \otimes dx^k \otimes dx^m \tag{A.12}$$

where

$$R^i_{jkm} = \frac{\partial \Gamma^i_{mj}}{\partial x^k} - \frac{\partial \Gamma^i_{kj}}{\partial x^m} + \sum_{a=1}^{n} (\Gamma^a_{mj}\Gamma^i_{ka} - \Gamma^a_{kj}\Gamma^i_{ma}) \tag{A.13}$$

If $X = \Sigma X^i \, \partial/\partial x^i$, $Y = \Sigma Y^i \, \partial/\partial x^i$, and $Z = \Sigma Z^i \, \partial/\partial x^i$, then

$$R(X,Y)Z = \sum_{i,j,k,m} R^i_{jkm} Z^j X^k Y^m \frac{\partial}{\partial x^i} \tag{A.14}$$

## A.2 PSEUDO-RIEMANNIAN MANIFOLDS

Let M be a smooth paracompact Hausdorff manifold and let $\pi : TM \longrightarrow M$ denote the tangent bundle of M. A *pseudo-Riemannian metric* g for M is a smooth symmetric tensor field of type (0,2) on M such that for each $p \in M$ the tensor $g|_p : T_pM \times T_pM \longrightarrow \mathbb{R}$ is a nondegenerate inner product of signature $(-, \ldots, -, +, \ldots, +)$. Here nondegenerate means that for each nontrivial vector $v \in T_pM$ there is some $w \in T_pM$ such that $g_p(v,w) \neq 0$. Although we consider only smooth metrics, some authors have studied more general pseudo-Riemannian metrics [cf. Parker (1979), Taub (1980)].

In local coordinates $(U,(x^1,\ldots,x^n))$ on M, the pseudo-Riemannian metric g can be represented as

$$g \mid U = \sum_{i,j=1}^{n} g_{ij}(x)\, dx^i \otimes dx^j$$

where $g_{ij} = g_{ji}$ and $\det g \neq 0$. If g has s negative eigenvalues and $r = n - s$ positive eigenvalues, then the signature of g will be denoted by (s,r). For each fixed point $p \in M$ there are local coordinates such that $g|_p$ can be represented as $\mathrm{diag}\{-1,\ldots,-1,+1,\ldots,+1\}$.

For each pseudo-Riemannian manifold (M,g) there is an associated pseudo-Riemannian manifold (M,-g) obtained by replacing g with -g. Aside from some minor changes in sign, there is no essential difference between (M,g) and (M,-g). Thus results for spaces of signature (s,r) may always be translated into corresponding results for spaces of signature (r,s) by appropriate sign changes and inequality reversals.

Given a pseudo-Riemannian manifold (M,g), there is a unique affine connection $\nabla$ on M such that

$$Z(g(X,Y)) = g(\nabla_Z X,Y) + g(X,\nabla_Z Y) \tag{A.15}$$

and

$$\nabla_X Y - \nabla_Y X = [X,Y] \tag{A.16}$$

for all $X,Y,Z \in \mathfrak{X}(M)$. This connection $\nabla$ is called the *Levi-Civita connection* of $(M,g)$. Equation (A.15) is the requirement that the connection $\nabla$ be compatible with g. Setting $Z = c'$ in (A.15), it is easily seen that parallel translation of vector fields along any curve c in M preserves inner products. Equation (A.16) is just the requirement that $\nabla$ be torsion free [cf. Eq. (A.8)]. The connection coefficients of $\nabla$ are given by

$$\Gamma^k_{ij} = \frac{1}{2} \sum_{a=1}^{n} g^{ak}\left(\frac{\partial g_{ia}}{\partial x^j} - \frac{\partial g_{ij}}{\partial x^a} + \frac{\partial g_{aj}}{\partial x^i}\right) \tag{A.17}$$

where $g^{ij}$ represents the (2,0) tensor defined by the requirement that

$$\sum_{a=1}^{n} g^{ia} g_{aj} = \delta^i_j$$

The local representations $g^{ij}$ and $g_{ij}$ may be used to raise and lower indices. For example, if the upper index of the curvature tensor is lowered, we obtain the components of the *Riemann-Christoffel tensor*

$$R_{ijkm} = \sum_{a=1}^{n} g_{ai} R^a_{jkm} \tag{A.18}$$

Alternatively, one may define the Riemann-Christoffel tensor $\tilde{R}$ as the (0,4) tensor such that

$$\tilde{R}(W,Z,X,Y) = g(W,R(X,Y)Z) \tag{A.19}$$

The trace of the curvature tensor is the *Ricci curvature*, a symmetric (0,2) tensor. For each $p \in M$, the Ricci curvature may be interpreted as a symmetric bilinear map $\mathrm{Ric}_p : T_pM \times T_pM \longrightarrow \mathbb{R}$. To evaluate Ric(v,w), let $e_1, \ldots, e_n$ be an orthonormal basis for $T_pM$. Then

$$\begin{aligned} \text{Ric}(v,w) &= \sum_{i=1}^{n} g(e_i,e_i)g(R(e_i,w)v,e_i) \\ &= \sum_{i=1}^{n} g(e_i,e_i)\tilde{R}(e_i,v,e_i,w) \end{aligned} \tag{A.20}$$

Alternatively, we may express v and w in the natural basis as $v = \Sigma v^i \partial/\partial x^i$, $w = \Sigma w^i \partial/\partial x^k$ and then write

$$\text{Ric}(v,w) = \sum_{i,j=1}^{n} R_{ij}v^iw^j \tag{A.21}$$

where

$$R_{ij} = \sum_{a=1}^{n} R^a_{iaj} \tag{A.22}$$

The *Ricci tensor* is the (1,1) tensor field which corresponds to the Ricci curvature. The components of the Ricci tensor may be obtained by raising one index of the Ricci curvature. Either component may be raised since the Ricci curvature is symmetric. Thus

$$R^i_j = \sum_{a=1}^{n} g^{ai}R_{aj} = \sum_{a=1}^{n} g^{ai}R_{ja} \tag{A.23}$$

The trace of the Ricci tensor is the *scalar curvature* $\tau$. Historically this function has been denoted by the much used symbol R. Accordingly,

$$\tau = R = \sum_{a=1}^{n} R^a_a \tag{A.24}$$

Thus if $e_1, \ldots, e_n$ is an orthonormal basis of $T_pM$, we have

$$\tau = R = \sum_{i=1}^{n} g(e_i,e_i)\text{Ric}(e_i,e_i) \tag{A.25}$$

A two-dimensional linear subspace E of $T_pM$ is called a *plane section*. The plane section E is said to be *nondegenerate* if for each nontrivial vector $v \in E$, there exists a vector $u \in E$ such that $g(v,u) \neq 0$. This is equivalent to the requirement that $g_p \mid E$ be a nondegenerate inner product on E. If v and w form a basis of the nondegenerate plane section E, then $g(v,v)g(w,w) - [g(v,w)]^2$ is a nonzero quantity which represents the square of the pseudo-Euclidean area of the parallelogram in E determined by v and w. The *sectional curvature* K(p,E) of the nondegenerate plane section E with basis {v,w} is then given by

$$K(p,E) = \frac{g(R(w,v)v,w)}{g(v,v)g(w,w) - [g(v,w)]^2} \tag{A.26}$$

This quantity is not defined on degenerate plane sections since the denominator in the above definition is always zero for degenerate plane sections.

If $v = \Sigma\, v^i\, \partial/\partial x^i$ and $w = \Sigma\, w^i\, \partial/\partial x^i$ are a basis of the nondegenerate plane section E, then equation (A.26) may be written as

$$K(p,E) = \frac{\Sigma\, R_{ijkm} w^i v^j w^k v^m}{\Sigma\, g_{ij} v^i v^j g_{km} w^k w^m - (\Sigma\, g_{ij} v^i w^j)^2} \tag{A.27}$$

Pseudo-Riemannian manifolds which have the same sectional curvature on all (nondegenerate) plane sections are said to have *constant curvature*. If (M,g) is pseudo-Riemannian of constant curvature c, then

$$R(X,Y)Z = c[g(Y,Z)X - g(X,Z)Y]$$

[cf. Graves and Nomizu (1978, p. 268)].

A nondegenerate plane E is said to be *timelike* if it is spanned by a spacelike and a timelike tangent vector. It may be shown that if the sectional curvatures of the timelike planes are bounded from both above and below and dim $M \geqslant 3$, then (M,g) has constant curvature [cf. Harris (1979a, Appendix A)]. Thus there is no apparent

Lorentzian analogue for the concept of a "pinched Riemannian manifold" [cf. Cheeger and Ebin (1975, p. 118)]. On the other hand, families of Lorentzian manifolds conformal to space-times of constant curvature may be constructed which have all timelike sectional curvatures bounded in one direction [cf. Harris (1979b)]. However if the sectional curvature of all nondegenerate planes is bounded from above (or from below), then the sectional curvature is constant [cf. Kulkarni (1979)]. Thus it is essential to restrict attention here to timelike planes only.

The *gradient* is defined for pseudo-Riemannian manifolds just as for Riemannian manifolds. If $f : M \longrightarrow \mathbb{R}$ is a smooth function, then df is a (0,1) tensor on M and grad f is the (1,0) tensor field which corresponds to df. Thus

$$Y(f) = (df, Y) = g(\text{grad } f, Y) \tag{A.28}$$

for an arbitrary vector field Y. In local coordinates the vector field grad f is represented by

$$\text{grad } f = \sum_{i,j=1}^{n} g^{ij} \frac{\partial f}{\partial x^i} \frac{\partial}{\partial x^j} \tag{A.29}$$

## A.3 NULL RICCI CURVATURE IN TWO-DIMENSIONAL MANIFOLDS

A pseudo-Riemannian manifold (M,g) of signature (1, n - 1) [i.e., signature (-, +, ..., +)] is a Lorentzian manifold. In this section we show that Ric(v,v) = 0 for any null vector v of a two-dimensional Lorentzian manifold.

PROPOSITION A.1 Let (M,g) be a two-dimensional Lorentzian manifold. If v is a null vector, then Ric(v,v) = 0.

*Proof.* Fix an arbitrary point p of M and let $e_1$, $e_2$ be an orthonormal basis of $T_pM$ with $g(e_1,e_1) = -1$ and $g(e_2,e_2) = +1$. The bilinearity of Ric(-,-) implies that it suffices to show that Ric(v,v) = Ric(w,w) = 0 for the linearly independent null vectors

$v = e_1 + e_2$ and $w = e_1 - e_2$. Since the proofs for v and w are similar, we will only give the proof that $\mathrm{Ric}(v,v) = 0$ for $v = e_1 + e_2$.

Using Eq. (A.20), $g(e_1,e_1) = -1$, and $g(e_2,e_2) = +1$, we obtain

$$\mathrm{Ric}(v,v) = -g(R(e_1, e_1 + e_2)(e_1 + e_2), e_1) + g(R(e_2, e_1 + e_2)(e_1 + e_2), e_2)$$

The trilinearity of the map $(X,Y,Z) \longrightarrow R(X,Y)Z$ and the curvature identity $R(X,Y) = -R(Y,X)$ yield

$$\mathrm{Ric}(v,v) = -g(R(e_1,e_2)e_1,e_1) - g(R(e_1,e_2)e_2,e_1) + g(R(e_2,e_1)e_1,e_2) + g(R(e_2,e_1)e_2,e_2)$$

The identities $\tilde{R}(W,Z,X,Y) = -\tilde{R}(Z,W,X,Y)$ and $\tilde{R}(W,Z,X,Y) = -\tilde{R}(W,Z,Y,X)$ of the Riemann-Christoffel tensor [cf. Eq. (A.19)], then imply the desired result that $\mathrm{Ric}(v,v) = 0$. □

Appendix B

# THE GENERIC CONDITION

In Section 11.2, we used the definition (cf. Definition 11.7) that a timelike geodesic c with tangent vector $W = c'$ satisfies the generic condition if for some $t_0$ in the domain of c, the curvature endomorphism

$$R(-,W)W\big|_{t_0} : V^{\perp}(c(t_0)) \longrightarrow V^{\perp}(c(t_0))$$

is not identically zero. That is, there exists $y \in V^{\perp}(c(t_0))$ such that $R(y,W(t_0))W(t_0) \neq 0$. We will show in this appendix that this condition is equivalent to the condition commonly used in general relativity that

$$W^c W^d W_{[a} R_{b]cd[e} W_{f]} \neq 0 \tag{B.1}$$

at the point $c(t_0)$. Here $V^{\perp}(c(t_0)) = \{v \in T_{c(t_0)}M : g(v,c'(t_0)) = 0\}$. We also used the definition (cf. Definition 11.7) in Chapter 11 that a null geodesic $\beta$ satisfies the generic condition if for some $t_0$ in the domain of $\beta$, the quotient bundle endomorphism

$$\overline{R}(-,\beta'(t_0))\beta'(t_0) : G(\beta(t_0)) \longrightarrow G(\beta(t_0))$$

is nonzero [cf. equation (9.25) of Section 9.3 for $G(\beta(t_0))$]. We will show that this condition is equivalent to (B.1) holding at $\beta(t_0)$ with $W = \beta'$. Thus the space-time (M,g) satisfies the generic condition iff for each inextendible nonspacelike geodesic $\gamma$ with tangent W there is some point along $\gamma$ such that (B.1) holds.

A sufficient condition for the generic condition to hold along the nonspacelike geodesic $\gamma$ is that $\mathrm{Ric}(\gamma',\gamma') \neq 0$ at some point of $\gamma$. It follows that a space-time with $\mathrm{Ric}(v,v) > 0$ for all nonzero, nonspacelike vectors v must satisfy both the generic and strong energy conditions.

We begin by showing that the generic condition for a timelike geodesic can be characterized as $R_{bnen} \neq 0$.

LEMMA B.1 Let c be a timelike geodesic and let $\{v_1, v_2, \ldots, v_n = c'(t_0)\}$ be a basis for $T_{c(t_0)}M$. If the components of the Riemann-Christoffel tensor are given with respect to this basis, then c satisfies the generic condition at $t_0$ iff $R_{bnen} \neq 0$ at $c(t_0)$ for some $1 \leqslant b, e \leqslant n - 1$.

*Proof.* If c satisfies the generic condition at $t_0$, then there is some $y_1 \in V^{\perp}(c(t_0))$ with $R(y_1,v_n)v_n \neq 0$. Since R is trilinear and $R(v_n,v_n) = 0$, the existence of $y_1$ implies that there is some y in the span of $v_1, \ldots, v_{n-1}$ such that $R(y,v_n)v_n \neq 0$. Using the nondegeneracy of g at $c(t_0)$, we obtain $x_1 \in T_{c(t_0)}M$ such that $\tilde{R}(x_1,v_n,y,v_n) = g(x_1,R(y,v_n)v_n) \neq 0$. Here $\tilde{R}$ denotes the Riemann-Christoffel tensor defined in Appendix A.2, formula (A.19). Since $\tilde{R}$ is multilinear and satisfies $\tilde{R}(v_n,v_n,y,v_n) = 0$, there exists an x in the span of $v_1, \ldots, v_{n-1}$ such that $\tilde{R}(x,v_n,y,v_n) \neq 0$. It follows that $R_{bnen} \neq 0$ for some $1 \leqslant b, e \leqslant n - 1$.

Conversely, if $R_{bnen} \neq 0$, then $\tilde{R}(v_b + \alpha v_n, v_n, v_e + \beta v_n, v_n) = \tilde{R}(v_b,v_n,v_e,v_n) \neq 0$ for all $\alpha,\beta \in \mathbb{R}$. Thus $R(v_e + \beta v_n, v_n)v_n \neq 0$ for all $\beta \in \mathbb{R}$ which implies that $R(y,v_n)v_n \neq 0$ for some $y \in V^{\perp}(c(t_0))$. □

We now show that Eq. (B.1) characterizes the generic condition along timelike geodesics.

PROPOSITION B.2 If c is a timelike geodesic with tangent $W = c'$, then c satisfies the generic condition at $t_0$ iff

$$W^c W^d W_{[a} R_{b]cd[e} W_{f]} = 0$$

at $c(t_0)$.

*Proof.* We may assume without loss of generality that c is a unit speed geodesic. We may thus use an orthonormal basis $\{e_1, \ldots, e_n = W = c'(t_0)\}$ at $c(t_0)$ to evaluate this tensor. With respect to this basis and dual co-basis, the components corresponding to W are given by $W^1 = \cdots = W^{n-1} = W_1 = \cdots = W_{n-1} = 0$, $W^n = 1$, and $W_n = -1$. Consequently,

$$\begin{aligned} W^c W^d W_{[a} R_{b]cd[e} W_{f]} &= \frac{1}{4}[W_a R_{bnne} W_f - W_b R_{anne} W_f - W_a R_{bnnf} W_e + W_b R_{annf} W_e] \\ &= \frac{1}{4}[\delta_a^n R_{bnne} \delta_f^n - \delta_b^n R_{anne} \delta_f^n - \delta_a^n R_{bnnf} \delta_e^n + \delta_b^n R_{annf} \delta_e^n] \end{aligned}$$

It is easily seen that this expression is nonzero iff $R_{bnne} \neq 0$ for some $1 \leqslant b, e \leqslant n - 1$. The result now follows from Lemma B.1 since $R_{bnne} = -R_{bnen}$. □

Using Lemma B.1, we will now show that if $\mathrm{Ric}(c',c') \neq 0$ at some point of the timelike geodesic c, then c satisfies the generic condition.

LEMMA B.3 Let $c : (a,b) \longrightarrow M$ be a timelike geodesic and assume that $\mathrm{Ric}(c',c') \neq 0$ at some point of c. Then c satisfies the generic condition.

*Proof.* We may assume without loss of generality that c is unit speed. Supposing that $\mathrm{Ric}(c'(t_0),c'(t_0)) \neq 0$, let $\{e_1, e_2, \ldots, e_n = c'(t_0)\}$ be an orthonormal basis for $T_{c(t_0)}M$. Then at $t_0$ we have

$$\mathrm{Ric}(c',c') = \sum_{i=1}^{n} g(e_i,e_i) g(R(e_i,c')c',e_i)$$

Since $R(c',c')c' = 0$, this reduces to

$$\mathrm{Ric}(c',c') = \sum_{i=1}^{n-1} g(R(e_i,c')c',e_i)$$

$$= \sum_{i=1}^{n-1} R_{inin}$$

Hence $\mathrm{Ric}(c',c') \neq 0$ implies $R_{inin} \neq 0$ for some value of $i \in \{1, \ldots, n-1\}$. The result now follows from Lemma B.1. □

We now consider the generic condition for a null geodesic $\beta : (a,b) \longrightarrow M$. For each point p of $\beta$, there is some parameterization of $\beta$, some orthonormal basis $\{e_1,\ldots,e_{n-2},\tilde{e}_{n-1},\tilde{e}_n\}$ for $T_pM$ and some null vector $N \in T_pM$ such that

$$\tilde{e}_n = \frac{\beta' + N}{\sqrt{2}} \qquad \tilde{e}_{n-1} = \frac{\beta' - N}{\sqrt{2}}$$

as p. Here we have let $\tilde{e}_n$ be the unit timelike vector in the basis $\{e_1,\ldots,e_{n-2},\tilde{e}_{n-1},\tilde{e}_n\}$. It is easily checked that $g(\beta',N) = g(\tilde{e}_n,\tilde{e}_n) = -1$ so that $\{e_1,e_2,\ldots,e_{n-1},N,\beta'\}$ is a pseudo-orthonormal basis for $T_pM$. In this pseudo-orthonormal basis, the metric tensor at p has the form

$$g_{ij} = \delta_{ij} \qquad \text{if } 1 \leq i,j \leq n-2$$

$$g_{nn-1} = g_{n-1n} = -1$$

$$g_{nn} = g_{n-1\,n-1} = 0$$

We now show that using this basis, the generic condition for $\beta$ at p is equivalent to $R_{bnne} \neq 0$ for some $1 \leq b,e \leq n-2$.

LEMMA B.4 Let $p = \beta(t_0)$ be a point of the null geodesic $\beta$ and let $\{e_1,\ldots,e_{n-2},N,\beta'\}$ be the pseudo-orthonormal basis for $T_pM$ defined above. If the components of the Riemann-Christoffel tensor are given with respect to this basis, then $\beta$ satisfies the generic condition at $t_0$ iff $R_{bnne} \neq 0$ at $\beta(t_0)$ for some $1 \leq b,e \leq n-2$.

*Proof.* If $N(\beta(t_0))$ denotes the $(n-1)$-dimensional linear space of vectors orthogonal to $\beta'(t_0)$, then $N(\beta(t_0))$ has the basis $\{e_1,\ldots,e_{n-2},\beta'\}$. Let $G(\beta(t_0)) = N(\beta(t_0))/[\beta'(t_0)]$, and let $\pi : N(\beta(t_0)) \longrightarrow G(\beta(t_0))$ denote the natural projection map. Recall that the curvature endomorphism $\overline{R}(-,\beta')\beta' : G(\beta(t_0)) \longrightarrow G(\beta(t_0))$ is defined by $\overline{R}(\overline{v},\beta')\beta' = \pi \circ R(v,\beta')\beta'$ where v is any vector in $\pi^{-1}(\overline{v})$ [cf. (9.34) of Section 9.3].

Now assume that the generic condition along $\beta$ is satisfied at $t_0$. Then there is some $v_1 \in N(\beta(t_0))$ with $\overline{R}(\pi(v_1),\beta')\beta' = \pi \circ R(v_1,\beta')\beta' \neq 0$. Using $R(\beta',\beta')\beta' = 0$ and the trilinearity of R, it follows that there is some $e_i$ for $1 \leq i \leq n-2$ such that $\pi \circ R(e_i,\beta')\beta' \neq 0$. Since the quotient metric $\overline{g}$ on $G(\beta(t_0))$ is nondegenerate, there exists $\overline{w} \in G(\beta(t_0))$ with $\overline{g}(\overline{w}, \pi \circ R(e_i,\beta')\beta') \neq 0$. Hence there is some $e_j$ with $1 \leq j \leq n-2$ such that $R_{jnin} = g(e_j,R(e_i,\beta')\beta') \neq 0$.

Conversely, assume that $R_{jnin} \neq 0$ for some $1 \leq i,j \leq n-2$. Then it is easily checked that $\overline{g}(\overline{R}(\pi(e_i),\beta')\beta',\pi(e_j)) \neq 0$ and hence $\overline{R}(-,\beta')\beta' : G(\beta'(t_0)) \longrightarrow G(\beta'(t_0))$ is not identically zero. $\square$

We now show that Eq. (B.1) characterizes the generic condition along null geodesics.

PROPOSITION B.5 If $\beta$ is a null geodesic with tangent $W = \beta'$, then $\beta$ satisfies the generic condition at $t_0$ iff

$$W^c W^d W_{[a} R_{b]cd[e} W_{f]} \neq 0$$

at $\beta(t_0)$.

*Proof.* We may assume without loss of generality that the components of the Riemann-Christoffel tensor are given with respect to the above pseudo-orthonormal basis $\{e_1,\ldots,e_{n-2},N,\beta'(t_0)\}$. If $W = \beta'$, then the components of W in this basis and corresponding dual co-basis are $W^1 = \cdots = W^{n-1} = W_1 = \cdots = W_{n-2} = W_n = 0$, $W^n = +1$, and $W_{n-1} = -1$. Consequently,

$$W^c W^d W_{[a}{}^R{}_{b]cd[e}{}^W{}_{f]} = \frac{1}{4}\Big[\delta_a^{n-1} R_{bnne}\delta_f^{n-1} - \delta_b^{n-1} R_{anne}\delta_f^{n-1} - \delta_a^{n-1} R_{bnnf}\delta_e^{n-1} + \delta_b^{n-1} R_{annf}\delta_e^{n-1}\Big]$$

This quantity is zero whenever one or more of a, b, e, or f equals n. Furthermore, if it is nonzero, exactly one of a or b must be n - 1 and exactly one of e or f must be n - 1. It follows that $W^c W^d W_{[a}{}^R{}_{b]cd[e}{}^W{}_{f]} \neq 0$ iff $R_{bnne} \neq 0$ for some $1 \leq b,e \leq n - 2$ and the proposition follows from Lemma B.4. □

We now prove the null analogue of Lemma B.3.

LEMMA B.6 Let β be a null geodesic and assume that $Ric(\beta'(t_0), \beta'(t_0)) \neq 0$. Then β satisfies the generic condition at $t_0$.

*Proof.* We first calculate $Ric(\beta'(t_0),\beta'(t_0))$ using the orthonormal basis $\{e_1,\ldots,e_{n-2},\tilde{e}_{n-1},\tilde{e}_n\}$ defined above. Now

$$Ric(\beta'(t_0),\beta'(t_0)) = \sum_{i=1}^{n-2} g(e_i,e_i)g(e_i,R(e_i,\beta')\beta') + \sum_{i=n-1}^{n} g(\tilde{e}_i,\tilde{e}_i)g(\tilde{e}_i,R(\tilde{e}_i,\beta')\beta')$$

The same arguments used in Proposition A.1 of Appendix A.3 may be used to show that

$$\sum_{i=n-1}^{n} g(\tilde{e}_i,\tilde{e}_i)g(\tilde{e}_i,R(\tilde{e}_i,\beta')\beta') = 0$$

Hence,

$$Ric(\beta'(t_0),\beta'(t_0)) = \sum_{i=1}^{n-2} g(e_i,R(e_i,\beta')\beta') = \sum_{i=1}^{n-2} R_{inin}$$

where we have used $R_{abcd}$ to represent the components of the Riemann-Christoffel tensor with respect to the pseudo-orthonormal basis $\{e_1,\ldots,e_{n-2},N,\beta'(t_0)\}$. Clearly, $Ric(\beta'(t_0),\beta'(t_0)) \neq 0$ implies $R_{inin} = - R_{inni} \neq 0$ for some $1 \leq i \leq n - 2$. The lemma now follows from Lemma B.4. □

Using Lemmas B.3 and B.6 together with the definition of the strong energy condition (cf. Definition 11.8), we obtain

PROPOSITION B.7 Let $(M,g)$ be a space-time. If $\mathrm{Ric}(v,v) > 0$ for all nonzero nonspacelike vectors of $(M,g)$, then $(M,g)$ satisfies both the generic condition and the strong energy condition.

## Appendix C

## THE EINSTEIN EQUATIONS

The purpose of this appendix is to give a brief description of the Einstein equations. A heuristic derivation of these equations may be found in Frankel (1979, Chapter 3). Since these equations are only assumed for manifolds of dimension four, we restrict our attention to this dimension.

The Einstein equations relate purely geometric quantities to the energy-momentum tensor T which is a physical quantity. They may thus be used to state the strong energy condition in terms of the energy-momentum tensor. In the case of a perfect fluid the tensor T takes a simple form. This is important in general relativity because the matter of the universe is assumed to behave like a perfect fluid in the standard cosmological models.

### C.1 THE ENERGY-MOMENTUM TENSOR AND THE EINSTEIN EQUATIONS

The physical motivation for studying Lorentzian manifolds is the assumption that a gravitational field may be effectively modeled by some Lorentzian metric g defined on a suitable four-dimensional manifold M. Since every manifold which admits one Lorentzian metric admits uncountably many Lorentzian metrics, it is necessary to decide which Lorentzian metric should be used to model a given gravitational problem. This question led to the Einstein equations which relate the metric tensor g, Ricci curvature Ric, and scalar curvature R to the energy-momentum tensor T. The tensor T is to be determined from physical considerations dealing with the distribution of matter and energy [cf. Hawking and Ellis (1973, Chapter 3), Misner, Thorne, and Wheeler (1973, Chapter 5)]. The *Einstein equations* may be written as

$$\mathrm{Ric} - \frac{1}{2} Rg + \Lambda g = 8\pi T \tag{C.1}$$

where $\Lambda$ is a constant, known as the cosmological constant. The constant factor of $8\pi$ is simply for scaling purposes. In local coordinates (C.1) becomes

$$R_{ij} - \frac{1}{2} Rg_{ij} + \Lambda g_{ij} = 8\pi T_{ij} \tag{C.2}$$

where $1 \leqslant i,j \leqslant 4$.

The Ricci curvature and scalar curvature involve the first and second partial derivatives of the components $g_{ij}$ of the metric tensor. Hence the Einstein equations represent (nonlinear) partial differential equations in the metric and its first two derivatives. These sixteen equations reduce to ten equations because all of the tensors in (C.1) are symmetric. There is a further reduction to six equations [cf. Misner, Thorne and Wheeler (1973, p. 409)] because of the curvature identity

$$\sum_{j=1}^{4} (R^{ij} - \frac{1}{2} Rg^{ij} + \Lambda g^{ij})_{;j} = 0 \tag{C.3}$$

which implies four conservation laws given by

$$\sum_{j=1}^{4} T^{ij}{}_{;j} = 0 \tag{C.4}$$

Here $;j$ denotes covariant differentiation in the $x^j$ direction.

The Einstein equations do not determine the metric on M without sufficient boundary conditions. For example, let $M = \mathbb{R}^2 \times S^2 = \{(t,r) : t \in \mathbb{R} \text{ and } r > 2m\} \times S^2$ and set $\Lambda = 0$, $T_{ij} = 0$ and $d\Omega^2 = d\theta^2 + \sin^2\theta\, d\phi^2$. Then M with this cosmological constant and energy-momentum tensor admits both the flat metric $ds^2 = -dt^2 + dr^2 + r^2 d\Omega^2$

and the Schwarzschild metric $ds^2 = -(1 - 2m/r)\, dt^2 + (1 - 2m/r)^{-1} dr^2 + r^2 d\Omega^2$ as solutions to the Einstein equations. Each of these metrics is asymptotically flat and each of these metrics is *Ricci flat* (i.e., Ric = 0). However, the Schwarzschild metric has a nonzero curvature tensor and hence these metrics are not isometric.

On the other hand, a counting argument shows that in general one expects the Einstein equations to determine the metric up to diffeomorphism [cf. Hawking and Ellis (1973, p. 74)]. For first note that the metric tensor has 16 components which by symmetry are reduced to 10 independent components. Furthermore, 4 of these 10 components can be accounted for by the dimension of M which allows 4 degrees of freedom. Thus the Einstein equations yield 6 independent equations to determine 6 essential components of the metric tensor. A rigorous approach to the problem of existence and uniqueness of solutions to the Einstein equations using Cauchy surfaces with initial data may be found in Hawking and Ellis (1973, Chapter 7); see also Marsden, Ebin, and Fischer (1972, pp. 233-264).

## C.2 THE STRONG ENERGY CONDITION AND THE ENERGY-MOMENTUM TENSOR

We now relate the strong energy condition $\mathrm{Ric}(v,v) \geqslant 0$ for nonspacelike vectors to the energy-momentum tensor. In order to evaluate the scalar curvature R in terms of T at $p \in M$, let $\{e_1, e_2, e_3, e_4\}$ be an orthonormal basis of $T_pM$ and use (C.1) to obtain

$$\Sigma\, g(e_i,e_i)[\mathrm{Ric}(e_i,e_i) - \frac{1}{2} Rg(e_i,e_i) + \Lambda g(e_i,e_i)]$$
$$= 8\pi\, \Sigma\, g(e_i,e_i)T(e_i,e_i)$$

Using dim M = 4 and the fact that the scalar curvature is the trace of the Ricci curvature, this equation becomes

$$R - 2R + 4\Lambda = 8\pi \operatorname{tr} T$$

Hence

$$R = -8\pi \operatorname{tr} T + 4\Lambda \tag{C.5}$$

The Einstein equations then become

$$\mathrm{Ric} - \frac{1}{2}(-\ 8\pi\ \mathrm{tr}\ T + 4\Lambda)g + \Lambda g = 8\pi\ T$$

Thus

$$\mathrm{Ric} = 8\pi\left(T - \frac{\mathrm{tr}\ T}{2}\ g + \frac{\Lambda}{8\pi}\ g\right) \tag{C.6}$$

This shows that $\mathrm{Ric}(v,v) \geqslant 0$ is equivalent to the inequality $T(v,v) \geqslant [(\mathrm{tr}\ T)/2 - \Lambda/8\pi]g(v,v)$. It follows that when $\Lambda = 0$ and $\dim M = 4$ our strong energy condition

$$\mathrm{Ric}(v,v) \geqslant 0 \qquad \text{for all nonspacelike } v$$

is equivalent to the condition [cf. Hawking and Ellis (1973, p. 95)]

$$T(v,v) \geqslant \frac{\mathrm{tr}\ T}{2}\ g(v,v) \qquad \text{for all nonspacelike } v$$

## C.3 PERFECT FLUIDS

Consider a fluid which moves through a space-time with a unit speed timelike tangent vector v. The flow lines of the fluid are then the integral curves of the vector field v. The fluid is said to be a *perfect fluid* if it has an energy density $\mu$, pressure p and energy-momentum tensor T such that

$$T = (\mu + p)\omega \otimes \omega + pg \tag{C.7}$$

which is

$$T_{ij} = (\mu + p)v_i v_j + pg_{ij} \tag{C.8}$$

in local coordinates. Here $\omega = \Sigma\ v_i dx^i$ is the 1-form corresponding to the vector field $v = \Sigma\ v^i\ \partial/\partial x^i$. It follows from the above form of T that a perfect fluid is an isotropic fluid which is free of shear and viscosity.

Recall that a Robertson-Walker space-time is a warped product of the form $(a,b) \times_f H$ where H is an isotropic Riemannian manifold (cf. Section 4.4). Robertson-Walker space-times are used to construct cosmological models in which the matter of the universe is taken to be a perfect fluid. In this case, the vector field v is given by $\partial/\partial t$ and the flow lines of the fluid are thus the curves $\gamma(t) = (t, y_0)$ for $y_0 \in H$. The energy-momentum tensor T has the required form (C.7) and the functions $\mu$ and p only depend on t. A discussion of the Robertson-Walker cosmological models may be found in Hawking and Ellis (1973, p. 134 ff); see also Misner, Thorne, and Wheeler (1973, Part VI).

We now consider the strong energy condition in space-times for which T has the form (C.7). If $\{e_1, e_2, e_3, e_4 = v\}$ is an orthonormal basis for $T_pM$, then the trace of T may be calculated by

$$\begin{aligned} \operatorname{tr} T &= \sum_{i=1}^{4} g(e_i, e_i) T(e_i, e_i) \\ &= -(\mu + p) + 4p \\ &= 3p - \mu \end{aligned}$$

Thus using (C.6), it follows that the strong energy condition is equivalent to

$$T(w,w) \geqslant \left(\frac{3p - \mu}{2} - \frac{\Lambda}{8\pi}\right) g(w,w)$$

for all nonspacelike w. From (C.7) we obtain

$$(\mu + p)[g(v,w)]^2 + pg(w,w) \geqslant \left(\frac{3p - \mu}{2} - \frac{\Lambda}{8\pi}\right) g(w,w)$$

which simplifies to

$$(\mu + p)[g(v,w)]^2 \geqslant \left(\frac{p - \mu}{2} - \frac{\Lambda}{8\pi}\right) g(w,w) \tag{C.9}$$

Since $g(w,w) \leqslant 0$, and $g(v,w) \neq 0$, Eq. (C.9) shows that a negative cosmological constant has the effect of making the strong energy condition more plausible and that a positive cosmological constant has the opposite effect. Tipler (1977b) has made a study of the geodesic incompleteness of space-times which have negative cosmological constants. Einstein originally introduced the cosmological constant because the Einstein equations (C.1) with $\Lambda = 0$ predict a universe which is either expanding or contracting and in the early part of this century it was believed that the universe was essentially static. After the discovery that the universe is expanding, the original motivation for the cosmological constant was removed. On the other hand, removing $\Lambda$ from the theory has been more difficult. While astronomical experiments have failed to detect a $\Lambda$ different from zero, one may argue that $\Lambda$ is so small that the experiments have not been sufficiently sensitive.

Appendix D

# JACOBI FIELDS AND TOPONOGOV'S THEOREM FOR LORENTZIAN MANIFOLDS[†]

One of the important consequences of the Rauch comparison theorem in Riemannian geometry is the Toponogov triangle comparison theorem [cf. Cheeger and Ebin (1975, pp. 42-49)]: Let M be a complete Riemannian manifold (metric here and elsewhere in this appendix, denoted by $\langle -,- \rangle$) with sectional curvature of all 2-planes $\sigma$ in M satisfying $K(\sigma) \geq H$ for some number H. Let $(\gamma_1, \gamma_2, \gamma_3)$ be geodesics in M forming a triangle: $\gamma_1(0) = \gamma_2(0)$, $\gamma_1(L_1) = \gamma_3(0)$, and $\gamma_2(L_2) = \gamma_3(L_3)$, where $L_i = L(\gamma_i)$. Suppose that $\gamma_2$ and $\gamma_3$ are minimal geodesics, and, in the case that $H = q^2$ $(q > 0)$, suppose that $L_i \leq \pi/q$, $i = 1, 2, 3$. Assume the geodesics $\gamma_i$ are parameterized by arc length, and define $\alpha_3 = \langle \gamma_1'(0), \gamma_2'(0) \rangle$ and $\alpha_2 = \langle -\gamma_1'(L_1), \gamma_3'(0) \rangle$. For a triangle $(\overline{\gamma}_1, \overline{\gamma}_2, \overline{\gamma}_3)$, possibly in another manifold, $\overline{\alpha}_2$ and $\overline{\alpha}_3$ are defined similarly.

(a) In the simply connected two-dimensional Riemannian manifold of constant curvature H, denoted by $M_H$, there is a geodesic triangle $(\overline{\gamma}_1, \overline{\gamma}_2, \overline{\gamma}_3)$ with $L(\overline{\gamma}_i) = L_i$, $i = 1, 2, 3$, and $\overline{\alpha}_2 \leq \alpha_2$ and $\overline{\alpha}_3 \leq \alpha_3$. This triangle is determined up to congruence if $H \leq 0$ or if $H > 0$ and all $L_i < \pi/q$.

(b) In $M_H$, let $\overline{\gamma}_1$ and $\overline{\gamma}_2$ be geodesics with $\overline{\gamma}_1(0) = \overline{\gamma}_2(0)$, $L(\overline{\gamma}_1) = L_1$, $L(\overline{\gamma}_2) = L_2$, and $\overline{\alpha}_3 = \alpha_3$. Let $\overline{\gamma}_3$ be a minimal geodesic between the endpoints of $\overline{\gamma}_1$ and $\overline{\gamma}_2$. Then $L(\overline{\gamma}_3) \geq L(\gamma_3)$.

[†]By Steven G. Harris, Department of Mathematics, University of Missouri, Columbia, Missouri.

It is possible to develop an analogous theorem for Lorentzian geometry; the program will be sketched here, although the proofs will be omitted. Details appear in Harris (1979a) and Harris (1979b, pp. 3-41).

The first step is to modify the timelike Rauch comparison theorem I (cf. Theorem 10.11) so that it applies to timelike geodesics, not without conjugate points, but without focal points: If N is a submanifold of a manifold M, then a point $q \in M$ will be said to be a *focal point of* N *from* p ($p \in N$) if q is the image of a critical point of exp in the normal bundle of N at p. For v a vector in the tangent bundle TM, N(v) will denote the submanifold of M which is the image under exp of a small enough neighborhood of the origin in the perpendicular space of v such that exp is an embedding on it. Thus N(v) is an (n - 1)-dimensional submanifold orthogonal to v.

In the following statement of the second Rauch theorem, and elsewhere in this appendix, $A \wedge B$ will denote the 2-plane spanned by the vectors A and B.

THEOREM D.1 (Timelike Rauch II) Let $V_i$ be a Jacobi field along a unit-speed timelike geodesic $\gamma_i$ in a space-time $M_i$, i = 1, 2, with $\gamma_i : [0,L] \longrightarrow M_i$; let $T_i = \gamma_i'$. Suppose that $\langle V_1,V_1\rangle_0 = \langle V_2,V_2\rangle_0$ and $\langle V_1,T_1\rangle_0 = \langle V_2,T_2\rangle_0$ and $(\nabla_{T_i} V_i)(0) = 0$. Further suppose that for any vectors $X_i$ at $\gamma_i(t)$,

$$K(X_1 \wedge T_1) \geqslant K(X_2 \wedge T_2)$$

and that $\gamma_2$ has no focal point of $N(\gamma_2'(0))$ from $\gamma_2(0)$. Then for all t in [0,L],

$$\langle V_1,V_1\rangle_t \geqslant \langle V_2,V_2\rangle_t$$

An important corollary of this theorem shows how curvature can affect the lengths of timelike curves.

COROLLARY D.2 Let $\gamma_i : [0,L] \longrightarrow M_i$, i = 1, 2, be two timelike (or two null or two spacelike) geodesics, and let $E_i$ be parallel unit timelike vector fields along $\gamma_i$ with $\langle E_1, T_1\rangle = \langle E_2, T_2\rangle$ ($T_i = \gamma_i'$). Let $f : [0,L] \longrightarrow \mathbb{R}$ be any smooth real-valued function. Suppose that for all t in [0,L], $\exp_{\gamma_i(t)}(f(t)E_i(t))$ is defined; call this $c_i(t)$. Suppose further that for all t in [0,L], the geodesic $\eta : s \longmapsto \exp_{\gamma_2(t)}(sE_2(t))$ has no focal point of $N(\eta'(0))$ from $\eta(0)$ for $s \leqslant f(t)$. Finally, suppose that for all timelike 2-planes $\sigma_i$ in $M_i$,

$$K(\sigma_1) \geqslant K(\sigma_2)$$

Then, for all t in [0,L],

$$\langle c_1', c_1'\rangle_t \geqslant \langle c_2', c_2'\rangle_t$$

Thus, if $c_1$ is a nonspacelike curve, so is $c_2$, and

$$L(c_1) \leqslant L(c_2)$$

This corollary makes possible a triangle comparison theorem for "thin" triangles. The model spaces used for comparison are the two-dimensional de Sitter and anti-de Sitter spaces of constant curvature [cf. Hawking and Ellis (1973, pp. 124-134), Wolf (1961, pp. 114-118)]. A triangle of timelike geodesics $(\gamma_1, \gamma_2, \gamma_3)$ is "thin" in this context if the following holds: For a given H, let $\overline{\gamma}_1$ and $\overline{\gamma}_2$ be timelike geodesics in the simply connected two-dimensional Lorentzian manifold of constant curvature H (denoted by $M_H$) with $\overline{\gamma}_1(0) = \overline{\gamma}_2(0)$, $L(\overline{\gamma}_1) = L_1$, $L(\overline{\gamma}_2) = L_2$, and $\overline{\alpha}_3 = \alpha_3$. First, suppose there is a timelike geodesic $\overline{\gamma}_3$ between the endpoints of $\overline{\gamma}_1$ and $\overline{\gamma}_2$. Let $\overline{E}$ be the parallel translate of $\overline{\gamma}_1'(0)$ along $\overline{\gamma}_2$. For each t in $[0,L_2]$, there is a smallest positive number f(t) such that $\exp(f(t)\overline{E}(t))$ lies on $\overline{\gamma}_3$. Second, suppose that for all such t, the geodesic $s \longmapsto \exp(s\overline{E}(t))$ has no focal point of $N(\overline{E}(t))$ from $\overline{\gamma}_2(t)$ for $s \leqslant f(t)$. If these two suppositions hold, and if $\gamma_3$ is maximal and all timelike planes $\sigma$ in M satisfy $K(\sigma) \leqslant H$, then $L(\gamma_3) \geqslant L(\overline{\gamma}_3)$.

The problem is to start with a more general timelike geodesic triangle, slice it up into "thin" triangles, apply the result just mentioned to each slice, and put them back together. In the Riemannian theorem, completeness is used to ensure that in any triangle $(\gamma_1,\gamma_2,\gamma_3)$, minimal geodesics can be extended from $\gamma_3(L_3)$ to $\gamma_1$, slicing up the original triangle. In the Lorentzian context global hyperbolicity succeeds just as well, so long as $H \geqslant 0$. For $H = -q^2$, however, a problem arises: Not even the model spaces $M_H$ are globally hyperbolic; indeed, by Proposition 10.8, no timelike geodesic of length greater than $\pi/q$ can be maximal in a Lorentzian manifold whose timelike planes $\sigma$ satisfy $K(\sigma) \leqslant -q^2$. A solution to this problem lies in a new concept, a sort of global hyperbolicity in the small: For x and y in a Lorentzian manifold M, let $C(x,y)$ denote the space of nonspacelike curves in M from x to y, modulo reparameterization, with the compact-open topology.

DEFINITION D.3 A Lorentzian manifold M is *globally hyperbolic of order* q $(q > 0)$ if M is strongly causal and for all points x and y in M with $\sup\{L(\gamma) : \gamma \in C(x,y)\} < \pi/q$, this space $C(x,y)$ is compact.

The Lorentzian analogue of Toponogov's theorem can now be stated (geodesics parameterized with unit speed):

THEOREM D.4 (Lorentzian Triangle Comparison Theorem) Let M be a space-time whose timelike planes $\sigma$ satisfy $K(\sigma) \leqslant H$ for some constant H; M is to be globally hyperbolic, or, in case $H = -q^2$, globally hyperbolic of order q. Let $(\gamma_1,\gamma_2,\gamma_3)$ be a triangle of timelike geodesics with $\gamma_2$ the future-directed side between the pastmost and futuremost of the three endpoints, $\gamma_1$ the other future-directed side from $\gamma_2(0)$, and $\gamma_3$ the remaining future-directed side. Suppose that $\gamma_2$ and $\gamma_3$ are maximal and, if $H = -q^2$, that $L_i < \pi/q$, i = 1, 2, 3. Then

(a) There is in $M_H$ a timelike geodesic triangle $(\bar\gamma_1,\bar\gamma_2,\bar\gamma_3)$ with $L(\bar\gamma_i) = L_i$, i = 1, 2, 3, and with $\bar\alpha_2 \geqslant \alpha_2$ and $\bar\alpha_3 \geqslant \alpha_3$.

(b) For timelike geodesics $\bar\gamma_1$ and $\bar\gamma_2$ constructed in $M_H$ with $\bar\gamma_1(0) = \bar\gamma_2(0)$, $L(\bar\gamma_1) = L_1$, $L(\bar\gamma_2) = L_2$, and $\bar\alpha_3 = \alpha_3$, if there is a timelike geodesic $\bar\gamma_3$ between the endpoints of $\bar\gamma_1$ and $\bar\gamma_2$, then $L(\bar\gamma_3) \leqslant L(\gamma_3)$.

The Toponogov triangle comparison theorem can be used to show how a bound on sectional curvature can rigidly determine a complete Riemannian manifold if the limits of the curvature-imposed constraints are attained. One such result is the maximal diameter theorem [cf. Cheeger and Ebin (1975, p. 110)]: If a complete Riemannian manifold $M^n$ satisfies $K(\sigma) \geqslant q^2 > 0$ for all planes $\sigma$, and if the diameter of M attains the maximum thus allowable, $\pi/q$, then M is isometric to the sphere $S^n$ of curvature $q^2$.

Theorem D.4 can be used similarly.

THEOREM D.5 Let $M^n$ be a space-time which is globally hyperbolic of order q and satisfies $K(\sigma) \leqslant -q^2$ for all timelike planes $\sigma$. Suppose that M possesses a complete timelike geodesic $\gamma$ which is maximal on all intervals of length $\pi/q$. Then M is isometric to the simply connected geodesically complete n-dimensional Lorentzian manifold of constant curvature $-q^2$.

In addition to timelike geodesics, it is possible to study the effects of curvature on Jacobi fields along null geodesics. Sectional curvature cannot be used for this, since it is undefined for null (singular) planes, but a similar concept, introduced here, will work.

DEFINITION D.6 If $\sigma$ is a null plane and N is any nonzero element of the one-dimensional space of null vectors in $\sigma$, then the *null sectional curvature of* $\sigma$ *with respect to* N, $K_N(\sigma)$, is defined by

$$K_N(\sigma) = \frac{\langle R(A,N)N,A\rangle}{\langle A,A\rangle}$$

where A is any nonnull vector in $\sigma$.

This expression for null sectional curvature is independent of the vector A and depends in a quadratic fashion on the vector N. This curvature quantity has certain interesting relations with ordinary sectional curvature. For instance [cf. Proposition 2.3 in Harris (1979a)].

PROPOSITION D.7 If at a single point in a Lorentzian manifold, the null sectional curvatures are all positive (resp., negative), then timelike sectional curvature at that point is unbounded below (resp., above); and if the null sectional curvatures all vanish, then the timelike and spacelike sectional curvatures are all equal. Thus, a Lorentzian manifold of dimension at least three has constant curvature iff it has null sectional curvature everywhere zero.

Null sectional curvature can be used much like timelike sectional curvature with regard to Jacobi fields.

PROPOSITION D.8 Let $\beta : [0,L] \longrightarrow M^n$ be a null (affinely parameterized) geodesic with $T = \beta'$. If for all nonnull vectors V perpendicular to T, $K_T(V \wedge T) \leqslant 0$, then $\beta$ has no conjugate points. If for some q, $K_T(V \wedge T) \geqslant q^2$ for all such V, or, more generally, if $\mathrm{Ric}(T,T) \geqslant (n-2)q^2$, then $L \geqslant \pi/q$ implies $\beta$ does have a conjugate point.

Null sectional curvature also lends itself to a Rauch-type theorem:

THEOREM D.9 (Rauch Comparison Theorem for Null Geodesics) Let $\beta_i : [0,L] \longrightarrow M_i$, $i = 1, 2$ be null geodesics; $T_i = \beta_i'$. Let $V_i$, $i = 1, 2$ be perpendicular Jacobi fields along $\beta_i$, not everywhere parallel to $T_i$ (and, thus, nowhere parallel to $T_i$), with $V_i(0) = 0$ and $\langle V_1', V_1' \rangle_0 = \langle V_2', V_2' \rangle_0$.

Suppose that for any t in [0,L] and for any nonnull vectors $X_i$ at $\beta_i(t)$, perpendicular to $T_i$,

$$K_{T_1}(X_1 \wedge T_1) \leqslant K_{T_2}(X_2 \wedge T_2)$$

and that $\beta_2$ has no points conjugate to $\beta_2(0)$. Then for all t in [0,L],

$$\langle V_1, V_1 \rangle_t \geqslant \langle V_2, V_2 \rangle_t$$

The necessity of using perpendicular vector fields makes this theory less tractable than that for timelike geodesics.

Details for Proposition D.8 and Theorem D.9 can be found in Harris (1979a).

# REFERENCES


Anderson, J. L. (1967). *Principles of Relativity Physics*, Academic Press, New York.

Auslander, L. and L. Marcus (1959). *Flat Lorentz 3-Manifolds*, Memoir *30*, Amer. Math. Soc.

Avez, A. (1963). Essais de géométrie riemannienne hyperbolique globale. Applicationes à la Relativité Générale, Ann. Inst. Fourier 132, 105-190.

Beem, J. K. (1976a). Conformal changes and geodesic completeness, Commun. Math. Phys. *49*, 179-186.

Beem, J. K. (1976b). Globally hyperbolic space-times which are timelike Cauchy complete, Gen. Rel. Grav. *7*, 339-344.

Beem, J. K. (1976c). Some examples of incomplete space-times, Gen. Rel. Grav. *7*, 501-509.

Beem, J. K. (1977). A metric topology for causally continuous completions, Gen. Rel. Grav. *8*, 245-257.

Beem, J. K. (1978a). Homothetic maps of the space-time distance function and differentiability, Gen. Rel. Grav. *9*, 793-799.

Beem, J. K. (1978b). Proper homothetic maps and fixed points, Lett. Math. Phys. *2*, 317-320.

Beem, J. K. (1980). Minkowski space-time is locally extendible, Commun. Math. Phys., *72*, 273-275.

Beem, J. K., and P. E. Ehrlich (1973). Stability of geodesic incompleteness for Robertson-Walker space-times, Gen. Rel. Grav., to appear.

Beem, J. K. and P. E. Ehrlich (1977). Distance lorentzienne finie et géodésiques f-causales incomplètes, C. R. Acad. Sci. Paris Ser. A *581*, 1129-1131.

Beem, J. K. and P. E. Ehrlich (1978). Conformal deformations, Ricci curvature and energy conditions on globally hyperbolic space-times, Math. Proc. Camb. Phil. Soc. *84*, 159-175.

Beem, J. K. and P. E. Ehrlich (1979a). Singularities, incompleteness and the Lorentzian distance function, Math. Proc. Camb. Phil. Soc. *85*, 161-178.

Beem, J. K. and P. E. Ehrlich (1979b). The space-time cut locus, Gen. Rel. Grav. *11*, 89-103.

Beem, J. K. and P. E. Ehrlich (1979c). Cut points, conjugate points and Lorentzian comparison theorems, Math. Proc. Camb. Phil. Soc. *86*, 365-384.

Beem, J. K. and P. E. Ehrlich (1979d). A Morse index theorem for null geodesics, Duke Math. J. *46*, 561-569.

Beem, J. K. and P. E. Ehrlich (1980). Stability of geodesic incompleteness for Robertson-Walker space-times, Gen. Rel. Grav., to appear.

Beem, J. K., P. E. Ehrlich and T. G. Powell (1980). Warped product manifolds in relativity, Einstein volume, Athens, Greece, to appear.

Beem, J. K. and P. Y. Woo (1969). *Doubly Timelike Surfaces*, Memoir *92*, Amer. Math. Soc.

Berger, M. (1960). Les variétés riemanniennes (1/4) - pincées, Annali della Scuola Normale Sup. di Pisa, Ser. III, *14*, 161-170.

Birkhoff, and G.-C. Rota (1969). *Ordinary Differential Equations*, second edition, Blaisdell, Waltham, Massachusetts.

Bishop, R. and R. Crittenden (1964). *Geometry of Manifolds*, Academic Press, New York.

Bishop, R. L. and B. O'Neill (1969). Manifolds of negative curvature, Trans. Amer. Math. Soc. *145*, 1-49.

Bölts, G. (1977). Existenz und Bedeutung von konjugierten Werten in der Raum-Zeit, Bonn Universität Diplomarbeit.

Bondi, H. (1968). *Cosmology*, second edition, Cambridge University Press, Cambridge.

Bosshard, B. (1976). On the b-boundary of the closed Friedmann model, Commun. Math. Phys. *46*, 263-268.

Boyer, R. H. and R. W. Lindquist (1967). Maximal analytic extension of the Kerr metric, J. Math. Phys. *8*, 265-281.

Boyer, R. H. and T. G. Price (1965). An interpretation of the Kerr metric in General Relativity, Proc. Camb. Phil. Soc. *61*, 531-534.

Brill, D. and F. Flaherty (1976). Isolated maximal surfaces in spacetime, Commun. Math. Phys. *50*, 157-165.

Budic, R. and R. K. Sachs (1974). Causal boundaries for general relativistic space-times, J. Math. Phys. *15*, 1302-1309.

Budic, R. and R. K. Sachs (1976). Scalar time functions: differentiability, in *Differential Geometry and Relativity*, eds. M. Cohen and M. Flato, Reidel: Dordrecht, 215-224.

Busemann, H. (1942). *Metric Methods in Finsler Spaces and in the Foundations of Geometry*, Annals of Math. Studies *8*, Princeton University Press, Princeton, New Jersey.

Busemann, H. (1955). *The Geometry of Geodesics*, Academic Press, New York.

Busemann, H. (1967). *Timelike Spaces*, Dissertationes Math. Rozprawy Mat. *53*.

Busemann, H. and J. K. Beem (1966). Axioms for indefinite metrics, Rnd. Cir. Math. Palermo *15*, 223-246.

Cahen, M. and M. Parker (1980). *Pseudo-riemannian Symmetric Spaces*, Memoir *229*, Amer. Math. Soc.

Carter, B. (1971a). Causal structure in space-time, Gen. Rel. Grav. *1*, 349-391.

Carter, B. (1971b). Axisymmetric black hole has only two degrees of freedom, Phys. Rev. Lett. *26*, 331-333.

Cheeger, J. and D. Ebin (1975). *Comparison Theorems in Riemannian Geometry*, North-Holland, Amsterdam.

Cheeger, J. and D. Gromoll (1971). The splitting theorem for manifolds of nonnegative Ricci curvature, J. Diff. Geo. *6*, 119-128.

Cheeger, J. and D. Gromoll (1972). On the structure of complete manifolds of nonnegative curvature, Ann. Math. *96*, 413-433.

Chicone, C. and P. Ehrlich (1980). Line integration of Ricci curvature and conjugate points in Lorentzian and Riemannian manifolds, Manuscripta Math. *31*, 297-316.

Choquet-Bruhat, Y., A. E. Fischer, and J. E. Marsden (1979). Maximal hypersurfaces and positivity of mass, Il nuovo cimento, to appear.

Clarke, C. J. S. (1970). On the global isometric embedding of pseudo-Riemannian manifolds, Proc. Roy. Soc. Lond. *A314*, 417-428.

Clarke, C. J. S. (1971). On the geodesic completeness of causal space-times, Proc. Camb. Phil. Soc. *69*, 319-324.

Clarke, C. J. S. (1973). Local extensions in singular space-times, Commun. Math. Phys. *32*, 205-214.

Clarke, C. J. S. (1975). Singularities in globally hyperbolic space-times, Commun. Math. Phys. *41*, 65-78.

Clarke, C. J. S. (1976). Space-time singularities, Commun. Math. Phys. *49*, 17-23.

Clarke, C. J. S. and B. G. Schmidt (1977). Singularities: the state of the art, Gen. Rel. and Grav. *8*, 129-137.

Cohn-Vossen, S. (1936). Total Krümmung und geodatische linien auf einfach zusammenhängenden offenen vollständigen Flächenstücken, Recueil Mathematique *1*, 139-163.

Crittenden, R. (1962). Minimum and conjugate points in symmetric spaces, Cand. J. Math. *14*, 320-328.

Dajczer, M. and K. Nomizu (1979). On the boundedness of Ricci curvature of an indefinite metric, Bol. Soc. Brazil Math., to appear.

Dajczer, M. and K. Nomizu (1980). On sectional curvature of indefinite metrics II, Math. Annalen *247*, 279-282.

Dodson, C. T. J. (1978). Space-time edge geometry, Int. J. Theor. Phys. *17*, 389-504.

Dodson, C. T. J. (1980). *Categories, Bundles and Spacetime Topology*, Shiva Math. Series *1*, Shiva, Kent.

Dodson, C. T. J. and T. Poston (1977). *Tensor Geometry: The Geometric Viewpoint and Its Uses*, Survey and Reference Works in Math. *1*, Pitman, San Francisco.

Dodson, C. T. J. and L. J. Sulley (1980). On bundle completion of parallelizable manifolds, Math. Proc. Camb. Phil. Soc. *87*, 523-525.

Eardley, D. M. and L. Smarr (1979). Time functions in numerical relativity: Marginally bound dust collapse, Phys. Rev. D *19*, 2239-2259.

Eberlein, P. (1972). Product manifolds that are not negative space forms, Mich. Math. J. *19*, 225-231.

Eberlein, P. (1973). When is a geodesic flow of Anosov type I, J. Diff. Geo. *8*, 437-463.

Eberlein, P. and B. O'Neill (1973). Visibility manifolds, Pacific J. Math. *46*, 45-109.

Ehresmann, C. (1951). Les connexions infinitésmiales dans un espace fibré différentiable, Colloque de Topologie (Espaces Fibrés) Bruxelles 1950, Masson, Paris, 29-55.

Ehrlich, P. E. (1974). Continuity properties of the injectivity radius function, Compositio Math. *29*, 151-178.

Ehrlich, P. E. (1976a). Metric deformations of curvature I: local convex deformations, Geometriae Dedicata *5*, 1-24.

Ehrlich, P. E. (1976b). Metric deformations of curvature II: compact 3-manifolds, Geometriae Dedicata *5*, 147-161.

Einstein, A. (1916). Die Grundlage der allgemeinen Relativitätstheorie, Annalen der Phys. *49*, 769-822.

Einstein, A. (1953). *The Meaning of Relativity*, fourth edition, Princeton University Press, Princeton, New Jersey.

Ellis, G. F. R. and B. G. Schmidt (1977). Singular space-times, Gen. Rel. Grav. *8*, 915-953.

Eschenburg, J.-H. (1975). Stabilitätsverhalten des Geodätischen Flüsses Riemannscher Mannigfaltigkeiten, Thesis, Bonn University, 1975.

Eschenburg, J.-H. and J. O'Sullivan (1976). Growth of Jacobi fields and divergence of geodesics, Math. Zeitschrift *150*, 221-237.

Everson, J. and C. J. Talbot (1976). Morse theory on timelike and causal curves, Gen. Rel. Grav. *7*, 609-622.

Everson, J. and C. J. Talbot (1978). Erratum: Morse theory on timelike and causal curves, Gen. Rel. Grav. *9*, 1047.

Fegan, H. and R. Millman (1978). Quadrants of Riemannian metrics, Mich. Math. J. *25*, 3-7.

Fischer, A. E. and J. E. Marsden (1972). The Einstein equations of evolution--a geometric approach, J. Math. Phys. *13*, 546-568.

Flaherty, F. (1975a). Lorentzian manifolds of nonpositive curvature, Proc. Symp. Pure Math. *27*, part 2, Amer. Math. Soc., 395-399.

Flaherty, F. (1975b). Lorentzian manifolds of nonpositive curvature II, Proc. Amer. Math. Soc. *48*, 199-202.

Frankel, T. (1979). *Gravitational Curvature*, W. H. Freeman, San Francisco.

Freudenthal, H. (1931). Über die enden topologischer Raüme und Gruppen, Math. Zeitschrift, *33*, 692-713.

Friedlander, F. G. (1975). *The Wave Equation on a Curved Space-time*, Cambridge University Press, Cambridge.

Galloway, G. (1977). Closure in antisotropic cosmological models, J. Math. Phys. *18*, 250-252.

Galloway, G. (1979). A generalization of Myers' Theorem and an application to relativistic cosmology, J. Diff. Geo. *14*, 105-116.

Geroch, R. P. (1968a). Spinor structure of space-times in general relativity I, J. Math. Phys. *9*, 1739-1744.

Geroch, R. P. (1968b). What is a singularity in general relativity, Ann. Phys. (N.Y.) *48*, 526-540.

Geroch, R. P. (1969). Limits of spacetimes, Commun. Math. Phys. *13*, 180-193.

Geroch, R. P. (1970). Domain of dependence, J. Math. Phys. *11*, 437-449.

Geroch, R. P. and G. T. Horowitz (1979). Global structure of space-time, in *General Relativity: An Einstein Centenary Survey*, eds. S. Hawking and W. Israel, Cambridge University Press, Cambridge, 212-293.

Geroch, R. P., E. H. Kronheimer and R. Penrose (1972). Ideal points in space-time, Proc. Roy. Soc. Lond. *A327*, 545-567.

Göbel, R. (1976). Zeeman topologies on space-times of general relativity theory, Commun. Math. Phys. *46*, 289-307.

Graves, L. (1979). Codimension one isometric immersions between Lorentz spaces, Trans. Amer. Math. Soc., *252*, 367-392.

Graves, L. and K. Nomizu (1978). On sectional curvature of indefinite metrics, Math. Ann. *232*, 267-272.

Green, L. W. (1958). A theorem of E. Hopf, Mich. Math. J. *5*, 31-34.

Green, R. E. (1970). *Isometric Embeddings of Riemannian and Pseudo-Riemannian Manifolds*, Memoir *97*, Amer. Math. Soc.

Gromoll, D., W. Klingenberg and W. Meyer (1975). *Riemannsche Geometrie im Grossen*, Lecture Notes in Math *55*, Springer-Verlag, Berlin.

Gromoll, D. and W. Meyer (1969). On complete open manifolds of positive curvature, Ann. of Math. *90*, 75-90.

Gulliver, R. (1975). On the variety of manifolds without conjugate points, Trans. Amer. Math. Soc. *210*, 185-201.

Harris, S. G. (1979a). A triangle comparison theorem for Lorentz manifolds, Indiana Math. J., to appear.

Harris, S. G. (1979b). Some comparison theorems in the geometry of Lorentz manifolds, Thesis, University of Chicago.

Hartman, P. (1964). *Ordinary Differential Equations*, Wiley, New York.

Hawking, S. W. (1967). The occurrence of singularities in cosmology III. Causality and singularities, Proc. Roy. Soc. Lond. *A300*, 187-201.

Hawking, S. W. (1968). The existence of cosmic time functions, Proc. Roy. Soc. Lond. *A308*, 433-435.

Hawking, S. W. (1971). Stable and generic properties in general relativity, Gen. Rel. Grav. *1*, 393-400.

Hawking, S. W. and G. F. R. Ellis (1973). *The Large Scale Structure of Space-time*, Cambridge University Press, Cambridge.

Hawking, S. W., A. R. King and P. J. McCarthy (1976). A new topology for curved space-time which incorporates the causal, differential and conformal structures, J. Math. Phys. *17*, 174-181.

Hawking, S. W. and R. Penrose (1970). The singularities of gravitational collapse and cosmology, Proc. Roy. Soc. Lond. *A314*, 529-548.

Hawking, S. W. and R. K. Sachs (1974). Causally continuous space-times, Commun. Math. Phys. *35*, 287-296.

Helgason, S. (1962). *Differential Geometry and Symmetric Spaces*, Academic Press, New York.

Hermann, R. (1968). *Differential Geometry and the Calculus of Variations*, Academic Press, New York.

Hicks, N. J. (1965). *Notes on Differential Geometry*, D. Van Nostrand, Princeton, New Jersey.

Hirsch, M. (1976). *Differential Topology*, Grad. Texts in Math. *33*, Springer-Verlag, New York.

Hopf, E. (1948). Closed surfaces without conjugate points, Proc. Nat. Acad. Sci. *34*, 47-51.

Hopf, H. and W. Rinow (1931). Über den Begriff des vollständigen differentialgeometrischen Fläche, Comment. Math. Helv. *3*, 209-225.

Johnson, R. A. (1977). The bundle boundary in some special cases, J. Math. Phys. *18*, 898-902.

Kelley, J. L. (1955). *General Topology*, Univ. Ser. in Higher Math., D. Van Nostrand, Princeton, New Jersey.

Klingenberg, W. (1959). Contributions to Riemannian geometry in the large, Ann. of Math. *69*, 654-666.

Klingenberg, W. (1961). Über Riemannsche Mannigfaltigkeiten mit positiver Krümmung, Comment. Math. Helv. *35*, 47-54.

Klingenberg, W. (1962). Über Riemannsche Mannigfaltigkeiten mit nach oben beschränkter Krümmung, Annali di Mat. *60*, 49-59.

Klingenberg, W. (1978). *Lectures on Closed Geodesics*, Grundlehren der mathematischen Wissenschaften *230*, Springer-Verlag, New York.

Kobayashi, S. (1967). On conjugate and cut loci, Studies in global geometry and analysis, MAA studies in Math. *4*, 96-122.

Kobayashi, S. and K. Nomizu (1963). *Foundations of Differential Geometry*, vol. I, Interscience Tracts in Pure and Applied Math. *15*, John Wiley, New York.

Kronheimer, E. H. and R. Penrose (1967). On the structure of causal spaces, Proc. Camb. Phil. Soc. *63*, 481-501.

Kruskal, M. D. (1960). Maximal extension of Schwarzschild metric, Phys. Rev. *119*, 1743-1745.

Kulkarni, R. S. (1978). Fundamental groups of homogeneous space-forms, Math. Ann. *234*, 51-60.

Kulkarni, R. S. (1979). The values of sectional curvature in indefinite metrics, Comment. Math. Helv. *54*, 173-176.

Kundt, W. (1963). Note on the completeness of spacetimes, Zs. für Phys. 172, 488-489.

Lee, K. K. (1975). Another possible abnormality of compact space-time, Canad. Math. Bul. *18*, 695-697.

Lerner, D. E. (1972). Techniques of topology and differential geometry in general relativity, Springer Lecture Notes in Phys. *14*, 1-44.

Lerner, D. E. (1973). The space of Lorentz metrics, Commun. Math. Phys. *32*, 19-38.

Marathe, K. (1972). A condition for paracompactness of a manifold, J. Diff. Geo. 7, 571-573.

Markus, L. (1955). Line element fields and Lorentz structures on differentiable manifolds, Ann. of Math. *62*, 411-417.

Marsden, J. E. (1973). On completeness of homogeneous pseudo-Riemannian manifolds, Indiana Univ. Math. J. *22*, 1065-1066.

Marsden, J. E., D. G. Ebin, and A. E. Fischer (1972). Diffeomorphism groups, hydrodynamics and relativity, in Proceedings of the thirteenth biennial seminar of the Canadian Mathematical Congress, ed. J. R. Vanstone, 135-279.

Miller, J. G. (1979). Bifurcate Killing horizons, J. Math. Phys. *20*, 1345-1348.

Milnor, J. (1963). *Morse Theory*, Ann. of Math. Studies *51*, Princeton University Press, Princeton, New Jersey.

Misner, C. W. (1967). Taub-NUT space as a counterexample to almost anything, in *Relativity and Astrophysics I. Relativity and Cosmology*, Amer. Math. Soc., ed. J. Ehlers, 160-169.

Misner, C. W. and A. H. Taub (1969). A singularity-free empty universe, Soc. Phys. J.E.T.P. *28*, 122-133.

Misner, C. W., K. S. Thorne and J. A. Wheeler (1973). *Gravitation*, W. H. Freeman, San Francisco.

Moncrief, V. (1975). Spacetime symmetries and linearization stability of the Einstein equations, J. Math. Phys. *16*, 493-498.

Morrow, J. (1970). The denseness of complete Riemannian metrics, J. Diff. Geo. *4*, 225-226.

Morse, M. (1934). *The Calculus of Variations in the Large*, Amer. Math. Soc. Colloq. Pub. vol. *18*.

Munkres, J. R. (1963). *Elementary Differential Topology*, Ann. of Math. Studies *54*, Princeton University Press, Princeton, New Jersey.

Munkres, J. R. (1975). *Topology: A First Course*, Prentice-Hall, Englewood Cliffs, New Jersey.

Myers, S. B. (1935). Riemannian manifolds in the large, Duke Math. J. *1*, 39-49.

Myers, S. B. and N. Steenrod (1939). The group of isometries of a Riemannian manifold, Ann. of Math. *40*, 400-416.

Nomizu, K. (1979). Left invariant Lorentz metrics on Lie groups, Osaka Math. J. *16*, 143-150.

Nomizu, K. and H. Ozeki (1961). The existence of complete Riemannian metrics, Proc. Amer. Math. Soc. *12*, 889-891.

Ohanian, H. C. (1976). *Gravitation and Spacetime*, W. W. Norton, New York.

O'Neill, B. (1966). *Elementary Differential Geometry*, Academic Press, New York.

O'Neill, B. (1981). *Semi-riemannian Manifolds*, Wiley, New York, to appear.

O'Sullivan, J. (1974). Manifolds without conjugate points, Math. Annalen *210*, 295-311.

Palais, R. S. (1957). On the differentiability of isometries, Proc. Amer. Math. Soc. *8*, 805-807.

Parker, P. E. (1979). Distributional geometry, J. Math. Phys. *20*, 1423-1426.

Pathria, R. K. (1974). *The Theory of Relativity*, second edition, Pergamon Press, Oxford.

Penrose, R. (1965). Gravitational collapse and space-time singularities, Phys. Rev. Lett. *14*, 57-59.

Penrose, R. (1968). Structure of space-time, in Battelle Recontres, Lectures in Mathematics and Physics, ed. by de Witt, C. M. and Wheeler, J. A., Benjamin, New York, 121-235.

Penrose, R. (1972). *Techniques of Differential Topology in Relativity*, Regional Conference Series in Applied Math. 7, SIAM, Philadelphia.

Petrov, A. Z. (1969). *Einstein Spaces*, Pergamon Press, Oxford.

Poincaré, H. (1905). Sur les lignes géodésiques des surfaces convexes, Trans. Amer. Math. Soc. *6*, 237-274.

Rauch, H. (1951). A contribution to differential geometry in the large, Ann. of Math. *54*, 38-55.

Retzloff, D. G., B. DeFacio, and P. Dennis (1981). A new mathematical formulation of accelerated observers in general relativity, J. Math. Phys., to appear.

Rinow, W. (1932). Über Zusammenhange der Differentialgeometrie im Grossen und im Kleinen, Math. Z. *35*, 512-528.

Ryan, M. P. and L. C. Shepley (1975). *Homogeneous Relativistic Cosmologies*, Princeton Series in Physics, Princeton University Press, Princeton, New Jersey.

Sachs, R. K. and H. Wu (1977a). *General Relativity for Mathematicians*, Grad. Texts in Math. *48*, Springer Verlag, New York.

Sachs, R. K. and H. Wu (1977b). General Relativity and cosmology, Bull. Amer. Math. Soc. *83*, 1101-1164.

Schmidt, B. G. (1971). A new definition of singular points in general relativity, Gen. Rel. Grav. *1*, 269-280.

Schmidt, B. G. (1973). The local b-completeness of space-times, Commun. Math. Phys. *29*, 49-54.

Seifert, H.-J. (1967). Global connectivity by timelike geodesics, Zs. f. Naturforsche. *22a*, 1356-1360.

Seifert, H.-J. (1971). The causal boundary of space-times, Gen. Rel. Grav. *1*, 247-259.

Seifert, H.-J. (1977). Smoothing and extending cosmic time functions, Gen. Rel. Grav. *8*, 815-831.

Serre, J. P. (1951). Homologie singulière des espaces fibrés, applications, Ann. Math. *54*, 425-505.

Smith, J. W. (1960a). Fundamental groups on a Lorentz manifold, Amer. J. Math. *82*, 873-890.

Smith, J. W. (1960b). Lorentz structures on the plane, Trans. Amer. Math. Soc. *95*, 226-237.

Spivak, M. (1970). *A Comprehensive Introduction to Differential Geometry*, Vol. II, Publish or Perish Press, Boston.

Steenrod, N. E. (1951). *The Topology of Fibre Bundles*, Princeton University Press, Princeton, New Jersey.

Taub, A. H. (1951). Empty space-times admitting a three parameter group of motions, Ann. of Math. *53*, 472-490.

Taub, A. H. (1980). Space-times with distribution valued curvature tensors, to appear.

Thorpe, J. (1969). Curvature and the Petrov canonical forms, J. Math. Phys. *10*, 1-7.

Thorpe, J. (1977a). Curvature invariants and space-time singularities, J. Math. Phys. *18*, 960-964.

Thorpe, J. (1977b). The observer bundle, Abstracts of contributed papers to the 8th International General Relativity Congress, 334.

Tipler, F. (1977a). Singularities and causality violation, Ann. of Phys. *108*, 1-36.

Tipler, F. (1977b). Singularities in universes with negative cosmological constant, Astrophys. J. *209*, 12-15.

Tipler, F. (1977c). Black holes in closed universes, Nature *270*, 500-501.

Tipler, F. (1977d). Causally symmetric space-times, J. Math. Phys. *18*, 1568-1573.

Tipler, F. (1978). General Relativity and conjugate ordinary differential equations, J. Differential Equations *30*, 165-174.

Tipler, F. (1979). Existence of closed timelike geodesics in Lorentz spaces, Proc. Amer. Math. Soc., *76*, 145-147.

Tipler, F., C. J. S. Clarke, and G. F. R. Ellis (1980). Singularities and horizons--a review article, in *General Relativity and Gravitation* vol. 2, ed. A. Held, Plenum Press, New York, 97-206.

Tits, J. (1955). *Sur certaines classes d'espaces homogenes de groups de Lie*, Memoir Belgian Academy of Sciences.

Tomimatsu, A. and H. Sato (1973). New series of exact solutions for gravitational fields of spinning mass, Prog. Theor. Phys. *50*, 95-110.

Uhlenbeck, K. (1975). A Morse theory for geodesics on a Lorentz manifold, Topology *14*, 69-90.

Walker, A. G. (1944). Completely symmetric spaces, J. Lond. Math. Soc. *19*, 219-226.

Wang, H.-C. (1951). Two theorems on metric spaces, Pacific J. Math. *1*, 473-480.

Wang, H.-C. (1952). Two-point homogeneous spaces, Ann. of Math. *55*, 177-191.

Weinberg, S. (1972). *Gravitation and Cosmology*, John Wiley, New York.

Wheeler, J. A. (1977). Singularity and unanimity, Gen. Rel. Grav. *8*, 713-715.

Whitehead, J. H. C. (1932). Convex regions in the geometry of paths, Quart. J. Math. Oxford Ser. *3*, 33-42.

Whitehead, J. H. C. (1933). Convex regions in the geometry of paths--Addendum, Quart. J. Math. Oxford Ser. *4*, 226-227.

Whitehead, J. H. C. (1935). On the covering of a complete space by the geodesics through a point, Ann. of Math. *36*, 679-704.

Wolf, J. A. (1961). Homogeneous manifolds of constant curvature, Comment. Math. Helv. *36*, 112-147.

Wolf, J. A. (1974). *Spaces of Constant Curvature*, 3rd edition, Publish or Perish Press, Boston.

Wolter, F.-E. (1979). Distance function and cut loci on a complete Riemannian manifold, Archiv. der Math. *32*, 92-96.

Woodhouse, N. M. J. (1973). The differentiable and causal structures of space-time, J. Math. Phys. *14*, 495-501.

Woodhouse, N. M. J. (1976). An application of Morse theory to space-time geometry, Commun. Math. Phys. *46*, 135-152.

Zeeman, E. C. (1964). Causality implies the Lorentz group, J. Math. Phys. *5*, 490-493.

Zeeman, E. C. (1967). The topology of Minkowski space, Topology 6, 161-170.

# NOTATION

Length

| | |
|---|---|
| $L_0(\gamma)$ | Riemannian arc length, 3 |
| $L(\gamma)$ | Lorentzian arc length, 82 |

Manifolds

| | |
|---|---|
| $\Delta(M)$ | diagonal of M, 102 |
| $\chi(M)$ | Euler characteristic of M, 5 |
| $\mathfrak{X}(M)$ | smooth vector fields on M, 402 |
| $M \times_f H$ | warped product of M with H, 56 |
| $\mathbb{R}^n_s$ | pseudo-Euclidean space of signature (s, n - s), 123 |
| $\mathbb{R}P^3$ | real projective 3-space, 131 |
| $T_{-1}M$ | unit observer bundle, 223 |

Maps

| | |
|---|---|
| $\exp_p$ | exponential map, 20 |
| $\tau_v$ | canonical isomorphism, 260 |

Metrics

| | |
|---|---|
| $C(M,g)$ | equivalence class of Lorentzian metrics conformal to g, 6 |
| $\mathrm{Con}(M)$ | collection of all conformal equivalence classes, 173 |
| $\bar{g}(v,w)$ | quotient metric, 296 |
| $g_1 <_A g_2$ | partial order on metrics, 29 |
| $\mathrm{Lor}(M)$ | space of all Lorentzian metrics on M, 28 |

Second fundamental forms

| | |
|---|---|
| $L_n$ | second fundamental form operator, 54 |
| $S$ | second fundamental form, 54 |
| $S_n$ | second fundamental form in direction n, 54 |

Space-times

| | |
|---|---|
| $\tilde{H}^n_1$ | anti-de Sitter space-time, 124 |
| $S^n_1$ | de Sitter space-time, 124 |

Tensors and curvatures

| | |
|---|---|
| $K(p,E)$ | sectional curvature, 409 |
| $R$ | curvature tensor, 405 |
| $R$, $\tau$ | scalar curvature, 408 |
| Ric | Ricci curvature, 408 |
| $T$ | energy-momentum tensor, 421, 422 |
| $T(X,Y)$ | torsion tensor, 404 |
| $W(A,B)$ | Wronskian, 312 |
| $\theta$, $\bar{\theta}$ | expansion, quotient expansion, 349, 351 |
| $\omega$, $\bar{\omega}$ | vorticity, quotient vorticity, 349, 351 |
| $\sigma$, $\bar{\sigma}$ | shear, quotient shear, 349, 351 |

Topologies

| | |
|---|---|
| $C^0$-topology | on curves, 40 |
| $C^0$-topology | on Lor(M), 28 |
| $C^r$-topology | on Lor(M), 28 |

Vector fields and curves

| | |
|---|---|
| $C_{(p,q)}$ | piecewise smooth future-directed timelike curves from p to q, 279 |
| $G(\beta)$ | quotient bundle of $N(\beta)$ with $[\beta']$, 295 |
| $J_t(c)$ | Jacobi fields along c which vanish at a and t, 267 |
| $J_t(\beta)$ | Jacobi fields along $\beta$ which vanish at a and t, 301 |
| $N(c(t))$ | vector space orthogonal to $c'(t)$, 250 |
| $N(\beta(t))$ | vector space orthogonal to $\beta'(t)$, 294 |
| $V^{\perp}(c)$ | piecewise smooth vector fields orthogonal to c, 250 |
| $V_0^{\perp}(c)$ | vector fields in $V^{\perp}(c)$ which vanish at the endpoints of curve c, 250 |
| $V^{\perp}(\beta)$ | piecewise smooth vector fields orthogonal to $\beta$, 294 |

| | |
|---|---|
| $V_0^{\perp}(\beta)$ | vector fields in $V^{\perp}(\beta)$ which vanish at the endpoints of $\beta$, 294 |
| $\mathfrak{X}(\beta)$ | piecewise smooth sections of the quotient bundle $G(\beta)$, 297 |
| $\mathfrak{X}_0(\beta)$ | sections in $\mathfrak{X}(\beta)$ which vanish at the endpoints of $\beta$, 297 |
| $\Delta_{t_i}(X')$ | change in $X'$ at $t_i$, 251, 252 |
| $\Omega_{p,q}$ | piecewise smooth curves from $p$ to $q$, 3 |

# INDEX